T0180188

Sustainable Textiles: Production, Processing, Manufacturing & Chemistry

Series Editor

Subramanian Senthilkannan Muthu, Head of Sustainability, SgT and API, Kowloon, Hong Kong

This series aims to address all issues related to sustainability through the lifecycles of textiles from manufacturing to consumer behavior through sustainable disposal. Potential topics include but are not limited to: Environmental Footprints of Textile manufacturing; Environmental Life Cycle Assessment of Textile production; Environmental impact models of Textiles and Clothing Supply Chain; Clothing Supply Chain Sustainability; Carbon, energy and water footprints of textile products and in the clothing manufacturing chain; Functional life and reusability of textile products; Biodegradable textile products and the assessment of biodegradability; Waste management in textile industry; Pollution abatement in textile sector; Recycled textile materials and the evaluation of recycling; Consumer behavior in Sustainable Textiles; Eco-design in Clothing & Apparels; Sustainable polymers & fibers in Textiles; Sustainable waste water treatments in Textile manufacturing; Sustainable Textile Chemicals in Textile manufacturing. Innovative fibres, processes, methods and technologies for Sustainable textiles; Development of sustainable, eco-friendly textile products and processes; Environmental standards for textile industry; Modelling of environmental impacts of textile products; Green Chemistry, clean technology and their applications to textiles and clothing sector; Eco-production of Apparels, Energy and Water Efficient textiles. Sustainable Smart textiles & polymers, Sustainable Nano fibers and Textiles; Sustainable Innovations in Textile Chemistry & Manufacturing; Circular Economy, Advances in Sustainable Textiles Manufacturing; Sustainable Luxury & Craftsmanship; Zero Waste Textiles.

More information about this series at https://link.springer.com/bookseries/16490

Subramanian Senthilkannan Muthu · Ali Khadir
Editors

Advanced Oxidation Processes in Dye-Containing Wastewater

Volume 2

Editors
Subramanian Senthilkannan Muthu
SgT Group and API
Hong Kong, Kowloon, Hong Kong

Ali Khadir
Western University
London Ontario, ON, Canada

ISSN 2662-7108 ISSN 2662-7116 (electronic)
Sustainable Textiles: Production, Processing, Manufacturing & Chemistry
ISBN 978-981-19-0884-2 ISBN 978-981-19-0882-8 (eBook)
https://doi.org/10.1007/978-981-19-0882-8

This Springer imprint is published by the registered company Springer Nature Singapore Pte Ltd.
The registered company address is: 152 Beach Road, #21-01/04 Gateway East, Singapore 189721, Singapore

Contents

About the Editors

Dr. Subramanian Senthilkannan Muthu currently works for SgT Group as Head of Sustainability, and is based out of Hong Kong. He earned his Ph.D. from The Hong Kong Polytechnic University, and is a renowned expert in the areas of Environmental Sustainability in Textiles and Clothing Supply Chain, Product Life Cycle Assessment (LCA), and Product Carbon Footprint Assessment (PCF) in various industrial sectors. He has 5 years of industrial experience in textile manufacturing, research and development, and textile testing, and over a decade of experience in life cycle assessment (LCA), and carbon and ecological footprints assessment of various consumer products. He has published more than 100 research publications, written numerous book chapters, and authored/edited over 100 books in the areas of Carbon Footprint, Recycling, Environmental Assessment, and Environmental Sustainability.

Dr. Ali Khadir is an environmental engineer and a member of the Young Researcher and Elite Club, Islamic Azad University of Shahre Rey Branch, Tehran, Iran. He has published several articles and book chapters in reputed international publishers, including Elsevier, Springer, Taylor & Francis, and Wiley. His articles have been published in journals with IF of greater than 4, including Journal of Environmental Chemical Engineering and International Journal of Biological Macromolecules. He also has been the reviewer of journals and international conferences. His research interests center on emerging pollutants, dyes, and pharmaceuticals in aquatic media, advanced water, and wastewater remediation techniques and technology.

Fenton Process in Dye Removal

Maicon S. N. dos Santos, Carolina E. D. Oro, João H. C. Wancura, Rogério M. Dallago, and Marcus V. Tres

Abstract The Fenton process is considered one of the main Advanced Oxidation Process (AOP) applied for the treatment of industrial effluents and wastewaters. Synthetic dyes are mainly discharged in the environment and are characterized as highly injurious components for human health and nature. In this context, a suitable treatment for the efficient removal of these elements is encouraging. Correspondingly, this chapter includes topics referring to the application of the Fenton process for dye removal in industrial effluents. Initially, the chapter presents general characteristics of Fenton processes, homogeneous and heterogeneous Fenton processes, and general properties and characteristics of dye removal. Furthermore, the main applications of Fenton processes for dye removal are included. In conclusion, future outlooks and technological challenges will be provided.

Keywords Dyes · Effluents · Environmental pollutants · Fenton process · Industrial effluents · Organic compounds · Wastewater · Wastewater treatment

1 Introduction

The intense industrial advance experienced in recent years resulted in large amounts of highly contaminating effluents directed to the environment. These contaminants originated from many fields of application, such as pharmaceuticals, cosmetics,

M. S. N. dos Santos · M. V. Tres (✉)
Laboratory of Agroindustrial Processes Engineering (LAPE), Federal University of Santa Maria (UFSM), 1080 Sete de Setembro St., Center DC, Cachoeira do Sul, RS 96,508-010, Brazil
e-mail: marcus.tres@ufsm.br

C. E. D. Oro · R. M. Dallago
Department of Food Engineering, Regional Integrated University of Alto Uruguai e das Missões (URI), 1621 Sete de Setembro St., Fátima DC, Erechim, RS 99,709-910, Brazil

J. H. C. Wancura
Department of Chemical Engineering, Federal University of Santa Maria (UFSM), 1000 Roraima Av., Camobi DC, Santa Maria, RS 97,105-900, Brazil

S. S. Muthu and A. Khadir (eds.), *Advanced Oxidation Processes in Dye-Containing Wastewater*, Sustainable Textiles: Production, Processing, Manufacturing & Chemistry, https://doi.org/10.1007/978-981-19-0882-8_1

textiles, health care, and pesticides, and are highly unfavorable to human and environmental health [10, 61]. Moreover, it is estimated that approximately 500 million tons of these contaminants are conducted to the environment annually [76]. These effluents present high concentrations of organic materials and essentially toxic compounds and represent a significant risk to the environment and to the maintenance of indispensable natural resources [20, 57]. The harmful potential of these contaminants is irrefutable due to the disintegration of natural resources, the loss of characteristic properties, and the mutation of different biochemical processes that occur in organisms and in the organism–environment association [60]. A high deal of emphasis is associated with a large volume of industrial organic pollutants, particularly synthetic dyes.

Synthetic dyes are characterized as highly soluble organic composite materials in water and wastewater from a range of industries, mainly related to textile manufacturing [36]. The scenario of the production of dyes is significant since more than 700,000 tons of dyes are produced annually and about 15% of this total is lost in the environment [26]. The removal of this type of material is arduous and, consequently, its permanence gives undesirable colorations to the water and presents a high potential for contamination. Effluents with high dye concentrations are extremely toxic to the edaphoclimatic balance of soils and surface waters [90]. Furthermore, continuous exposure to this type of element can cause irreparable damage to human health, such as cancer of different classifications, such as lung, bladder, breast, etc. [67, 70]. Nonetheless, high concentrations of dyes in water bodies drastically affect the potential for water reoxygenation and light penetration, making the aquatic balance impossible, and minimizing the physicochemical quality of these resources [26]. This perturbing scenario indicated the necessity to adopt degradation measures for these materials and alternatives for maximum elimination of dyes during effluents treatment.

The adoption of wastewater treatment practices aimed at removing dyes has been widely explored in the literature. Treatment protocols require the application of certain strategies that involve different types of treatment (primary, secondary, and tertiary treatments) and processes (biological, physical, and chemical treatment processes) [58]. These strategies comprise widely used methods such as adsorption, ozonation, anaerobic treatment, and coagulation/flocculation [47, 80, 86]. Nevertheless, the implementation of only one type of treatment process is insufficient for the adequate removal of dyes and an efficient elimination can be achieved through the investigation of a range of integrated methods [79]. However, the performance of these methods is restricted due to the high costs generated and the sludge produced, which is a high pollutant and toxic product to nature [5]. Furthermore, biological treatments are limited due to their non-biodegradability in the environment [57]. Thus, processes that provide an eco-friendly approach and economic viability are encouraging for the removal of toxic compounds from industrial effluents [5].

AOPs, such as the application of the Fenton reaction (H_2O_2/Fe^{2+}), are replaceable strategies to enhance the degradation of organic compounds due to the high reactivity of hydroxyl radicals (OH-), originating from the catalytic decomposition of the H_2O_2 compound, in reaction with Fe^{2+} (Eq. 1) [10]:

$$Fe^{2+} + H_2O_2 \rightarrow Fe^{3+} + OH^{\bullet} + OH^{-} \quad (1)$$

The free radicals have a strong oxidizing capacity, which provides removal of most pollutants present in industrial effluents [43]. The high efficiency in the color removal and mineralization process of the compounds combined with the availability and facility of handling provide Fenton with an excellent role in wastewater treatment [5]. Furthermore, the Fenton process is economically advantageous, since Fe^{2+} can be easily eliminated from the solution and the H_2O_2 is decomposed into non-toxic substances (H_2O and O_2) [1]. However, the high degradation of the compounds is significantly influenced by the operating conditions (temperature, pressure, pH, etc.), which causes an increase in the cost of the process [89].

An acidic medium with pH 3 becomes the best condition for the generation of hydroxyl radicals from the Fenton reaction [27]. Furthermore, kinetic studies have shown that the maximum potential of the Fenton process can be reached at temperatures of 60 °C, with removal of up to 99% color and 65% COD (Chemical Oxygen Demand) [12]. Correspondingly, it is verified that many scientific studies have been performed with the objective of optimizing the operational conditions of the process and potentiating the dye removal through the Fenton method. Thus, the application of Fenton's reaction in a range of production lines has been explored with highly promising results, such as carpet industry (93% color removal and 98% COD removal) [33], pharmaceutical wastewater (up to 82% COD removal) [81], cosmetic wastewater (72% COD removal) [44], and textile wastewater (95% color removal) [7], (up to 92%, up to 89%, and up to 94.3%, for color, COD, and turbidity, respectively) [24].

Additionally, it is pertinent to evaluate the prospect of studies aimed at the application of the Fenton reaction in the elimination of dyes, especially in industrial effluents. The Fenton process is widely explored for wastewater treatment and is described as an innovative strategy and efficient and environmentally friendly approach. Based on the presupposed, basic research to quantify scientific publications published in the last 5 years was conducted in scientific platforms that are widely investigated by the academy: Scopus®, Science Direct®, Web of Science®, and Science Research®. To perform the data research, the following keywords were adopted: *Fenton's reaction*; *Fenton's reaction/dye removal*; and *Fenton's reaction / dye removal/wastewater treatment* (Fig. 1).

According to the application of the Fenton reaction for the elimination of dyes, the scientific scenario is successful and has improved over the years. As reported by the Science Research® database, 104 manuscripts were published in 2019, significantly higher than the 36 articles published in 2016. Even though 2020 shows a slight decline compared to 2019 on most scientific platforms, the scenario when exploring the Fenton process to the removal of dyes in effluents is promising and has been potentialized for application in the next few years worldwide. Accordingly, an investigation of the scientific production in a geographical context, demonstrating the worldwide distribution of the Fenton's reaction applied to the dye removal treatment, is relevant (Fig. 2).

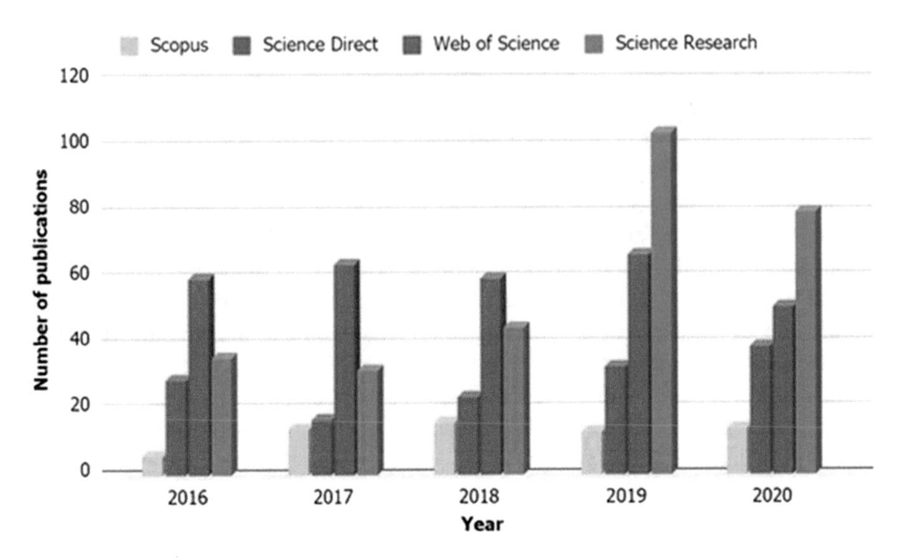

Fig. 1 Worldwide quantification of scientific publications on the Fenton's reaction applied to the dye removal in wastewater treatments distributed by year from 2016 to 2020 according to the Scopus®, Science Direct®, Web of Science®, and Science Research® scientific platforms according to the following keywords: *Fenton's reaction*; *Fenton's reaction/dye removal*; and *Fenton's reaction/dye removal/wastewater treatment*

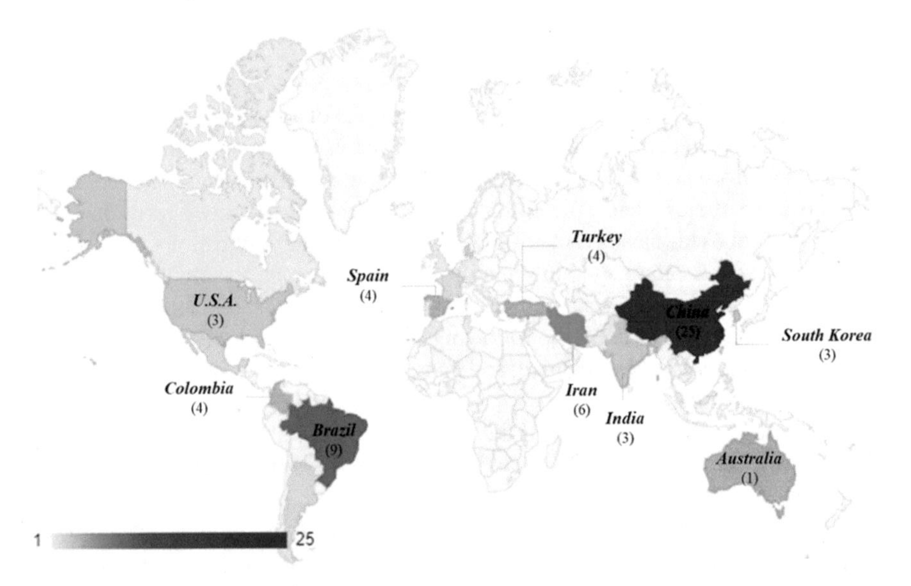

Fig. 2 Geographical dispersion of the worldwide scientific publications on the Fenton's reaction applied to dye removal in wastewater treatments distributed by year from 2016 to 2020 according to the Scopus®, Science Direct®, Web of Science®, and Science Research® scientific platforms according to the following keywords: *Fenton's reaction*; *Fenton's reaction/dye removal*; and *Fenton's reaction/dye removal/wastewater treatment*

Considering the intense volume of effluents produced over the years, the adoption of treatment practices for these materials is widely investigated in countries with high potential and industrial and economic leadership, such as China, the U.S.A., Brazil, and India. The industrial projection and population expansion of these countries motivated a scenario in which the demand for effluent treatments has become excellent environmental and economic strategies. Furthermore, essentially desert countries, such as Iran, or with extensive territories characterized by unequal access and distribution of water, such as Turkey and Spain, the necessity for an adequate configuration regarding the treatment of residual and industrial effluents on a large scale. Appropriately, a large proportion of scientific studies on this subject has been examined recently.

Overall, the aim of this study is to perform a comprehensive and encouraging review on Fenton process applied to dye removal for wastewater treatment. Additionally, general characteristics of processes involving the Fenton reaction, broad application of the Fenton process in dye removal, and future perspectives are presented.

2 General Characteristics of Fenton Processes

Wastewater effluents generated in various industrial segments are usually characterized and differentiated by the diversity and quantity of their pollutants, and by the presence of numerous compounds, which can be organic and/or inorganic. One of the main problems faced by conventional effluent treatment plants, that is, based on the use of primary (physicochemical) and secondary (biological) processes, is the presence of non-biodegradable organic compounds, such as pesticides, drugs, dyes, among other compounds. These recalcitrant/non-biodegradable compounds are characterized by their persistence when subjected to conventional biological treatments and by phase transfer when using adsorption processes and/or chemical coagulation, leading to their concentration [45, 54].

In order to remedy these problems, oxidative systems have been increasingly highlighted in recent years. They are classified as tertiary treatment and their success is mainly due to their mineralization process, which converts pollutants into water, carbon dioxide, and inorganic ions, leading to low levels of contaminants and residues at the end of the process [84, 92].

In this context, AOPs have been arousing interest. AOPs are defined as processes based on the employment and in situ generations of strong oxidants for the degradation of organic compounds. Traditionally, the vast majority of AOPs are based on the formation of the hydroxyl radical (•OH). More recently, oxidation processes based on sulfate and chlorine radicals have also been studied [9, 45].

The hydroxyl radical with a standard reduction potential (E°) of 2.8 V is the oxidizing agent with the highest known potential, being only inferior to fluorine, the top element of the electrochemical series ($E^\circ = 3.0$ V). The other oxidizing agents commonly used in the chemical oxidative treatment of wastewater are ozone ($E^\circ =$

2.1 V), hydrogen peroxide (E° = 1.77 V), hypochlorite (E° = 1.5 V), and chlorine gas (E° = 1.4 V) [6, 52]. In this context, the potential use of the hydroxyl radical in oxidative treatment processes is justified, especially concerning the degradation of refractory/persistent compounds, promoting their total mineralization [9].

Miklos et al. [45] classified AOPs into five categories, namely based on reactions with ozone, based on processes with UV radiation, electrochemical, catalytic, and physical. However, this is a general classification, and it is important to emphasize that some AOPs may fall into more than one category, as they use combined technologies for the treatment of water with the presence of various pollutants, such as antibiotics, herbicides, insecticides, endocrine disruptors, products of personal care, pharmaceuticals, dyes, and organic matter [14]. One of the advantages of AOPs is that they usually operate at ambient temperature and pressure. Furthermore, they are considered versatile processes since different systems of reagents can be used to generate the oxidant, allowing for a better adaptation with the specific requirements for each type of treatment intended.

Among the AOPs, the Fenton process is widely studied and is based on the dissociation of hydrogen peroxide (H_2O_2) into hydroxyl radicals due to the catalytic effect of ferrous ion (Fe^{2+}) or metallic iron (Fe^0) [64]. Fenton's reaction was discovered and reported by H.J.H. Fenton in 1894 on the oxidation of tartaric acid in the presence of iron. The author observed that the ferrous ion acts as a catalyst and can activate H_2O_2 in the tartaric acid oxidation reaction [15].

The Fenton process is based on the reaction between ferrous ions and hydrogen peroxide. Ferrous ions catalyze the decomposition of hydrogen peroxide through a redox reaction, in which Fe^{2+} ions act as an anode (Eq. 2), providing electrons, and H_2O_2 as a cathode (Eq. 3), consuming electrons, resulting in the generation of hydroxyl radicals (Eq. 4) [31, 84].

$$2Fe^{2+} \rightarrow 2Fe^{3+} + 2e^{-} \text{Anode reaction} \tag{2}$$

$$H_2O_2 + 2e^{-} \rightarrow \bullet OH + OH^{-} \text{Cathode reaction} \tag{3}$$

$$2Fe^{2+} + H_2O_2 \rightarrow 2Fe^{3+} + \bullet OH + OH^{-} \text{Global reaction} \tag{4}$$

The spontaneity of this redox reaction can be proven by the electromotive force of the redox system involved (Eq. 3), where Fe^{2+}, with an E°= 0.77 V, acts as the anode and H_2O_2, with an E°= 1.77 V as the cathode, whose final global reaction value is E°= +1.0 V.

The concepts of electrochemistry determine that, for a redox reaction to occur spontaneously, the element/compound that oxidizes, acting as an anode (that is, as a supplier of the electrons that hydrogen peroxide needs to decompose) must present reduction potential inferior to the element/compound that is reduced (in this case the hydrogen peroxide). Thus, any element or compound with a reduction potential lower than that of H_2O_2 (E°= 1.77 V), such as ferrous ions (E°= 0.77 V), is electrochemically enabled, in its reduced form, as a possible promoter of radical decomposition of hydrogen peroxide.

The success of the Fenton process can be attributed, in part, to the ease of operation and flexibility for industrial implementation, in addition to the high efficiency in the mineralization of pollutants [6].

The Fenton process, when compared to other AOPs, has some advantages, such as (1) energy is not required to activate the chemical reaction to form the hydroxyl radical (the use of energy is only for pumping the effluent to the treatment), (2) the reaction takes place at room temperature and atmospheric pressure, (3) relatively short reaction times, and (4) the operation is simple, flexible, and easily industrially implemented [56, 84].

2.1 Variables that Affect the Process

In Fenton systems, the main parameters to be considered are the pH and concentrations of H_2O_2 and reduced iron in the reaction medium [3], as well as the use of promoting agents, such as short-chain carboxylic acids, like formic acid (HCOOH) and acetic acid (H_3CCOOH), as all these parameters can significantly affect the efficiency of the process [17].

Regarding pH, the literature recommends, for a better performance of the process, the use of values between 2.5 and 3.5 [35, 46]. pH values lower than 2.5, due to the excess of $H^+_{(aq)}$ ions (protons) in the reaction medium, can lead to the consumption of hydroxyl radicals (Eq. 5). At pHs > 4.0, ferrous ions tend to hydrolyze the water, which occurs concomitantly with its oxidation, leading to their precipitation in the form of ferric hydroxide ($Fe(OH)_3$). In both cases, the interference is negative for the efficiency of the process [84].

$$\bullet OH + H^+ + e^- \rightarrow H_2O \quad (5)$$

Considering that the pH adjustment of the reaction medium, when necessary, tends to be carried out by the addition of an acid and that the active form of iron is the reduced one, some considerations must be observed, mainly concerning the character (oxidizing, reducing, or dehydrating) of the acid employed. Nitric acid (HNO_3), with an oxidizing character, can negatively interfere in the process since it can lead the oxidation of ferrous ions (Fe^{2+}) to ferric ions (Fe^{3+}), without activity for the catalytic decomposition of H_2O_2 into hydroxyl radical. Hydrochloric acid (HCl), with a reducing character, would be the most recommended. However, its ability to oxidize to chlorine gas when in the presence of a strong oxidizing agent, such as dichromate ($Cr_2\,O_7^{2-}$) and hydroxyl radical, can also negatively interfere in the process. In this context, the most recommended acid for pH correction is sulfuric acid (H_2SO_4), an acid with dehydrating characteristics, whose oxidative action is only manifested in the presence of high temperatures (>150 °C) [71].

Regarding the ferrous ions used as catalyst, when present in excess (above their optimal concentration, which should be obtained experimentally), due to their reducing character, they tend to have a negative effect on the Fenton process due to the consumption of hydroxyl radicals (Eq. 6), of oxidizing character. This effect is also evidenced when there is no substrate to be oxidized [3, 84].

$$Fe^{2+} + \bullet OH \rightarrow Fe^{3+} + OH^- \ K = 2,5 - 5 \text{ x } 10^8 \left(L.\ mol^{-1}.s^{-1}\right) \quad (6)$$

During the Fenton process, parallel reactions of radical–radical and radical–hydrogen peroxide types can also occur [3]. The concentration of hydrogen peroxide used as a hydroxyl radical precursor must be optimized experimentally, as in excess it will act as a hydroxyl radical scavenger [3, 18, 84] (Eqs. 7–9), eliminating it from the reaction environment, negatively interfering in the efficiency of the degradation process of organic pollutants.

$$\bullet OH + \bullet OH \rightarrow H_2O_2 K = 5 - 8 \times 10^9 \left(L.\ mol^{-1}.s^{-1}\right) \quad (7)$$

$$H_2O_2 + \bullet OH \rightarrow HO_2 \bullet + H_2O K = 1,7 - 4,5 \times 10^7 \left(L.\ mol^{-1}.s^{-1}\right) \quad (8)$$

$$HO_2 \bullet + \bullet OH \rightarrow H_2O + O_2 \quad K = 1,4 \times 10^{10} \left(L.\ mol^{-1}.s^{-1}\right) \quad (9)$$

Other factors that influence the Fenton process are the type of effluent to be treated, the concentration of organic contaminants and inorganic salts, reaction time, and precursor of the iron catalyst used in heterogeneous systems [46], in addition to the promoting effect provided by the presence of short-chain carboxylic acids, such as formic and acetic acids [17], and by phenolic compounds [2].

The promoting effect of carboxylic acids was demonstrated experimentally by Ferras et al. (2007) for the degradation of dyes with H_2O_2 and iron (II) supported as a catalyst. The tests were conducted in the absence and presence of formic acid, at a concentration corresponding to 5% (mol/mol) in relation to the peroxide, showing, respectively, 30% and 100% removal of the color from the medium with 5 min of reaction. Computational calculations in Quantum Mechanics have shown that the energy bound to the hydroxyl radical resulting from the peracid (resulting from the reaction between hydrogen peroxide and formic acid), with 64.7 kcal/mol, is approximately 2.5 times higher than that of the hydroxyl radical produced from hydrogen peroxide, with 25.4 kcal/mol [17].

A similar effect on degradation speed was observed by [2] for other classes of compounds, such as phenols. The favorable effect observed was linked to the ability of phenols to reduce iron (III) ions to iron (II), considered the most active species for the Fenton reaction. The complexing effect of some compounds to iron can also favor the Fenton reaction, delaying its oxidation from Fe^{2+} to Fe^{3+}.

Other factors that influence the Fenton process are the type of effluent to be treated, the concentration of organic contaminants and inorganic salts, reaction time, production of other compounds (by-products) due to the non-selective character of the hydroxyl radical, and the temperature. Furthermore, the type of catalyst to be used in the heterogeneous Fenton can also influence the process [46].

2.2 *Homogeneous Versus Heterogeneous Fenton Processes*

The Fenton process has peculiarities inherent to the homogeneous and heterogeneous processes that differentiate them, such as the number of phases and the form (in solution or solid) of the iron precursor used. In addition, the mode of addition of hydrogen peroxide must be conducted differently between the systems and will determine the efficiency of the process.

2.2.1 Homogeneous Fenton Processes

The homogeneous system, which is characterized by having a single-phase, demands that the active iron be found soluble in solution, that is, as ferrous ions ($Fe^{2+}_{(aq)}$). This feature provides a uniform distribution of ferrous ions in the reaction medium and favors the catalytic decomposition reaction of H_2O_2 into •OH radicals. The decomposition reaction manifests itself instantly and requires contact with the pollutant to conduct its degradation, because of the oxidative character of •OH. In the absence of pollutants and excess H_2O_2, •OH tends to self-oxidize, through parallel reactions, leading to the generation of water and carbon dioxide, providing a competitive effect [14, 32, 56].

When in excess, H_2O_2 can act as a hydroxyl radical scavenger in parallel reactions, forming with it the hydroperoxyl radical (HO_2•), with lower reduction potential ($E°=$ 1.42 V) (Eq. 7), which can react with another HO• molecule, generating water and molecular oxygen, according to Eq. 8.

In this context, the addition of hydrogen peroxide in a single step is avoided (Fig. 1), as this condition would contribute to the competitive effect, since it leads to the generation of point environments within the reaction medium containing a high concentration of hydroxyl radicals and excess H_2O_2 in relation to pollutants, thus reducing the efficiency of the process. Figure 3 illustrates the Fenton process and the result of the addition of hydrogen peroxide in a single step in the removal of color from an effluent.

For illustrative purposes, it is considered that the volume (1V) of H_2O_2 added, in theory, would be enough to remove all the color of the effluent. However, this is not what is observed due to the competitive and negative effect of the excess of H_2O_2 in a specific region of the reaction medium, provided by the addition in a single step of all the H_2O_2 (1V). Due to the instantaneity of the hydroxyl radical generation reaction associated with its reactivity, this form of addition prevents the H_2O_2 from accessing the pollutant molecules (dye) in the reaction medium. That is, the molecules located at the extremes of the reactor (a region not contemplated with the addition of H_2O_2) and the efficiency of the process is impaired.

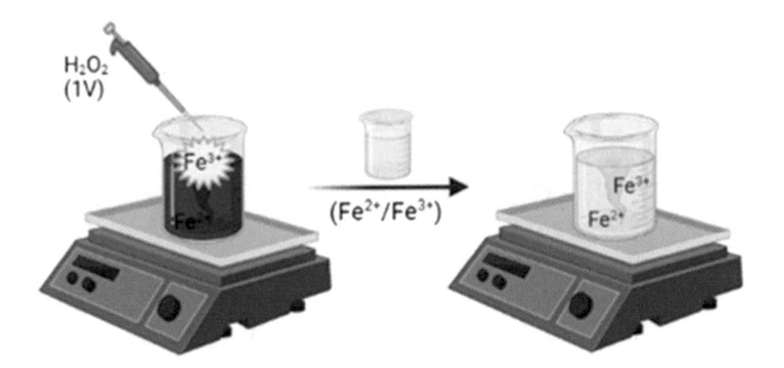

Fig. 3 One-step addition of hydrogen peroxide (1V) in effluent containing blue dye

To overcome this problem, the addition of hydrogen peroxide in homogeneous Fenton systems must be carried out continuously, by dripping or spraying, to improve its distribution in the reaction medium, enhancing the process. Figure 4 illustrates the same test described in Fig. 3, with the difference that the volume of H_2O_2 (1V) is added fractionally in four steps, each corresponding to ¼ of the volume (¼V).

The fractional addition of H_2O_2, by allowing a homogenization of the reaction medium before each new addition, allows pollutant/dye molecules that were present in solution locations not covered with H_2O_2 in the first addition, to access the hydroxyl radicals generated in the addition subsequent, thus enhancing the efficiency of the process. This improvement in efficiency is illustrated in Fig. 2 by the decrease in the intensity of the blue color between additions (¼V) and the appearance of an orange coloration at the end of the process, as a consequence of the oxidation of ferrous ions (Fe^{2+}) to ferric ions (Fe^{3+}). The minimization of the competitive effect of H_2O_2 also contributes to improving the efficiency of the process, since the fractional addition employs smaller volumes in each addition, and leads to the generation of point regions in the reaction medium with lower concentrations of H_2O_2 in relation to the concentration of pollutants.

The presence of ferric ions in solution after the Fenton process is one of the main drawbacks to its application, as it leads to an effluent with a high concentration of iron in solution, normally above 10.0 mg/L. For this reason, the Fenton homogeneous process is normally used in association with additional treatment, such as chemical treatment. For this, it is necessary to alkalinize the reaction medium, intended for the removal of ferric ions in the form of ferric hydroxide ($Fe(OH)_3$) (Fig. 5), at values

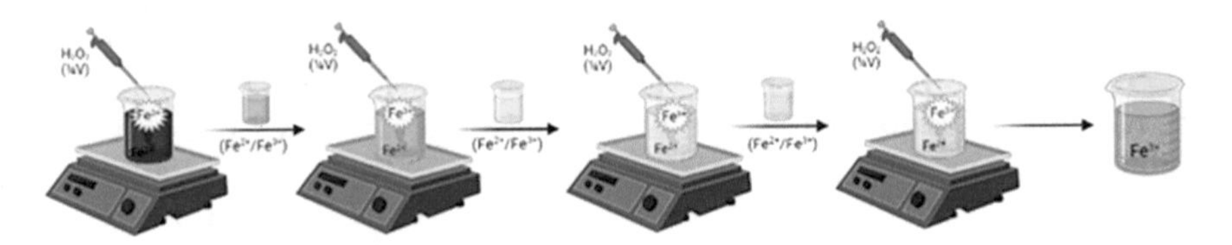

Fig. 4 Stepwise addition of hydrogen peroxide (¼V) to effluent containing blue dye

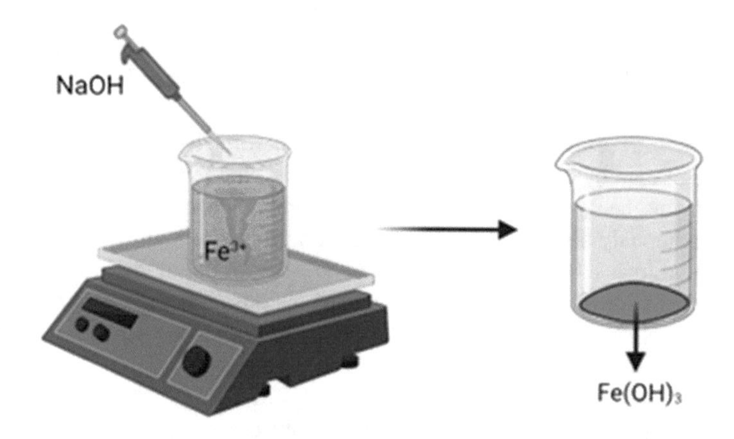

Fig. 5 Addition of NaOH for precipitation of Fe3+ in the form of sludge to obtain a clean effluent

that meet the legislation, resulting in the generation of sludge and additional costs to the treatment [31, 46].

2.2.2 Heterogeneous Fenton Processes

Heterogeneous Fenton processes are characterized by the presence of two phases: one solid and one liquid. Of these, the solid phase corresponds to the iron catalyst, which must be in one of its reduced states (Fe^0 or Fe^{2+}), whether as metallic iron, an alloy, an oxide or supported.

Of these, the use of iron oxides (magnetite, ferrihydrite, hematite, goethite, schwertmannite, lepidocrocite, and maghemite) derived from mining processes, as potential precursors of iron, in Heterogeneous Fenton processes are confirmed by the literature [77]. It should be noted that its direct applications are ineffective, due to the iron being oxidized in the form of ferric ions. For this reason, a pre-treatment step is needed, usually with H_2 and heat, aimed at reducing the iron (III), present in its structures, to Fe(II) or Fe^0. The degree of reduction is directly associated with temperature, time of concentration of H_2 used [17].

The heterogeneous system, for having two phases, allows a more effective separation of the catalyst from the reaction medium, through filtration. It also allows its use in continuous flow systems, in which H_2O_2 is added to the effluent before contact with Fe catalytic precursor contained inside the reactor. It should be noted that the reaction will occur until all the reduced iron present in the reactor is oxidized.

Another important feature in the heterogeneous process that deserves to be highlighted is the way of adding peroxide in batch mode, which can vary according to the order of addition of the reactive (iron and H_2O_2). When H_2O_2 is added to the effluent before the iron, its addition must be in a single step, leading to the generation of a solution containing the pollutant and H_2O_2 homogeneously distributed. This contributes to an effective contact between them and will favor the degradation of the pollutant in the generation of the hydroxyl radical by the subsequent addition of the reduced iron catalyst in a solid state to the reaction medium, generating a heterogeneous system.

When the iron catalyst is added to the effluent before H_2O_2, it can be added in a single step or also in a fractional way. In both cases, the hydroxyl radical will only manifest itself when in contact with the reduced iron on the catalyst surface, which is not homogeneously distributed, but rather in certain places in the reaction medium. This feature allows a better distribution of H_2O_2 temporally in the reaction medium before contact with the iron catalyst, minimizing the possibility of generating regions with high concentrations of H_2O_2 in relation to the pollutant and, consequently, its competitive effect.

Figure 6 illustrates the addition of H_2O_2 (1V) in a single time, in two distinct sequences concerning heterogeneous iron. In both cases, the total removal of the blue color is observed over time, since the volume of peroxide (1V) used, in theory, is enough for this to occur, only if the competitive effect is not significant, as observed in the homogeneous system for this same volume (Fig. 6).

Depending on the Fe catalyst, it may or may not leach from the solid phase to the reaction medium in the form of ferric ions. When this occurs, after the iron precursor removal step, the solution tends to present an orange color. In these cases, the precipitation and removal of the iron present in the solution must be carried out as well as in the homogeneous process (Fig. 6). Therefore, the choice of iron precursor is important. In addition to ensuring reaction efficiency, the right catalyst avoids further precipitation and separation steps. However, when precipitation is necessary, as little sodium hydroxide (NaOH) as possible should be used.

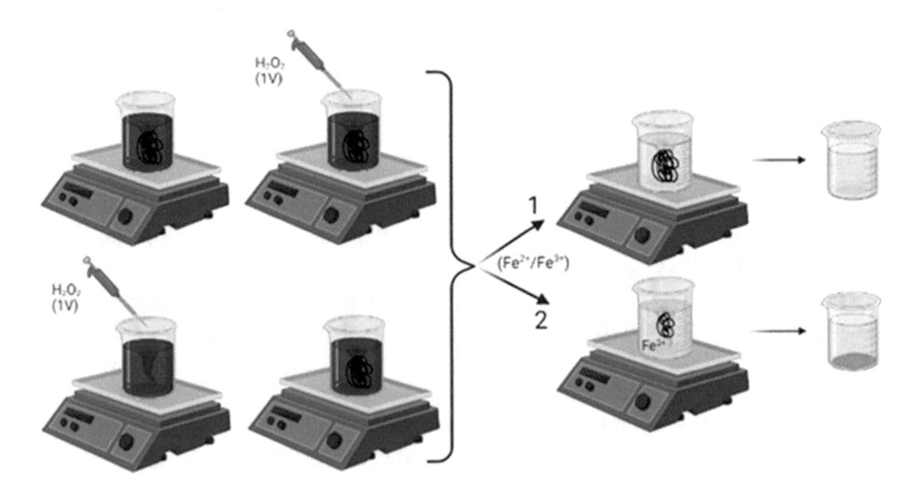

Fig. 6 Addition of hydrogen peroxide in one step (1V) in effluent containing blue dye by the heterogeneous Fenton process

2.2.3 Applications of the Fenton Processes

Usually, the Fenton process, as well as the other AOPs, are used in association with other treatment processes (physical-chemical and biological) with the purpose of enhancing the removal of undesirable characteristics in wastewater. This optimizes effluent treatment to meet the parameters of current environmental legislation.

This associated application can occur as *(i) first stage*: before a biological treatment when the effluent is recalcitrant, toxic, or non-biodegradable. In this way, an increase in its biodegradability is obtained, to remove the organic load; or *(ii) final stage:* after the primary and/or secondary treatments, as a final polishing to meet the legislation parameters.

2.3 General Properties and Characteristics of Dye Removal

Several industrial segments use synthetic dyes in their processes, such as the textile, leather, food, cosmetics, paper, pharmaceutical industries, among others. Of these, the textile sector stands out as the main consumer of inks and dyes, and consequently, as one of the main generators of liquid effluents with pigmentation due to the high volume of water used during the various stages of the fabric process [51]. Estimates show that to produce 1 kg of fabric approximately 200 L of high-quality water is needed to avoid interfering with the stability of the process [69]. It is estimated that 10–20% of the inks and pigments are lost during the preparation, dyeing, and rinsing phases of the fabric pieces [25]. After the dyeing process, textile wastewater must undergo some treatment capable of removing or degrading complex organic substances before its disposal. In addition to coloring, wastewater has a high pH and suspended solids [73]. The presence of dyes in water is easily visible (less than 1 $mg.L^{-1}$ for some dyes) and undesirable. In addition to the impact caused by the change in color, dyes contribute to the increase in the non-biodegradable organic load content of the effluent. Furthermore, the strong colors of these effluents negatively affect the aquatic environment, causing a reduction in the ability of ultraviolet radiation to penetrate, harming the aquatic biota [51].

Dyes and inks are classified by their application, structure, color, or particle charge in solution, and are further separated into the categories of synthetic and natural dyes [73, 93]. Of these, synthetic dyes are still the most used due to their high dyeing power and chemical stability (to light, heat, and pH variation), providing uniformity and color stability to the products, in addition to low cost [18]. However, they are characterized by presenting greater difficulty in terms of removal, especially in biological systems as they are not biodegradable. In some cases, they present toxicity against microorganisms present in the activated sludge destined for their treatments, inhibiting their actions [50]. Therefore, the Fenton process can contribute to improving biodegradability in biological processes, being used in an associated way.

Considering the literature reports, this improvement in biodegradability must be experimentally evaluated for each effluent. In some cases, the by-products of the oxidative process can show higher toxicity compared to the pre-treated effluent [16].

In this context, to evaluate the efficiency of the Fenton process in removing dyes, the use of UV–VIS spectrophotometric techniques is not recommended, since the coloration of a compound is directly linked to its chemical structure. Changes in this structure, no matter how simple it is, for example, a simple rupture of a double bond, can lead to loss of color due to changes in the resonance structure of the molecule (responsible for its capacity to absorb radiation), caused by the decrease the number of conjugated double bonds. This means that the loss of color does not necessarily represent the mineralization of the molecule [71].

In this context, the parallel use of other parameters is recommended, such as analysis of Total Organic Carbon (TOC), Chemical Oxygen Demand (COD), Biochemical Oxygen Demand (BOD), and toxicity, which allow an assessment of the alteration or removal of organic load, as well as its biodegradability and possible toxicity. For example, the dye degradation rate (k_{dye}) and the mineralization rate (k_{TOC}) can be calculated according to Eqs. 10 and 11 [37].

$$ln\frac{C_0}{C_t} = k_{dye}.t \quad (10)$$

$$ln\frac{TOC_0}{TOC_t} = k_{TOC}.t \quad (11)$$

where C_0 and C_t are the dye concentrations at time 0 and reaction time t, respectively. TOC_0 and TOC_t are the TOC concentrations at time 0 and reaction time t, respectively.

3 Applications of Fenton Processes in Dye Removal

3.1 Recent Proposals of Fenton Processes for Dye Removal

Dyes are coloring agents widely applied in diverse types of chemical processes at the industrial level such as textile, food, leather, plastic, cosmetic, paper, pharmaceutical, paint, etc.

Nidheesh et al. [49, 50] cite that up to 15% of the dyes used annually in industries enter the environment as wastes. With a typical chemical oxygen demand ranging from 150 to 12,000 mg·L^{-1}, total suspended solids between 2,900 and 3,100 mg·L^{-1} and biochemical oxygen demand ranging from 80 to 6,000 mg·L^{-1}, these compounds impede the entrance of light and oxygen in aquatic systems, causing water eutrophication, under-oxygenation, and odor alteration of the medium [55, 83].

Among the different techniques currently used to remove dyes from wastewater, Fenton process has gained attention in recent years of researchers of the area. Considering that the homogeneous Fenton process is a well-established technology and in virtue of the inconveniences associated with it, mainly in relation to the sludge formed, the heterogeneous Fenton process has been the main topic of researches recently published about the subject. Such researches have focused on the development of novelties Fe-based catalysts anchored at distinct supports, returning interesting results. For example, [41] prepared a catalyst compounded by Fe^{2+} anchored at two-dimensional molybdenite (from natural bulk molybdenite) and used as heterogeneous Fenton catalyst to degrade methylene blue in the presence of light. According to the results presented, the catalyst demonstrated high catalytic activity on dye elimination, showing 100% of degradation efficiency in five cycles with a slight leaching amount of Mo and Fe ions, suggesting the solid stability and reusability in the H_2O_2 system. In another interesting research, [94] investigated the performance of $Fe_{80}Si_{10}B_{10}$ and $Fe_{83}Si_5B_8P_4$ amorphous ribbons for the degradation of methylene blue and Rhodamine B. At 298 $_K$, pH 3, H_2O_2 concentration of 1 mmo·L^{-1}, 0.5 ^{g}·L^{-1} of catalyst, and dye concentration of 20 mg·L^{-1}, the authors reported for both amorphous ribbon samples 95% of dyes decomposition within 11 min. Moreover, additional assays demonstrated the high stability and reusability of the compounds, where both catalysts could be reutilized up to 26 times without significant losses on catalytic efficiency. In summary, Table 1 presents some recently published works about dye removal from water solutions using the Fenton technique.

3.2 *Fenton Process Applied to Treatment of Real Wastewater Containing Dyes*

In terms of the Fenton process, it is usual the publication of researches where developed catalysts are tested in real industrial wastewater instead of only synthetic effluents (dyes solutions, for example). Such researches aim to prove the efficiency of a given compound/process in a real effluent, increasing the industrial appeal of the study. Ribeiro and Nunes [63] describe that the researches in this field are primarily focused on synthesizing novelty materials to substitute the conventional iron salt Fenton catalysts and, for heterogeneous process, regenerate and reutilize the catalyst in subsequent reactions.

In one of these researches, [11] investigated the treatment of textile effluents by the heterogeneous Fenton process in a fixed-bed reactor filled with a catalyst based on activated carbon impregnated with 7 wt.% of iron. Firstly, a solution of Alcian Blue-tetrakis chloride dye was used as a model: at 50 °C, 30.0 mM of H_2O_2, pH 2.5, and a catalyst to wastewater ratio of 3.3 g catalyst·(min·ml)$^{-1}$, 93.2% of discoloration was achieved and 54.1% of total organic carbon was removed. In a second moment, these best conditions found were applied to the treatment of a real textile effluent used on cotton dyeing, with a TOC content, COD content, and BOD_5 content of

Table 1 Some catalysts proposed to dye removal using Fenton processes recently published on literature

Pollutant	Catalyst	Catalyst Load	H_2O_2 Load	System pH	Duration	Efficiency	Observations	References
Methylene Blue	Basalt powder	1.0 g·L^{-1}	5.0 mM	2.0	60 min	87% degradation	100% degradation when assisted by visible light	Saleh et al. [66]
Basic Red 18		1.0 g·L^{-1}	2.5 mM	2.0	60 min	28% degradation	74% degradation when assisted by visible light	
AR14	$FeSO_4 \cdot 7H_2O$	0.2 g·L^{-1}	194 mMol L^{-1}	3.0	20 min	100% degradation		Wakrim et al. [82]
Acid Blue 29	Electric arc furnace steel activated by sulfuric acid	0.15 g·L^{-1}	6.0 mM	4.5	–	95% degradation	The catalyst kept the oxidative degradation for seven cycles	Nasuha et al. [48]
Methylene Blue		0.15 g·L^{-1}	8.0 mM	4.5	–	82% degradation		
Rhodamine B	$CuCr_2O_4/CeO_2$	10 wt%	4.0 mM	–	15 min	100% degradation		Ghorai et al. [19]
Methylene Blue		10 wt%	4.0 mM	–	20 min	98% degradation		

(continued)

Table 1 (continued)

Pollutant	Catalyst	Catalyst Load	H_2O_2 Load	System pH	Duration	Efficiency	Observations	References
Methyl Orange		10 wt%	4.0 mM	–	30 min	96% degradation		
Methylene Blue	Fe_3O_4/Natural-Zeolite	5.0 $g{\cdot}L^{-1}$	42.4 mM	6.3	1.440 min	97% removal	Kinetics showed that the reaction order changes from zero to second order according H_2O_2 load	Kuntubek et al. [34]
Rhodamine B	Ni-Cu/MWCNTs	0.01 mM	0.12 mM	3.0	50 min	86% degradation		Tariq et al. [75]
Methylene Blue	Magnetic nanoparticles	2.0 $g{\cdot}L^{-1}$	560 mM	3.5	90 min	100% removal	Assisted by magnetic heating induction	Rivera et al. [65]
Methylene Blue	PCN-250 (Fe_2Mn)	0.5 $g{\cdot}L^{-1}$	6.100 mM	-	720 min	100% removal		Kirchon et al. [30]
Amaranth	$CoFe_2O_4$/ZSM-5	0.5 $g{\cdot}L^{-1}$	8.0 mMol L^{-1}	3.0	60 min	95% removal	Assisted by visible light	Oliveira et al. [53]

(continued)

Table 1 (continued)

Pollutant	Catalyst	Catalyst Load	H_2O_2 Load	System pH	Duration	Efficiency	Observations	References
Acid Black 1	zero-valent iron-based catalyst/kaolinite	0.3 $g \cdot L^{-1}$	4.0 mM	2.0	120 min	98% decolorization	After four cycles, decolorization rate remains 72.5%	Kakavandi et al. [29]
Methyl Orange	Fe_3O_4/reduced graphene oxide	1.0 $g \cdot L^{-1}$	9.69 mM	3.0	60 min	94% removal	The process followed a first-order kinetics model	Xu et al. [88]
Methylene Blue	Mesoporous α-Fe_2O_3/SiO_2	1.0 $g \cdot L^{-1}$	1.76 mM	7.0	60 h	100% removal	Catalyst also has a good performance on pH of 3.0–11.0	Wu et al. [87]
Rhodamine B	Fe_3O_4/cellulose aerogel nanocomposite	3.0 $g \cdot L^{-1}$	99 mM	3.0	64 h	100% removal	Catalyst presented high dye degradation (97%–100%) for six reuses	Jiao et al. [28]
Acid Orange II	Fe_0-Fe_2O_3 embedded ordered mesoporous carbon	0.5 $g \cdot L^{-1}$	109.6 mM	7.0	24 h	100% removal	Kept good activity after five recycles	Jing Wang et al. [85]
Rhodamine B	Fe/Fe_2O_3	0.2 $g \cdot L^{-1}$	0.08 mM	4.3	30 min	80% removal		Sun et al. [74]
Methylene Blue	$Cu/CuFe_2O_4$	0.1 $g \cdot L^{-1}$	15 mM	2.5	4 min	99% removal		Li et al. [38]

(continued)

Table 1 (continued)

Pollutant	Catalyst	Catalyst Load	H_2O_2 Load	System pH	Duration	Efficiency	Observations	References
Methylene Blue	Reduced $CuFe_2O_4$	0.1 $g{\cdot}L^{-1}$	0.5 mM	3.2	25 min	74% removal	Kept good activity after five recycles	Qin et al. [59]
Methylene Blue	TiO_2 nanoparticles/NH_2-MIL88B(Fe)	0.2 $g{\cdot}L^{-1}$	20 mM	7.0	150 min	100% degradation	Assisted by visible light. Catalyst efficiency was not reduced after five reuses	Li et al. [39]
Acid Red 73	Fe-g-C_3N_4 graphitized mesoporous carbon	0.8 $g{\cdot}L^{-1}$	40 mM	4.0 – 10.0	40 min	99.2% removal	Assisted by visible light	Ma et al. [42]

174.7 mg·L^{-1}, 495 mg·L^{-1}, and 127.5 mg·L^{-1}, respectively. Using the synthetized catalyst, the authors reported 96.7% of discoloration and an interesting reduction of the TOC (73.6%), COD (66.3%), and BOD_5 (72.5%). The authors also tested the stability of the catalyst in five consecutive runs, without less than 5% of activity losses. Another promising result described by the authors was the low iron leaching observed, with only 1.25 wt.% of the initial Fe load lost after 60 h of operation.

Bae et al. [4] proposed a hybrid process combining biological and chemical treatment to degradation of refractory compounds present in real textile wastewater, with a flow rate of approximately 100,000 m^3 by day, generated from more than sixty dyeing factories in South Korea. The chemical treatment choose was homogeneous Fenton oxidation. The characteristics of the textile wastewater were pH 10–12, temperature of 40 °C, COD of 1,150 mg·L^{-1}, soluble COD of 1,100 mg·L^{-1}, and color of approximately 1,180 ADMI units. The authors observed that that the Fenton process only reduced the color-imparting bonds in the dye materials instead of completely degrading them (operating conditions: 4.0 mM of H_2O_2, 4.2 mM of Fe^{2+}, pH of 3.5, and 30 min). However, the Fenton process significantly reduces the soluble chemical oxygen demand by 66% and color by 73% in the raw wastewater. The authors describe that the biological treatment reduced the degradable compounds considerably, in such a way that the Fenton process could successfully eliminate recalcitrant compounds, making the hybrid system more economical.

Aiming to solve sludge management issues during dyeing wastewater treatment, [91] proposed a technique for use the excess biological sludge and ferric sludge produced in the system and to convert it into a Fenton catalyst (magnetic biochar composite) under hydrothermal carbonization conditions. The best hydrothermal conditions were evaluated by the efficacy of the obtained products in degrade methylene blue via Fenton process. The sludge was collected from a secondary settling tank of a textile and dyeing facility and the ferric sludge was collected from a Fenton pretreatment unit of the same industry. At ambient temperature, 8.8 mM of H_2O_2, 1.0 g·L^{-1} of magnetic biochar composite, pH 3.0, and 30 min, the Fenton process was able to treat the dyeing wastewater, removing the initial chemical oxygen demand and total organic carbon removal in 47% and 49%, respectively.

3.3 Aspects in Large-Sclae Applications

Homogeneous Fenton processes have been extensively evaluated and employed in dye removal treatments [72]. The processes are cost-effective, reagents are easily available and the systems can be efficiently scaled up for practical applications [40]. However, the high amounts of effluent that are precipitated as ferric hydroxide sludge when the reaction is neutralized in the post-treatment require technical intervention and maintenance, increasing the process costs [62].

Alternatively, the heterogeneous Fenton process has attracted the attention of researchers in the last years due to its capacity to eliminate non-biodegradable organic compounds, such as synthetic dyes, from wastewater solutions due to its advantage of work under a wider pH range and less generation of sludge containing iron. However, [40] cited that corrosion issues associated with the catalyst in Fenton systems remain as a drawback for large-scale industrial applications.

Although the number of studies using the Fenton process at large scales is still scarce, the focus on the development of alternatives to implement the technique at the industrial level had been explored in some researches recently published. The main issue here is associated with the typical large amounts of commercial H_2O_2 required in the Fenton process that impact the operational costs and hinders its application at the industrial level [8]. Thomas et al. [78] point out that researches about the Fenton process should be focusing on strategies for improving the utilization of H_2O_2 and not only on ideas for enhancing the efficiency of Fenton catalysts. Once Fenton processes are employed for treating wastewater with substantial pollutants loads, usually, it is necessary additional amounts of H_2O_2 in the reaction, making the process expensive. In this sense, optimization of the H_2O_2 loads could make Fenton reactions significantly cheaper [78].

Nevertheless, some recent researches that envision the use of Fenton processes to remove dyes are discussed in this section. In an interesting work, [68] investigate a hybrid technique combining Fenton and microbiological consortia technique for biodegradation of tannery effluent containing toxic azo dye (acid blue 113). The process consists of an effluent pre-treatment through the Fenton process aiming to guarantee the dye degradation regardless of its initial load in the subsequent biological step. The pre-treatment of dye effluent with Fenton, using H_2O_2 to chemical oxygen demand mass ratio of 0.5:1 and 70 mg/L of Fe^{+2}, reduced the dye concentration by 40%, and a maximum dye degradation of 85% was achieved. The results obtained by the authors highlight the potential of the hybrid process to be applied to large-scale industrial effluent treatment.

Taking advantage of the Chilean potential in terms of solar energy, [13] investigate the degradation of acid yellow 42, a textile dye widely used in the Chilean textile industry using an 8 L-solar photoelectron-Fenton pilot plant. Using an aqueous solution containing 100 $mg{\cdot}L^{-1}$ of total organic carbon of dye, a complete degradation was achieved by the authors after 270 min of process using 1.0 mM of Fe^{+2} at pH 3.0, assisted by a current density of 50 $^{mA}{\cdot}cm^{-2}$.

Gu et al. [21] evaluated the degradation of high-toxic dye intermediates from naphthalene compounds, such as 6-nitro-1,2,4-acid (with initial chemical oxygen demand of 7.300 $mg{\cdot}L^{-1}$), applying a two-stage Fenton model in a 2 m^3-pilot plant with an inner circulation system aiming recycling the treated effluent back to the influent distribution system. With a total load of H_2O_2 applied of 4.9 $g{\cdot}L^{-1}$, Fe^{2+} to H_2O_2 molar ratio of 1:20 and initial pH of 2.5, 93% of the chemical oxygen demand could be removed with 62% of oxidation efficiency when a flow to influent inner circulation ratio of 5:1 was applied. The system efficiency could be proved when only 41% of reduction to the chemical oxygen demand was obtained when the single-stage Fenton without circulation was used.

Hammad et al. [22] proposed a low-cost technique for large-scale production of iron oxide/graphene nanostructures with a controllable graphene load for a heterogeneous Fenton reaction. Degradation efficiencies higher than 99% were achieved after

60 min for methylene blue, rhodamine B, acid orange 7, and phenol at a concentration of 60 mg·mL^{-1} under UV-A assistance of 1.6 mW·cm^{-2}. The authors also reported good stability of the iron oxide/graphene nanocomposites to a pH range from 3 to 9, which allowing the application of these compounds for degradation of a wide range of azo dye contaminants and benzene-based pollutants as cost-effective treatments of wastewater.

In another research, [23] developed a large-scale and cost-effective method to prepare a pure spinel-type $(Mg,Ni)(Fe,Al)_2O_4$ from saprolite laterite ore via acid leaching and co-precipitation. The compound was posteriorly tested on the Fenton catalytic process for degradation of distinct dyes (rhodamine B, methylene blue, methyl orange, malachite green, and reactive yellow 3), assisted by oxalic acid and sunlight. The authors describe that a combination of iron oxides with polycarboxylic acids, like oxalic acid, form a $[Fe(C_2O_4)_3]^{3-}$ complex at the surface of the iron oxide that boosts the quantum efficiency of the compound in relation to ≡Fe-OH complex ions, improving the overall effectiveness of the system. Under optimized conditions, yields higher than 95% were obtained for concentrations of dyes ranging from 10 to 50 mg·mL^{-1} within 180 min of process. It is interesting to mention that the yield and total organic carbon removal for rhodamine B were over 98% and 46%, respectively, even after five cycles.

4 Future Outlooks and Technological Challenges

Fenton processes have been widely performed and studies involving operational optimization and enhancement of organic compounds removal capacity have been performed for wastewater treatment. The expansion process of this method is closely associated to the massive industrial expansion verified over the years. A potential disintegration of pollutants in industrial effluents through innovative formulas has been a key point for obtaining eco-friendly and economically favorable methods. However, some challenges have been explored to optimize Fenton processes in order to increase the removal efficiency of industrial synthetic dyes and provide an equilibrium in terms of operating costs and appropriate from an environmental panorama.

One of the main challenges pointed out with strongly promising perspectives for the coming years is the combination of Fenton with other processes that have shown some efficacy in the elimination of organic compounds. Textile production has increased at significant levels, mainly to supply the growing demand. The extensive use of technologies that permit to improve the performance of the Fenton reaction through the dissociation of Fenton elements and enable higher amounts of OH- radicals is the focus of several scientific studies published in recent years and that have been performed. Additionally, a large part of the effluents produced from the degradation of these elements can be recycled and applied for other purposes or in other stages of the production process. Thus, the integration of the Fenton process with secondary and tertiary processes and biological treatments can indicate expressive operational gains and a symbol of progress for a suggestive degradation pathway.

Furthermore, targeting AOPs from a large-scale perspective is one of the biggest ambitions for industrial processing. This context is closely related to the strong rigidity of environmental regulations in several countries regarding the discharge and treatment of wastewater. However, high costs and the demand for an appropriate technical operational handling that meets a continuous variable legislative demand become important limitations and can threaten the development of AOPs on a larger scale.

In this context, a background of economic costs, especially to verify the operational conditions and parameterization of indispensable variables for a higher economic benefit, is interesting. The Fenton process does not need the energy to activate H_2O_2, and its performance is considered superior to several other integrated methods. In this case, detailed studies aimed at optimizing the process can provide a broad field of study for the coming years. The efficiency of dye removal performance under unspecified conditions or in a set of integrated methods can increase the development of different types of reactors and different processes can be observed. Current studies have highlighted that this background is closely associated to reaction factors, such as pH, initial dye, H_2O_2, and Fe^{2+} concentrations, temperature, and pressure. Thus, it is concluded that the prospecting of scientific studies correlated with the parameterization of reactions is promising and encouraging for the future.

References

1. Adar E (2020) Optimization of triple dye mixture removal by oxidation with Fenton. Int J Environ Sci Technol 17(11):4431–4440. https://doi.org/10.1007/s13762-020-02782-1
2. Aguiar A, Ferraz A, Contreras D, Rodríguez J (2007) Mechanism and applications of the fenton reaction assisted by iron-reducing phenolic compounds. Quim Nova 30(3):623–628
3. Babuponnusami A, Muthukumar K (2014) A review on Fenton and improvements to the Fenton process for wastewater treatment. J Environ Chem Eng 2(1):557–572
4. Bae W, Won H, Hwang B, de Toledo RA, Chung J, Kwon K, Shim H (2015) Characterization of refractory matters in dyeing wastewater during a full-scale Fenton process following pure-oxygen activated sludge treatment. J Hazard Mater 287:421–428
5. Bahmani P, Kalantary RR, Esrafili A, Gholami M, Jafari AJ (2013) Evaluation of Fenton oxidation process coupled with biological treatment for the removal of reactive black 5 from aqueous solution. J Environ Health Sci Eng 11(1)
6. Bokare AD, Choi W (2014) Review of iron-free Fenton-like systems for activating H2O2 in advanced oxidation processes. Elsevier B.V
7. Cetinkaya SG, Morcali MH, Akarsu S, Ziba CA, Dolaz M (2018) Comparison of classic Fenton with ultrasound Fenton processes on industrial textile wastewater. Sustain Environ Res 28(4):165–170. https://doi.org/10.1016/j.serj.2018.02.001
8. Comninellis C, Kapalka A, Malato S, Parsons SA, Poulios I, Mantzavinos D (2008) Advanced oxidation processes for water treatment: advances and trends for R&D. J Chem Technol Biotechnol 83(6):769–776
9. Dewil R, Mantzavinos D, Poulios I, Rodrigo MA (2017) New perspectives for advanced oxidation processes. J Environ Manage 195:93–99

10. Dong C, Xing M, Zhang J (2020) Recent progress of photocatalytic fenton-like process for environmental remediation. Front Environ Chem 1(September):1–21
11. Duarte F, Morais V, Maldonado-Hódar FJ, Madeira LM (2013) Treatment of textile effluents by the heterogeneous Fenton process in a continuous packed-bed reactor using Fe/activated carbon as catalyst. Chem Eng J 232:34–41
12. Ertugay N, Acar FN (2017) Removal of COD and color from Direct Blue 71 azo dye wastewater by Fenton's oxidation: kinetic study. Arabian J Chem 10:S1158–S1163. https://doi.org/10.1016/j.arabjc.2013.02.009
13. Espinoza C, Romero J, Villegas L, Cornejo-Ponce L, Salazar R (2016) Mineralization of the textile dye acid yellow 42 by solar photoelectro-Fenton in a lab-pilot plant. J Hazard Mater 319:24–33
14. Fast SA, Gude VG, Truax DD, Martin J, Magbanua BS (2017) A critical evaluation of advanced oxidation processes for emerging contaminants removal. Environ Process 4(1):283–302
15. Fenton HJH (1894) Oxidation of tartaric acid in presence of iron. J Chem Soc Trans 65:899–910
16. Fernandes NC, Brito LB, Costa GG, Taveira SF, Cunha-Filho MSS, Oliveira GAR, Marreto RN (2018) Removal of azo dye using Fenton and Fenton-like processes: Evaluation of process factors by Box-Behnken design and ecotoxicity tests. Chem Biol Interact 291:47–54
17. Ferraz W, Oliveira LCA, Dallago R, da Conceição L (2007) Effect of organic acid to enhance the oxidative power of the fenton-like system: computational and empirical evidences. Catal Commun 8(2):131–134
18. Garcia-Segura S, Brillas E (2017) Applied photoelectrocatalysis on the degradation of organic pollutants in wastewaters. J Photochem Photobiol C 31:1–35
19. Ghorai K, Panda A, Bhattacharjee M, et al (2021) Facile synthesis of $CuCr_2O_4/CeO_2$ nanocomposite: a new Fenton like catalyst with domestic LED light assisted improved photocatalytic activity for the degradation of RhB, MB and MO dyes. Appl Surf Sci 536:147604. https://doi.org/10.1016/j.apsusc.2020.147604
20. Gomes IB, Maillard JY, Simões LC, Simões M (2020) Emerging contaminants affect the microbiome of water systems—strategies for their mitigation. npj Clean Water 3(1). https://doi.org/10.1038/s41545-020-00086-y
21. Gu L, Nie J-Y, Zhu N, Wang L, Yuan H-P, Shou Z (2012) Enhanced Fenton's degradation of real naphthalene dye intermediate wastewater containing 6-nitro-1-diazo-2-naphthol-4-sulfonic acid: A pilot scale study. Chem Eng J 189–190:108–116
22. Hammad M, Fortugno P, Hardt S, Kim C, Salamon S, Schmidt TC, Wende H, Schulz C, Wiggers H (2021) Large-scale synthesis of iron oxide/graphene hybrid materials as highly efficient photo-Fenton catalyst for water remediation. Environ Technol Innov 21:101239
23. Han X, Liu S, Huo X, Cheng F, Zhang M, Guo M (2020) Facile and large-scale fabrication of (Mg,Ni)(Fe,Al)2O4 heterogeneous photo-Fenton-like catalyst from saprolite laterite ore for effective removal of organic contaminants. J Hazard Mat 392:122295
24. Hayat H, Mahmood Q, Pervez A, Bhatti ZA, Baig SA (2015) Comparative decolorization of dyes in textile wastewater using biological and chemical treatment. Sep Purif Technol 154:149–153. https://doi.org/10.1016/j.seppur.2015.09.025
25. Innocenzi V, Prisciandaro M, Centofanti M, Vegliò F (2019) Comparison of performances of hydrodynamic cavitation in combined treatments based on hybrid induced advanced Fenton process for degradation of azo-dyes. J Environ Chem Eng 7(3):103171
26. Ismail M, Akhtar K, Khan MI, Kamal T, Khan MA, M Asiri A, Seo J, KhanSB (2019) Pollution, toxicity and carcinogenicity of organic dyes and their catalytic bio-remediation. Curr Pharm Des 25(34):3645-3663
27. Javaid R, Qazi UY (2019) Catalytic oxidation process for the degradation of synthetic dyes: an overview. Int J Environ Res Public Health 16(11):1–27
28. Jiao Y, Wan C, Bao W, et al (2018) Facile hydrothermal synthesis of Fe_3O_4@cellulose aerogel nanocomposite and its application in Fenton-like degradation of Rhodamine B. Carbohydr Polym 189:371–378. https://doi.org/10.1016/j.carbpol.2018.02.028
29. Kakavandi B, Takdastan A, Pourfadakari S, et al (2019) Heterogeneous catalytic degradation of organic compounds using nanoscale zero-valent iron supported on kaolinite: mechanism,

kinetic and feasibility studies. J Taiwan Inst Chem Eng 96:329–340. https://doi.org/10.1016/j.jtice.2018.11.027
30. Kirchon A, Zhang P, Li J, et al (2020) Effect of isomorphic metal substitution on the fenton and photo-fenton degradation of methylene blue using Fe-based metal–organic frameworks. ACS Appl Mater Interfaces 12:9292–9299. https://doi.org/10.1021/acsami.9b21408
31. Krüger RL, Dallago RM, Di Luccio M (2009) Degradation of dimethyl disulfide using homogeneous Fenton's reaction. J Hazard Mater 169(1–3):443–447
32. Kumar A, Rana A, Sharma G, Naushad M, Dhiman P, Kumari A, Stadler FJ (2019) Recent advances in nano-Fenton catalytic degradation of emerging pharmaceutical contaminants. J Mol Liq 290:111177
33. Kumar P, Teng TT, Chand S, Wasewar KL (2011) Fenton oxidation of carpet dyeing wastewater for removal of cod and color. Desalin Water Treat 28(1–3):260–264
34. Kuntubek A, Kinayat N, Meiramkulova K, et al (2020) Catalytic oxidation of methylene blue by use of natural zeolite-based silver and magnetite nanocomposites. Processes 8:471. https://doi.org/10.3390/pr8040471
35. Lee YY, Fan C (2020) Mechanistic exploration of the catalytic modification by co-dissolved organic molecules for micropollutant degradation during fenton process. Chemosphere 258:127338
36. Lellis B, Fávaro-Polonio CZ, Pamphile JA, Polonio JC (2019) Effects of textile dyes on health and the environment and bioremediation potential of living organisms. Biotechnol Res Innov 3(2):275–290
37. Li X, Jin X, Zhao N, Angelidaki I, Zhang Y (2017) Novel bio-electro-Fenton technology for azo dye wastewater treatment using microbial reverse-electrodialysis electrolysis cell. Biores Technol 228:322–329
38. Li Y, Jiang J, Fang Y, et al (2018a) TiO_2 Nanoparticles anchored onto the metal–organic framework NH_2 -MIL-88B(Fe) as an adsorptive photocatalyst with enhanced fenton-like degradation of organic pollutants under visible light irradiation. ACS Sustain Chem Eng 6:16186–16197. https://doi.org/10.1021/acssuschemeng.8b02968
39. Li Z, Lyu J, Ge M (2018b) Synthesis of magnetic $Cu/CuFe_2O_4$ nanocomposite as a highly efficient Fenton-like catalyst for methylene blue degradation. J Mater Sci 53:15081–15095. https://doi.org/10.1007/s10853-018-2699-0
40. Litter MI, Slodowicz M (2017) An overview on heterogeneous Fenton and photoFenton reactions using zerovalent iron materials. J Adv Oxid Technol 20(1)
41. Liu C, Yang B, Chen J, Jia F, Song S (2022) Synergetic degradation of Methylene Blue through photocatalysis and Fenton reaction on two-dimensional molybdenite-Fe. J Environ Sci 111:11–23
42. Ma J, Yang Q, Wen Y, Liu W (2017) Fe-g-C_3N_4/graphitized mesoporous carbon composite as an effective Fenton-like catalyst in a wide pH range. Appl Catal B Environ 201:232–240. https://doi.org/10.1016/j.apcatb.2016.08.048
43. Manaa Z, Chebli D, Bouguettoucha A, Atout H, Amrane A (2019) Low-cost photo-fenton-like process for the removal of synthetic dye in aqueous solution at circumneutral pH. Arab J Sci Eng 44(12):9859–9867. https://doi.org/10.1007/s13369-019-04101-4
44. Marcinowski PP, Bogacki JP, Naumczyk JH (2014) Cosmetic wastewater treatment using the Fenton, photo-Fenton and H 2O2/UV processes. J Environ Sci Health Part A Toxic/Hazard Subst Environ Eng 49(13):1531–1541
45. Miklos DB, Remy C, Jekel M, Linden KG, Drewes JE, Hübner U (2018) Evaluation of advanced oxidation processes for water and wastewater treatment–a critical review. Water Res 139:118–131
46. Mirzaei A, Chen Z, Haghighat F, Yerushalmi L (2017) Removal of pharmaceuticals from water by homo/heterogonous Fenton-type processes–a review. Chemosphere 174:665–688
47. Mohammed Redha Z (2020) Multi-response optimization of the coagulation process of real textile wastewater using a natural coagulant. Arab J Basic Appl Sci 27(1):406–422. https://doi.org/10.1080/25765299.2020.1833509

48. Nasuha N, Hameed BH, Okoye PU (2021) Dark-Fenton oxidative degradation of methylene blue and acid blue 29 dyes using sulfuric acid-activated slag of the steel-making process. J Environ Chem Eng 9:104831. https://doi.org/10.1016/j.jece.2020.104831
49. Nidheesh PV, Gandhimathi R, Ramesh ST (2013a) Degradation of dyes from aqueous solution by Fenton processes: a review. Environ Sci Pollut Res 20(4):2099–2132
50. Nidheesh PV, Gandhimathi R, Ramesh ST (2013b) Degradation of dyes from aqueous solution by Fenton processes: a review. Environ Sci Pollut Res 20(4):2099–2132
51. Nidheesh PV, Zhou M, Oturan MA (2018) An overview on the removal of synthetic dyes from water by electrochemical advanced oxidation processes. Chemosphere 197:210–227
52. Nogueira RFP, Jardim WF (1998) A fotocatálise heterogênea e sua aplicação ambiental. Quim Nova 21(1):69–72
53. Oliveira JS de, Halmenschlager F da C, Jahn SL, Foletto EL (2019) $CoFe_2O_4$ synthesis on $MgAl_2O_4$ and ZSM-5 supports for use in pollutant degradation by heterogeneous photo-Fenton process under visible and solar irradiation. Matéria (Rio Janeiro) 24. https://doi.org/10.1590/s1517-707620190004.0823
54. Oller I, Malato S, Sánchez-Pérez JA (2011a) Combination of advanced oxidation processes and biological treatments for wastewater decontamination-a review. Sci Total Environ 409(20):4141–4166
55. Oller I, Malato S, Sánchez-Pérez JA (2011) Combination of advanced oxidation processes and biological treatments for wastewater decontamination—a review. Sci Total Environ 409(20):4141–4166
56. Oturan MA, Aaron JJ (2014) Advanced oxidation processes in water/wastewater treatment: principles and applications: a review. Crit Rev Environ Sci Technol 44(23):2577–2641
57. Patel SK, Patel SG, Patel GV (2020) Degradation of reactive dye in aqueous solution by fenton, photo-fenton process and combination process with activated Charcoal and TiO2. Proc Nat Acad Sci India Sect A Phys Sci 90(4):579–591. https://doi.org/10.1007/s40010-019-00618-3
58. Piaskowski K, Świderska-Dąbrowska R, Zarzycki PK (2018) Dye removal from water and wastewater using various physical, chemical, and biological processes. J AOAC Int 101(5):1371–1384
59. Qin Q, Liu Y, Li X, et al (2018) Enhanced heterogeneous Fenton-like degradation of methylene blue by reduced $CuFe_2O_4$. RSC Adv 8:1071–1077. https://doi.org/10.1039/C7RA12488K
60. Ramírez-Malule H, Quiñones-Murillo DH, Manotas-Duque D (2020) Emerging contaminants as global environmental hazards a bibliometric analysis. . Emerg Contam 6:179–193
61. Rasheed T, Bilal M, Nabeel F, Adeel M, Iqbal HMN (2019) Environmentally-related contaminants of high concern: potential sources and analytical modalities for detection, quantification, and treatment. Environ Int 122:52–66
62. Rial JB, Ferreira ML (2021) Challenges of dye removal treatments based on IONzymes: beyond heterogeneous Fenton. J Water Process Eng 41:102065
63. Ribeiro JP, Nunes MI (2021) Recent trends and developments in Fenton processes for industrial wastewater treatment–a critical review. Environ Res 197:110957
64. Ribeiro JP, Marques CC, Portugal I, Nunes MI (2020) Fenton processes for AOX removal from a kraft pulp bleaching industrial wastewater: optimisation of operating conditions and cost assessment. J Environ Chem Eng 8(4):104032
65. Rivera FL, Recio FJ, Palomares FJ, et al (2020) Fenton-like degradation enhancement of methylene blue dye with magnetic heating induction. J Electroanal Chem 879:114773. https://doi.org/10.1016/j.jelechem.2020.114773
66. Saleh M, Bilici Z, Kaya M, et al (2021) The use of basalt powder as a natural heterogeneous catalyst in the Fenton and Photo-Fenton oxidation of cationic dyes. Adv Powder Technol 32:1264–1275. https://doi.org/10.1016/j.apt.2021.02.025
67. Sanad MF, Shalan AE, Bazid SM, Abdelbasir SM (2018) Pollutant degradation of different organic dyes using the photocatalytic activity of ZnO@ZnS nanocomposite materials. J Environ Chem Eng 6(4):3981–3990. https://doi.org/10.1016/j.jece.2018.05.035.

68. Shanmugam BK, Easwaran SN, Mohanakrishnan AS, Kalyanaraman C, Mahadevan S (2019) Biodegradation of tannery dye effluent using Fenton's reagent and bacterial consortium: a biocalorimetric investigation. J Environ Manage 242:106–113
69. Silva LGM, Moreira FC, Cechinel MAP, Mazur LP, de Souza AAU, Souza SMAGU, Boaventura RAR, Vilar VJP (2020) Integration of Fenton's reaction based processes and cation exchange processes in textile wastewater treatment as a strategy for water reuse. J Environ Manag 272(July)
70. Singh Z, Chadha P (2016) Textile industry and occupational cancer. J Occupat Med Toxicol 11(1):1–6. https://doi.org/10.1186/s12995-016-0128-3
71. Skoog DA, Holler FJ, Nieman TA (2002) Princípios de análise instrumental. Bookman, Porto Alegre
72. Soon AN, Hameed BH (2011) Heterogeneous catalytic treatment of synthetic dyes in aqueous media using Fenton and photo-assisted Fenton process. Desalination 269(1–3):1–16
73. Stupar SL, Grgur BN, Radišić MM, Onjia AE, Ivanković ND, Tomašević AV, Mijin D (2020) Oxidative degradation of Acid Blue 111 by electro-assisted Fenton process. J Water Process Eng 36(May)
74. Sun M, Chu C, Geng F, et al (2018) Reinventing Fenton chemistry: iron oxychloride nanosheet for pH-insensitive H_2O_2 activation. Environ Sci Technol Lett 5:186–191. https://doi.org/10.1021/acs.estlett.8b00065
75. Tariq M, Muhammad M, Khan J, et al (2020) Removal of Rhodamine B dye from aqueous solutions using photo-Fenton processes and novel Ni-Cu@MWCNTs photocatalyst. J Mol Liq 312:113399. https://doi.org/10.1016/j.molliq.2020.113399
76. Thomaidis NS, Asimakopoulos AG, Bletsou AA (2012) Emerging contaminants: a tutorial mini-review. Global NEST J 14(1):72–79
77. Thomas N, Dionysiou DD, Pillai SC (2021a) Heterogeneous Fenton catalysts: a review of recent advances. J Hazard Mater 404(PB):124082
78. Thomas N, Dionysiou DD, Pillai SC (2021b) Heterogeneous Fenton catalysts: a review of recent advances. J Hazard Mater 404(July 2020)
79. Tkaczyk A, Mitrowska K, Posyniak A (2020) Synthetic organic dyes as contaminants of the aquatic environment and their implications for ecosystems: a review. Sci Total Environ 717:137222. https://doi.org/10.1016/j.scitotenv.2020.137222.
80. Venkatesh S, Venkatesh K, Quaff AR (2017) Dye decomposition by combined ozonation and anaerobic treatment: cost effective technology. J Appl Res Technol 15(4):340–345. https://doi.org/10.1016/j.jart.2017.02.006
81. Vergili I, Gencdal S (2015) Applicability of combined Fenton oxidation and nanofiltration to pharmaceutical wastewater. Desalin Water Treat 56(13):3501–3509
82. Wakrim A, Dassaa A, Zaroual Z, et al (2021) Mechanistic study of carmoisine dye degradation in aqueous solution by Fenton process. Mater Today Proc 37:3847–3853. https://doi.org/10.1016/j.matpr.2020.08.405
83. Wang J, Tang J (2021) Fe-based Fenton-like catalysts for water treatment: Catalytic mechanisms and applications. J Mol Liq Molecular 332:115755
84. Wang N, Zheng T, Zhang G, Wang P (2016) A review on Fenton-like processes for organic wastewater treatment. J Environ Chem Eng 4(1):762–787
85. Wang J, Liu C, Qi J, et al (2018) Enhanced heterogeneous Fenton-like systems based on highly dispersed Fe0-Fe_2O_3 nanoparticles embedded ordered mesoporous carbon composite catalyst. Environ Pollut 243:1068–1077. https://doi.org/10.1016/j.envpol.2018.09.057
86. Wong S, Ghafar NA, Ngadi N, Razmi FA, Inuwa IM, Mat R, Amin NAS (2020) Effective removal of anionic textile dyes using adsorbent synthesized from coffee waste. Sci Rep 10(1):1–13
87. Wu Z, Zhu W, Zhang M, et al (2018) Adsorption and Synergetic Fenton-like Degradation of Methylene Blue by a Novel Mesoporous α-Fe_2O_3 /SiO_2 at Neutral pH. Ind Eng Chem Res 57:5539–5549. https://doi.org/10.1021/acs.iecr.8b00077
88. Xu H-Y, Li B, Shi T-N, et al (2018) Nanoparticles of magnetite anchored onto few-layer graphene: a highly efficient Fenton-like nanocomposite catalyst. J Colloid Interface Sci 532:161–170. https://doi.org/10.1016/j.jcis.2018.07.128

89. Yan Q, Zhang J, Xing M (2020) Cocatalytic Fenton reaction for pollutant control. Cell Rep Phys Sci 1(8):100149. https://doi.org/10.1016/j.xcrp.2020.100149
90. Yaseen DA, Scholz M (2019) Textile dye wastewater characteristics and constituents of synthetic effluents: a critical review. Springer, Berlin, Heidelberg. https://doi.org/10.1007/s13762-018-2130-z
91. Zhang H, Xue G, Chen H, Li X (2018) Magnetic biochar catalyst derived from biological sludge and ferric sludge using hydrothermal carbonization: Preparation, characterization and its circulation in Fenton process for dyeing wastewater treatment. Chemosphere 191:64–71
92. Zheng J, Gao Z, He H, Yang S, Sun C (2016) Efficient degradation of Acid Orange 7 in aqueous solution by iron ore tailing Fenton-like process. Chemosphere 150:40–48
93. Zhou Y, Lu J, Zhou Y, Liu Y (2019) Recent advances for dyes removal using novel adsorbents: a review. Environ Pollut 252:352–365
94. Zuo M, Yi S, Choi J (2021) Excellent dye degradation performance of FeSiBP amorphous alloys by Fenton-like process. J Environ Sci 105:116–127

Fenton-Like Processes for the Removal of Cationic Dyes

Md. Saddam Hossain, Md. Yeasin Pabel, and Md. Mominul Islam

Abstract Due to a rapid industrial development in last few decades, large amounts of synthetic dyes have been discharged into water bodies that cause water pollution and substantially damage the aquatic environments. Consequently, the treatment of dyestuffs in the effluents discharged especially from the textile industries are crucial to protect our environments from pollutions. In this regard, advanced oxidation processes (AOPs) have been considered as a promising method of which Fenton process is well-known from earlier. In this chapter, a comprehensive discussion on the removal of textile dyes, mainly cationic dyes, from wastewaters through degradation with Fenton-like processes is presented. The fundamentals and techniques associated with these processes are elaborated. The features and efficiency of degradation of Fenton-like processes developed based on various catalysts are compared by an extensive review on recent literatures. Advantages and limitations of these methods in practical applications are especially highlighted. All of these attempts altogether create a good knowledge base for the researchers working in the textile and environmental fields.

Keywords Fenton process · Fenton-like processes · Hydroxyl radical · Organic dyes · Textile dyes · Efficiency · Decolorization · Mineralization

1 Introduction

In the last few decades, the huge industrialization has caused a massive stress on the environment. Contamination of surface water with organic dyes is one of the major environmental pollutions. Dyes especially synthetic organic dyestuffs have been widely used in textile, pulp, rubber, plastic, printing, pharmaceutical, and food

Md. S. Hossain
Department of Chemistry, Khulna University of Engineering and Technology, Khulna 9203, Bangladesh

Md. Y. Pabel · Md. M. Islam (✉)
Department of Chemistry, Faculty of Science, University of Dhaka, Dhaka 1000, Bangladesh
e-mail: mominul@du.ac.bd

S. S. Muthu and A. Khadir (eds.), *Advanced Oxidation Processes in Dye-Containing Wastewater*, Sustainable Textiles: Production, Processing, Manufacturing & Chemistry, https://doi.org/10.1007/978-981-19-0882-8_2

industries as coloring agents [1–175, 177–179]. Basically more than 30,000 commercial dyes based on 8,000 different chemical structures are used in these industries, and a part of them are discharged with the industrial wastewaters [61, 69, 89]. Chemical structures of some common cationic dyes are presented in Scheme 1. Most of them are persistent and being detected in ground and surface waters. Nowadays, the contamination of water has attracted attentions of the scientific community because it has serious consequences for the humans as well as for the ecosystem [103]. Consequently, the treatment of wastewaters containing dyestuffs discharged especially from the textile industries is crucial to protect our environments.

Over the different methods practiced [5, 8, 20, 21, 60, 62, 73, 81, 98, 101, 102, 108, 113, 118, 119, 125, 127, 132, 158, 160, 167, 179], advanced oxidation processes (AOPs) have showed an exciting potential for the treatment of textile wastewater containing organic dyes due to its important inherent advantages. In AOPs, organic dyes are basically degraded into different gaseous and ionic or neutral fragments that are considered to be non-toxic to the aquatic environments. Among the several oxidizing agents used for treatment wastewaters, hydroxyl radical ($OH^{\bullet}$) is much

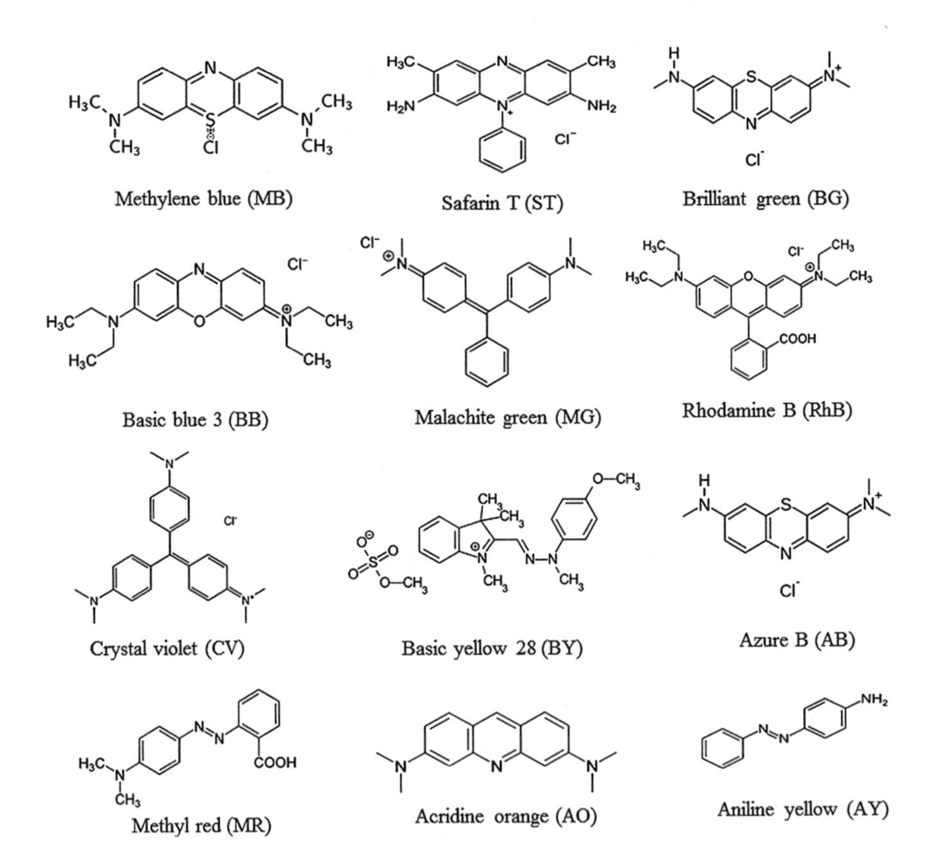

Scheme 1 Chemical structures of commonly studied dyes

more lucrative for its different inherent features as summarized in Table 1. $OH^{\bullet}$ radical generated through AOPs oxidizes complex organic species to smaller organic fragments or completely mineralized to CO_2 and H_2O [73, 81, 119, 160].

Fenton process is one of the most frequently used AOPs, where H_2O_2 and Fe^{2+} are used as the oxidant and catalyst, respectively. The major advantages of Fenton process are (i) it exhibits high performance and simplicity in use, since it can be operated at ambient conditions [21, 125] and (ii) it is non-toxic [60], because H_2O_2 can break down into eco-friendly species like H_2O and O_2. The Fenton process has been employed to treat the wastewaters containing organic dyes discharged from olive-oil mill [20], textile [8], laboratory [19], pesticide [13], cosmetic [18], dyeing

Table 1 Features of some important oxidants used in AOPs [109, 114, 135, 162]

Oxidants	Chemical formula	E^0 (V)	Products leave	Hazard level	Method of generation	Used for dye degradation
Molecular oxygen	O_2	1.23	ROS, H_2O	±	EC	[91]
Chlorine	Cl_2	1.36	Cl^-	++	EF	[48]
Hypoiodous acid	HOI	1.45	I^-, IO_3^-	+		
Hypochlorous acid	HOCl	1.49	OCl^-	+	EF	[48]
Chlorine dioxide	ClO_2	1.57	ClO_2^-	++		
Hypobromous acid	HOBr	1.59	OBr^-	+		
Permanganate	MnO_4^-	1.68	Mn^{2+}, H_2O	±		
Perhydroxyl radical	$^{\bullet}OOH$	1.70	$O_2^{\bullet-}$, H_2O_2	+	SF	[143]
Hydrogen peroxide	H_2O_2	1.78	H_2O, O_2	±	EF	[33, 133]
Ozone	O_3	2.07	O_2	±	CDP	[92]
Peroxodisulfate	$S_2O_8^{2-}$	2.12	$SO_4^{\bullet-}$, SO_4^{2-}	+	EF, CA	[140, 174]
Atomic oxygen	O	2.42	O^{2-}	±	EU	[143]
Sulfate radical	$SO_4^{\bullet-}$	2.60	SO_4^{2-}	+	CA	[80, 83, 131]
Hydroxyl radical	$^{\bullet}OH$	2.80	H_2O, O_2	±	F, EC, PF, EF, SF	[6, 30, 77, 89, 143]
Positive hole on TiO_2	TiO_2^+	3.20	TiO_2	+	PF	[147]

E^0 = Standard reduction potential at 25 °C; F: Conventional Fenton; PF: Photo-Fenton; EF: Electro-Fenton; SF: Sono-Fenton; CDP: Corona discharge process; CA: Catalytic activation; EC: Electrochemical; EU: Electrolysis with ultrasound. The symbols ‘ + ’ and ‘ ± ’ signify as hazardous and safe character, respectively, of the product(s) generated in a particular process

[105, 152], fermentation brine from green olives [154], pharmaceutical [169], cork cooking [146], pulp mill [38], and phenol [117] industries. However, it has several disadvantages [186] and some of them are listed as follows:

(1) It has a relatively high operational cost and risks for the storage, transportation, and handling of H_2O_2.
(2) It requires high quantities of chemicals to acidify effluents to become pH 2–4 and hence neutralization of the treated solutions is needed before disposal.
(3) It leaves iron sludge that creates secondary management problems.
(4) It is not fully efficient for an overall mineralization because Fe^{3+} ion forms complexes with organic dye species directly or its fragmented parts that are difficult to be destroyed with generated $OH^{\bullet}$ species.
(5) It has difficulties in recycling the catalyst, Fe^{2+} ions dissolved in the medium.

Later, in order to overcome these problems, solid material-support Fenton process and some other Fenton-like processes (vide infra) have been developed [11, 27, 53, 68, 69, 71, 89, 90, 120, 121, 123, 151, 157, 158, 164, 165, 182]. Moreover, some other kinds of homo-/hetero-geneous catalysts have been employed to replace Fe^{2+} in so-called Fenton process [11, 53, 68, 120, 182]. Zero-valent iron (Fe^0), different forms of iron-based minerals such as magnetite (Fe_3O_4), goethite (α-FeOOH), maghemite (γ-Fe_2O_3), hematite (α-Fe_2O_3), ferrihydrite, lepidocrocite (γ-FeOOH), and pyrite (FeS_2) have been attempted as efficient Fenton catalysts [144].

Fenton-like processes are basically the AOPs treatment wherein the generation of different reactive oxygen species (ROS) including $OH^{\bullet}$ species take place via different physical and chemical methods coupled with the conventional Fenton process [11, 53, 27, 68, 69, 71, 89, 91, 120, 121, 123, 151, 157, 158, 164, 165, 183]. Ozonation [5], O_3/H_2O_2 [108], the effects of different physical fields such as photo-field, electric-field [41, 62, 64, 177], microwave-field [9, 37], cavitation effect [55, 72, 149] on the efficiency of both homogeneous and heterogeneous Fenton processes [127], e.g., Fe^{2+}/H_2O_2/UV, H_2O_2/UV [118], TiO_2/UV [62], ZnO/UV [101] are the examples of Fenton-like processes. In the recent years, the generation of ROS by electrochemical means for the application in the treatment of textile wastewaters [4, 15, 27] as a Fenton-like process has been attracting attention and becoming popular. All of these Fenton-like processes are, however, known after the technique hybridized with the conventional Fenton process, i.e., photo-, electro-, cavitation-, microwave-Fenton processes, and so on.

This chapter focuses on the degradation of cationic dyes with Fenton-like processes. The discussion is commenced with the fundamentals associated with Fenton and Fenton-like processes. The catalysts used and their preparation, and characterization are described in short. The degradation scenarios of several textiles dyes with different Fenton-like processes are extensively discussed and summarized with the view of recent literatures. Effectiveness and limitations of these methods in practical applications for the removal of dyes from wastewaters are compared. The future prospects of Fenton-like processes for cationic dye degradation are also highlighted.

2 Characteristics of Fenton and Different Fenton-Like Processes

2.1 Fenton Process

Fenton process basically involves the use of H_2O_2 and Fe^{2+}, known as Fenton's reagent, with notorious application for the treatment of wastewaters. H_2O_2 has been deemed as a green chemical because it gives rise to gaseous O_2 and H_2O as by-products. H_2O_2 is actually a mild oxidant and its oxidation potential (Table 1) varies depending on the solution pH [27]. The reaction of H_2O_2 with Fe^{2+} ion originates a very strong oxidizing and unstable ROS, $OH^{\bullet}$ species. It is the second strongest oxidizing agent known with a potential, $E°$ ($OH^{\bullet}/H_2O$) of 2.8 V and hence can degrade the organic and organometallic pollutants in a non-selective manner to a complete mineralized products. Owing to its very short lifetime, as a few nanoseconds in water, $OH^{\bullet}$ species is to generate in situ in the reaction medium.

In Fenton process, in general, the $OH^{\bullet}$ species generated in situ react with organic molecules via three possible attacking modes: (i) Dehydrogenation or abstraction of a hydrogen atom from alkanes and alcohols to form water with a high rate (Eqs. 1 and 2), (ii) Hydroxylation or electrophilic addition to a double bond (i.e., alkenes or alkynes) or aromatic or heterocyclic rings (Eq. 2) with a relatively slower rate [36], and (iii) Electron transfer or redox reactions:

$$OH^{\bullet} + RH \rightarrow H_2O + [*] \tag{1}$$

$$OH^{\bullet} + [*] \rightarrow H_2O + CO_2 \tag{2}$$

where [*] represents intermediates, for example,

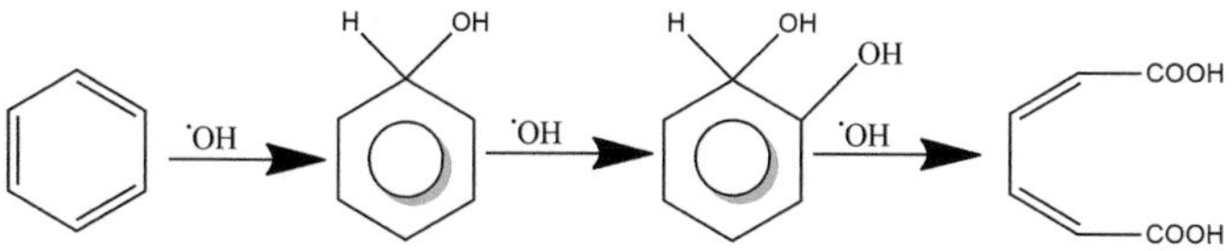

In the accepted mechanism of the Fenton process, the rate of formation of $OH^{\bullet}$ (Eq. 3) is very fast. The second-order rate constant (k) of various reactions involving Fenton process are compared in Fig. 1.

$$Fe^{2+} + H_2O_2 \rightarrow Fe^{3+} + OH^{\bullet} + OH^{-} \tag{3}$$

However, the formation of $OH^{\bullet}$ species (Eq. 3) becomes operative at an optimum pH range of 2.8–3.0, wherein the Fe^{3+}/Fe^{2+} redox couple regulates the reaction in a catalytic manner with a high number of cycles up to 2200 [34]. In the catalytic chain,

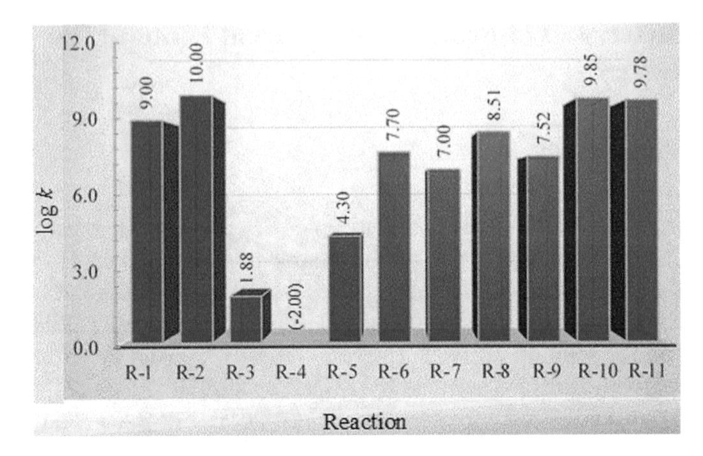

Fig. 1 The second-order rate constant (k) of different reactions involving Fenton process. R-1, R-2, R-3, ..., and R-11 signify the reactions expressed by Eqs. (1), (2), (3),...., and (11), respectively. Data of k plotted are the average values collected from different Refs. [14, 36, 136, 141, 150, 178]

Fe^{2+} is regenerated slowly ($k = 0.001–0.01\ M^{-1}\,s^{-1}$) by the reaction between Fe^{3+} and H_2O_2 yielding in perhydroxyl radical ($HOO^{\bullet}$) as follows:

$$Fe^{3+} + H_2O_2 \rightarrow Fe^{2+} + HOO^{\bullet} + H^{+} \tag{4}$$

The generated $HOO^{\bullet}$ species exhibits a low oxidation power compared to $OH^{\bullet}$ that, in practice, it is quite unreactive towards organic matter [27]. However, Fe^{2+} is regenerated more rapidly upon reduction of Fe^{3+} with $HOO^{\bullet}$ and superoxide ion ($O_2^{\bullet-}$) as follows:

$$Fe^{3+} + HOO^{\bullet} \rightarrow Fe^{2+} + O_2 + H^{+} \tag{5}$$

$$Fe^{3+} + O_2^{\bullet-} \rightarrow Fe^{2+} + O_2 \tag{6}$$

$$Fe^{3+} + O_2^{\bullet-} + 2H_2O \rightarrow Fe^{2+} + H_2O_2 \tag{7}$$

The enhanced reactivity and non-specificity in the reaction of $OH^{\bullet}$ towards both organic and inorganic substrates ($k > 10^{8}\ M^{-1}\,s^{-1}$ [14] result in various competitive processes that negatively affect the oxidation process of organic species [36, 136, 142, 150, 178]. Some of these may be noted as follows:

$$Fe^{2+} + OH^{\bullet} \rightarrow Fe^{3+} + OH^{-} \tag{8}$$

$$H_2O_2 + OH^{\bullet} \rightarrow HOO^{\bullet} + H_2O \tag{9}$$

$$HOO^{\bullet} + OH^{\bullet} \rightarrow H_2O + O_2 \quad (10)$$

$$OH^{\bullet} + OH^{\bullet} \rightarrow H_2O_2 \quad (11)$$

In addition, these inhibition reactions promote the scavenging of ROS and are thus competitive with the main reactions of the destruction of organic compounds (Eq. 1) and eventually restrict experimental design. However, the reaction shown in Eq. (8) is important, since it determines the optimum Fe^{2+} content in the medium to minimize the consumption of $OH^{\bullet}$ species. Other associated reactions shown by Eqs. (9)–(11) play a relatively minor role despite their quite high rates (Fig. 1), because the concentrations of ROS generated in the bulk via these reactions are relatively low. It is noteworthy to mention that the presence of inorganic ions like chloride (Cl^-), sulfate (SO_4^{2-}), nitrate (NO^-_3), carbonate (CO_3^{2-}), and hydrogen carbonate (HCO_3^-) strongly affect the rate of the reaction shown by Eq. (3) due to their radical scavenging role [22].

The overall Fenton chemistry has been simplified by considering the dissociation of H_2O_2 to form H_2O as follows:

$$2Fe^{2+} + H_2O_2 + 2H^+ \rightarrow 2Fe^{3+} + 2H_2O \quad (12)$$

It is clear that H^+ is required in the decomposition of H_2O_2, indicating the need for an acid environment to maximize the production of $OH^{\bullet}$ species. On the other hand, Fe^{3+} ions generated through reactions expressed by Eq. (3) react with OH^- ions present in neutral and near-neutral solutions to form insoluble $Fe(OH)_3$ [111]. This further reduces the overall oxidation efficiency of Fenton's reagent and requires continuous addition of Fe^{2+} salts. These disadvantages of Fenton process (Scheme 2) limit its whispered application for the treatment of wastewater although $OH^{\bullet}$ species generated has a great potential for the effective destruction of a large number of hazardous organic pollutants. These facts lead to the modification of the conventional

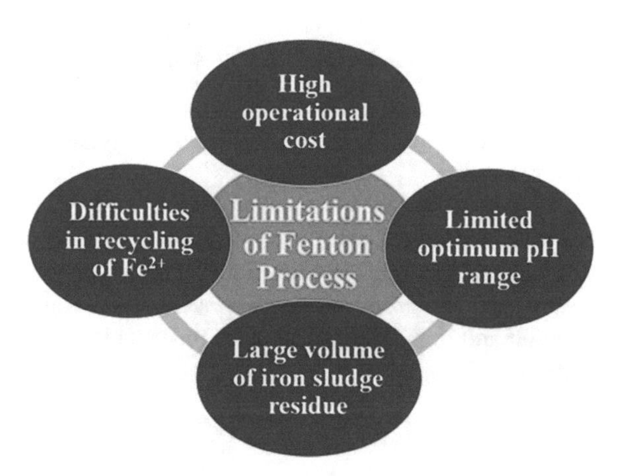

Scheme 2 Major limitations of the Fenton process

Fenton process in different ways that are collectively classified as "Fenton-like" reactions or Fenton-like processes [86, 100, 132]. Hence, the phenomenon in "Fenton-like process" is used to indicate any process that has similar chemistry to the conventional Fenton process, but $OH^{\bullet}$ species is in situ generated from different sources other than H_2O_2 oxidant and Fe^{2+} catalyst [126, 176]. These are namely photo-, electro-, photo-electro-, sono-, sono-photo-, sono-electro-, bio-, or microwave-Fenton processes, and so on.

2.2 *Different Fenton-Like Processes*

The chemistry of Fenton-like processes is essentially similar to that of conventional Fenton process. However, based on the nature of catalyst, the Fenton-like processes can also be classified into as heterogeneous and homogeneous processes. The catalysts exist in aqueous phase as the oxidant in homogeneous Fenton-like processes, while in heterogeneous processes the catalysts play their role from the solid phase [78, 130, 170, 173, 176]. Generally, the heterogeneous catalytic processes are considered as more efficient and environmentally benign for practical applications, indeed. The details on the catalysts used are discussed later. In this section, the fundamentals of different types of Fenton-like processes are described briefly.

2.2.1 Photo-Fenton Process

Electromagnetic radiations ($h\upsilon$) of the UV–visible ranges play an important role in Fenton process. Photo-Fenton (PF) is actually the combination of conventional Fenton reagent and light (Fig. 2a). The use of light minimizes the catalyst loading and enhances the efficiency of catalyst in principle [35] and hence helps to generate additional $OH^{\bullet}$ species in the systems through the reactions shown by Eqs. (13) and (14).

$$H_2O_2 + h\upsilon \rightarrow 2OH^{\bullet} + O_2(\lambda < 310\ \text{nm}) \tag{13}$$

$$Fe(OH)^{2+} + h\upsilon \rightarrow Fe^{2+} + OH^{\bullet}(\lambda < 580\ \text{nm}) \tag{14}$$

$[Fe(OH)]^{2+}$ is the predominant species of Fe^{3+} in acidic solutions with pH of 2.5–4.0. In addition, photo-generated Fe^{2+} Eq. (14) continues to involve in Fenton reaction and generates additional $OH^{\bullet}$ radicals. Hence, it enhances the reaction rate and reduces the iron dosage. However, the use of UV–visible light sources and periodic addition of H_2O_2 increase the establishment and operational costs of PF method.

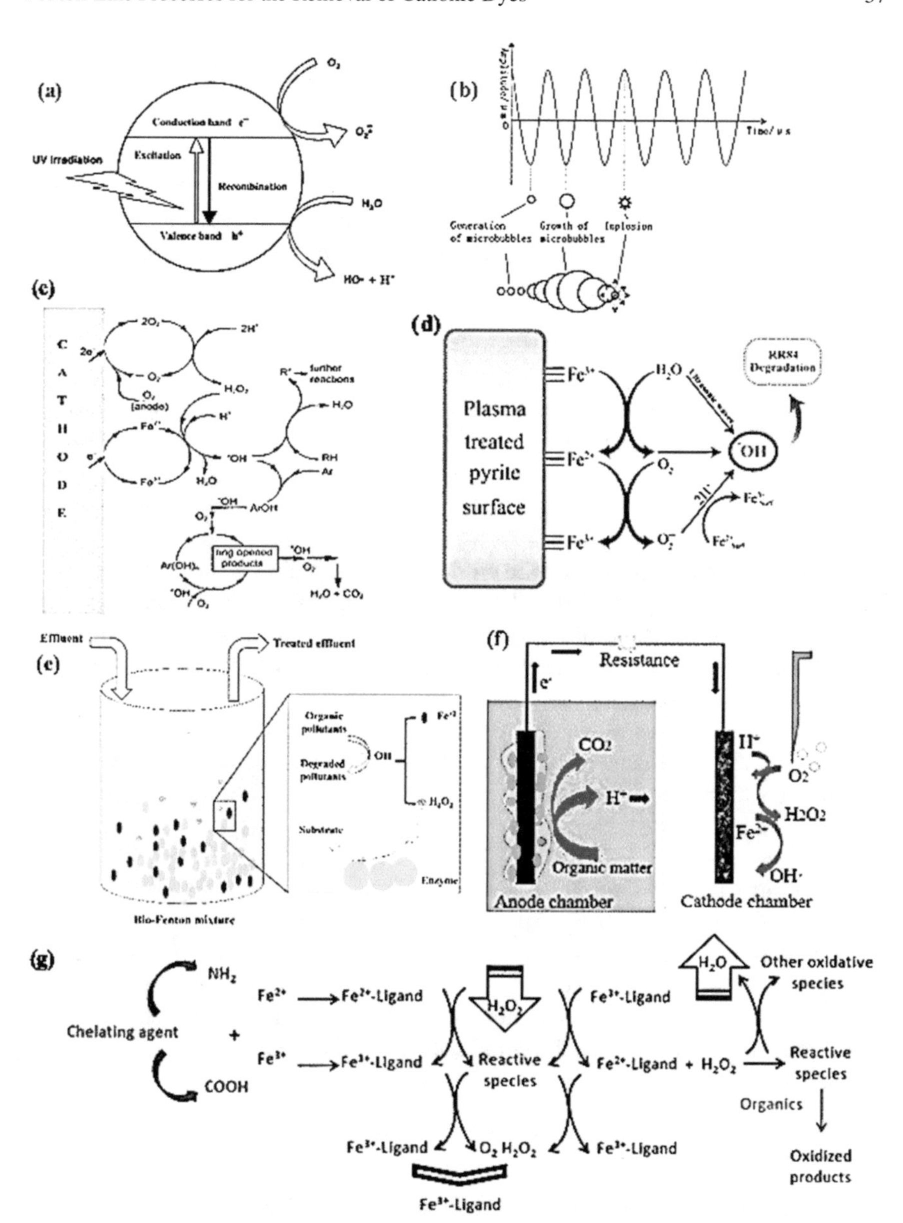

Fig. 2 Illustration of different Fenton-like processes: **a** Photo-, **b** Cavitation-, **c**, Electro-, **d** Sono-, **e** Bio-, **f** Bio-Electro-, and **g** Chelating ligand-assisted Fenton processes [57, 136, 138, 140, 176, 189]

2.2.2 Electro-Fenton Process

Electro-Fenton (EF) is an indirect electrochemical treatment process involving in situ generation of the reagents, i.e., H_2O_2 and/or Fe^{2+} ion used for the Fenton reaction (Fig. 2c). In this process, H_2O_2 is electrochemically generated at cathode surface by the two-electron reduction of O_2 ($2e^-$ ORR) shown in Eq. (15), in acidic medium on graphite, various forms of carbon including glassy carbon (GC), mercury pool, and oxygen-diffusion cathodes [31, 137, 138].

$$O_2(g) + 2H^+(aq) + 2e^- \rightarrow H_2O_2(aq) \tag{15}$$

On the other hand, Fe^{2+} ion can be generated by the $2e^-$ oxidation of a sacrificial metal Fe electrode as an anode:

$$Fe\ (s) \rightarrow Fe^{2+}(aq) + 2e^- \tag{16}$$

Later the produced species react to produce the $OH^\bullet$ radical of the Fenton reaction. Meanwhile, at sufficient applied voltage, water is oxidized at the anode. Controlled and continuous generations of H_2O_2 and Fe^{2+} are the advantages of this EF process. In addition, this method has many advantages over the chemical Fenton reaction, e.g., no need to store H_2O_2 and less sludge is formed from the iron species and other formed complexes. However, this method has some limitations such as gradual corrosion of electrodes and electrical power consumption, which affect directly the cost of the wastewater treatment [97, 148].

2.2.3 Photo-Electro-Fenton Process

Combination of PF and EF as described above processes can enhance the degradation rate of organic dyes called photo-electro-Fenton (PEF) process. Simply, when irradiation of electromagnetic radiation by an UV–visible light source occurs, in situ $OH^\bullet$ species are generated as the consequences of the reactions expressed by Eqs. (14) and (17).

$$Fe(OOCR)^{2+} + h\upsilon \rightarrow Fe^{2+} + CO_2 + R^\bullet \tag{17}$$

The reaction shown by Eq. (14) is the photoreduction of Fe^{3+} in $Fe(OH)^{2+}$ ion to form Fe^{2+}, but the photodecomposition of iron carboxylic acid complex occurs in case of the reaction expressed by Eq. (17) [28, 76, 129, 163]. H_2O_2 can also be electrochemically generated at cathode surface by the $2e^-$ ORR (Eq. 15) and Fe^{2+} ion can be generated by the $2e^-$ oxidation of sacrificial metal Fe anode as represented in Eq. (16). Production of additional $OH^\bullet$ species and regeneration of Fe^{2+} ion from the photodegradation of iron complexes shown in Eq. (13) and ferric carboxylates Eq. (17) are the advantages of PEF over conventional EF process, in general [28, 27].

2.2.4 Peroxi-Coagulation Process

Peroxi-coagulation is a modified EF process that is the combination of EF and electrocoagulation. Its function is rather complex. In this process, Fe or stainless steel is used as the anode to supply Fe^{2+} ions in an aqueous medium instead of the external addition of Fe^{2+} ion required to operate the Fenton reaction producing $OH^{\bullet}$ species. Thus, Fe^{2+} ions are continuously generated from the anode by the oxidation of a sacrificial anode (e.g., Fe) according to Eq. (16). At the cathode, which is necessarily selected to be the appropriate one for carrying on the $2e^-$ ORR expressed Eq. (15), H_2O_2 is generated. Ultimately chemical species generated at both anode and cathode react to form $OH^{\bullet}$ species in the system. With increasing the electrolysis time, Fe^{3+} ions accumulate in the aqueous medium leading to the formation of $Fe(OH)_3$ precipitate. Hence, organic pollutants are degraded by the simultaneous attack of $OH^{\bullet}$ species regulating in the degradation reaction and $Fe(OH)_3$ precipitate leading to the coagulation and precipitation processes [29].

$$2H_2O + 2e^- \rightarrow H_2 + 2OH^- \tag{18}$$

The increase in pH of the solution via the H_2O reduction at the cathode surface (Eq. 18) [58] increases the rate of formation of $Fe(OH)_3$ and causes electrocoagulative removal of pollutants along with oxidative action of Fenton's reaction. Although the peroxi-coagulation exhibits a higher efficiency in the removal of a pollutant than EF process, it produces a higher amount of sludge in the electrolytic cell [29]. This is one of the major disadvantages of peroxi-coagulation process. The other problem is associated with the supply of H_2O_2 produced at the cathode surface as Fe^{2+} ion in a competitive manner. Fundamentally, the rate of formation of Fe^{2+} ion from the sacrificial Fe anode in the medium is very fast, while the production rate of H_2O_2 via ORR at the cathode is rather slow because the solubility of O_2, chemical feed for the cathode, is too low (*ca.* 1 mM) and the electrode reaction follows a sluggish kinetics. Moreover, the results of higher sludge formation may cause the scavenging reactions expressed by Eqs. (9) and (19) [27].

$$Fe^{3+} + H_2O_2 \rightarrow Fe(OOH)^{2+} + H^+ \rightleftharpoons Fe^{2+} + HO_2^{\bullet} + H^+ \tag{19}$$

2.2.5 Sono- or Cavitation-Fenton Process

The coupling of ultrasound with the Fenton process (Fig. 2b) renders two-fold advantages over the conventional Fenton process. In sono-Fenton (SF) process, H_2O and O_2 are dissociated by ultrasonic waves (Eqs. (20) and (21)) that enhances not only the generation of $OH^{\bullet}$ species but also in situ generates H_2O_2 in the system (Eq. 22) [3, 133, 145]. The ultrasound obviously enhances the rate of degradation and increases the extent of mineralization of organic compounds [3, 133, 145, 153]. Moreover, H_2O and O_2 are used as additional precursors for $OH^{\bullet}$ generation and Fe^{2+} catalyst is easily regenerated in this process [3, 133, 145].

$$H_2O +))) \rightarrow H^\bullet + OH^\bullet \tag{20}$$

$$O_2 +))) + H_2O \rightarrow 2OH^\bullet \tag{21}$$

$$2H^\bullet + O_2 \rightarrow H_2O_2 \tag{22}$$

2.2.6 Sono-Photo-Fenton Process

This process is principally a hybrid of two Fenton-like processes such as SF (Fig. 2b) and PF (Fig. 2d). The scenario of generation of $OH^\bullet$ species is common as of PF process (Eqs. 13 and 14). However, sono-photo-Fenton (SPF) system appears to be promising for the treatment of wastewaters containing organic dyes over the PF and SF processes [193]. The most common problem in PF process is the decrease in efficiency of the photocatalyst for a continuous operation. Gogate and Pandit have pointed out that passivation of the active sites on the solid catalyst surface possibly occurs by the contaminants present in the solution, resulting in the decrease of efficiency of the catalyst used [82]. To solve this problem, ultrasound irradiation has been adopted that take part in a continuous cleaning up the catalyst surface during the operation [82].

2.2.7 Sono-Electro-Fenton Process

Sono-Electro-Fenton (SEF) process is the combination of SF and EF processes. Combined effect of ultrasound irradiation and the in situ electrogeneration of Fenton's reagent exhibits a high performance of this process [139]. Ultrasound irradiation continuously cleans up and activates the electrode surface at which redox reactions take place [112]. The synergistic effect of SEF process towards the rate of dye degradation appears due to (i) the high rate of mixing and enhanced mass transfer of Fe^{3+} and O_2 towards the cathode for the electrogeneration of Fe^{2+} and H_2O_2 (Eq. 15), (ii) the additional generation of $OH^\bullet$ radicals by sonolysis (Eq. 20–22) and (iii) continuous cleaning of the electrode surface by dissolution of inhibiting layers [27]. SEF is considered to be a promising, efficient, and environmentally friendly method for dye removal [27, 112].

2.2.8 Bio-Fenton Process

Bio-Fenton (BF) process refers to an in situ sustainable production of H_2O_2 that mitigates the danger of transporting and storing of H_2O_2 and decreases thus the chance of accidents. It is achieved by enzymatic activity in the presence of bio-sourced materials and a source of iron ions (Fig. 2e). The enzyme used should be able to catalyze a reaction that produces H_2O_2 as a main or by-product from bio-source materials used as a Fenton reagent [63]. Alcohol oxidase or glucose oxidase acts as the catalyst for the production of H_2O_2 from bio-sources. However, the most commonly used enzyme is glucose oxidase (GOX), which is well-known to catalyze the oxidation reaction of D-glucose to form D-glucono-1,5-lactone and H_2O_2 as a by-product (Eq. 23). The thus-generated H_2O_2 in the solution is attacked by Fe^{2+} ions catalyst to produce $OH^•$ species as expressed in Eq. (3).

$$C_6H_{12}O_6 + H_2O + O_2 \xrightarrow{GOX} C_6H_{12}O_7 + H_2O_2 \tag{23}$$

This method has been considered as a promising alternative to the conventional Fenton process, because of its important advantages, e.g., it is a sustainable, consumes low power, and can produce H_2O_2 onsite at a lower overall cost [63, 66, 99].

2.2.9 Bio-Electro-Fenton Process

Bio-Electro-Fenton Process (BEF) is a modified process of BF process, wherein the reduction of Fe^{3+} ion to form Fe^{2+} ion is operated via a coupled electrochemical method. In the anodic compartment, electrons are liberated due to biological activity (see Fig. 2f) by consuming acetate, as an example, substrate (Eq. 24) [59]. Then the generated electrons are migrated via the anode to the cathode by an external circuit as occurred in conventional microbial fuel cell [22, 75]. H_2O_2 is continuously generated in the cathodic compartment by a $2e^-$ ORR (see Eq. 15). In addition, Fe^{2+} ions are generated in situ by the reduction of iron containing substances present in the cathode chamber or directly on the cathode (Eq. 16).

$$CH_3COO^- + 2H_2O \rightarrow 2CO_2 + 7H^+ + 8e^- \tag{24}$$

Hence, $OH^•$ species are sustainably generated for the treatment of wastewaters from the in situ formed H_2O_2 in cathodic compartment of the cell.

Importantly, Fe^{2+} ions are required to be generated in situ in BEF process that can be included in the composite cathodes to reduce the amounts of excess metals and additives as well as the impurities in the treated water [75, 181]. These facts lead to avoid filtration or precipitation of these materials to recycle the catalysts. Moreover, the additional power supply is not required because it is self-dependent and the generation of electricity is achieved by the activity of microbial reactions in the anodic compartment.

2.2.10 Ligand-Assisted-Fenton Process

The PF process described above is efficient well in the solution of pH 3.0 at which the stability of Fe^{3+} hydroxyl complexes are maximum and $Fe(OH)^{2+}$ ion is highly photoactive [11, 70, 107]. In solution of pH > 3, a ligand, L is necessary to form a complex with Fe^{3+} ions (Eq. 25). Chelating ligands facilitate the dissolution of Fe^{3+} ions at pH ranging from 3 to 8 (see Fig. 2g) [188]. The most effective ligands that have been commonly used include, but are not limited to, citrate, oxalate, ethylenediaminetetraacetic acid (EDTA), humic acid, ethylenediamine disuccinic acid, and catechin.

$$Fe^{3+}L + h\upsilon \rightarrow Fe^{2+} + L^{+\bullet} \tag{25}$$

In the presence of ligands the high yield of Fe^{2+} could be achieved in solutions of pH ranging from 3 to 8. This allows to generate $OH^{\bullet}$ species with a high yield even under near neutral conditions and hence improves the removal efficiency of the organic pollutants dissolved in wastewaters.

Similarly Cu^{2+}-complex (Cu^{2+}-L) and Cu^{2+} ion itself can also act as the catalyst to in situ generate $OH^{\bullet}$ species under neutral conditions [124].

$$Cu^{2+}L + h\upsilon \rightarrow Cu^{+} + L^{+\bullet} \tag{26}$$

$$Cu^{+} + H_2O_2 \rightarrow Cu^{2+} + OH^{\bullet} + OH^{-} \tag{27}$$

It is mentioned that organic ligands, L used can also be degraded by the $OH^{\bullet}$ species owing to their instability as the treatment proceeds [166]. Hence, these ligands cannot act as a catalyst in this type of Fenton process. One of the most important limitations of this method is that the chelating agents unexpectedly introduces an additional organic components into the wastewater treatment process, i.e., Fenton/PF process, although these can be destroyed by the $OH^{\bullet}$ radicals.

2.2.11 Microwave-Fenton Process

The combination of microwave (MW) radiation with the EF process has shown a positive effect due to the acceleration of the reduction of Fe^{3+} to Fe^{2+} ions [181, 184]. This is because under microwave irradiation "hot spots" could be generated on the surface of the electrode. These high temperatures promote the transduction of electrons to the cathode surface and then accelerate the reduction of Fe^{3+} ions.

MW radiation is an electromagnetic wave with a frequency of 0.3 GHz–300 GHz [181, 184]. When MW irradiation is applied on dipole molecules dipole polarisation is occurred. The change in direction of microwave-field causes a rotation of the dipole molecules. But the rate at which the dipole rotates cannot correlate to the rate at which the MW field shifts direction. It causes "internal friction" among molecules, which

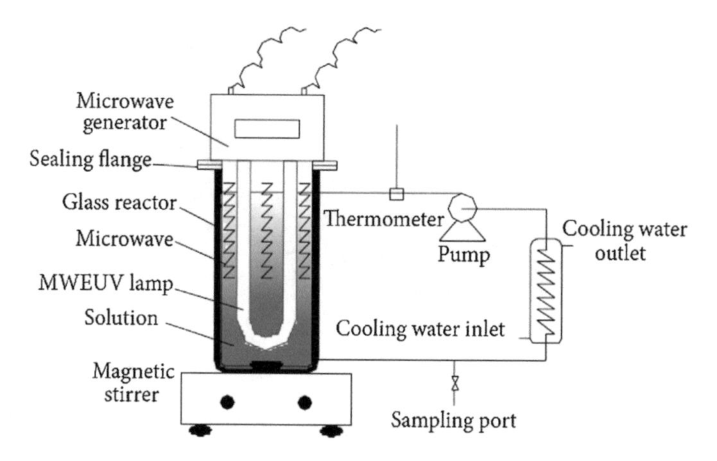

Fig. 3 A typical setup of MW-UV-Fenton-like system [44]

result in heating of the reaction mixture. But reflections and refractions of irradiation at local boundaries between phases causes to the formation of so-called "hot spots"; chemical reactivity can be enhanced by the "hot spots" effect [181, 184].

Another combination of MW irradiation with the PF processes has also shown a positive effect; the homogeneous MW-UV system could generate more $OH^{\bullet}$ radicals (Fig. 3). The reason is that H_2O_2 molecule can be activated by the absorption of MW radiation and then produced two $OH^{\bullet}$ radicals more readily. In heterogeneous processes, the photo-catalytic activity of a catalyst is enhanced by absorption of MW irradiation [44, 181, 184, 192]. However, the MW irradiation exhibits an obvious superiority compared to the Fenton process alone [44, 181, 184]. MW irradiation employed to decompose contaminants has numerous advantages due to mentioned thermal and nonthermal effects of the irradiation [44, 180, 184, 192].

3 Catalysis and Catalysts Used

Based on nature of catalysts, Fenton-like processes can further be classified into heterogeneous and homogeneous processes. In homogeneous process, the catalysts used exist in the aqueous phase as oxidants, while in heterogeneous process (Scheme 3) the catalysts are in solid beds [78, 130, 170, 173, 176]. Most commonly used catalysts in homogeneous Fenton-like process are a wide variety of metals such as Cu^{+}, Mn^{2+}, Ti^{3+}, and Co^{2+} [7, 17, 182] that have the potential to react with H_2O_2, hypochlorite (OCl^{-}), peroxydisulfate ($S_2O_8^{2-}$), or even organic hydroperoxides (ROOH) [45, 175] to produce reactive oxidants.

Solid catalysts can be grouped into two categories such as (i) Bulk catalysts: When the entire catalyst consists of the catalytically active substance, the solid catalyst is

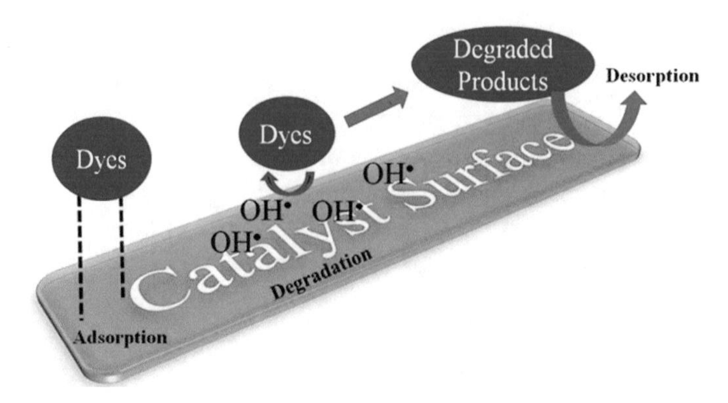

Scheme 3 Heterogeneous Fenton-like process

called a bulk catalyst, and (ii) Supported catalysts: In supported catalysts, the catalytically active materials are dispersed over the high surface area supported material. This type of catalyst is the most often used one for wastewater treatment.

3.1 Homogenous Process versus Heterogeneous Process

The researchers have focused on the heterogeneous Fenton-like catalysts to increase the catalytic activity placed in solid sources without leaching solid catalyst to the aqueous environment. Three major steps are mainly involved in heterogeneous Fenton-like process (see Scheme 3):

Step 1. Adsorption of dyes on the surface of catalyst,
Step 2. Generation of $OH^{\bullet}$ species from oxidants and reaction with dyes [79, 87], and
Step 3. Desorption of degraded products from the surface of catalyst.

In heterogeneous Fenton-like process, most popular catalyst is the Fe^{3+} in the insoluble forms that are naturally occurring minerals such as magnetite, hematite, pyrite and maghemite [74, 144], magnetite/$S_2O_8^{2-}$, etc. In other studies, Cu-loaded clay/UV and H_2O_2 system have been successfully applied to degrade trichloroethylene and azo-dye acid black 1, respectively [156, 185]. However, the scientists have been trying to employ various metals [93] to develop heterogeneous catalysts because the heterogeneous Fenton-like processes have several advantages over homogeneous Fenton reactions [78, 130, 168, 170, 173, 176]. Some of them are as it (i) inhibits the release of Fe^{3+} ion into the discharged water, (ii) leads to a higher removal efficiency of the organic pollutants, (iii) offers its application over a wide operational pH range [10, 67], (iv) consumes less energy and (v) allows an ease in recycling and reusing the catalysts employed.

3.2 Preparations of Some Important Catalysts

Catalysis is a surface phenomenon occurring on the surface or interface of the catalyst. The activity of a catalytic can be enhanced by increasing active sites that can be regulated by tuning surface area and porosity of solid catalysts. There are different processes for the synthesis of catalysts, for example, solid state reaction, precipitation, co-precipitation, gelation, crystallization, sol–gel and hydrothermal methods, and so on. Fourier transform infrared spectroscopy (FT-IR), scanning electron microscopy (SEM), transmission electron microscopy (TEM), energy dispersive X-ray analysis (EDX), BET surface area analysis, X-ray diffraction (XRD), X-ray fluorescence spectroscopy (XPS), etc. have been commonly used to characterize the catalysts. The methods of preparation and characterization of some important catalysts used in Fenton-like processes are described below.

3.2.1 $LiFe(WO_4)_2$

$LiFe(WO_4)_2$ catalyst is basically the supplement of Fe^{2+} ion required to operate the main Fenton reaction expressed by Eqs. (1) and (2) and has been used to degrade of MB in wastewaters [96]. It can be synthesized using solid state reaction through four different steps (Scheme 4): These steps are (i) Powders of Li_2CO_3, Fe_2O_3, and WO_3 with a molar ratio of 1:1:4 are mixed uniformly in a mortar, (ii) Mixture is then mechanically milled with hydrous ethanol, followed by a drying to remove the ethanol, and (iii) The obtained mass is calcined to produce a well-crystallized $LiFe(WO_4)_2$ [96].

The morphology and microstate of $LiFe(WO_4)_2$ have been studied with SEM and XRD (Fig. 4) [96]. The $LiFe(WO_4)_2$ particles are rod-like and the length of rod varies in the range of 300–500 nm with average width of 200 nm. $LiFe(WO_4)_2$ has been evaluated to be a highly crystalline and single-phase structure and exists with its pure form.

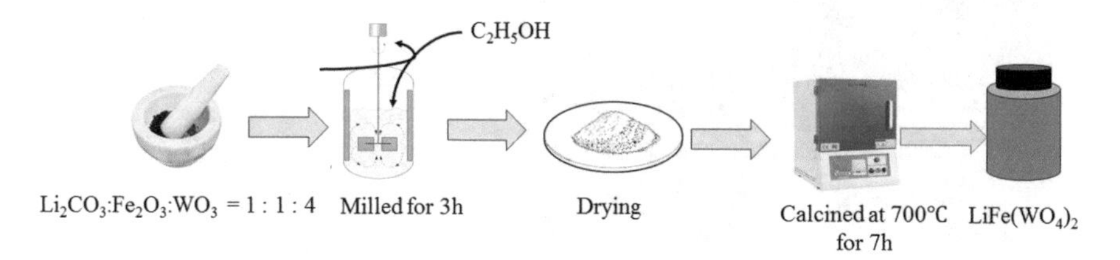

Scheme 4 Preparation of $LiFe(WO_4)_2$ by solid state reaction [96]

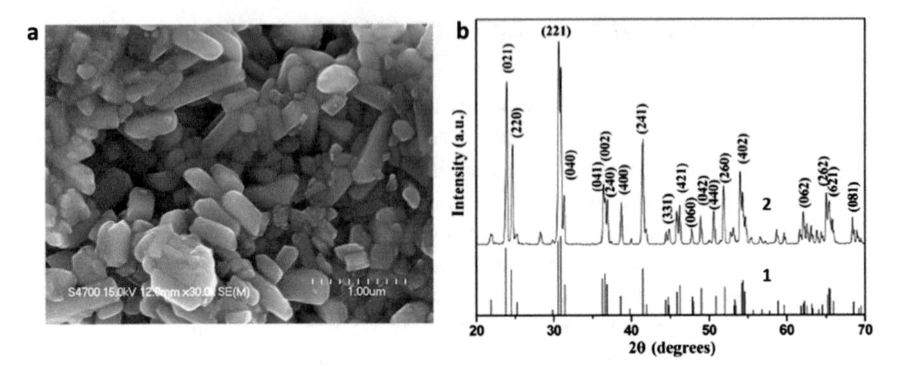

Fig. 4 **a** The typical SEM image of the $LiFe(WO_4)_2$ nanoparticles; **b** The XRD pattern of (1) theoretical $LiFe(WO_4)_2$ and (2) the synthesized $LiFe(WO_4)_2$ particles [96]

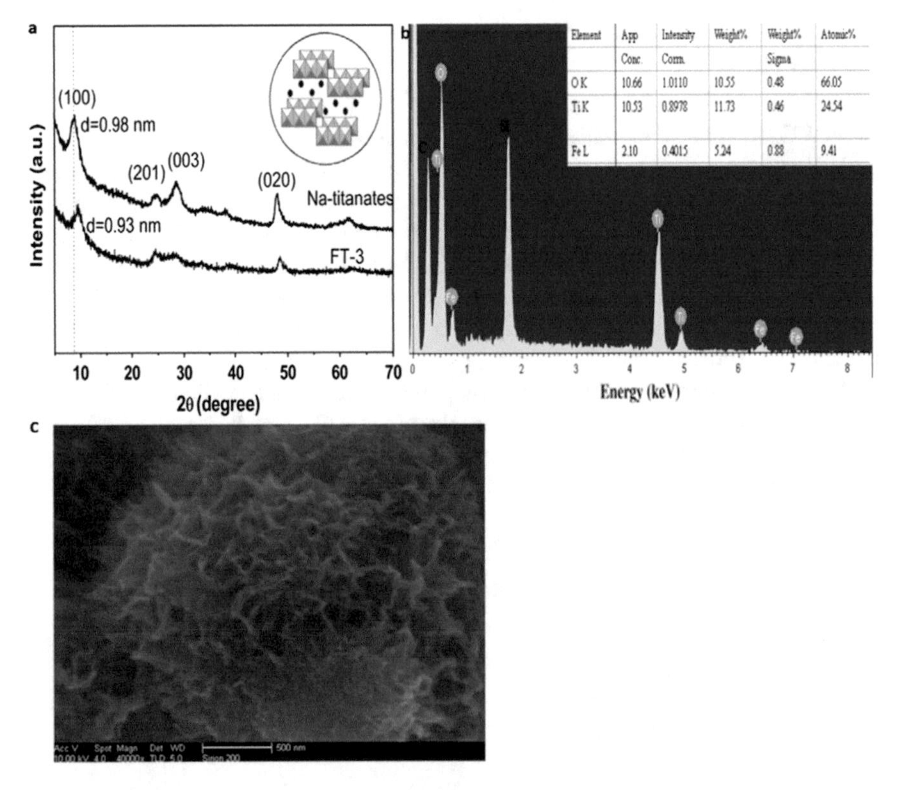

Fig. 5 **a** X-ray diffraction patterns of Na-titanates and Fe-titanates. Inset: schematic diagram of the structures, **b** EDX spectrum of Fe-titanates. Insert: the data obtained excluding data of C and Si which due to the substrate and **c** SEM morphology of Na-titanates [42]

3.2.2 Fe_3O_4/ZnO/Graphene Composite

It is one of the most widely studied catalysts. It has been used in Fenton, PF, SF, and SPF methods to degrade MB and congo-red (CR). Fe_3O_4/ZnO/graphene nanocomposites can be prepared by two successive synthetic methods, i.e., sol-gel method followed by hydrothermal method. In fact, Fe_3O_4/ZnO nanocomposites are firstly synthesized with sol-gel method, and their subsequent incorporation with graphene is performed by following a simple hydrothermal method [159]. Briefly, a specified amount of graphene is dissolved in a solution of water: ethanol (80: 20) through ultrasonication for 2 h, followed by the addition of around 10% of Fe_3O_4/ZnO with a stirring for 2 h to achieve a homogeneous suspension. The suspension is then put into a Teflon-sealed autoclave and heated at 120 °C for 3 h. The resulting nanocomposite is isolated by centrifugation and dried at 70 °C for 12 h.

The XRD patterns for the Fe_3O_4/ZnO/graphene nanocomposites can be indexed as the typical cubic spinel Fe_3O_4 and hexagonal wurtzites ZnO. The diffraction patterns for cubic spinel Fe_3O_4 have six main peaks at 2θ of 30.14°, 35.49°, 43.28°, 53.76°, 57.20°, and 62.83° that correspond to (2 2 0), (3 1 1), (4 0 0), (4 2 2), (5 1 1), and (4 4 0) planes, respectively. On the other hand, the hexagonal wurtzites ZnO exhibited peaks at 2θ of 34.47°, 36.26°, 47.50°, 56.70°, 68.06°, and 69.21° for (1 0 0), (0 0 2), (1 0 1), (1 0 2), (1 1 0), and (1 0 3) planes, respectively. However, the peak for graphene has not been seen in the measured XRD patterns. This has been considered due to the growth of Fe_3O_4/ZnO on the graphene sheets and the low graphene incorporation in the composites. The EDX spectra show that the composites without incorporation of graphene include the peaks for Fe at 0.67, 6.43, and 7.05 keV, the peaks for Zn at 1.02, 8.66, and 9.58 keV, and the peaks for oxygen at 0.52 keV. These results indicate the successful formation of Fe_3O_4/ZnO composites. In the XRD response of Fe_3O_4/ZnO/graphene, the presence of elemental carbon at 0.30 keV reveals that the Fe_3O_4/ZnO composites are grown on the surface of graphene [159].

3.2.3 Layered Fe-Titanate

Layered Fe-titanate is a typical photo-catalysts used in PF process to remove various textiles dyes from wastewaters [42]. Fe-containing titanate catalyst has been prepared by following two successive methods: alkali hydrothermal treatment and ion exchange process. In a typical synthetic procedure, meta-titanic acid and NaOH solution are mixed at proper ratio and stirred in a magnetic stirrer for several hours to form a milk-like suspension. Then this mixture is sealed in an autoclave with an internal substrate of polytetrafluoroethylene (PTFE) and heated at 80 °C for 24 h. Then the precipitate is filtered and washed until pH neutral with deionized water and dried at 80 °C for 24 h. The obtained products are characterized by different techniques. The Na-titanates are subsequently treated in Fe^{3+} solutions to obtain Fe-titanates through ion exchange process.

The XRD patterns show that the prominent peaks of both Na-titanates and Fe-titanates are typical of layered titanates [42]. In Na-titanates, the interlayer distance

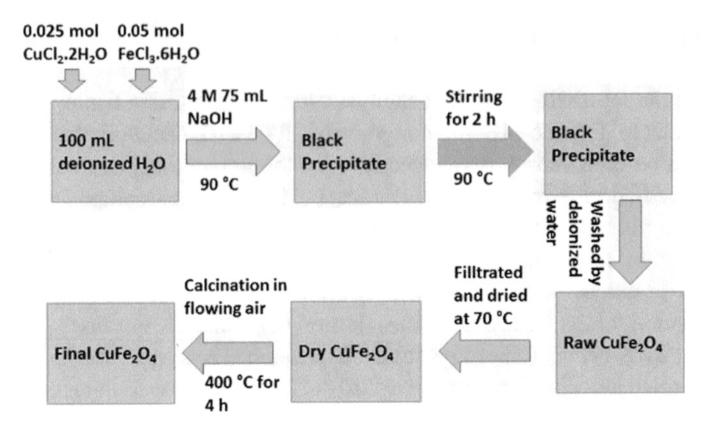

Scheme 5 Preparation of $CuFe_2O_4$ in classical alkaline medium [151]

of Na-titanates is about 0.98 nm. But the interlayer distance of Fe-titanates after the ion exchange treatment decreases to 0.93 nm, which indicates that all Na^+ ions are replaced by Fe^{3+} ions successfully. This result is consistent with EDX results of Fe-titanates where EDX response clearly shows that in Fe-titanates only Fe, Ti, and O elements exist, no Na element is observed. SEM analysis evidences the formation of layered Na-titanate that possesses the flower-like morphology.

3.2.4 Reduced Copper Ferrite-$CuFe_2O_4$

Reduced copper ferrite ($CuFe_2O_4$) has shown higher catalytic activity compared to raw $CuFe_2O_4$ for the removal of MB by heterogeneous Fenton-like process [151]. Preparation of $CuFe_2O_4$ in a classical alkaline medium is shown in Scheme 5.

The reduced $CuFe_2O_4$ is obtained by thermal treatment of prepared $CuFe_2O_4$ at 400 °C in a quartz tube under hydrogen gas flow (30 mL min^{-1}) for 4 h with a heating rate of 10 °C min^{-1} [151]. The results of typical characterizations of $CuFe_2O_4$ and reduced $CuFe_2O_4$ are shown in Fig. 6. The TEM images show that $CuFe_2O_4$ have a smooth surface and reduced $CuFe_2O_4$ have plumy look. The XRD patterns evidence that the prepared catalysts have acceptable purity. In the XPS spectra, the presence of a Fe^0 peak with weak intensity at 706.1 eV is suggestive of the loading of Fe^0 in the reduced $CuFe_2O_4$. For the XPS of Cu 2*p* regions, the peak at binding energy of 932.5 eV for the reduced $CuFe_2O_4$ is assigned to Cu^0 in the reduced $CuFe_2O_4$.

4 Removal of Cationic Dyes Contaminants via Degradation

The scenarios of oxidative degradation of various cationic dyes along with their applications and toxic features are summarized in Table 2. The efficiency of a particular

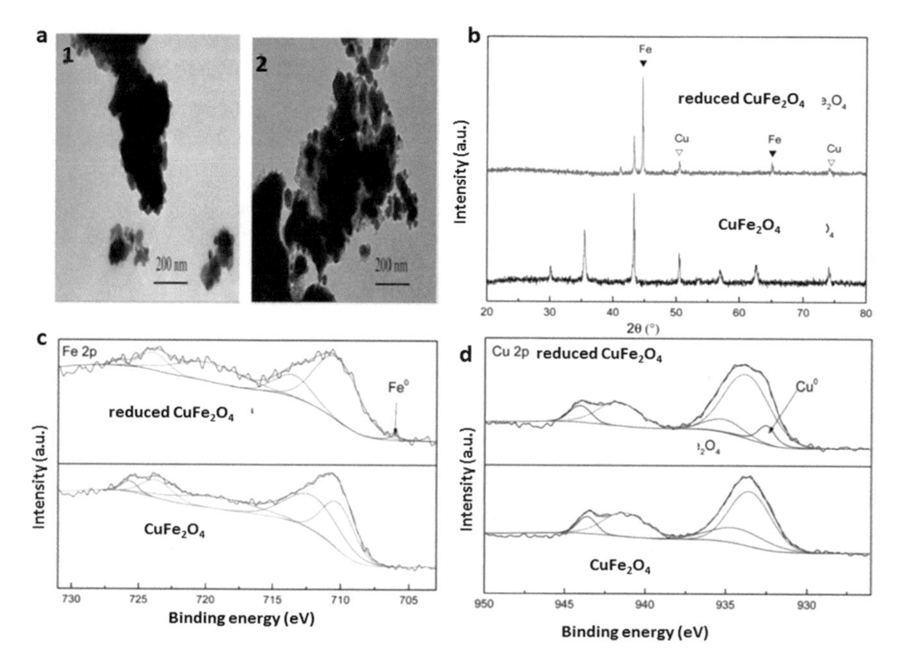

Fig. 6 **a** TEM images of (1) $CuFe_2O_4$ and (2) reduced $CuFe_2O_4$, **b** XRD patterns of $CuFe_2O_4$ and reduced $CuFe_2O_4$, and XPS spectra for Fe 2*p* (**c**) and Cu 2*p* (**d**) of $CuFe_2O_4$ and reduced $CuFe_2O_4$ [151]

method for decolorization or mineralization of model dyes with optimum conditions is especially highlighted here. Turn over number (TON) and turn over frequency (TOF) that are the characteristic parameters tell about the potential of a catalyst. It is known that catalyst having higher TON and TOF is considered to be more efficient. The recycle ability of the catalyst is a very important factor to judge the applicability of the method. In general, the cost is less in case of the catalysts that can be recycled for the repeated uses. In the following sub-sections, the detailed study on the oxidative degradation of common cationic dyes with different Fenton-like processes is systematically discussed.

4.1 Acridine Orange

Acridine orange (AO) is a cationic dye that serves as a nucleic acid-selective fluorescent dye used for the determination of cell cycle. AO has been used in many different areas including epifluorescence microscopy, and the assessment of sperm chromatin quality. A complete decolorization of AO has been achieved with the conventional Fenton process in an aqueous solution [43]. The effects of different

Table 2 Features of different textiles dyes and their degradations

Dye	Features		Degradation			References
	Hazardous effects	Applications	Method	Optimum conditions	Efficiency (%)	
AO[a]	Toxic to aquatic organisms	To assessment of sperm chromatin quality; in epi-fluorescene microscopy	Fenton [i]	$[H_2O_2] = 2.0$ mM; $[Fe^{2+}] = 0.4$ mM; pH = 3	[d]95.8 [i]	[i] [43]
A-O[b]	Toxic to aquatic organism; may cause nausea and vomiting	To stain acidic vacuoles; as a nucleic acid-selective fluorescent	EF [ii]	$[AO^b] = 50$ mg L^{-1}; $[FeSO_4 \cdot 7H_2O] = 0.22$ mM; $j = 10$ mA cm^{-2}; pH = 3.0, $V = 3$ L min^{-1}; $[Na_2SO_4] = 0.05$ M	[d]97.8 [ii] [m]92.5 [ii]	[ii] [165]

(continued)

Table 2 (continued)

Dye	Features		Degradation			References
	Hazardous effects	Applications	Method	Optimum conditions	Efficiency (%)	
AB	Intercalation into DNA and partitioning into the membrane lipid of biological cells; phytotoxic; cytotoxic	As a cardioprotective agent; histological dye, drug metabolite etc.	PF, SPF [iii], EF [iv]	PF: [AB] = 1.0×10^{-4} M; [Fe^{3+}] = 6.67×10^{-4} M; pH = 2.2; [H_2O_2] = 1.5 mL; I = 75.5 mW cm^{-2} [iii] SPF: [AB] = 1.33×10^{-4} M; [Fe^{3+}] = 5.0×10^{-4} M; pH = 2.1; [H_2O_2] = 0.5 mL; I = 75.5 mW cm^{-2}; f = 40 kHz [iii], [cat.] = 8.69 g; V = 1 L min^{-1}; pH = 2.0; working volume of 0.15 L; potential drop 14.19 V [iv]	100 [iii] [d]98-100 [iv] [m]89	[iii] [172] [iv] [155]

(continued)

Table 2 (continued)

Dye	Features		Degradation			References
	Hazardous effects	Applications	Method	Optimum conditions	Efficiency (%)	
BB-3	Causes irritation to the eye; skin; very toxic to aquatic life; environmental hazard	As a popular textile dye	EF [v]	i = 100 mA (5.6 mA cm^{-2}); T = 15 °C; [BB-3] = 0.2 mM; [$NaNO_3$] = 0.1 M; [Fe^{3+}] = 0.1 mM; pH = 3; V = 100 mL min^{-1}	$^{m/t}$91.6 ^{mce}34.84	[v] [142]
BB-41	Very toxic to aquatic life; causes irritation to the skin; eye; respiratory tract	As a textile dye	PF [vi, vii]	[BB-41] = 0.05 mM; [H_2O_2]/[Fe^{2+}] = 20; [Fe^{3+}] = 0.6 mM; [H_2O_2] = 12 mM; pH = 3 [vi], [BB-41] = 0.05 mM; pH = 3.0; [H_2O_2] = 2 mM; [H_2O_2]/[Fe^{3+}] = 10; [H_2O_2]/[BB-41] = 40 [vii]	^{t}93.2 ^{d}100	[vi] [25] [vii] [26]

(continued)

Table 2 (continued)

Dye	Features		Degradation			References
	Hazardous effects	Applications	Method	Optimum conditions	Efficiency (%)	
BR-46 or Cationic Red X-GRL	Causes foot dermatitis; highly allergic	To dye acrylic fiber	PF [vi] EF [viii]	[BB46] = 0.05 mM; $[H_2O_2]/[Fe^{2+}] = 20$; $[Fe^{3+}] = 0.6$ mM, $[H_2O_2] = 12$ mM; pH = 3 [vi], $j = 8.89$ mA cm^{-2}; pH = 3; $[Fe^{2+}] = 5$ mM [viii]	[m]93 [vi] [d]100 [vi] [d]97.62 [viii] [t]67.89 [viii]	[viii] [110]
BY-28	Adverse effects on human health causing cancer; tumors	In textile industry; printing industry	PF [vi] PEF, PF, UV/ZnO, PEF/ZnO, UV-C [ix]	[BY28] = 0.05 mM; $[H_2O_2]/[Fe^{2+}] = 20$; $[Fe^{3+}] = 0.6$ mM; $[H_2O_2] = 12$ mM; pH = 3 [vi], [BY28] = 10 mg L^{-1}; pH = 3.0; $[Fe^{3+}] = 0.1$ mM; $\lambda = 100–280$ nm [ix]	[m]93 [vi] [d]100 [vi] [d]98.8 [ix] [m]94.7 [ix]	[ix] [94]

(continued)

Table 2 (continued)

Dye	Features		Degradation			References
	Hazardous effects	Applications	Method	Optimum conditions	Efficiency (%)	
CV	Very toxic to aquatic life; corrosive; acute toxic; causes eye irritation	As a textile dye; in ball-point pens	OZ, PO, EPOP, ECP, EPO, [x, xi]	$[H_2O_2] = 15$ mM; [CV] = 50 mg L^{-1}; $[O_3] = 2$ mg L^{-1} min^{-1}; pH = 9 [x], $[H_2O_2] = 7.5$ mM; pH = 3.5; [CV] = 0.005 g L^{-1} [xi]	> [d]90 [x] [d]94.1 [xi]	[x] [1] [xi] [171]
MG	Causes irritation to the gastrointestinal tract and even cancer upon ingestion; skin irritation redness and pain	As a coloring agent on silk; wool; jute; leather; cotton; paper; acrylic; etc,	F [xii] FL [xiii]	$[Fe^{2+}] = 0.08$ mM; $[H_2O_2] = 12.5$ mM; pH = 3.4 [xii] $[MG] = 3 \times 10^{-5}$ M; $[Fe^{3+}] = 1.0 \times 10^{-3}$ M; $[H_2O_2] = 5 \times 10^{-2}$ M; pH = 3; T = 30 °C [xiii]	[d]99 [xii] [d]95.5, [m]70 [xiii]	[xii] [191] [xiii] [85]

(continued)

Table 2 (continued)

Dye	Features		Degradation			References
	Hazardous effects	Applications	Method	Optimum conditions	Efficiency (%)	
MB	Increases heart rate; vomiting; shock; Heinz body formation; cyanosis; jaundice; quadriplegia and tissue necrosis in human	To stain cells printing calico; to print cotton and tannin; as an antiseptic	FL, PF [xiv], FL [xv], EF [xvi] FL [xvii]	[MB] = 10 mg L^{-1}; [cat.] = 20 g L^{-1}; [H_2O_2] = 174.4 mM; I = 10 010 lx; pH = 12 [xiv] [MB] = 50 mg L^{-1}; [cat.] = 0.1 g L^{-1}; [H_2O_2] = 0.5 mM; pH = 3.2 [xv] pH = 3.0; [cat.] = 1.0 g L^{-1}; [MB] = 50 mg L^{-1} $[MB]_o$ = 500 mg L^{-1}; [cat.] = 10 g L^{-1}; pH = 3; $[H_2O_2]_o$ = 0.2 M; T = 50 °C [xvii]	[d]96 [xiv] [d]74 [xv] [m]96.6 [xvi] [d]99.7 [xvii]	[xiv] [121] [xv] [151] [xvi] [74] [xvii] [96]

(continued)

Table 2 (continued)

Dye	Features		Degradation			References
	Hazardous effects	Applications	Method	Optimum conditions	Efficiency (%)	
MR	Causes cancer; health and ecological hazards	As an indicator for acid–base titrations	EF [xviii]	$[MR]_o = 100$ mg L^{-1}; $E_c = -0.55$ V; $V = 0.4$ L min^{-1}; $[Fe^{2+}] = 0.2$ mM; pH = 3; $[Na_2SO_4] = 0.1$ M [xviii]	[d]80	[xviii] [194]
RhB	Causes teratogenic effects on public health; a carcinogenic agent	To dye silk; wool; jute; leather; and cotton in the textile industry	3D-E-Fenton, EF [xix]	$[Fe^{2+}] = 3$ mM; $[Fe^{2+}]/[H_2O_2] < 1$; $E = 2.0$ V; pH = 6.2 [xix]	[d]99	[xix] [115]
ST	Causes irritation to the eye; skin; and respiratory tract	As coloring agent in food industry; for dyeing tannin; cotton; bast fibers; wool; silk; leather and paper	FL [xx]	pH = 3; $[Fe^{2+}]/[ST] < 1$; $[H_2O_2]/[Fe^{2+}] > 3$ [xx]	-	[xx] [122]

a = Acridine orange; b = Auramine-O; d = Decolorization; m = Mineralization; t = Total organic carbon; V = Air flow; T = Temperature; λ = Wavelength of light; cat. = Catalyst; V = Cell voltage; l = Electrode gap; I = Intensity of light; i = Applied current; j = Current density; f = Frequency; ACF = Activated carbon fiber; PTEF= Polytetrafluoroethylene; EPOP = Electrolysis/peroxene/H_2O_2, E = Applied potential, FL = Fenton-Like operational parameters, for instances, the concentrations of dye, Fe^{2+}, H_2O_2, solution pH, and the presence of Cl^- ion on the oxidative degradation of AO have been clarified. More than 95.8% efficiency of decolorization of AO has been achieved within 10 min at an optimum operating condition of 0.2, 2.0, and 0.4 mM of dye, H_2O_2, and Fe^{2+} ion, respectively, and the solution pH of 3.0. Adverse effects of Cl^- ion on the decolorization efficiency of AO has been noticed, since Cl^- ions exhibit scavenging effects towards $OH^•$ radicals [43].

The pathway of degradation of AO has been explored through identifying the intermediates with the help of high performance liquid chromatography (HPLC) coupled with a photodiode array detector and electrospray ionization (ESI) mass spectrometer (see Fig. 7). The results of analysis indicate that the degradation of AO occurs in a stepwise manner to yield *mono-*, *di-*, *tri-*, and *tetra-N*-de-methylated AO species generated. The methyl groups are removed one by one as is confirmed by the observation of the gradual shifting of wavelength of the band towards the blue region (Fig. 7A–b). Hence, the $OH^•$ radicals generated in the Fenton process are efficient enough to completely decolorize and partially mineralize the AO dye in an aqueous solution in a short reaction time [43].

4.2 *Auramine-O*

Auramine-O (A-O) appears as yellow needle crystals in their pure form. It is a cationic dye widely used as a fluorescent stain [104]. Degradation of A-O through EF process using Ti/PbO_2-EF anode in which H_2O_2 generated in situ by the ORR has been carried out [165]. The degradation of AO has been analyzed from the view of total organic carbon (TOC), mineralization current efficiency (MCE), energy consumption (E_c), and degradation mechanism and kinetics (Fig. 8). It has been observed that TOC removal depends on the electrode materials, for instance, after 180 min TOC removal has been found to be 92.54, 66.3, and 36.88% at Ti/PbO_2-EF, Ti/PbO_2-AO, and Pt-EF anodes, respectively. The similar dependence in MCE and E_c have also been noticed.

It has been observed that the efficiency of degradation increases as the current density increases. This effect can interpret with the excess production of H_2O_2, and regeneration of Fe^{2+} ion that promotes the generation of $OH^•$ species [141]. In addition, when the applied current intensity exceeds 10 mA cm^{-2}, this effect on the kinetics of degradation becomes negligible. The association of applied current density on the enhanced degradation has been considered to be due to the fact that H_2O produces a side reaction (Eq. 28) that reduces the production of H_2O_2 [56].

$$2H_2O \rightarrow O_2 + 4H^+ + 4e^- \quad (28)$$

The cost of disposing of pollutants is an important indicator of a good method. This can be estimated from the profile of energy consumption per TOC mass removed. Under the optimal conditions, the MCE and E_c for the removal of AO using Ti/PbO_2-carbon felt operated for 180 min have been found to be 92.5% and 0.9 kWh $(TOC)^{-1}$,

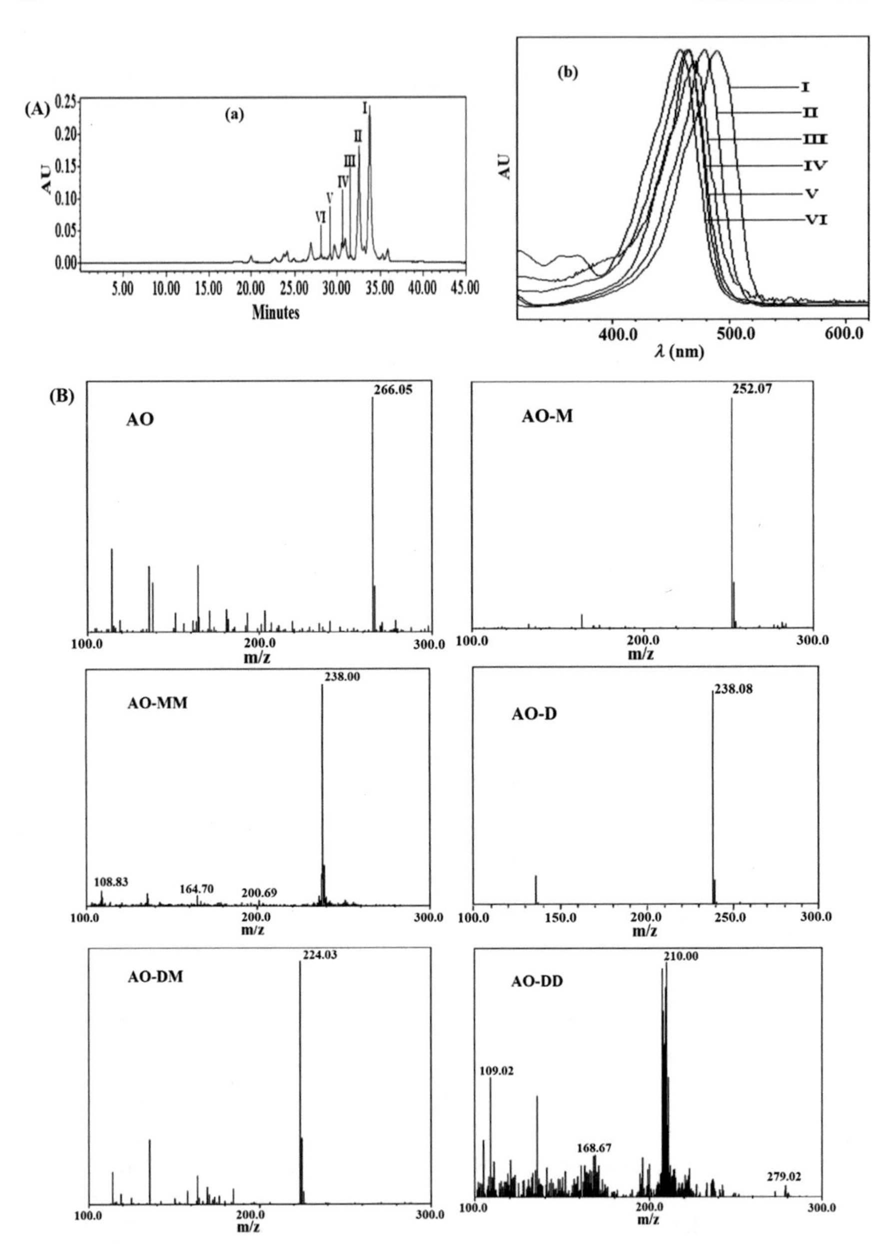

Fig. 7 **A** HPLC chromatogram of the reacted solution after 30 min of Fenton reaction (**a**), and the absorption spectra of the N-de-methylated intermediates formed during the Fenton process of the AO dye corresponding to the peaks in the HPLC chromatogram (**b**). Spectra have been recorded using the photodiode array detector. The spectra denoted by I-VI correspond to the peaks I-VI in Fig. (**A–a**), respectively. **B** ESI mass spectra of the N-de-methylated intermediates separated by HPLC-ESI-mass spectrometry (MS) method [43]

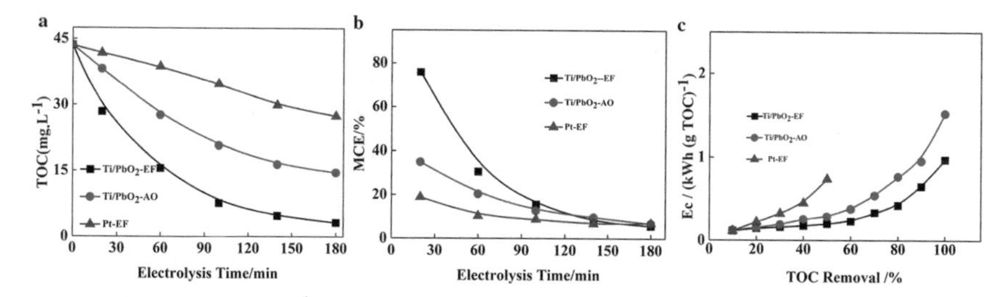

Fig. 8 Comparison between Ti/PbO$_2$-EF, Ti/PbO$_2$-A-O and Pt-EF for evolution of TOC removal (**a**), MCE (**b**), and E_c (**c**) during degradation of 50 mgL^{-1} AO solutions. Conditions: Pt-EF and Ti/PbO$_2$-EF: FeSO$_4$·7H$_2$O is 0.22 mM; pH = 3; aeration rates is 3 Lmin^{-1}; current density is 10 mAcm^{-2}; addition amount of Na$_2$SO$_4$ is 0.05 M; Ti/PbO$_2$-A-O without FeSO$_4$·7H$_2$O and the aeration) [165]

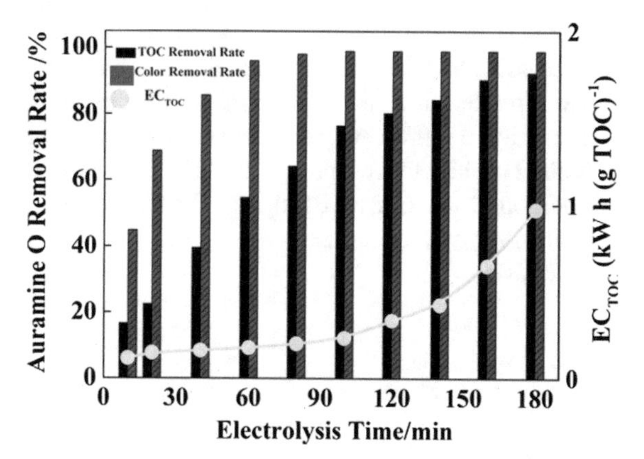

Fig. 9 Optimal conditions of TOC removal rate, energy consumption and the decolorization rate by the EF process using Ti/PbO$_2$-carbon felt. The condition is FeSO$_4$•7H$_2$O = 0.22 mM; pH = 3; initial concentration = 50 mg L^{-1}; aeration rate = 3 L min^{-1}; current density = 10 mA cm^{-2}; addition amount of Na$_2$SO$_4$ = 0.05 M [165]

respectively (Fig. 9). However, after 60 min, the MCE and E_c become about 64.27% and 0.114–0.2305 kWh (TOC)$^{-1}$), respectively. The observed increase in the rate of TOC removal and slow growth of E_c against time may result from a complicated mechanism occurring during the dye degradation process [106]. However, the EF process using Ti/PbO$_2$-carbon felt process is revealed to be highly effective method in removing organic dyes from wastewaters.

4.3 *Azure-B*

Azure-B (AB) is an organic chloride salt having 3-(dimethylamino)-7-(methylamino)phenothiazin-5-ium as the counterion. It is used as a fluorochrome, a histological dye, a drug metabolite, a cardioprotective agent, and an antidepressant. The oxidative degradation of AB has been studied by PF, SPF, and EF processes (see Table 2). A complete mineralization of AB to form CO_2 and H_2O has been achieved using PF and SPF. Multi-methods have been employed [172] to study the oxidative degradation of AB. In homogeneous aqueous solution, the operating conditions of PF and SPF for AB degradation have been optimized. The rate of degradation has been revealed to increase in the presence of ultrasound and it follows a pseudo-first-order kinetics monitored with spectroscopic techniques (Fig. 10). A tentative mechanism for the decomposition of AB via SPF reaction has been proposed [172]. However, the regeneration of Fe^{2+} ions from Fe^{3+} ion is highlighted to be the main advantage of SPF process, since further separation of Fe^{3+} ions is not needed after wastewater treatment.

The applicability of EF technique to degrade AB using Fe alginate gel beads has been evaluated in a comparative study with Fenton process [155]. The effect of pH on the process based on Fe alginate beads and on the EF process with free Fe and Fe alginate bead have been studied. Fe alginate beads show a physical integrity in a wide range of pH (2–8) and hence increase the efficiency of the process. Almost a complete decolorization (~98–100%) of dye has been obtained by EF process in successive batches without operational problems, e.g., clogging, bead breakage, or overpressure at a residence time of 30 min. At low pH of 2, the efficiency of decolorization significantly decreases with free Fe, mainly because Fe^{3+} precipitates at high pH values that eventually leads to a decease in the concentration of iron ions in solution [155]. Hence, Fe alginate beads exhibiting a stable performance opens a promising route for a fast and cost-effective treatment of wastewater containing organic dyes.

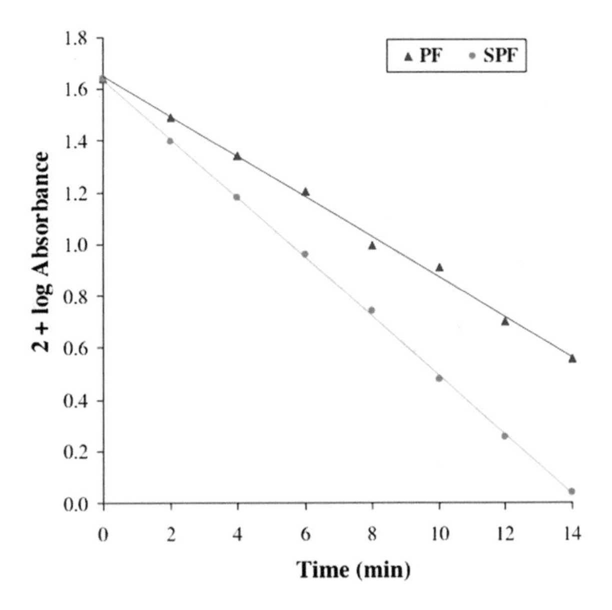

Fig. 10 A typical run; PF: [AB] = 1.0×10^{-4} M, pH = 2.2, $[Fe^{3+}] = 6.67\times10^{-4}$ M, H_2O_2 = 1.5 mL and light intensity = 75.5 mW cm^{-2}, SPF: [AB] = 1.33×10^{-4} M, pH = 2.1, $[Fe^{3+}] = 5.0\times10^{-4}$ M, H_2O_2 = 0.5 mL, light intensity = 75.5 mW cm^{-2} and frequency = 40 kHz [172]

4.4 Basic Blue 3

Basic blue-3 (BB-3) appears as bronze powder in its pure form. It is readily soluble in water and widely used for dyeing wool and acrylic blended fabric graft copolymerization. A novel cathode material, carbon sponge (CS)-based EF process has been employed for treating BB-3 contaminated wastewaters [142]. It has been observed that CS electrode exhibits a greater efficiency than carbon felt (CF) electrode for degradation of BB-3. For instance, the TOC have been achieved to be 91.6 and 50.8% with CS and CF electrodes, respectively, after 8 h electrolysis. This difference is revealed to be associated with the variation in the production of electrogenerated H_2O_2. The amount of H_2O_2 determined with iodide method [142] at CS electrode is about three times higher than that produced at CF electrode. The effects of some experimental conditions, e.g.; applied current value, type of supporting electrolyte, O_2 flow rate, pH, and temperature on the efficiency of degradation of BB-3 have been clarified. The conditions of degradation and mineralization for CS electrode-based EF process have been optimized as: [BB-3] = 0.2 mM; [$NaNO_3$] = 0.1 M; [Fe^{3+}] = 0.1 mM; pH = 3; $i = 5.6$ mA cm^{-2}; $T = 15$ °C, O_2 flow rate = 100 mL min^{-1}. Temperature has been found to exhibit a significant adverse effect on electrosynthesis of H_2O_2. This is reasonable, because the solubility of O_2 in water as the stability of H_2O_2 decrease with increasing temperature. The degradation of BB-3 follows a pseudo-first-order kinetics in both cases. MCE values have been obtained to be 34.84 and 9.13% for the CS and CF electrodes in the first 30 min, respectively, that is, the efficiency of CS electrode is four-times higher than that of CF electrode. Therefore, CS electrode in EF system is a promising in the field of electrochemical generation of H_2O_2 as well as degradation of organic pollutants.

4.5 Basic Blue 41

Basic blue-41 (BB-41) looks like bright blue at normal temperature. When it dissolves in water the solution becomes blue. It has been used for dyeing acrylic fiber, scattered the wool, the knitting flocking, and knitted fabrics. PF process has been attempted to degrade BB-41 using the experimental setup represented in Fig. 11 [25]. The mineralization efficiency depends on the ratio of [H_2O_2]/[BB-41] and the optimum ratio is 40 at which a complete mineralization with a 93.2% of TOC removal could be achieved. The degradation of BB-41 follows first-order kinetics [25].

4.6 Basic Red 46 or Cationic Red X-GRL

Basic red-46 (BB-46) is a dark red powdery cationic dye used in wool, silk, acrylic/cellulosic (cotton or linen) fiber blends, polyester, and acrylic textile printing

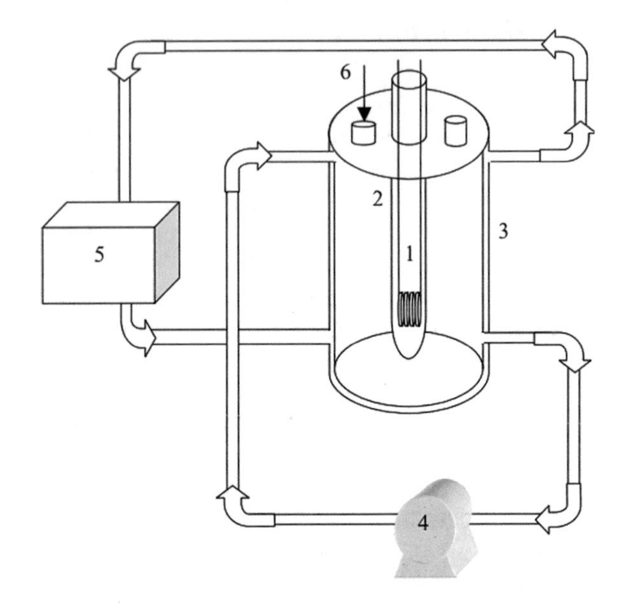

Fig. 11 Typical experimental setup of PE employed. UV lamp (1), Pyrex jacket (2), reaction vessel (3), pump (4), thermostat (5), sampling (6). The optimum conditions for the degradation of BB: 0.052 mM of BB and [H_2O_2], respectively and [H_2O_2]/[Fe^{3+}] = 10 at solution pH = 3.0 [25]

(sweaters, shirts, socks, etc.). The EF process using an activated carbon fiber (ACF) cathode has been adopted for the degradation of the azo dye, BB-46 [110]. At the experimental optimized condition, the efficiency of decolorization could be achieved to be 97.62% within 180 min. The rate of degradation can be enhanced by increasing temperature of the system and it follows a pseudo-first-order kinetics. These results suggest that in case of EF processes the reactions expressed by Eqs. (12) and (15), and the reaction of $OH^{\bullet}$ radicals oxidizing the dye species (Eqs. (1) and (2)) could be accelerated by increasing temperature. The efficiencies of decolorization and TOC removal have been found to decrease with increasing initial concentration of dye feed in the system. The effect of Cu^{2+} and Mn^{2+} ions on the degradation profile has been tested to evaluate their catalytic performance as a substitute for Fe^{2+} ion. Cu^{2+} and Mn^{2+} ions have been found to be more efficient than Fe^{2+} in this regard. In the mechanism it has been proposed that (i) a fast destruction of complexes of Cu^{2+} with the intermediates formed (see Eq. 29) during degradation of dye species takes place [32, 162, 195] and (ii) the greater production of $OH^{\bullet}$ species in the medium of the Cu^{2+}/Cu^{+} catalytic system (Eqs. 29 and 30) contributes to the observed enhancement [54, 76]. On the contrary, in case of Mn^{2+} ion, Mn^{3+} possessing a higher standard reduction potential accepts electrons at a faster rate than that of Fe^{3+} (Eq. 31) that ultimately catalyze the regeneration reaction of Mn^{2+} that ultimately catalyzes the disproportionation reaction of H_2O_2 to produce $OH^{\bullet}$ radicals at a faster rate [95].

$$Cu^{2+} + HO_2^{\bullet} \rightarrow Cu^{+} + H^{+} + O_2 \quad (29)$$

$$Cu^{+} + H_2O_2 \rightarrow Cu^{2+} + OH^{\bullet} + OH^{-} \quad (30)$$

$$Mn^{3+} + e^- \rightarrow Mn^{2+}(E^0 = 1.51 \text{ V } vs. \text{NHE}) \quad (31)$$

4.7 *Basic Yellow 28*

Basic yellow-28 (BY-28) is a yellow color cationic. It is easily soluble in water at room temperature. It has been mainly used in acrylic fiber, scattered article fiber dyeing and can also be employed for a direct printing acrylic and silk. BY-28 has been attempted to degrade by photocatalytic process using immobilized ZnO nanoparticles that are combined with EF process at CNT–PTFE cathode [94]. The excellence of this study has been revealed by a comparative study with ultraviolet-C (UV-C), UV/ZnO, EF, PEF, and PEF/ZnO for decolorization of BY-28 in water. The process at PEF/ZnO composite exhibits the highest decolorization efficiency of 91.3% at the first 40 min, while PEF, UV/ZnO, EF, and UV-C processes led to the decolorization efficiencies of only 51.8, 17.1, 6.9, and 6.13%, respectively. It has been further clarified that removal efficiency increases with increasing applied current that might be due to the production of a large amount of H_2O_2 through reaction shown by Eq. (15). Moreover, the increase in the concentration of BY-28, the efficiencies of removal gradually decrease to 97.1, 73.3, and 54.5% for 20, 30, and 40 mg L^{-1}, respectively. This may be happened due to the competitive consumption of $OH^{\bullet}$ radical by the generated intermediates at high initial concentration of BY-28. On the other hand, UV-C (100–280 nm) light irradiation exhibits the highest decolorization efficiency compared to other UV-A (315–380 nm) and UV-B (280–315 nm) light irradiations. This is obvious since UV-C can generate more $OH^{\bullet}$ species by photolysis of the produced H_2O_2 (Eq. 15) [51, 50]. The maximum TOC removal has been found to be 94.7% with PEF/ZnO electrode. Six different compounds namely 1,2,3,3-tetramethyl indoline, 4-methoxy benzenamine, hydroquinone, butenedioic acid, propanoic acid, and glyoxylic acid as by-products accompanying BY-28 degradation have successfully been detected by GC–MS analysis.

4.8 *Basic Blue 41, Basic Red 46, and Basic Yellow 28*

A complete decolorization of a mixture of three basic dyes, viz., BB-41, BR-46, and BY-28 has been investigated using PF process, $UV/Fe^{3+}/H_2O_2$ [26]. The rate of degradation and efficiency of mineralization has been found to depend on the initial concentration of H_2O_2 and Fe^{3+} ion as well as their ratio. The degradation of the dye mixture follows a pseudo-first-order kinetics. The optimum initial concentration of Fe^{3+} has been found to be 0.6 mM and further increases in the concentration of Fe^{3+} adversely affect the mineralization efficiency. This fact would be associated with the brown turbidity of the medium that hinders the degree of absorption of light required for the PF as well as leads so-called scavenging effects of OH$^{\bullet}$ radicals (Eqs. 5 and 8) [46]. A complete decolorization of the mixture and the TOC removal of BB-41, BR-46, and BY-28 have been found to be 93, 85, and 95%, respectively.

4.9 Crystal Violet

Crystal violet (CV), also known as methyl violet 10B, is a triarylmethane cationic dye used as a histological stain and in Gram's method of classifying bacteria [2]. It is used as a textile dye, to dye paper and as a component of ball-point pens, and inkjet printers [16]. The degradation of CV has been investigated by applying different AOPs, e.g., ozonization, peroxone (PO), electrolysis, electrolysis/H_2O_2 (ECP), electroperoxone (EPO), and electrolysis/peroxene/H_2O_2 (EPOP) methods [1]. All processes are capable of decolorization of CV indeed, while PO and EPOP are more efficient in this regard (Fig. 12). A significant decrease in oxidation rate of CV has been observed owing to the addition of methanol as scavenger and EDTA since they are prone to react with in situ generated OH$^{\bullet}$ radicals. The toxicity test and growth of *Escherichia coli* colonies confirms that the treated effluent can be discharged to the environment.

Iron-loaded ZSM-5 zeolite catalyst has been found to be efficient and robust in catalyzing degradation for CV [171]. At the optimum condition the efficiency of decolorizaiton of CV has been determined to be 94.1%. In addition acidic condition favors the decolorization process. Doubling the amount of the catalyst enhanced the decolorization from 94.1 to 99.6%, while the chemical oxygen demand (COD) changes from 50 to 58.8%. The stability of the catalyst has been found to be maintained even after using the catalyst for three cycles, and a small iron leaching also

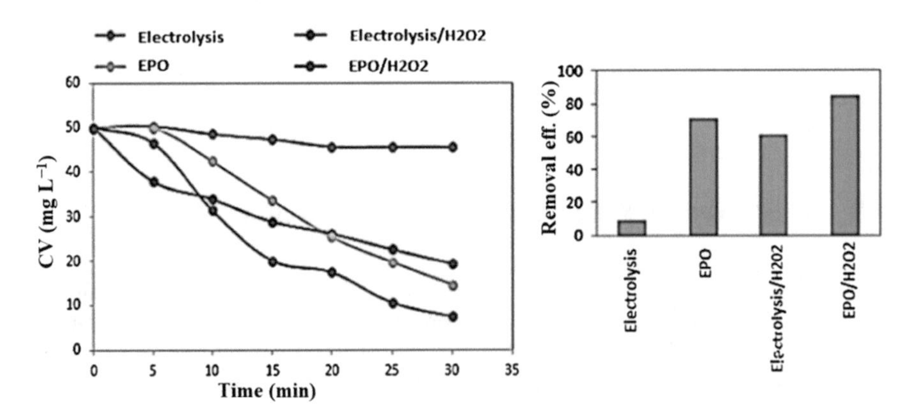

Fig. 12 Effect of electrolysis and its combinations with ozonation and peroxone on CV decolorization [1]

proves the stability of the catalyst. Thus, iron-loaded ZSM-5 zeolite seems to be an efficient for dye removal from textile wastewaters.

4.10 Malachite Green

Malachite green (MG) is an organic compound that is used as a cationic dyestuff. MG is traditionally used as a dye for materials such as silk, leather, and paper [49]. MW-Fenton process has been employed for the degradation of MG [191]. Basically, MW irradiation is not able to de-colorization MG separately and only 23.5% de-colorization of MG has been found with the conventional Fenton process. Interestingly, the combined approach of both methods, i.e., MW-Fenton offers *ca.* decolorization efficiency of 95.4%. The rates of COD removal and de-colorization of MG have been observed to be 82.0 and 99.0%, respectively, within 5 min under the optimum condition. However, a new peak appears in the UV range of the spectra, which indicates that the dye species are not completely decomposed into CO_2 and H_2O, rather to some small fragments. The apparent kinetics equation of $-dC/dt = 0.0337 \times [MG]^{0.9860} \times [Fe^{2+}]^{0.8234} \times [H_2O_2]^{0.1663}$ for MG decolorization has been established. This observation suggests that the concentration of Fe^{2+} ion is the dominant factor in achieving acceptable removal efficiency with MW-Fenton process [191].

In another study, the mineralization of MG dye using Fenton-like reaction, Fe^{3+}/H_2O_2 has been studied [85]. The effects of different operational parameters like solution pH, the initial concentrations of Fe^{3+}, H_2O_2, and dye, temperature, and added electrolytes (Cl^- and SO_4^{2-}) on the oxidation of dye have been clarified. Cl^- ion affects the degradation of MG negligibly, whereas SO_4^{2-} ion exhibits a significant effect. The overall efficiencies of degradation and mineralization have been found to be *ca.* 95 on and 70%, respectively, under the optimum conditions and 15 min operation. The degradation of MG follows a pseudo-second-order kinetics and the reaction is spontaneous and endothermic in nature.

4.11 Methylene Blue

Methylene blue (MB) has been commonly studied using different Fenton-like systems. In fact, MB has been used as a model cationic dye for evaluating efficiencies of other dye removal techniques, probably because of its availability and its intense blue color. Co^{2+} loaded surfactant-modified alumina (SMA), designated as Co-SMA, a heterogeneous catalyst has a great influence on the degradation MB compared to homogeneous cobalt catalyzed Fenton reaction [121]. Adsorption of Co^{2+} on SMA is *ca.* 6.4-times higher than that on normal alumina. MB has been adsolubilized on the Co-SMA surface for further degradation and about 100% adsorption occurs within the first 10 min (Fig. 13). On the other hand, the affinity of adsorption of MB towards

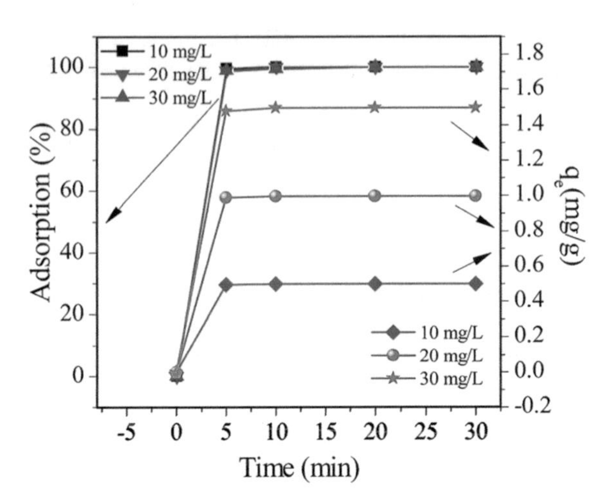

Fig.13 Kinetics of adsolubilization of MB on Co-SMA surface at different initial concentrations [121]

alumina surface is very low and only *ca.* 5% adsorption has been found to occur in 1 h. This is because no surfactant bilayer is present on the alumina surface to host MB.

The degradation of MB has been carried out in fully adsolubilized state on Co-SMA surface using PF process. The decolorization reaction follows a zero-order kinetics, which depicts a true surface catalyzed reaction. Co^{2+} ion and MB are firmly attached to the adsorption layer and $OH^\bullet$ radicals are generated from the reaction of Co^{2+} and H_2O_2 leading to degradation reaction in an efficient way [121]. PF degradation of MB on Co-SMA surface follows a Langmuir–Hinshelwood mechanism [65].

Fluorescence measurement has confirmed that $OH^\bullet$ radicals are generated and utilized in the Fenton reaction. The production of $OH^\bullet$ radical has been found to be much slower (Eq. 32) compared to its consumption (Eq. 36). So, the rate of degradation of MB, finally, only depends on the rate of ROS formation. Since MB is absent in the slower step (or steps), the kinetics of MB degradation follows zero-order reaction mechanism with respect to MB [121].

$$\mathrm{Co^{2+}(adsorbed\ on\ SMA) + H_2O_2 \rightarrow\ Co^{3+} + OH^- + OH^\bullet(slow)} \tag{32}$$

$$\mathrm{Co^{3+} + H_2O_2 \rightarrow\ Co^{2+} + HOO^\bullet + H^+} \tag{33}$$

$$\mathrm{Co^{3+} + H_2O + visible\ light \rightarrow\ Co^{2+} + OH^\bullet + H^+} \tag{34}$$

$$\mathrm{OH^\bullet + H_2O_2 \rightarrow\ H_2O + HOO^\bullet} \tag{35}$$

$$\mathrm{OH^\bullet + MB \rightarrow\ Products\ (fast)} \tag{36}$$

$$\text{Products} \rightarrow \text{Released from the SMA surface} \quad (37)$$

The degradation reaction is facilitated in presence of visible light. The decolorization efficiency of almost 96% has been observed at pH 12, whereas 46% at pH 3 under the optimized experimental conditions. This is because heterogeneous Co^{2+} system exhibits enhanced reactivity in alkaline condition. The degradation products are analyzed by ESI–MS shown in Fig. 14. It is clear that MB turns into smaller fragments upon degradation. TON and TOF of the catalyst are found to be 6.82×10^{20} molecules g^{-1} and 1.89×10^{17} molecule g^{-1} s^{-1}, respectively [120].

Moreover, the Co-SMA catalyst exhibits a good performance up to third cycle (Fig. 15c). Therefore, Co-SMA-based catalyst is much more efficient to degrade MB as compared to a homogeneous cobalt-catalyzed Fenton reaction, because of the presence of the admicellar surface on SMA that accommodates MB more effectively [121].

Reduced $CuFe_2O_4$ shows superior catalytic activity in the presence of H_2O_2 compared to untreated $CuFe_2O_4$ for the removal of MB [151]. This is because reduced catalysts possess a high surface area (Fig. 6a) that provides more active sites for H_2O_2 decomposition and produces more reactive oxidants such as $OH^\bullet$ species. A small amount of metal ions leached from the reduced $CuFe_2O_4$ and these leached ions could act as homogeneous Fenton catalysts in MB degradation. Hence, the removal of MB is attributed to both the homogeneous and heterogeneous Fenton-like reactions [151].

Loading Fe^0/Cu^0 bimetallic particles could facilitate the decomposition of H_2O_2 into $OH^\bullet$ and accelerate electron transfer from Fe^0 and/or Cu^0 to $CuFe_2O_4$ [47, 128]. The degradation of MB follows a pseudo-first-order kinetics. Effect of *tert*-butyl alcohol as $OH^\bullet$ scavenger reveals that the main reactive species is$OH^\bullet$ radical. Therefore, reduced $CuFe_2O_4$ could effectively activate H_2O_2 to generate $OH^\bullet$ in the heterogeneous Fenton-like reaction. A possible enhanced reaction mechanism of MB degradation by reduced $CuFe_2O_4$ is illustrated in Scheme 6.

Moreover, the magnetic hysteresis behavior of $CuFe_2O_4$ has revealed that reduced $CuFe_2O_4$ is ferromagnetic and has a magnetic saturation of about 6.3 emu g^{-1}, which ensure that the catalyst could be easily separated by a magnet and reused from an aqueous solution. It also exhibits a good activity after being recycled five times (Fig. 15a).

The degradation of MB has also been studied using a heterogeneous EF process based on a sepiolite/pyrite (Sep/FeS_2) nanocomposite [71]. The EF process with Sep/FeS_2 composite is found to be more efficient in MB removal than FeS_2 catalyst alone. Thus Sep/FeS_2 nanocomposite offers more sites for producing more $OH^\bullet$ species by which a 96.6% mineralization is achieved [71].

The MB degradation follows a pseudo-first-order reaction kinetic at graphite cathode. LC–MS results reveal that aromatic compounds are formed during the treatment of MB using EF process based on the Sep/FeS_2 composite [71]. The catalytic performance of the Sep/FeS_2 nanocomposite has been found to be slightly reduced after five cycles from 96.6 to 87.8% (Fig. 15b). However, one disadvantage of the proposed EF method here is that it needs to be operated under acidic conditions, so it

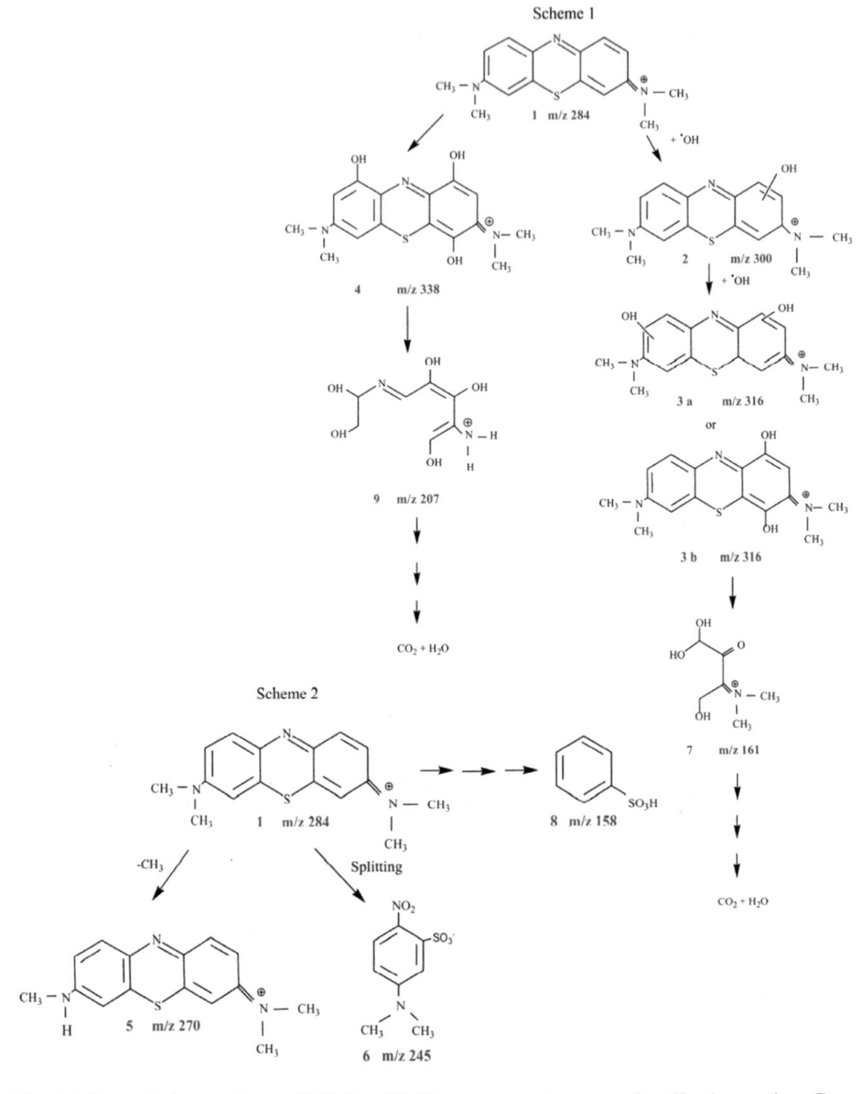

Fig. 14 Degradation products of MB identified by mass spectroscopy after 60 min reaction. Conc. of MB: 10 mg L^{-1}; dose of Co-SMA: 20 g L^{-1}; light intensity: 10 010 lx; [H_2O_2]: 348.8 mM) [121]

is best suited for treating acidic industrial wastewaters, or the wastewater pH would need to be adjusted [71].

$LiFe(WO_4)_2$ particles as a heterogeneous catalyst have been tested to exhibit the catalytic performance to decompose H_2O_2 to $OH^\bullet$ radicals which can destroy MB efficiently [96]. The initial dye concentration, catalyst dosage, H_2O_2 concentration, pH, and temperature have shown great influences on decolorization efficiency.

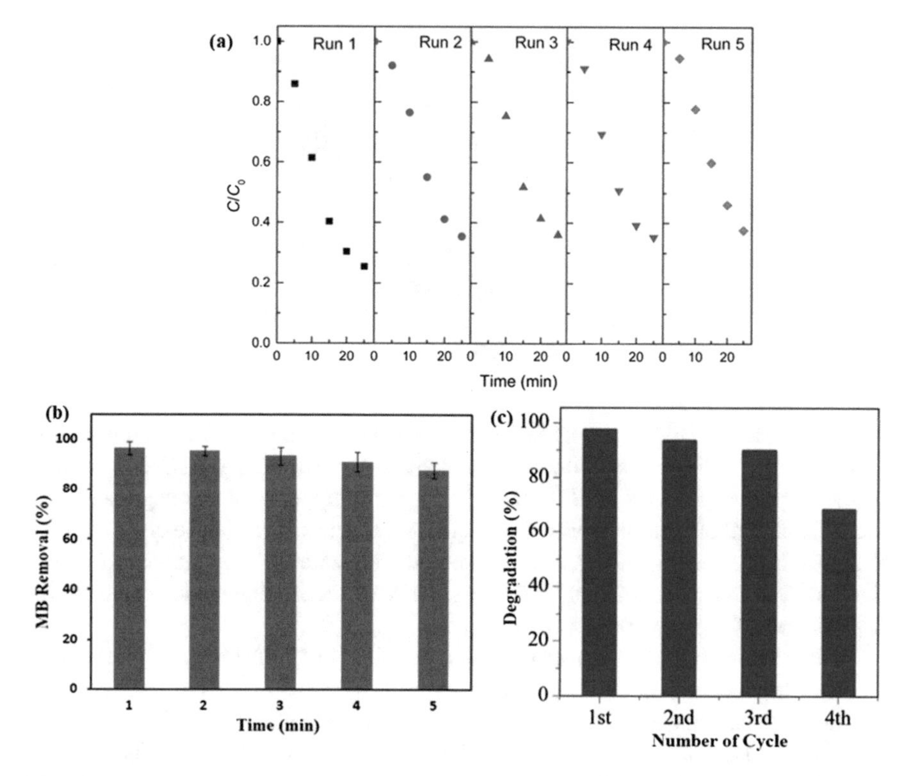

Fig. 15 **a** Degradation of MB in different batch runs in the reduced $CuFe_2O_4/H_2O_2$ system [151]. **b** Recycling tests for the MB degradation in the Sep/FeS_2-EF process system [71]. **c** Recycle ability of Co-SMA for MB degradation (initial conc. of MB: 10 mg L^{-1}; catalyst dose: 20 g L^{-1}; light intensity: 10 010 lx; [H_2O_2]: 348.8 mM) [121]

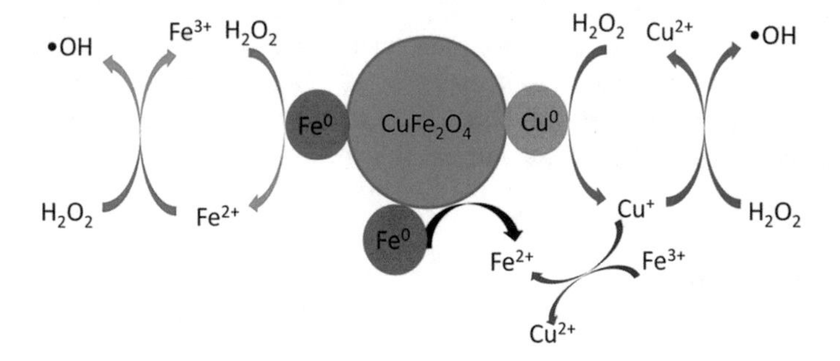

Scheme 6 Schematic diagram of MB degradation mechanism by reduced $CuFe_2O_4$ [151]

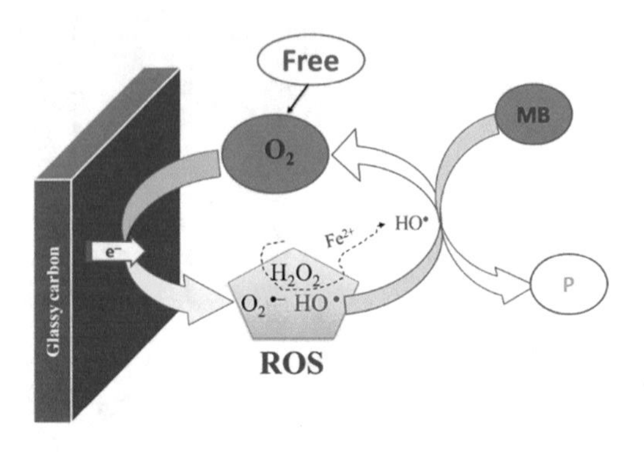

Scheme 7 Schematic diagram of electrochemical oxygen reduction reaction for dye degradation [89]

The decolorization efficiency of MB has been achieved to be 99.7% within 60 min at optimum condition. Gas chromatography-mass spectrometry (GC–MS) results of the degraded products confirm that the major component of the degraded fragments is phenol and aniline. Moreover, $LiFe(WO_4)_2$ exhibits to have good stability and reusability (see Fig. 4), indicating it could be a suitable heterogeneous Fenton-like catalyst and could overcome the drawback of the homogeneous catalysts for degradation of organic pollutants.

In another study, MB has been degraded by ROS ($O_2^{\bullet-}$, $OH^{\bullet}$, H_2O_2, etc.) generated via electrochemical ORR at a glassy carbon (GC) electrode in an acidic aqueous solution (see Scheme 7). Among these species, $OH^{\bullet}$ radical is revealed to be responsible for the degradation of MB species [89]. The pH of the solution has been found to be decreased upon degradation of MB and caused a consequent increase in electrical conductivity. That is, the degraded products are found to be ionic in character and both aliphatic and aromatic amine residues have been found to form upon the degradation of MB species by comparing experimental FR-IR and theoretical IR constructed by DFT shown in Fig. 16 [89]. This study provides essential information for the generation of ROS with electrochemical ORR that can efficiently degrade organic dye species to small fragments by consuming only air (O_2) and electricity. Therefore, this opens up a novel route to replace the conventional AOP method for the treatment of wastewater containing textile dyes.

4.12 Methyl Red

Methyl red (MR), also called C.I. acid red 2, is a cationic dye that turns into red in acidic solutions. It is used as an indication in acid–base titration. It is a dark red crystalline powder. MR is detectably fluorescent in 1:1 water: methanol at pH 7.0, with a maximum emission at 375 nm [52]. Degradation of MR dye has been carried out in an efficient EF system (Fig. 17) based on a graphite-polytetrafluoroethylene (PTFE) cathode with high production rate of H_2O_2 and current efficiency under wider pH ranges [194]. The effects of different experimental parameters on the removal

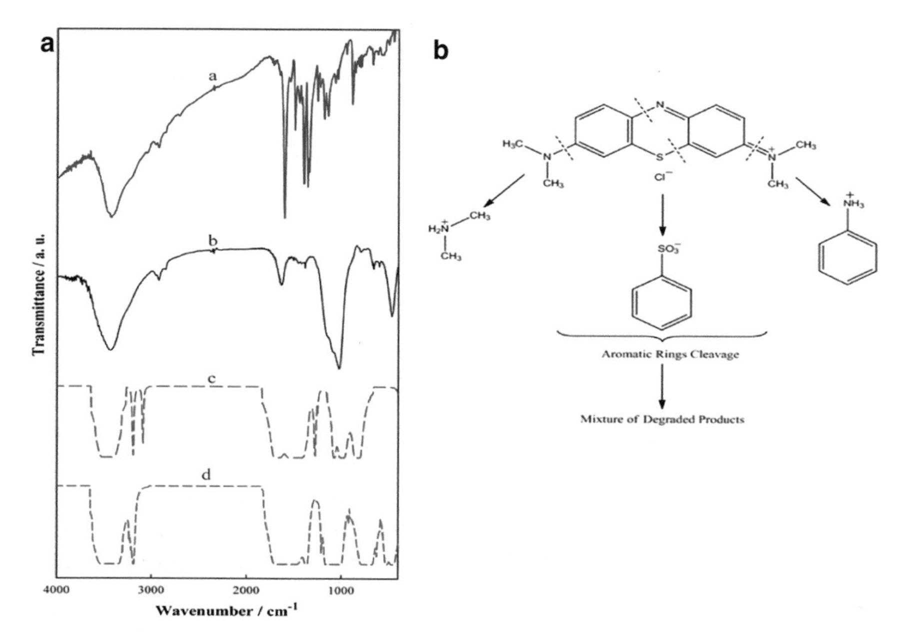

Fig. 16 **a** FT-IR spectra of MB before (**b**) and after degradation (**b**) and theoretical IR spectra of dimethylammonium (**c**) and anilinium (**d**) constructed using density functional theory. **b** Probable cleavage of bonds present in MB by OH• radical in aqueous solution. Dash-dash (——) line denotes the bond cleavage [89]

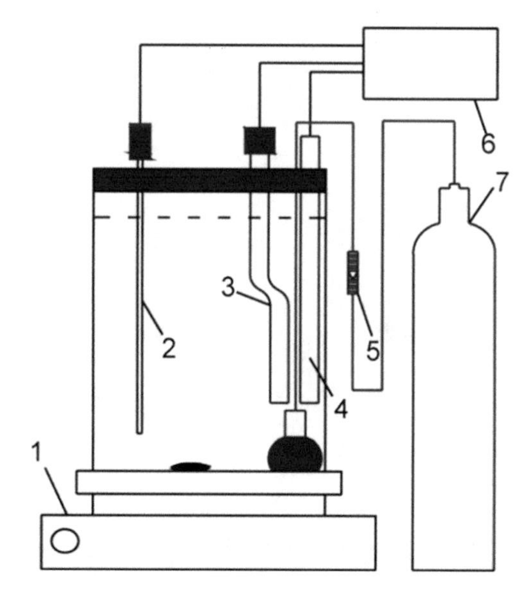

Fig. 17 Schematic diagram of the experimental setup of EF. 1, magnetic stirrer; 2, platinum wire; 3, SCE; 4, gas diffusion electrode; 5, gas flow meter; 6, potentiostat; 7, oxygen tank [194]

efficiency have been optimized and 80% of MR can be removed at an applied potential of −0.55 V versus SCE within 20 min. The degradation of MR follows two steps mechanism wherein the oxidation is faster initially and slows down with time. This is because a decrease in the concentration of Fe^{2+} ion and the formation of hard-to-treat intermediates occur during treatment.

4.13 Rhodamine B

Rhodamine B (RhB) is a cationic dye. It exists in equilibrium between two forms: an open (fluorescent) form and a closed (non-fluorescent spirolactam) form shown in Scheme 8. The open form dominates in acidic condition while the closed form in basic condition [23].

It is widely used as a tracer dye within water to determine the rate and direction of flow and transport. RhB is tunable around 610 nm when used as a laser dye. A novel three-dimensional EF (3D-EF) based on foam nickel, activated carbon fiber as the cathode and Ti/RuO_2–IrO_2 as the anode have been developed for a complete removal of RhB from wastewaters [115]. A comparison of removal efficacies of RhB by different methods is shown in Fig. 18. The higher efficiency of 3D-EF system is attributed to much more $OH^{\bullet}$ radicals generated at the foam nickel particle electrodes [115].

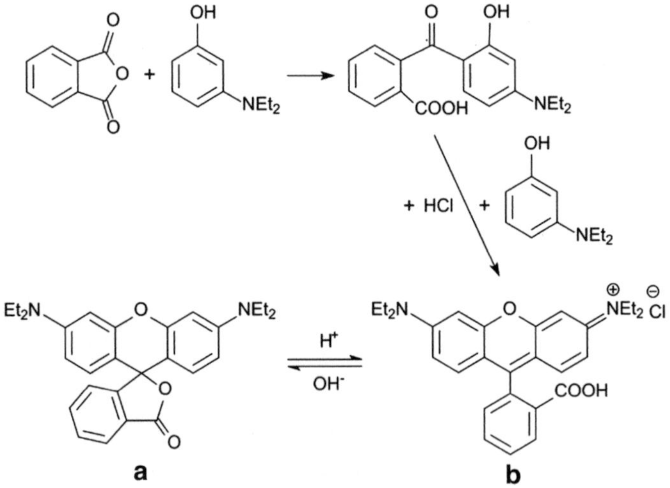

Scheme 8 Rhodamine B closed form (**a**) and open form (**b**)

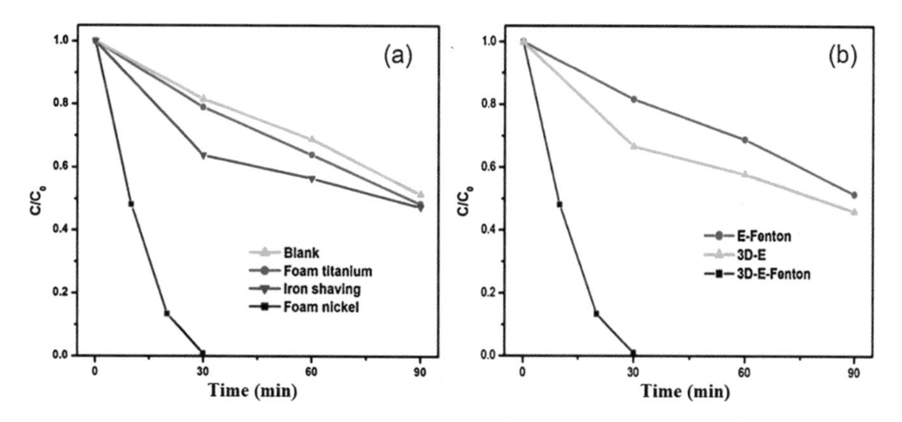

Fig. 18 **a** The degradation of RhB in 3D-E-Fenton system with different particle electrodes; **b** the degradation of RhB in different systems. Experimental conditions: pH = 6.2, applied voltage = 2.0 V and $[Fe^{2+}]$ = 3 M [115]

4.14 Safranin T

Safranin-T (ST), a cationic dye, is used in histology and cytology as a biological stain. It can also be used as a food dye in flavoring and coloring candies and cookies, used for dyeing tannin, cotton, wool, silk, leather, and paper [84, 187]. Fenton's reagent coupled with Fe^{3+}-chelate to catalyze the oxidation of H_2O_2 has been adopted for the degradation of ST [122]. This system is different to some extent from the classic Fenton's chemistry in aqueous solution. In fact, this Fe^{3+}-organic complex is highly active in catalytic decomposition of H_2O_2 to produce $OH^\bullet$ radicals and/or ferryl species capable of oxidizing the ST. This reaction also follows a pseudo-first-order kinetics. The activation energy for the Fe^{3+} chelate catalyzed degradation of ST has been found to be 75.91 kJ mol^{-1}.

5 Comparative Analysis of Different Fenton-Like Processes

The essence of applicability of the Fenton-like processes for the treatments of dye-contaminated wastewaters can be extracted from Tables 1 and 2. In the preceding sections, the advantages and limitations along with the operational efficiency of different methods in degrading known textile dyes are described extensively. In general, the lower capital and operational costs, and higher reaction rate are the main advantages of Fenton-Like processes over most of the contemporary AOPs applied in this purpose [40, 39].

The nutshell of major advantages and disadvantages of the Fenton-like processes are depicted in Scheme 9. Among these processes EF has attracted increased interest in recent years, not only for its capability of a high degradation rates, but also for its

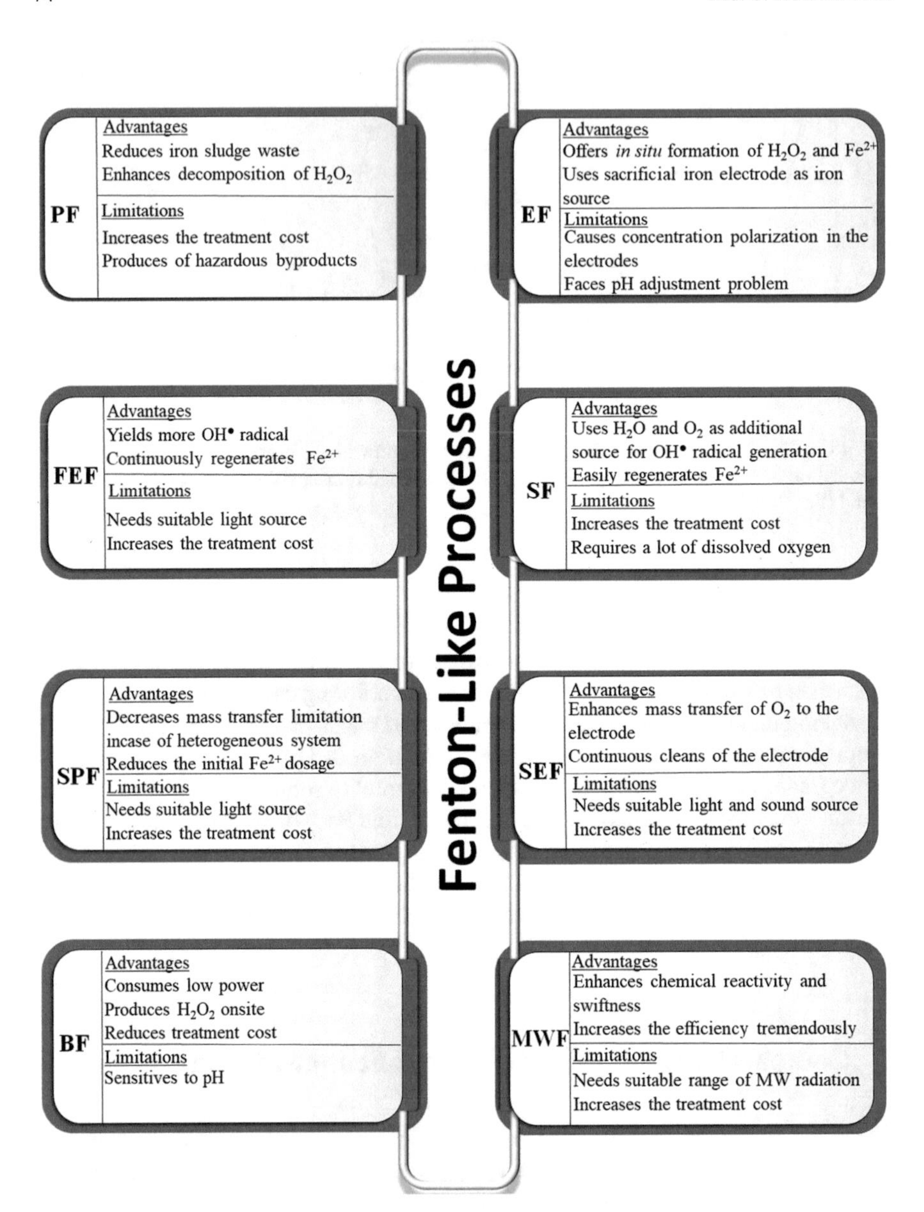

Scheme 9 Advantages and disadvantages of various processes

simplicity in installation and operation (Scheme 7). In addition, it is environmentally friendly, since the electron, the main reagent, is a clean species. The advanced addition in EF is the use of light in so-called PEF system (see Sect. 2.2.2) that actually reduces the operational cost, because the sunlight acts as a renewable energy source required to regulate the classical Fenton reaction.

Another most important point concerning degradation of dyes in water is the quality and quantity of its residual parts leaving to the water bodies. It has been generally taken into account that a complete mineralization of organic compounds, i.e., converting them to CO_2, H_2O, and inorganic ions could be achieved by Fenton-like processes. In order to provide an overall idea on the aptness of these method in practical application, a comparison on the performance of some typical Fenton-like methods employed for degradation of a model dye, MB, is given in Table 3.

At a glance, from the summary represented in Table 3 one may presume that none of the methods is efficient enough to degrade dye species to produce desired CO_2 and H_2O or any other gaseous material completely. Several residues whether ionic or neutral are produced that become the part of the water and obviously reduce the quality of water concerning pH, electrical conductivity, chemical oxygen demand (COD), biological oxygen demand (BOD), etc. Although all of the methods have some constraints in the application, some of them may be utilized commercially for their advantages regarding faster rate of degradation of dyes.

6 Concluding Remarks and Future Prospects

The treatment of dye-stuffs in the effluents discharged especially from the textile industries with AOPs is apparently a lucrative chemical module over the dozen methods practiced in this field [5, 8, 20, 21, 60, 62, 73, 81, 98, 101, 102, 108, 113, 118, 119, 125, 127, 132, 158, 160, 165, 179]. The standard reduction potential of $OH^{\bullet}$ species that is a clean chemical and is preferably used in AOPs takes the position just below the positive hole of TiO_2 semiconductor (see Table 1). Some of the known Fenton-like processes offer a clean production of $OH^{\bullet}$ species even without using any chemical, for example, via electrochemical ORR by consuming O_2 (i.e., air) and electricity (see Scheme 7).

The functional rate, handling, operation, supplies required and management of any device are important from industrial perspectives. On the other hand, the life time of $OH^{\bullet}$ species is very short in water but is highly reactive to organic compounds and hence it has to be prepared *onsite* for application in the degradation of textile dyes. Besides the conventional Fenton process, various Fenton-like processes discussed are truly excellent initiatives developed by the talented researchers around the globe for eco-friendly, renewable production of $OH^{\bullet}$ species with a competitive manner as industries demand. To enhance the rate of production of $OH^{\bullet}$ species sometimes catalysts have to be used that requires additional steps to be followed in handling and manipulating chemicals related to their synthesis.

The electrochemical two-electron ORR at the pristine solid surface of carbon or metal electrode producing in $OH^{\bullet}$ species might solve this practical problem associated with handling of the chemicals. Unfortunately, the rate of formation of $OH^{\bullet}$ species via ORR does not meet the criteria for an industrial scale production, since at the electrode surface faces a kinetic hurdle and hence the rate is slow, the mechanism is rather complicated and produces a series of ROS including H_2O_2,

Table 3 Features of degradation of MB with Fenton and different Fenton-like processes

Process	Catalyst used	Efficiency (%)	Products leave	Refs.	Limitations
Conventional Fenton	Reduced $CuFe_2O_4$ supported	74 in 25 min	nd	[151]	Secondary waste; Excess chemicals; Recycling problem
	$LiFe(WO_4)_2$ supported	^{d}99.7 in 60 min	Phenol and aniline	[96]	
	Fe_3O_4/ZnO/graphene composite supported	^{d}89 in 120 min	nd	[159]	
	Haematite supported	^{d}70 in >120 min	nd	[123]	
	Magnetite and MCM-41 supported	^{d}100; ^{m}43 in 180 min	nd	[134]	
PF	Co-alumina	~^{d}29 in 90 min	nd	[121]	Secondary waste; Excess chemicals; Recycling problem; Light source
	Co-SMA	~ ^{d}97.4 in 60 min	Hydroxylated amine, aromatic compounds containing sulphonic acid, nitro- amine groups, etc	[121]	
	Layered Fe-titanate	^{d}98.5 in 15 min	nd	[42]	
	Haematite	^{d}70 in 90 min	nd	[123]	
	Fe_3O_4/ZnO/graphene composite	^{d}100 in 120 min	nd	[159]	
	Fe_3O_4/ZnO/graphene composite	^{d}93 in 120 min	nd	[159]	
EF	Sep/FeS_2 supported	^{m}96.6 in 25 min	Aromatic compounds	[71]	Electrode cleaning problem; Slow kinetics
	No catalyst	~ ^{d}100	Dimethylammonium and anilinium	[89]	
SF	Nano zero-valent metals	^{d}100 in 30 min	nd	[161]	Light source; High capital cost; Recycling problem

(continued)

Table 3 (continued)

Process	Catalyst used	Efficiency (%)	Products leave	Refs.	Limitations
BEF	No catalyst	[d]97 in 8 h [m]99.6 in 16 h	nd	[190]	Sensitive to pH; High capital cost; Recycling problem
MWF	No catalyst	[d]93 in 1 min	nd	[116]	Light source; Complicated reaction with catalyst; High capital cost
MW-PF	TiO_2	[d]96 in 15 min [m]45 in 15 min	De-methylated MB	[88]	Light source; Interaction between catalyst and light; High capital cost

d = Decolorization; m = Mineralization; t = Total organic carbon; nd = Not detected

the solubility of O_2 feed in water is low. The attempts for a sole production of $OH^{\bullet}$ species in the electrochemical ORR using some clean pathways that assist to decompose H_2O_2 and lineup other ROS generated to form $OH^{\bullet}$ species are yet to be taken.

The advantages of oxidative degradation of dyes via AOPs over other methods employed for treatment of wastewaters such as adsorption, filtration, coagulation, and so on would only be beneficiary from the environmental point of views, whenever a complete mineralization to form CO_2 and H_2O or other non-toxic fragments is possible. Moreover, the mineralization to some ionic species would reduce water quality parameters concerning pH, COD, BOD, electrical conductivity from the standard. Although the efficiency of different Fenton-like methods described towards decolorization of dyes is satisfactory, a complete mineralization in an acceptable water quality parameter could be rarely achieved (Tables 2 and 3). All of these challenges are to be met to develop a smart technology for the treatment of dye-contaminated textile wastewaters in commercial and eco-friendly manners. Hence, a plenty of works on designing the materials especially for the electrochemical ORR in producing $OH^{\bullet}$ species in a clean and renewable manner that would light up a hope to meet the challenges in devising a smart, cost-effective dye-contaminated wastewater treatment technology are to be carried out.

Acknowledgements The financial support of University Grants Commission of Bangladesh is greatly appreciated. M. Y. P. acknowledges the fellowship of the Bose Center for Advanced Study and Research in Natural Sciences, University of Dhaka, Bangladesh.

References

1. Abdi M, Balagabri M, Karimi H, Hossini H, Rastegar SO (2020) Degradation of crystal violet (CV) from aqueous solutions using ozone, peroxone, electroperoxone, and electrolysis processes: a comparison study. Appl Water Sci 168(10):1–10. https://doi.org/10.1007/s13201-020-01252-w
2. Adak A, Bandyopadhyay M, Pal A (2005) Removal of crystal violet dye from wastewater by surfactant-modified alumina. Sep Purif Technol 44(2):139–144. https://doi.org/10.1016/j.seppur.2005.01.002
3. Adewuyi YG (2005) Sonochemistry in environmental remediation 1: Combinative and hybrid sonophotochemical oxidation processes for the treatment of pollutants in water. Environ Sci Technol 39(10):3409–3420. https://doi.org/10.1021/es049138y
4. Alfaya E, Iglesias O, Pazos M, Sanromán MA (2015) Environmental application of an industrial waste as catalyst for the electro-Fenton-like treatment of organic pollutants. RSC Adv 5(19):14416–14424. https://doi.org/10.1039/c4ra15934a
5. Amat AM, Arques A, Beneyto H, García A, Miranda MA, Seguí S (2003) Ozonisation coupled with biological degradation for treatment of phenolic pollutants: a mechanistically based study. Chemosphere 53(1):79–86. https://doi.org/10.1016/S0045-6535(03)00450-8
6. Ameta R, Chohadia AK, Jain A, Punjabi PB (2018) Fenton and photo-Fenton processes. In: Advanced oxidation processes for wastewater treatment: emerging green chemical technology. Academic Press 49–87
7. Armstrong WA (1969) Relative rate constants for reactions of hydroxyl radicals from the reaction of Fe(II) or Ti(III) with H_2O_2. Can J Chem 47(20):3737–3744. https://doi.org/10.1139/v69-623
8. Arslan-Alaton I (2007) Degradation of a commercial textile biocide with advanced oxidation processes and ozone. J Environ Manage 82(2):145–154. https://doi.org/10.1016/j.jenvman.2005.12.021
9. Atta AY, Jibril BY, Al-Waheibi TK, Al-Waheibi YM (2012) Microwave-enhanced catalytic degradation of 2-nitrophenol on alumina-supported copper oxides. Catal Commun 26:112–116. https://doi.org/10.1016/j.catcom.2012.04.033
10. Azabou S, Najjar W, Gargoubi A, Ghorbel A, Sayadi S (2007) Catalytic wet peroxide photo-oxidation of phenolic olive oil mill wastewater contaminants. Appl Catal B Environ 77(1–2):166–174. https://doi.org/10.1016/j.apcatb.2007.07.008
11. Babuponnusami A, Muthukumar K (2012) Advanced oxidation of phenol: a comparison between Fenton, electro-Fenton, sono-electro-Fenton and photo-electro-Fenton processes. Chem Eng J 183:1–9. https://doi.org/10.1016/j.jece.2013.10.011
12. Babuponnusami A, Muthukumar K (2014) A review on Fenton and improvements to the Fenton process for wastewater treatment. J Environ Chem Eng 2(1):557–572. https://doi.org/10.1016/j.cej.2011.12.010
13. Badawy MI, Ghaly MY, Gad-Allah TA (2006) Advanced oxidation processes for the removal of organophosphorus pesticides from wastewater. Desalination 194(1–3):166–175. https://doi.org/10.1016/j.desal.2005.09.027
14. Badellino C, Rodrigues CA, Bertazzoli R (2006) Oxidation of pesticides by in situ electrogenerated hydrogen peroxide: study for the degradation of 2,4-dichlorophenoxyacetic acid. J Hazard Mater 137(2):856–864. https://doi.org/10.1016/j.jhazmat.2006.03.035

15. Balci B, Oturan MA, Oturan N, Sires I (2009) Decontamination of aqueous glyphosate, (aminomethyl) phosphonic acid, and glufosinate solutions by electro-Fenton-like process with Mn^{2+} as the catalyst. J Agric Food Chem 57(11):4888–4894. https://doi.org/10.1021/jf900876x
16. Bale MS (1981) Management of the umbilicus with crystal violet solution. Can Med Assoc J 124(4):372
17. Bandala ER, Peláez MA, Dionysiou DD, Gelover S, Garcia J, Macías D (2007) Degradation of 2,4-dichlorophenoxyacetic acid (2,4-D) using cobalt-peroxymonosulfate in Fenton-like process. J Photochem Photobiol A Chem 186(2–3):357–363. https://doi.org/10.1016/j.jphotochem.2006.09.005
18. Bautista P, Mohedano AF, Gilarranz MA, Casas JA, Rodriguez JJ (2007) Application of Fenton oxidation to cosmetic wastewaters treatment. J Hazard Mater 143(1–2):128–134. https://doi.org/10.1016/j.jhazmat.2006.09.004
19. Benatti CT, Tavares CRG, Guedes TA (2006) Optimization of Fenton's oxidation of chemical laboratory wastewaters using the response surface methodology. J Environ Manage 80(1):66–74. https://doi.org/10.1016/j.jenvman.2005.08.014
20. Bianco B, De Michelis I, Vegliò F (2011) Fenton treatment of complex industrial wastewater: optimization of process conditions by surface response method. J Hazard Mater 186(2–3):1733–1738. https://doi.org/10.1016/j.jhazmat.2010.12.054
21. Bigda RJ (1995) Consider Fenton's chemistry for wastewater treatment. Chem Eng Prog 91(12)
22. Birjandi N, Younesi H, Ghoreyshi AA, Rahimnejad M (2016) Electricity generation, ethanol fermentation and enhanced glucose degradation in a bio-electro-Fenton system driven by a microbial fuel cell. J Chem Technol Biotechnol 91(6):1868–1876. https://doi.org/10.1002/jctb.4780
23. Birtalan E, Rudat B, Kölmel DK, Fritz D, Vollrath SB, Schepers U, Bräse S (2011) Investigating rhodamine B-labeled peptoids: scopes and limitations of its applications. J Pept Sci 96(5):694–701. https://doi.org/10.1002/bip.21617
24. Bossmann SH, Oliveros E, Göb S, Siegwart S, Dahlen EP, Payawan L, Straub M, Wörner M, Braun AM (1998) New evidence against hydroxyl radicals as reactive intermediates in the thermal and photochemically enhanced Fenton reactions. J Phys Chem A 102(28):5542–5550. https://doi.org/10.1021/jp980129j
25. Bouafia-Chergui S, Oturan N, Khalaf H, Oturan MA (2010) Parametric study on the effect of the ratios $[H_2O_2]/[Fe^{3+}]$ and $[H_2O_2]$/[substrate] on the photo-Fenton degradation of cationic azo dye Basic Blue 41. J Environ Sci Heal Part A 45(5):622–629. https://doi.org/10.1080/10934521003595746
26. Bouafia-Chergui S, Oturan N, Khalaf H, Oturan MA (2012) A photo-Fenton treatment of a mixture of three cationic dyes. Proc Eng 33:181–187. https://doi.org/10.1016/j.proeng.2012.01.1192
27. Brillas E, Sirés I, Oturan MA (2009) Electro-Fenton process and related electrochemical technologies based on Fenton's reaction chemistry. Chem Rev 109(12):6570–6631. https://doi.org/10.1021/cr900136g
28. Brillas E (2014) A review on the degradation of organic pollutants in waters by UV photoelectro-Fenton and solar photoelectro-Fenton. J Braz Chem Soc 25(3):393-417. http://doi.org/10.5935/0103-5053.20130257
29. Brillas E, Casado J (2002) Aniline degradation by Electro-Fenton® and peroxi-coagulation processes using a flow reactor for wastewater treatment. Chemosphere 47(3):241–248. https://doi.org/10.1016/S0045-6535(01)00221-1
30. Brillas E, Calpe JC, Casado J (2000) Mineralization of 2,4-D by advanced electrochemical oxidation processes. Water Res 34(8):2253–2262. https://doi.org/10.1016/S0043-1354(99)00396-6
31. Brillas E, Baños MÁ, Garrido JA (2003) Mineralization of herbicide 3,6-dichloro-2-methoxybenzoic acid in aqueous medium by anodic oxidation, electro-Fenton and photoelectro-Fenton. Electrochim Acta 48(12):1697–1705. https://doi.org/10.1016/S0013-4686(03)00142-7

32. Brillas E, Baños MÁ, Camps S, Arias C, Cabot PL, Garrido JA, Rodríguez RM (2004) Catalytic effect of Fe^{2+}, Cu^{2+} and UVA light on the electrochemical degradation of nitrobenzene using an oxygen-diffusion cathode. New J Chem 28(2):314–322. https://doi.org/10.1039/b312445b
33. Brillas E, Baños MÁ, Skoumal M, Cabot PL, Garrido JA, Rodríguez RM (2007) Degradation of the herbicide 2,4-DP by anodic oxidation, electro-Fenton and photoelectro-Fenton using platinum and boron-doped diamond anodes. Chemosphere 68(2):199–209. https://doi.org/10.1016/j.chemosphere.2007.01.038
34. Burns JM, Craig PS, Shaw TJ, Ferry JL (2010) Multivariate examination of Fe(II)/Fe(III) cycling and consequent hydroxyl radical generation. Environ Sci Technol 44(19):7226–7231. https://doi.org/10.1021/es903519m
35. Burrows HD, Santaballa JA, Steenken S (2002) Reaction pathways and mechanisms of photodegradation of pesticides. J Photochem Photobiol B Biol 67(2):71–108. https://doi.org/10.1016/S1011-1344(02)00277-4
36. Buxton G V., Greenstock CL, Helman WP, Ross AB (1988) Critical review of rate constants for reactions of hydrated electrons, hydrogen atoms and hydroxyl radicals ($^{\bullet}OH/O^{\bullet -}$) in Aqueous Solution. J Phys Chem Ref Data 17(2):513–886. https://doi.org/10.1063/1.555805
37. Carta R, Desogus F (2013) The enhancing effect of low power microwaves on phenol oxidation by the Fenton process. J Environ Chem Eng 1(4):1292–1300. https://doi.org/10.1016/j.jece.2013.09.022
38. Catalkaya EC, Kargi F (2007) Color, TOC and AOX removals from pulp mill effluent by advanced oxidation processes: a comparative study. J Hazard Mater 139(2):244–253. https://doi.org/10.1016/j.jhazmat.2006.06.023
39. Cañizares P, Paz R, Sáez C, Rodrigo MA (2009) Costs of the electrochemical oxidation of wastewaters: a comparison with ozonation and Fenton oxidation processes. J Environ Manage 90(1):410–420. https://doi.org/10.1016/j.jenvman.2007.10.010
40. Cañizares P, Lobato J, Paz R, Rodrigo MA, Sáez C (2007) Advanced oxidation processes for the treatment of olive-oil mills wastewater. Chemosphere 67(4):832–838. https://doi.org/10.1016/j.chemosphere.2006.10.064
41. Chand R, Ince NH, Gogate PR, Bremner DH (2009) Phenol degradation using 20, 300 and 520 kHz ultrasonic reactors with hydrogen peroxide, ozone and zero valent metals. Sep Purif Technol 67(1):103–109. https://doi.org/10.1016/j.seppur.2009.03.035
42. Chen Y, Li N, Zhang Y, Zhang L (2014) Novel low-cost Fenton-like layered Fe-titanate catalyst: preparation, characterization and application for degradation of organic colorants. J Colloid Interface Sci 422:9–15. https://doi.org/10.1016/j.jcis.2014.01.013
43. Chen CC, Wu RJ, Tzeng YY, Lu CS (2009) Chemical oxidative degradation of acridine orange dye in aqueous solution by Fenton's reagent. J Chinese Chem Soc 56(6):1147–1155. https://doi.org/10.1002/jccs.200900165
44. Cheng G, Lin J, Lu J, Zhao X, Cai Z, Fu J (2015) Advanced treatment of pesticide-containing wastewater using Fenton reagent enhanced by microwave electrodeless ultraviolet. Biomed Res Int 2015:1–8. https://doi.org/10.1155/2015/205903
45. Chevallier E, Jolibois RD, Meunier N, Carlier P, Monod A (2004) "Fenton-like" reactions of methylhydroperoxide and ethylhydroperoxide with Fe^{2+} in liquid aerosols under tropospheric conditions. Atmos Environ 38(6):921–933. https://doi.org/10.1016/j.atmosenv.2003.10.027
46. Chiou CS, Chen YH, Chang CT, Chang CY, Shie JL, Li YS (2006) Photochemical mineralization of di-n-butyl phthalate with H_2O_2/Fe^{3+}. J Hazard Mater 135(1–3):344–349. https://doi.org/10.1016/j.jhazmat.2005.11.072
47. Costa RC, Moura FC, Ardisson JD, Fabris JD, Lago RM (2008) Highly active heterogeneous Fenton-like systems based on Fe^0/Fe_3O_4 composites prepared by controlled reduction of iron oxides. Appl Catal B Environ 83(1–2):131–139. https://doi.org/10.1016/J.APCATB.2008.01.039
48. Cuerda-Correa EM, Alexandre-Franco MF, Fernández-González C (2020) Advanced oxidation processes for the removal of antibiotics from water: an overview. Water 12(1):102. https://doi.org/10.3390/w12010102

49. Culp SJ, Beland FA (1996) Malachite green: a toxicological review. Int J Toxicol 15(3):219–238. https://doi.org/10.3109/10915819609008715
50. Daneshvar N, Aleboyeh A, Khataee AR (2005) The evaluation of electrical energy per order (EEo) for photooxidative decolorization of four textile dye solutions by the kinetic model. Chemosphere 59(6):761–767. https://doi.org/10.1016/J.CHEMOSPHERE.2004.11.012
51. Daneshvar N, Salari D, Khataee AR (2004) Photocatalytic degradation of azo dye acid red 14 in water on ZnO as an alternative catalyst to TiO_2. J Photochem Photobiol A Chem 162(2–3):317–322. https://doi.org/10.1016/S1010-6030(03)00378-2
52. Das DK, Goswami P, Barman C, Das B (2012) Methyl red: a fluorescent sensor for Hg^{2+} over Na^+, K^+, Ca^{2+}, Mg^{2+}, Zn^{2+}, and Cd^{2+}. Environ Eng Res 17(S1):75–78. https://doi.org/10.4491/eer.2012.17.S1.S75
53. de Haan SB (1991) A review of the rate of pyrite oxidation in aqueous systems at low temperature. Earth-Sci Rev 31(1):1–10. https://doi.org/10.1016/0012-8252(91)90039-I
54. De Laat J, Gallard H (1999) Catalytic decomposition of hydrogen peroxide by Fe(III) in homogeneous aqueous solution: mechanism and kinetic modeling. Environ Sci Technol 33(16):2726–2732. https://doi.org/10.1021/es981171v
55. Dehghani M, Shahsavani E, Farzadkia M, Samaei MR (2014) Optimizing photo-Fenton like process for the removal of diesel fuel from the aqueous phase. J Environ Heal Sci Eng 12(1):1–7. https://doi.org/10.1186/2052-336X-12-87
56. Dirany A, Sirés I, Oturan N, Özcan A, Oturan MA (2012) Electrochemical treatment of the antibiotic sulfachloropyridazine: kinetics, reaction pathways, and toxicity evolution. Environ Sci Technol 46(7):4074–4082. https://doi.org/10.1021/es204621q
57. Divyapriya G, Nambi IM, Senthilnathan J (2016) Nanocatalysts in Fenton based advanced oxidation process for water and wastewater treatment. J Bionanosci 10(5):356–368. https://doi.org/10.1166/jbns.2016.1387
58. Drogui P, Asselin M, Brar SK, Benmoussa H, Blais JF (2008) Electrochemical removal of pollutants from agro-industry wastewaters. Sep Purif Technol 61(3):301–310. https://doi.org/10.1016/j.seppur.2007.10.013
59. Du Z, Li H, Gu T (2007) A state of the art review on microbial fuel cells: a promising technology for wastewater treatment and bioenergy. Biotechnol Adv 25(5):464–482. https://doi.org/10.1016/j.biotechadv.2007.05.004
60. Duarte F, Maldonado-Hódar FJ, Madeira LM (2011) Influence of the characteristics of carbon materials on their behaviour as heterogeneous Fenton catalysts for the elimination of the azo dye Orange II from aqueous solutions. Appl Catal B Environ 103(1–2):109–115. https://doi.org/10.1016/j.apcatb.2011.01.016
61. Ehrampoush M, Moussavi GH, Ghaneian M, Rahimi S, Ahmadian M (2011) Removal of methylene blue dye from textile simulated sample using tubular reactor and TiO_2/UV-C photocatalytic process. J Environ Heal Sci Eng 8(1):34–40.
62. El Hajjouji H, Barje F, Pinelli E, Bailly JR, Richard C, Winterton P, Revel JC, Hafidi M (2008) Photochemical UV/TiO_2 treatment of olive mill wastewater (OMW). Bioresour Technol 99(15):7264–7269. https://doi.org/10.1016/j.biortech.2007.12.054
63. Elhami V, Karimi A, Aghbolaghy M (2015) Preparation of heterogeneous bio-Fenton catalyst for decolorization of malachite green. J Taiwan Inst Chem Eng 56:154–159. https://doi.org/10.1016/j.jtice.2015.05.006
64. Elshafei GM, Yehia FZ, Dimitry OIH, Badawi AM, Eshaq G (2014) Ultrasonic assisted-Fenton-like degradation of nitrobenzene at neutral pH using nanosized oxides of Fe and Cu. Ultrason Sonochem 21(4):1358–1365. https://doi.org/10.1016/j.ultsonch.2013.12.019
65. Ertl G (1994) Reactions at well-defined surfaces. Surf Sci 299:742–754. https://doi.org/10.1016/0039-6028(94)90694-7
66. Eskandarian M, Mahdizadeh F, Ghalamchi L, Naghavi S (2014) Bio-Fenton process for Acid Blue 113 textile azo dye decolorization: characteristics and neural network modeling. Desalin Water Treat 52(25–27):4990–4998. https://doi.org/10.1080/19443994.2013.810325
67. Fajerwerg K, Debellefontaine H (1996) Wet oxidation of phenol by hydrogen peroxide using heterogeneous catalysis Fe-ZSM-5: a promising catalyst. Appl Catal B Environ 10(4):L229–L235. https://doi.org/10.1016/S0926-3373(96)00041-0

68. Fan X, Hao H, Wang Y, Chen F, Zhang J (2013) Fenton-like degradation of nalidixic acid with Fe^{3+}/H_2O_2. Environ Sci Pollut Res 20(6):3649–3656. https://doi.org/10.1007/s11356-012-1279-0
69. Farhana TI, Mollah MYA, Susan MABH, Islam MM (2014) Catalytic degradation of an organic dye through electroreduction of dioxygen in aqueous solution. Electrochim Acta 139:244–249. https://doi.org/10.1016/j.electacta.2014.06.145
70. Fassi S, Djebbar K, Bousnoubra I, Chenini H, Sehili T (2014) Oxidation of bromocresol green by different advanced oxidation processes: Fenton, Fenton-like, photo-Fenton, photo-Fenton-like and solar light comparative study. Desalin Water Treat 52(25–27):4982–4989. https://doi.org/10.1080/19443994.2013.809971
71. Fayazi M, Ghanei-Motlagh M (2020) Electrochemical mineralization of methylene blue dye using electro-Fenton oxidation catalyzed by a novel sepiolite/pyrite nanocomposite. Int J Environ Sci Technol 17(11):4541–4548. https://doi.org/10.1007/s13762-020-02749-2
72. Fei BL, Yan QL, Wang JH, Liu QB, Long JY, Li YG, Shao KZ, Su ZM, Sun WY (2014) Green oxidative degradation of Methyl Orange with Copper(II) Schiff base complexes as photo-Fenton-like catalysts. Zeitschrift fur Anorg und Allg Chemie 640(10):2035–2040. https://doi.org/10.1002/zaac.201300562
73. Feng J, Hu X, Yue PL (2006) Effect of initial solution pH on the degradation of Orange II using clay-based Fe nanocomposites as heterogeneous photo-Fenton catalyst. Water Res 40(4):641–646. https://doi.org/10.1016/J.WATRES.2005.12.021
74. Feng Y, Wu DL, Duan D, Lu MM (2012) Fenton-like oxidation of refractory chemical wastewater using pyrite. Adv Mat Res 518:2518–2525. https://doi.org/10.4028/www.scientific.net/AMR.518-523.2518
75. Feng CH, Li FB, Mai HJ, Li XZ (2010) Bio-electro-Fenton process driven by microbial fuel cell for wastewater treatment. Environ Sci Technol 44(5):1875–1880. https://doi.org/10.1021/es9032925
76. Flox C, Ammar S, Arias C, Brillas E, Vargas-Zavala AV, Abdelhedi R (2006) Electro-Fenton and photoelectro-Fenton degradation of indigo carmine in acidic aqueous medium. Appl Catal B Environ 67(1–2):93–104. https://doi.org/10.1016/j.apcatb.2006.04.020
77. Fu L, You SJ, Zhang GQ, Yang FL, Fang XH (2010) Degradation of azo dyes using in-situ Fenton reaction incorporated into H_2O_2-producing microbial fuel cell. Chem Eng J 160(1):164–169. https://doi.org/10.1016/j.cej.2010.03.032
78. Garrido-Ramírez EG, Theng BK, Mora ML (2010) Clays and oxide minerals as catalysts and nanocatalysts in Fenton-like reactions—a review. Appl Clay Sci 47(3–4):182–192. https://doi.org/10.1016/j.clay.2009.11.044
79. Gemeay AH, Mansour IA, El-Sharkawy RG, Zaki AB (2003) Kinetics and mechanism of the heterogeneous catalyzed oxidative degradation of indigo carmine. J Mol Catal A Chem 193(1–2):109–120. https://doi.org/10.1016/S1381-1169(02)00477-6
80. Ghanbari F, Moradi M (2017) Application of peroxymonosulfate and its activation methods for degradation of environmental organic pollutants: review. Chem Eng J 310:41–62. https://doi.org/10.1016/j.cej.2016.10.064
81. Ghatak HR (2014) Advanced oxidation processes for the treatment of biorecalcitrant organics in wastewater. Crit Rev Environ Sci Technol 44(11):1167–1219. https://doi.org/10.1080/10643389.2013.763581
82. Gogate PR, Pandit AB (2004) Sonophotocatalytic reactors for wastewater Treatment: a critical review. AIChE J 50(5):1051–1079. https://doi.org/10.1002/aic.10079
83. Guerra-Rodríguez S, Rodríguez E, Singh DN, Rodríguez-Chueca J (2018) Assessment of sulfate radical-based advanced oxidation processes for water and wastewater treatment: a review. Water 10(12):1828. https://doi.org/10.3390/w10121828
84. Gupta VK, Mittal A, Jain R, Mathur M, Sikarwar S (2006) Adsorption of Safranin-T from wastewater using waste materials-activated carbon and activated rice husks. J Colloid Interface Sci 303(1):80–86. https://doi.org/10.1016/j.jcis.2006.07.036
85. Hashemian S (2013) Fenton-like oxidation of malachite green solutions: kinetic and thermodynamic study. J Chem 2013. https://doi.org/10.1155/2013/809318

86. Hassan H, Hammed B (2011) Decolorization of Acid Red 1 by heterogeneous Fenton-like reaction using Fe-ball clay catalyst. Int Conf Environ Sci Eng IPCBEE 8:232–236.
87. He J, Yang X, Men B, Wang D (2016) Interfacial mechanisms of heterogeneous Fenton reactions catalyzed by iron-based materials: a review. J Environ Sci 39:97–109. https://doi.org/10.1016/j.jes.2015.12.003
88. Hong J, Sun C, Yang SG, Liu YZ (2006) Photocatalytic degradation of methylene blue in TiO_2 aqueous suspensions using microwave powered electrodeless discharge lamps. J Hazard Mater 133(1–3):162–166. https://doi.org/10.1016/j.jhazmat.2005.10.004
89. Hossain MS, Mollah MYA, Susan MABH, Islam MM (2020) Role of in situ electrogenerated reactive oxygen species towards degradation of organic dye in aqueous solution. Electrochim Acta 344:136146. https://doi.org/10.1016/j.electacta.2020.136146
90. Hossain MS, Nixon AHR, Susan MABH, Mollah MYA, Islam MM (2015) Electrochemical approach for treatment of textile effluents. TRC Book of Papers. 19–21
91. Hossain MS, Sahed A, Jahan N, Mollah MYA, Susan MABH, Islam MM (2021) Micelle core as a nest for residence of molecular oxygen–An electrochemical study. J Electroanal Chem 894:115361. https://doi.org/10.1016/j.jelechem.2021.115361
92. Huang X, Xu Y, Shan C, Li X, Zhang W, Pan B (2016) Coupled Cu(II)-EDTA degradation and Cu(II) removal from acidic wastewater by ozonation: performance, products and pathways. Chem Eng J 299:23–29. https://doi.org/10.1016/j.cej.2016.04.044
93. Hussain S, Aneggi E, Goi D (2021) Catalytic activity of metals in heterogeneous Fenton-like oxidation of wastewater contaminants: a review. Environ Chem Lett 19(3):2405–2424. https://doi.org/10.1007/s10311-021-01185-z
94. Iranifam M, Zarei M, Khataee AR (2011) Decolorization of C.I. Basic Yellow 28 solution using supported ZnO nanoparticles coupled with photoelectro-Fenton process. J Electroanal Chem 659(1):107–112. https://doi.org/10.1016/j.jelechem.2011.05.010
95. Irmak S, Yavuz HI, Erbatur O (2006) Degradation of 4-chloro-2-methylphenol in aqueous solution by electro-Fenton and photoelectro-Fenton processes. Appl Catal B Environ 63(3–4):243–248. https://doi.org/10.1016/j.apcatb.2005.10.008
96. Ji F, Li C, Zhang J, Deng L (2011) Efficient decolorization of dye pollutants with $LiFe(WO_4)_2$ as a reusable heterogeneous Fenton-like catalyst. Desalination 269(1–3):284–290. https://doi.org/10.1016/j.desal.2010.11.015
97. Jiang CC, Zhang JF (2007) Progress and prospect in electro-Fenton process for wastewater treatment. J Zhejiang Univ Sci A 8(7):1118–1125. https://doi.org/10.1631/jzus.2007.A1118
98. Kallel M, Belaid C, Mechichi T, Ksibi M, Elleuch B (2009) Removal of organic load and phenolic compounds from olive mill wastewater by Fenton oxidation with zero-valent iron. Chem Eng J 150(2–3):391–395. https://doi.org/10.1016/j.cej.2009.01.017
99. Karimi A, Aghbolaghy M, Khataee A, Shoa Bargh S (2012) Use of enzymatic bio-Fenton as a new approach in decolorization of malachite green. Sci World J 2012. https://doi.org/10.1100/2012/691569
100. Karthikeyan S, Titus A, Gnanamani A, Mandal AB, Sekaran G (2011) Treatment of textile wastewater by homogeneous and heterogeneous Fenton oxidation processes. Desalination 281:438–445. https://doi.org/10.1016/j.desal.2011.08.019
101. Karunakaran C, Anilkumar P (2007) Semiconductor-catalyzed solar photooxidation of iodide ion. J Mol Catal A Chem 265(1–2):153–158. https://doi.org/10.1016/j.molcata.2006.10.016
102. Kim JR, Santiano B, Kim H, Kan E (2013) Heterogeneous oxidation of methylene blue with surface-modified iron-amended activated carbon. Am J Anal Chem 4:115–122. https://doi.org/10.4236/ajac.2013.47a016
103. Kolpin DW, Furlong ET, Meyer MT, Thurman EM, Zaugg SD, Barber LB, Buxton HT (2002) Pharmaceuticals, hormones, and other organic wastewater contaminants in U.S. streams, 1999–2000: a national reconnaissance. Environ Sci Technol 36(6):1202–1211. https://doi.org/10.1021/es011055j
104. Kommareddi S, Abramowsky CR, Swinehart GL, Hrabak L (1984) Nontuberculous mycobacterial infections: comparison of the fluorescent auramine-o and Ziehl-Neelsen techniques in tissue diagnosis. Hum Pathol 15(11):1085–1089. https://doi.org/10.1016/S0046-8177(84)80253-1

105. Kušić H, Božić AL, Koprivanac N (2007) Fenton type processes for minimization of organic content in coloured wastewaters: Part I: processes optimization. Dye Pigm 74(2):380–387 https://doi.org/10.1016/j.dyepig.2006.02.022
106. Labiadh L, Oturan MA, Panizza M, Hamadi NB, Ammar S (2015) Complete removal of AHPS synthetic dye from water using new electro-Fenton oxidation catalyzed by natural pyrite as heterogeneous catalyst. J Hazard Mater 297:34–41. https://doi.org/10.1016/j.jhazmat.2015.04.062
107. Lee HJ, Lee H, Lee C (2014) Degradation of diclofenac and carbamazepine by the copper(II)-catalyzed dark and photo-assisted Fenton-like systems. Chem Eng J 245:258–264. https://doi.org/10.1016/j.cej.2014.02.037
108. Lee E, Lee H, Kim YK, Sohn K, Lee K (2011) Hydrogen peroxide interference in chemical oxygen demand during ozone based advanced oxidation of anaerobically digested livestock wastewater. Int J Environ Heal Sci Eng 8(2):381–388. https://doi.org/10.1007/BF03326225
109. Legrini O, Oliveros E, Braun AM (1993) Photochemical processes for water treatment. Chem Rev 93(2):671–698. https://doi.org/10.1021/cr00018a003
110. Lei H, Li H, Li Z, Li Z, Chen K, Zhang X, Wang H (2010) Electro-Fenton degradation of cationic red X-GRL using an activated carbon fiber cathode. Process Saf Environ Prot 88(6):431–438. https://doi.org/10.1016/j.psep.2010.06.005
111. Letaïef S, Casal B, Aranda P, Martín-Luengo MA, Ruiz-Hitzky E (2003) Fe-containing pillared clays as catalysts for phenol hydroxylation. Appl Clay Sci 22(6):263–277. https://doi.org/10.1016/S0169-1317(03)00079-6
112. Li H, Lei H, Yu Q, Li Z, Feng X, Yang B (2010) Effect of low frequency ultrasonic irradiation on the sonoelectro-Fenton degradation of cationic red X-GRL. Chem Eng J 160(2):417–422. https://doi.org/10.1016/j.cej.2010.03.027
113. Li X, Li X, Yang W, Chen X, Li W, Luo B, Wang K (2014) Preparation of 3D PbO_2 nanospheres@SnO_2 nanowires/Ti electrode and its application in methyl orange degradation. Electrochim Acta 146:15–22. https://doi.org/10.1016/j.electacta.2014.08.150
114. Litter MI (2005) Introduction to photochemical advanced oxidation processes for water treatment. Environ Photochem Part II:325–366. https://doi.org/10.1007/b138188
115. Liu W, Ai Z, Zhang L (2012) Design of a neutral three-dimensional electro-Fenton system with foam nickel as particle electrodes for wastewater treatment. J Hazard Mater 243:257–264. https://doi.org/10.1016/j.jhazmat.2012.10.024
116. Liu ST, Huang J, Ye Y, Zhang AB, Pan L, Chen XG (2013) Microwave enhanced Fenton process for the removal of methylene blue from aqueous solution. Chem Eng J 215:586–590. https://doi.org/10.1016/j.cej.2012.11.003
117. Lopez A, Mascolo G, Detomaso A, Lovecchio G, Villani G (2005) Temperature activated degradation (mineralization) of 4-chloro-3-methyl phenol by Fenton's reagent. Chemosphere 59(3):397–403. https://doi.org/10.1016/j.chemosphere.2004.10.060
118. Lucas MS, Peres JA (2009) Removal of COD from olive mill wastewater by Fenton's reagent: kinetic study. J Hazard Mater 168(2–3):1253–1259. https://doi.org/10.1016/j.jhazmat.2009.03.002
119. Madhavan J, Kumar PSS, Anandan S, Grieser F, Ashokkumar M (2010) Sonophotocatalytic degradation of monocrotophos using TiO_2 and Fe^{3+}. J Hazard Mater 177(1–3):944–949. https://doi.org/10.1016/J.JHAZMAT.2010.01.009
120. Maekawa J, Mae K, Nakagawa H (2014) Fenton-Cu^{2+} system for phenol mineralization. J Environ Chem Eng 2(3):1275–1280. https://doi.org/10.1016/j.jece.2014.05.009
121. Mahamallik P, Pal A (2016) Photo-Fenton process in a Co(II)-adsorbed micellar soft-template on an alumina support for rapid methylene blue degradation. RSC Adv 6(103):100876–100890. https://doi.org/10.1039/c6ra19857k
122. Malik PK (2004) Oxidation of Safranine T in aqueous solution using Fenton's Reagent: involvement of an Fe(III) Chelate in the catalytic hydrogen peroxide oxidation of Safranine T. J Phys Chem A 108(14):2675–2681. https://doi.org/10.1021/jp031082r
123. Mansour SA, Tony MA, Tayeb AM (2019) Photocatalytic performance and photodegradation kinetics of Fenton-like process based on haematite nanocrystals for basic dye removal. SN Appl Sci 1(3):1–8. https://doi.org/10.1007/s42452-019-0286-x

124. Masarwa M, Cohen H, Meyerstein D, Hickman DL, Bakac A, Espenson JH (1988) Reactions of low-valent transition-metal complexes with hydrogen peroxide. Are they "Fenton-like" or Not? 1. The case of Cu^+ aq. and Cr^{2+} aq. J Am Chem Soc 110(13):4293–4297. https://doi.org/10.1021/ja00221a031
125. Mesquita I, Matos LC, Duarte F, Maldonado-Hódar FJ, Mendes A, Madeira LM (2012) Treatment of azo dye-containing wastewater by a Fenton-like process in a continuous packed-bed reactor filled with activated carbon. J Hazard Mater 237:30–37. https://doi.org/10.1016/j.jhazmat.2012.07.066
126. Miklos DB, Remy C, Jekel M, Linden KG, Drewes JE, Hübner U (2018) Evaluation of advanced oxidation processes for water and wastewater treatment—a critical review. Water Res 139:118–131. https://doi.org/10.1016/j.watres.2018.03.042
127. Molina R, Martínez F, Melero JA, Bremner DH, Chakinala AG (2006) Mineralization of phenol by a heterogeneous ultrasound/Fe-SBA-15/H_2O_2 process: multivariate study by factorial design of experiments. Appl Catal B Environ 66(3–4):198–207. https://doi.org/10.1016/j.apcatb.2006.03.015
128. Munoz M, De Pedro ZM, Casas JA, Rodriguez JJ (2015) Preparation of magnetite-based catalysts and their application in heterogeneous Fenton oxidation—a review. Appl Catal B Environ 176:249–265. https://doi.org/10.1016/j.apcatb.2015.04.003
129. Nava JL, Sirés I, Brillas E (2014) Electrochemical incineration of indigo: a comparative study between 2D (plate) and 3D (mesh) BDD anodes fitted into a filter-press reactor. Environ Sci Pollut Res 21(14):8485–8492. https://doi.org/10.1007/s11356-014-2781-3
130. Navalon S, Alvaro M, Garcia H (2010) Heterogeneous Fenton catalysts based on clays, silicas and zeolites. Appl Catal B Environ 99:1–26. https://doi.org/10.1016/j.apcatb.2010.07.006
131. Neta P, Madhavan V, Zemel H, Fessenden RW (1977) Rate constants and mechanism of reaction of sulfate radical anion with aromatic compounds. J Am Chem Soc 99(1):163–164. https://doi.org/10.1021/ja00443a030
132. Neyens E, Baeyens J (2003) A review of classic Fenton's peroxidation as an advanced oxidation technique. J Hazard Mater 98(1–3):33–50. https://doi.org/10.1016/S0304-3894(02)00282-0
133. Nidheesh PV, Gandhimathi R, Ramesh ST (2013) Degradation of dyes from aqueous solution by Fenton processes: a review. Environ Sci Pollut Res 20(4):2099–2132. https://doi.org/10.1007/s11356-012-1385-z
134. Nogueira AE, Castro IA, Giroto AS, Magriotis ZM (2014) Heterogeneous Fenton-like catalytic removal of methylene blue dye in water using magnetic nanocomposite (MCM-41/Magnetite). J Catal 2014. https://doi.org/10.1155/2014/712067
135. Oh WD, Dong Z, Lim TT (2016) Generation of sulfate radical through heterogeneous catalysis for organic contaminants removal: current development, challenges and prospects. Appl Catal B Environ 194:169–201. https://doi.org/10.1016/j.apcatb.2016.04.003
136. Oturan MA (2000) Ecologically effective water treatment technique using electrochemically generated hydroxyl radicals for in situ destruction of organic pollutants: application to herbicide 2,4-D. J Appl Electrochem 30(4):475–482. https://doi.org/10.1023/A:1003994428571
137. Oturan MA, Pinson J (1995) Hydroxylation by electrochemically generated $OH^{\bullet}$ radicals. Mono- and polyhydroxylation of benzoic acid: products and isomer distribution. J Phys Chem 99(38):13948–13954. https://doi.org/10.1021/j100038a029
138. Oturan MA, Aaron JJ, Oturan N, Pinson J (1999) Degradation of chlorophenoxyacid herbicides in aqueous media, using a novel electrochemical method. Pestic Sci. 55(5):558–562. https://doi.org/10.1002/(SICI)1096-9063(199905)55:5<558::AID-PS968>3.0.CO;2-H
139. Oturan MA, Sirés I, Oturan N, Pérocheau S, Laborde JL, Trévin S (2008) Sonoelectro-Fenton process: a novel hybrid technique for the destruction of organic pollutants in water. J Electroanal Chem 624(1–2):329–332. https://doi.org/10.1016/j.jelechem.2008.08.005
140. Oturan N, Hamza M, Ammar S, Abdelhédi R, Oturan MA (2011) Oxidation/mineralization of 2-Nitrophenol in aqueous medium by electrochemical advanced oxidation processes using Pt/carbon-felt and BDD/carbon-felt cells. J Electroanal Chem 661(1):66–71. https://doi.org/10.1016/j.jelechem.2011.07.017

141. Özcan A, Şahin Y, Koparal AS, Oturan MA (2008) Degradation of picloram by the electro-Fenton process. J Hazard Mater 153(1–2):718–727. https://doi.org/10.1016/j.jhazmat.2007.09.015
142. Özcan A, Şahin Y, Savaş Koparal A, Oturan MA (2008) Carbon sponge as a new cathode material for the electro-Fenton process: comparison with carbon felt cathode and application to degradation of synthetic dye basic blue 3 in aqueous medium. J Electroanal Chem 616(1–2):71–78. https://doi.org/10.1016/j.jelechem.2008.01.002
143. Pang YL, Abdullah AZ, Bhatia S (2011) Review on sonochemical methods in the presence of catalysts and chemical additives for treatment of organic pollutants in wastewater. Desalination 277(1–3):1–14. https://doi.org/10.1016/j.desal.2011.04.049
144. Pereira MC, Oliveira LCA, Murad E (2012) Iron oxide catalysts: Fenton and Fenton-like reactions—a review. Clay Miner 47(3):285–302. https://doi.org/10.1180/claymin.2012.047.3.01
145. Pham TTH, Brar SK, Tyagi RD, Surampalli RY (2010) Influence of ultrasonication and Fenton oxidation pre-treatment on rheological characteristics of wastewater sludge. Ultrason Sonochem 17(1):38–45. https://doi.org/10.1016/j.ultsonch.2009.06.007
146. Pintor AM, Vilar VJ, Boaventura RA (2011) Decontamination of cork wastewaters by solar-photo-Fenton process using cork bleaching wastewater as H_2O_2 source. Sol Energy 85(3):579–587. https://doi.org/10.1016/j.solener.2011.01.003
147. Pirkanniemi K, Sillanpää M (2002) Heterogeneous water phase catalysis as an environmental application: a review. Chemosphere 48(10):1047–1060. https://doi.org/10.1016/S0045-6535(02)00168-6
148. Pliego G, Zazo JA, Garcia-Muñoz P, Munoz M, Casas JA, Rodriguez JJ (2015) Trends in the intensification of the Fenton process for wastewater treatment: an overview. Crit Rev Environ Sci Technol 45(24):2611–2692. https://doi.org/10.1080/10643389.2015.1025646
149. Pouran SR, Aziz AA, Daud WMAW (2015) Review on the main advances in photo-Fenton oxidation system for recalcitrant wastewaters. J Ind Eng Chem 21:53–69. https://doi.org/10.1016/j.jiec.2014.05.005
150. Qiang Z, Chang JH, Huang CP (2003) Electrochemical regeneration of Fe^{2+} in Fenton oxidation processes. Water Res 37(6):1308–1319. https://doi.org/10.1016/S0043-1354(02)00461-X
151. Qin Q, Liu Y, Li X, Sun T, Xu Y (2018) Enhanced heterogeneous Fenton-like degradation of methylene blue by reduced $CuFe_2O_4$. RSC Adv 8(2):1071–1077. https://doi.org/10.1039/c7ra12488k
152. Ramirez JH, Costa CA, Madeira LM (2005) Experimental design to optimize the degradation of the synthetic dye Orange II using Fenton's reagent. Catal Today 107:68–76. https://doi.org/10.1016/j.cattod.2005.07.060
153. Ranjit PJD, Palanivelu K, Lee CS (2008) Degradation of 2,4-dichlorophenol in aqueous solution by sono-Fenton method. Korean J Chem Eng 25(1):112–117. https://doi.org/10.1007/s11814-008-0020-7
154. Rivas FJ, Beltrán FJ, Gimeno O, Alvarez P (2003) Optimisation of Fenton's reagent usage as a pre-treatment for fermentation brines. J Hazard Mater 96(2–3):277–290. https://doi.org/10.1016/S0304-3894(02)00217-0
155. Rosales E, Iglesias O, Pazos M, Sanromán MA (2012) Decolourisation of dyes under electro-Fenton process using Fe alginate gel beads. J Hazard Mater 213:369–377. https://doi.org/10.1016/j.jhazmat.2012.02.005
156. Ruan X, Gu X, Lu S, Qiu Z, Sui Q (2015) Trichloroethylene degradation by persulphate with magnetite as a heterogeneous activator in aqueous solution. Environ Technol 36(11):1389–1397. https://doi.org/10.1080/09593330.2014.991353
157. Sabhi S, Kiwi J (2001) Degradation of 2,4-dichlorophenol by immobilized iron catalysts. Water Res 35(8):1994–2002. https://doi.org/10.1016/S0043-1354(00)00460-7
158. Saha S, Mollah MYA, Susan MABH, Islam MM (2015) Treatment of wastewater containing organic dyes: recovery of dye adsorbed on starch-based materials through conversion of adsorbent into alcohol. Dhaka Univ J Sci 63(2):119–124. https://doi.org/10.3329/dujs.v63i2.24446

159. Saleh R, Taufik A (2019) Degradation of methylene blue and congo-red dyes using Fenton, photo-Fenton, sono-Fenton, and sonophoto-Fenton methods in the presence of iron (II , III) oxide/zinc oxide/graphene (Fe_3O_4/ZnO/graphene). Sep Purif Technol 210:563–573. https://doi.org/10.1016/j.seppur.2018.08.030
160. Sharma S, Mukhopadhyay M, Murthy ZVP (2013) Treatment of chlorophenols from wastewaters by advanced oxidation processes. Sep Purif Rev 42(4):263–295. https://doi.org/10.1080/15422119.2012.669804
161. Singh J, Yang JK, Chang YY, Koduru JR (2017) Fenton-Like degradation of methylene blue by ultrasonically dispersed nano zero-valent metals. Environ Process 4(1):169–182. https://doi.org/10.1007/s40710-016-0199-2
162. Sirés I, Garrido JA, Rodríguez RM, Centellas F, Arias C, Brillas E (2005) Electrochemical degradation of paracetamol from water by catalytic action of Fe^{2+}, Cu^{2+}, and UVA light on electrogenerated hydrogen peroxide. J Electrochem Soc 153(1):D1. https://doi.org/10.1149/1.2130568
163. Sirés I, Brillas E, Oturan MA, Rodrigo MA, Panizza M (2014) Electrochemical advanced oxidation processes: today and tomorrow. A Rev Environ Sci Pollut Res 21(14):8336–8367. https://doi.org/10.1007/s11356-014-2783-1
164. Sultana S, Hossain MS, Susan MABH, Islam MM (2020) Electrosorption of heavy metal from aqueous solution on polyaniline modified graphite electrode. Bangladesh J Sci Res 31:1–6
165. Sun W, Yao Y (2021) Degradation of Auramine-O in aqueous solution by Ti/PbO_2-Electro-Fenton process by hydrogen peroxide produced in situ. Iran J Sci Technol Trans A Sci 45(1):145–154. https://doi.org/10.1007/s40995-020-00975-4
166. Sun Y, Pignatello JJ (1992) Chemical treatment of pesticide wastes: evaluation of iron(III) chelates for catalytic hydrogen peroxide oxidation of 2,4-D at circumneutral pH. J Agric Food Chem 40(2):322–327. https://doi.org/10.1021/jf00014a031
167. Sun SP, Zeng X, Li C, Lemley AT (2014) Enhanced heterogeneous and homogeneous Fenton-like degradation of carbamazepine by nano-Fe_3O_4/H_2O_2 with nitrilotriacetic acid. Chem Eng J 244:44–49 https://doi.org/10.1016/j.cej.2014.01.039
168. Tabet D, Saidi M, Houari M, Pichat P, Khalaf H (2006) Fe-pillared clay as a Fenton-type heterogeneous catalyst for cinnamic acid degradation. J Environ Manage 80(4):342–346. https://doi.org/10.1016/j.jenvman.2005.10.003
169. Tekin H, Bilkay O, Ataberk SS, Balta TH, Ceribasi IH, Sanin FD, Dilek FB, Yetis U (2006) Use of Fenton oxidation to improve the biodegradability of a pharmaceutical wastewater. J Hazard Mater 136(2): 258–265. https://doi.org/10.1016/j.jhazmat.2005.12.012
170. Thomas N, Dionysiou DD, Pillai SC (2021) Heterogeneous Fenton catalysts: a review of recent advances. J Hazard Mater 404:124082. https://doi.org/10.1016/j.jhazmat.2020.124082
171. Ünnü BA, Gündüz G, Dükkancı M (2016) Heterogeneous Fenton-like oxidation of crystal violet using an iron loaded ZSM-5 zeolite. Desalin Water Treat 57(25):11835–11849. https://doi.org/10.1080/19443994.2015.1044915
172. Vaishnave P, Kumar A, Ameta R, Punjabi PB, Ameta SC (2014) Photo oxidative degradation of azure-B by sono-photo-Fenton and photo-Fenton reagents. Arab J Chem 7(6):981–985. https://doi.org/10.1016/j.arabjc.2010.12.019
173. Vicente MA, Trujillano R, Ciuffi KJ, Nassar EJ, Korili SA, Gil A (2010) Pillared clays and related catalysts in green oxidation reactions. In: Pillared clays and related catalysts. 301–318. Springer, New York, NY. https://doi.org/10.1007/978-1-4419-6670-4_11
174. Wacławek S, Lutze HV, Grübel K, Padil VV, Černík M, Dionysiou DD (2017) Chemistry of persulfates in water and wastewater treatment: a review. Chem Eng J 330:44–62. https://doi.org/10.1016/j.cej.2017.07.132
175. Walling C, Camaioni DM (1978) Role of silver(II) in silver-catalyzed oxidations by peroxydisulfate. J Org Chem 43(17):3266–3271. https://doi.org/10.1021/jo00411a003
176. Wang N, Zheng T, Zhang G, Wang P (2016) A review on Fenton-like processes for organic wastewater treatment. J Environ Chem Eng 4(1):762–787. https://doi.org/10.1016/j.jece.2015.12.016

177. Wang C, Shih Y (2015) Degradation and detoxification of diazinon by sono-Fenton and sono-Fenton-like processes. Sep Purif Technol 140:6–12. https://doi.org/10.1016/j.seppur.2014.11.005
178. Wang A, Qu J, Ru J, Liu H, Ge J (2005) Mineralization of an azo dye acid red 14 by electro-Fenton's reagent using an activated carbon fiber cathode. Dye Pigm 65(3):227–233. https://doi.org/10.1016/j.dyepig.2004.07.019
179. Wang X, Hu X, Wang H, Hu C (2012) Synergistic effect of the sequential use of UV irradiation and chlorine to disinfect reclaimed water. Water Res 46(4):1225–1232. https://doi.org/10.1016/j.watres.2011.12.027
180. Wang N, Zheng T, Jiang J, Lung WS, Miao X, Wang P (2014) Pilot-scale treatment of p-nitrophenol wastewater by microwave-enhanced Fenton oxidation process: effects of system parameters and kinetics study. Chem Eng J 239:351–359. https://doi.org/10.1016/j.cej.2013.11.038
181. Wang XQ, Liu CP, Yuan Y, Li FB (2014) Arsenite oxidation and removal driven by a bio-electro-Fenton process under neutral pH conditions. J Hazard Mater 275:200–209. https://doi.org/10.1016/j.jhazmat.2014.05.003
182. Watts RJ, Sarasa J, Loge FJ, Teel AL (2005) Oxidative and reductive pathways in manganese-catalyzed Fenton's reactions. J Environ Eng 131(1):158–164. https://doi.org/10.1061/(asce)0733-9372(2005)131:1(158)
183. Xu HY, Liu WC, Qi SY, Li Y, Zhao Y, Li JW (2014) Kinetics and optimization of the decoloration of dyeing wastewater by a schorl-catalyzed Fenton-like reaction. J Serbian Chem Soc 79(3):361–377. https://doi.org/10.2298/JSC130225075X
184. Yang Y, Wang P, Shi S, Liu Y (2009) Microwave enhanced Fenton-like process for the treatment of high concentration pharmaceutical wastewater. J Hazard Mater 168(1):238–245. https://doi.org/10.1016/j.jhazmat.2009.02.038
185. Yip ACK, Lam FLY, Hu X (2005) Chemical-vapor-deposited copper on acid-activated bentonite clay as an applicable heterogeneous catalyst for the photo-fenton-like oxidation of textile organic pollutants. Ind Eng Chem Res 44(21):7983–7990. https://doi.org/10.1021/ie050647y
186. Yuan S, Gou N, Alshawabkeh AN, Gu AZ (2013) Efficient degradation of contaminants of emerging concerns by a new electro-Fenton process with Ti/MMO cathode. Chemosphere 93(11):2796–2804. https://doi.org/10.1016/j.chemosphere.2013.09.051
187. Zaghbani N, Hafiane A, Dhahbi M (2008) Removal of Safranin T from wastewater using micellar enhanced ultrafiltration. Desalination 222(1–3):348–356. https://doi.org/10.1016/j.desal.2007.01.148
188. Zepp RG, Faust BC, Hoigne J (1992) Hydroxyl radical formation in aqueous reactions (pH 3–8) of iron(II) with hydrogen peroxide: the photo-Fenton reaction. Environ Sci Technol 26(2):313–319. https://doi.org/10.1021/es00026a011
189. Zhang Y, Zhou M (2019) A critical review of the application of chelating agents to enable Fenton and Fenton-like reactions at high pH values. J Hazard Mater 362:436–450. https://doi.org/10.1016/j.jhazmat.2018.09.035
190. Zhang Y, Wang Y, Angelidaki I (2015) Alternate switching between microbial fuel cell and microbial electrolysis cell operation as a new method to control H_2O_2 level in bioelectro-Fenton system. J Power Sources 291:108–116. https://doi.org/10.1016/j.jpowsour.2015.05.020
191. Zheng H, Zhang H, Sun X, Zhang P, Tshukudu T, Zhu G (2010) The catalytic oxidation of malachite green by the microwave-Fenton processes. Water Sci Technol 62(6):1304–1311. https://doi.org/10.2166/wst.2010.411
192. Zhihui A, Peng Y, Xiaohua L (2005) Degradation of 4-Chlorophenol by microwave irradiation enhanced advanced oxidation processes. Chemosphere 60(6):824–827. https://doi.org/10.1016/j.chemosphere.2005.04.027
193. Zhong X, Royer S, Zhang H, Huang Q, Xiang L, Valange S, Barrault J (2011) Mesoporous silica iron-doped as stable and efficient heterogeneous catalyst for the degradation of C.I. acid orange 7 using sono-photo-Fenton process. Sep Purif Technol 80(1):163–171. https://doi.org/10.1016/j.seppur.2011.04.024

194. Zhou M, Yu Q, Lei L, Barton G (2007) Electro-Fenton method for the removal of methyl red in an efficient electrochemical system. Sep Purif Technol 57(2):380–387. https://doi.org/10.1016/j.seppur.2007.04.021
195. Zhou M, He J (2008) Degradation of cationic red X-GRL by electrochemical oxidation on modified PbO_2 electrode. J Hazard Mater 153(1–2):357–363. https://doi.org/10.1016/j.jhazmat.2007.08.056

Plasma Degradation of Synthetic Dyes

Batool Saeed, Shazia Shukrullah, Muhammad Yasin Naz, and Fareeda Zaheer

Abstract Most of the dyes are harmful to the environment due to their high toxicity. The elimination of synthetic dyes is the most difficult task before discharging waste from textile industries. The level of chemical oxygen is higher in wastewaters containing several organic components, particularly synthetic dyes. The important feature of plasma technique over traditional physical methods for the removal of pollutants is that pollutants are mineralized. The degradation of pollutants in plasma treatment process occurs through the production of hydroxyl radicals (°OH), which oxidize the majority of the organic contaminants in the water body. Furthermore, several other oxidative species (O_3, H_2O_2, and HO_2^-) can also be introduced in water due to plasma-based catalytic reactions. These species react with pollutants to degrade the toxic contaminants, particularly dyes. The deterioration of the dye specimen is more pronounced than the other specimens during plasma processing. This chapter covers all aspects of catalytic and non-catalytic plasma degradation of synthetic dyes from the industrial wastewaters.

Keywords Synthetic dyes · Plasma oxidation · Synthetic dyes · Wastewater treatment

1 Introduction

Organic compounds are the major component of contaminants in wastewater that are causing serious issues for the atmosphere and for human health all around the world [13]. Polluting substances affect the natural functioning of ecosystems and water quality. The discharge of industrial effluents and heavy metals in waterbodies raises concerns about human health, damages wildlife, and affects the long-term ecology [1]. Toxic industrial effluents cause reproductive failure, acute poisoning, and immune suppression [52]. Hence, in the situation of uncontrollable pollutant emissions, these emerging molecules must be handled before being discharged into

B. Saeed · S. Shukrullah (✉) · M. Y. Naz · F. Zaheer
Department of Physics, University of Agriculture, Faisalabad 38040, Pakistan
e-mail: zshukrullah@gmail.com

S. S. Muthu and A. Khadir (eds.), *Advanced Oxidation Processes in Dye-Containing Wastewater*, Sustainable Textiles: Production, Processing, Manufacturing & Chemistry, https://doi.org/10.1007/978-981-19-0882-8_3

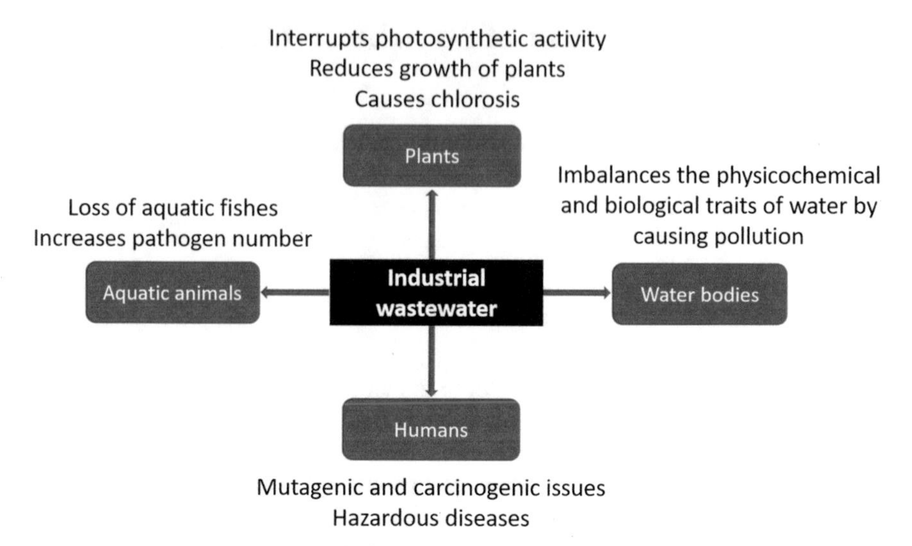

Fig. 1 Effects of textile wastewater on different elements of environment

aquatic life to meet stringent water quality regulations [6]. Figure 1 illustrates the effect of textile wastewater on different elements of the environment [39].

Various techniques are being developed to avoid pollutants from entering water bodies and to eliminate impurities from wastewater [8]. It is difficult to remove all adverse contaminants from water using traditional water treatment techniques, including chlorination, adsorption, coagulation, ultrasonication, and so on. For the remediation of contaminated water, advanced oxidation methodologies such as ozone oxidation, irradiation of high energy electrons, ionizing radiation exposure, plasma exposure, sonolysis, and carbon adsorption have recently been used. Plasma technology has recently been tested for the treatment of polluted water [52]. Plasma constitutes more 99% of our universe. Plasma is categorized into thermal and non-thermal segments depending on electron temperature and density [54]. One of the most effective oxidation methods for treating polluted water is a non-thermal plasma jet. A non-thermal plasma jet is a gas that has been electrically energized by passing it through a strong electric field [41]. Non-thermal plasmas (NTPs) have many novel applications, including material processing, ozone production, contaminants removal, nanomaterial synthesis, and medical treatments [50]. Plasma generates oxidizing species, charged species, UV radiations, and radicals, all of which are thought to AOPs [41]. As illustrated in Fig. 2, electromagnetic radiations, high-energy electrons, stable chemical species, and intermediate species are all produced by plasma discharge.

AOPs are important techniques that should be preferred over other methods, which result in unsafe sedimentation or the transition of pollutants from one form to another [20]. The oxidizing species generated by plasma do not emit any harmful by-products that are harmful to humans. Apart from chlorination and other conventional

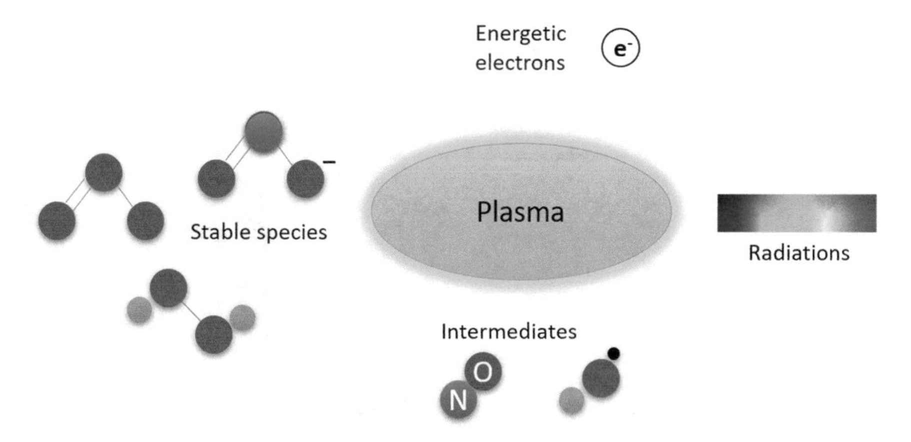

Fig. 2 Composition of a plasma discharge consisting of electromagnetic radiations, high energy electrons, stable chemical species, and intermediate species

processes, the plasma treatment method does not require any additional active ingredients. Depending on their oxidizing effect, reactive neutral species will typically disintegrate microorganism cell membrane, sheaths, and coverings. Compounds that cannot be oxidized by normal oxidants are decomposed by reactive nitrogen-based species (NO, NO_2) and oxygen-based species (O, O_2, O_3, .OH). Several authors addressed the application of plasma for pathogens disinfection [41]. When a suitable catalyst is present, non-thermal plasma can be quite effective for degrading pollutants, especially dyes. Silica gel, activated carbon, Fe^{2+}, TiO_2, and Fe_3O_4 are some of the identified catalysts for treatment of textile effluents [17] .

2 Introduction to Plasma Technology

2.1 Plasma Treatment Setup

Figure 3 illustrates a model of a DC plasma processing device that operates at atmospheric pressure. A high-voltage DC supply, fluctuating resistance chamber, tungsten anode, copper nozzle with an orifice diameter of 2 mm, sample holder, and gas supply connectors make up the setup. A 50 mm of water specimen was collected in the breaker and positioned on a sample holder. The tungsten electrode (anode) was immersed in water (sample). The copper nozzle, which served as cathode, was positioned 10 mm above the level of water. The gap between the electrodes was set to 20 mm. A 100 K safety resistance was added to the circuit to prevent a failure if the electrodes come into direct contact. When the circuit has been finalized, the argon gas was managed to pass out through the plasma nozzle at the fixed rate of flow of 5 LPM to generate a plasma jet, which made contact with a polluted water sample to initiate a chemical reaction.

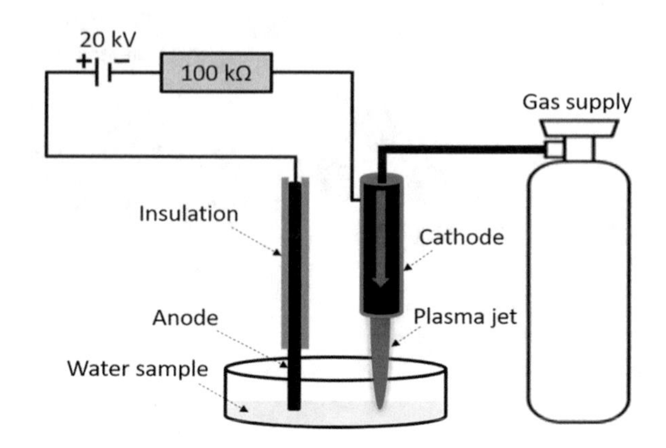

Fig. 3 Atmospheric pressure DC plasma treatment system

2.2 *Plasma Chemistry*

As the plasma jet was created by disposing of an argon gas mixture, Fig. 4 depicts the possible plasma chemistry. Neutrals, ionized particles, excited particles, and electrons may all be present in the plasma jet. The discharge was expected to contain exciting argon particles (Ar*), neutral argon (Ar), argon ions (Ar^+), nitrogen (N_2), oxygen (O_2), ozone O_3, CO_2, N_2O, and NO. Depending on the local environment, other species may also present in the plasma plume.

Contaminants were expected to degrade due to chemical interactions between plasma species with water molecules and contaminants, H_2O, OH, H, H_2O^+, and O_3 are assumed to be the reactive species in these interactions. Table 1 shows the oxidation potentials of the above species. The wastewater treatment procedure was performed at room temperature. Each sample was individually exposed to a plasma jet for 10 min without the use of a catalyst (TiO_2). The procedure was then repeated by inducing a catalyst (TiO_2) into the reaction only 0.5 g of catalyst was mixed into the water, which was then exposed to plasma for 10 min. The plasma-treated wastewater was filtered to extract the solid residue and catalyst. Filtered water was tested for removal efficiency after catalytic and non-catalytic treatment in the presence of plasma.

Several analyses on water samples were performed to determine the appropriateness of the plasma method for the purification of water discharge from industrial effluents. pH, turbidity, TDS conductivity, alkalinity, color, and other parameters are

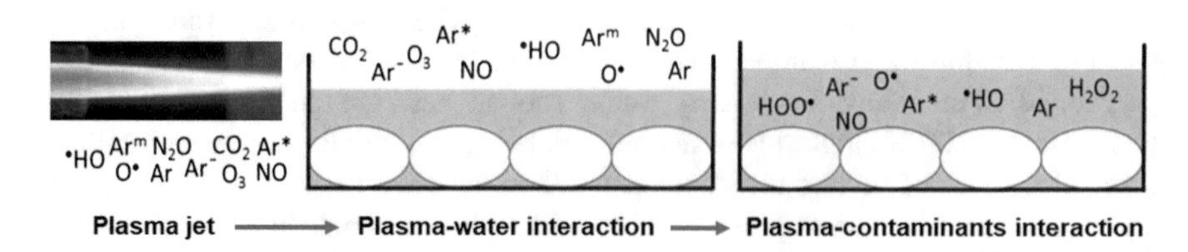

Fig. 4 The expected radicals and species in argon–air mixture plasma jet

Table 1 Oxidation potential of different chemical oxidizers

Oxidizer	Oxidation potential (V)
O*	2.42
Ozone(O3)	2.07
Hydroxyl radical (OH*)	2.80
Hydrogen peroxide (H2O2)	1.77
Permanganate ion	1.67
Chlorine	1.36
Chlorine oxide	1.50

tested as a part of these analyses. Prior to the plasma procedure, non-solid particles were detected in the water samples. During the plasma process, however, a white solid residue appears in the water. The residue obtained was removed from the liquid by centrifuge and dried in an oven at 80 °C for 6 h. Fourier transform infrared spectroscopy and x-ray diffraction were used to characterize the dry residue.

2.3 *Plasma Inactivation of Bacteria*

In addition to heavy metals, other pollutants and dyes, industrial effluents may contain gram-negative bacteria. As a result, the plasma treatment on the activation of Gram-negative bacteria in the water has also been investigated. Bacteria Escherichia coli (E.coli) were cultured and confined to the argon plasma jet under conditions similar to those used for water sample treatment. Under normal conditions, the bacteria cultured were exposed to plasma for 3 min. Counting colony forming units (CFUS) in the specimen afterward when plasma exposure was used to analyze the influence of plasma procedure and bacterial inactivation [52].

2.4 *Plasma–Water Interaction*

The plasma jet might include positive ions, negative ions, neutral species, electrons, and excited species based on environmental conditions. Furthermore, the plasma jet emits ultraviolent radiations, which play a role in the catalyst's photocatalytic activity. Typically, O_2, N_2, Co_2O_3, argon ions, neutral argon atoms, excited argon atoms, nitric oxide, and nitrogen dioxide were expected in the plasma jet. When reactive plasma species come into contact with water, waste is removed from it and pollutants are degraded. The existence of reactive species was determined using optical emission spectroscopy (OES) of the jet. The jet glow signals were collected using optical fiber and transmitted to an ocean optic spectrometer. The spectrometer was connected to the computer, which displayed the optical emission spectrum of the plasma jet. The electrical signal was generated by converting light signals. Electrical signals were captured in the pattern of the spectrum. With an optical resolution of 0.025 nm, the spectrum was listed in the range of 300–750 nm. It has been used to treat wastewater samples after the jet's OES. Water samples were taken from various locations and resources.

2.5 Optical Emission Spectroscopy of Plasma Jet

Figure 5 demonstrates a standard optical emission spectrum of the argon plasma jet containing OH, oxygen radical's and excited nitrogen from the air. The presence of OH, excited argon, excited oxygen, excited nitrogen, nitric oxide, and ozone in a plasma jet was confirmed by its emission spectrum. The jets energetic electrons excite and ionize the nitrogen and oxygen in the surrounding air, through a three-body reaction, oxygen molecules split into the atomic oxygen to produce ozone. Nitrogen within its ground state, on the other side, becomes excited because of the various collision with the electrons.

$$e + N_2\,(X^1 \sum_{g}^{+})\,v = 0 \rightarrow N_2\,(C^3 \prod_{u})\,v' = 0 + e$$

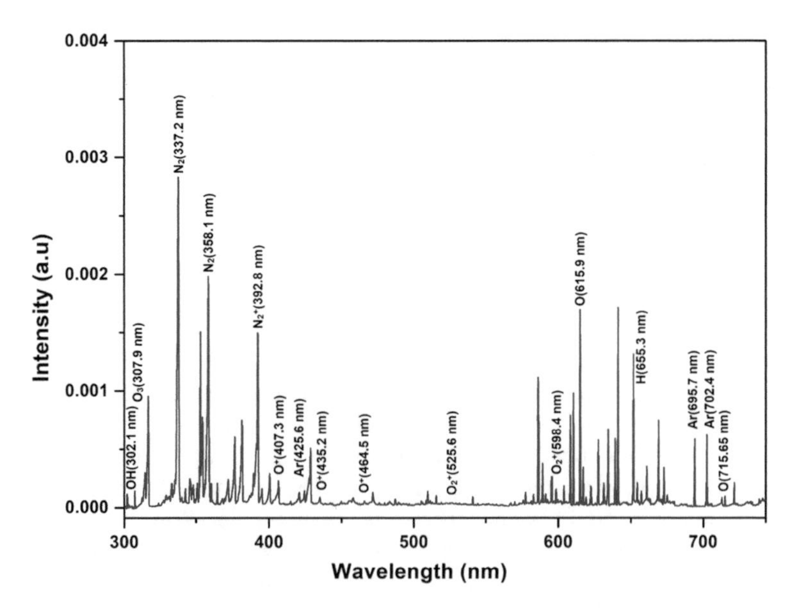

Fig. 5 Optical emission spectrum of argon plasma jet in open atmosphere

Electrons impact excitations of the $N_2(X_1\sum g^+)$ ground state and the $N_2(A_3\sum u^+)$ metastable state populates the $N_2(C_3\prod u)$ state; aside from the electron impacts excitations, the $N_2(C_3\prod u)$ state is also populated by associate excitation, pooling reaction, penning excitation, and energy transfer among colliding particles. By emitting a characteristics photon of the (0–0) band, this excited state decomposes into the second positive system of nitrogen.

$$N_2\,(C^3\prod_u)\,v' \rightarrow N_2\,(B^3\prod_g)\,v' = 0 + hv$$

When the second positive system interacts with the oxygen molecules, nitrogen oxide, ozone, and oxygen radicals are formed. During direct impact ionization of nitrogen in the ground state $N_2(X_1\sum g^+)$, the excited $N^{2+}(B^2\sum u^+)$ state is populated once more. The populated excited energy levels decompose into a first negative system by releasing a characteristic photo in the (0–0) band. The strength of radiations that are emitted is always directly proportionate to the excited state population density. The existence of excited NO, Ar, O_3, N_2, O^+, O^{2+}, N^{2+}, and OH- species in open atmospheric plasma jet was confirmed from the optical spectrum emission line intensities at 254.3 nm, 695–740, 307.9 nm, 330–380 nm, 715.6 nm, 400–500 nm, 500–600 nm, 390–415 nm, and 302–310 nm. The emission intensity proportions of the identified species and the second positive nitrogen system are significantly greater at the start of plasma jet activation. It demonstrates that the atmospheric air rapidly diffuses towards the jet and that the nitrogen concentration rises across the jet's path. The plasma jet was impacted on the water's surface and chemical species were identified by constructing FTIR spectra just at the water–plasma interface. A characteristic FTIR spectrum was produced by the researcher throughout the water–plasma interaction in atmospheric air. The availability of HNO_2, HNO_3, NO_2, N_2O, NO reactive species within plasma exposed water was confirmed using FTIR absorption peaks. Ozone formation had also been observed during water–plasma interactions in both nitrogen-rich environments and ambient air [42].

3 Different Dye Degradation Techniques

Coagulation, adsorption, biodegradation, membrane processes, and advanced oxidation processes (AOPs) were all investigated as methods of dye removal from wastewater. There are some benefits and drawbacks of these processes [35]. For instance, coagulation, the method of flocculation, involves an extensive amount of chemicals and generates a massive amount of sludge, drainage of which is a major issue. Although, land drainage is not acceptable because of the thrash metal and H_2O concentration in sludge, that demand treatment and de-watering individually. In addition, specific transportation is required for sludge, this raises the cost along with additional operations [33]. The maximum reduction was achieved by adsorption that uses activated charcoal 90%, 80%, and 87.6% in color, BOD_5, and COD, respectively, with an 11 g per liter adsorbent dose. However, this operation is regulated by numerous parameters, including temperature, a dosage of adsorbent, contact period, and speed of agitator [2]. Both coagulation and adsorption do not cause the destruction of dyes, rather, they contribute to the phase transition of the contaminants. Thus, the treatment of sludge remains a serious unsolved problem [25]. The membrane is very costly and has a very short life span, which demands periodic replacement that further increases costs [51]. Anaerobic degradation of azo dyes via Aeromonas Spp., pleobacterium Spp., and pseudomonas Spp. bacteria seems to be quicker than aerobic degradation, yet it generates aromatic amines that are highly harmful as compared to actual coloring agent [49].

Numerous analyses have been conducted to measure the efficiency of CGDE technique for the decolorization of contaminated water, which contain colorants. CGDE is the method of electrochemistry in which plasma is maintained through the discharge of DC glow within the electrolyte and electrodes. In the mechanism, the broad potential gradient in the plasma induces gaseous H_2O^+ ions to move towards the interface of plasma liquid and obtain sufficient energy through ionization and activation. They dissociate many water molecules, which produce a significant amount of •OH hydroxyl radical. Figure 6 shows some of the unique properties of •OH that make AOPs an effective technique for removing refractory compound [12]. Hydroxyl radicals (•OH) are very reactive species with a high capacity to oxidize (potential of oxidation is about 2.33 V) [22]. However, hydroxyl radical has a quiet shorter life span (3.7×10^{-9}s), as they are sensitive, they react easily with other compounds or with one another and therefore form H_2O_2 that is quite stable [28, 46].

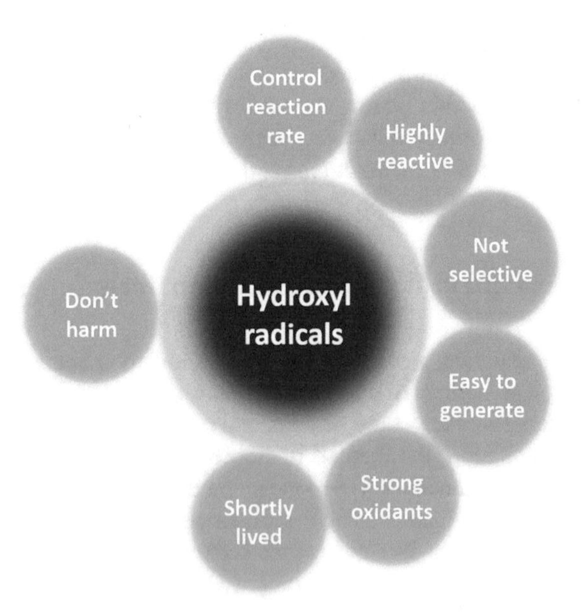

Fig. 6 Some key characteristics of hydroxyl radicals

Furthermore, gas injection inside the plasma region could facilitate the production of higher energy electrons, which produce many hydroxyl radicals. Furthermore, hydroxyl radicals produced in this procedure are observed to be highly reactive with all other existing species, as a result they can interact with the oxygen in the injected air generating numerous reactive species. This oxygen would also respond to H_2O molecules to provide O_3 and H_2O_2. The following equations describe the response of oxygen to hydroxyl radicals [12]:

$$e + O_2 \rightarrow O^{\bullet} + O^{\cdot} + e \tag{1}$$

$$O^{\bullet} + O_2 \rightarrow O_3 \tag{2}$$

$$O^{\bullet} + H_2O \rightarrow OH^{\bullet} + OH^{\bullet} \tag{3}$$

$$e + H_2O \rightarrow H^{\bullet} + OH^{\bullet} + e \tag{4}$$

$$^{\bullet}OH + O_2 \rightarrow HO_2. + O^{\bullet} \tag{5}$$

Equations (1)–(4) show that in addition to the percentage of OH, air injection generates a variety of reactive species (percentage of H_2O and O_2 radicals). The reduction potential for the percentage of O_2 and H_2O is 2.42 V and 1.00 V, respectively. It demonstrates that the dye degradation was high in the presence of these additional reactive species. Jamroz et al. [28] also reported similar process. They found that in a plasma reactor with the injection, active species like hydroxyl and water trigger the organic dye decolonization via Fenton reactions [11].

4 Fenton Reaction

In 1894, Fenton published the renowned Fenton reaction, which demonstrated an oxidation process using H_2O_2 (Hydrogen peroxide) as the oxidant and Fe (iron) as catalyst in acid (H^+) medium [21]. Many publications have comprehensively described the science of the Fenton reaction [61]. The reaction mechanism of the Fenton oxidation is somewhat complicated, and different factors influence the overall efficiency of the process. Fenton oxidation reactions, in general, begin with the formation of a Hydroxyl Free radical [19]. Hydroxyl radicals are among the aggressive oxidants, reacting to 10^6–10^{12} times quicker than O_3 (ozone) based on a substrate being degraded [37]. The Fenton step reactions are explained below Eqs. 6–12, with Eq. 6 describing the chain initiation process [58].

$$Fe^{2+} + H_2O_2 \rightarrow Fe^{3+} + OH^{\bullet} + OH^{-} \rightarrow \text{[Chain initiation process]} \tag{6}$$

$$Fe^{3+} + H_2O_2 \rightarrow Fe - OOH^{2+} + H^+ \tag{7}$$

$$Fe - OOH^{2+} \rightarrow Fe^{3+} + {}^{\bullet}O_2H \tag{8}$$

$$Fe^{3+} + {}^{\bullet}O_2H \rightarrow Fe^{2+} + O_2 + H^+ \tag{9}$$

$$Fe^{2+} + {}^{\bullet}OH \rightarrow Fe^{3+} + OH^- \rightarrow \text{[Chain termination process]} \tag{10}$$

$$H_2O_2 + {}^{\bullet}OH \rightarrow H_2O + {}^{\bullet}O_2H \tag{11}$$

$$\text{Organic toxic waste} + {}^{\bullet}OH \rightarrow \text{Degraded products} \tag{12}$$

The Fenton oxidation reaction requires an acidic medium to generate hydroxyl radical (•OH) and pH 3 is generally considered to be the optimized condition [7]. Above pH 4, a large amount of ferric hydroxide is found, which reduces the efficiency of the degradation process [53]. There are two broad categories of Fenton reaction heterogeneous and homogeneous process. Iron species will be in the same mode as the reactants in the homogeneous process, and mass transfer is not limited. The homogeneous method has many flaws, including sludge formation containing high iron concentration iron deactivation due to complex formation and a pH range (2.0–4.0) sensitivity. The functionalization of the heterogeneous catalysis, allowing it to effectively stimulate the deterioration of recalcitrant compounds while avoiding the production of ferric hydroxide sludge.

The plausible mechanistic paths have been proposed to explain heterogeneous catalytic Fenton reactions depending on recent research development and investigations [62]. The first route involves the iron percolating into the reaction solution and stimulating H_2O_2 through the homogeneous pathway. The depletion of H_2O_2 into Hydroxyl radical is the third method, which involves the adsorption of the investigating molecules on the catalyst surface. The heterogeneous catalyst seems to be more effective than the homogeneous catalyst for the removal of synthetic dyes in wastewater [30], according to the latest research development reports. Heterogeneous catalyst is often more advantageous because:

1. The catalyst is simple to use, precisely recoverable and reusable.
2. It could be used at broad pH range.
3. It avoids the formation of ferric hydroxide precipitation.

Complexing agents including zeolite, nafion, activated charcoal, resin, clay, and silica were being used as sustaining materials for Fe to improve efficiency. There are several Fenton processes that can be used, including electro-Fenton, photo-Fenton, photo-electro-Fenton, sono-Fenton, sono-electro-Fenton, and sono-photo-Fenton process [30]. Some of them are mentioned in Fig. 7 [44].

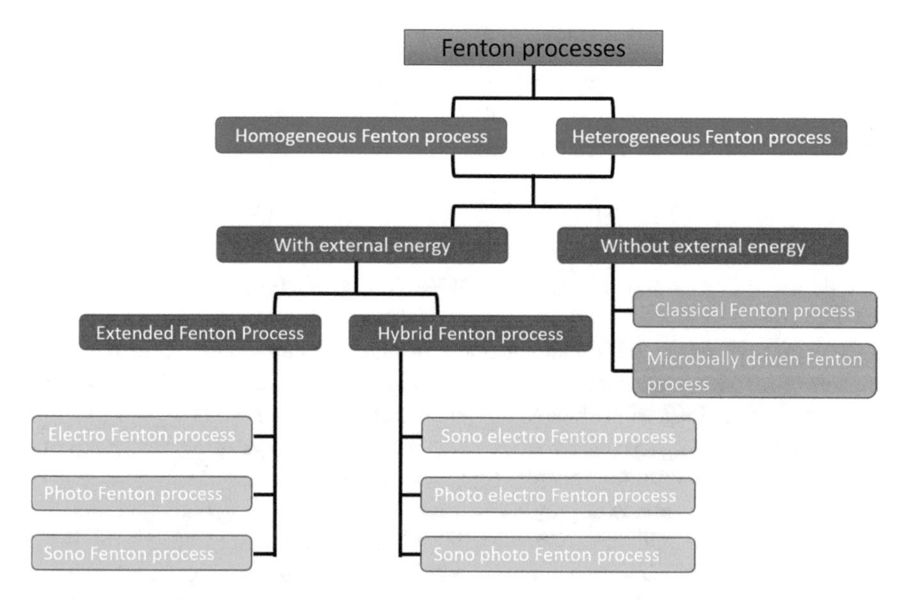

Fig. 7 Types of Fenton processes/reactions

4.1 Photo-Fenton Reaction

The photo-Fenton reaction is an improved version of the traditional. Fenton oxidation process in the gaze of UV–vis light with a wavelength less than 600 nm. UV–vis light involvement gives two additional paths for the discharge of hydroxyl radicals, increasing the degradation percentage of dye pollutants photoreduction of Fe^{2+} and Fe^{3+} ions shown in Eq. (13) and peroxide photolysis through shorter wavelength in Eq. (14). As illustrated in Fig. 8, the photo-generated ferrous ions join the Fenton

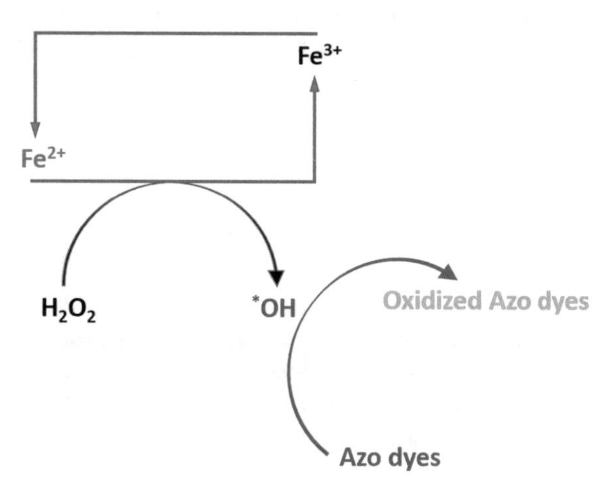

Fig. 8 Performance of the Photo-Fenton process in the degradation of azo dye [38]

reaction, which produces hydroxyl radicals. As a result, the photo-Fenton process has a higher oxidation percentage than the Fenton process without UV–vis light. Furthermore, the iron utilization and the resulting study formation have been significantly scaled back in the photo Fenton reaction.

$$[Fe(OH)]^{2+} \rightarrow Fe^{2+} + {}^{\bullet}OH \quad \lambda < 580\,nm \tag{13}$$

$$H_2O_2 + hv \rightarrow 2\,{}^{\bullet}OH \quad \lambda < 310\,nm \tag{14}$$

Several reviews on the implementations of photo-Fenton methods for removing various types of organic contaminants present in the wastewater have been authored in recent decades [30]. Many factors influencing the performance of the photo-Fenton method have been described by researchers, such as the type of light origin, lamp power, metal concentration, reactor structures, H_2O_2, and so on. Among these variables, light source and power play a significant role in evaluating reaction efficiency [15]. As the light irradiation source, traditional UV lamps are available in high and low-pressure mercury lamps. Mercury lamps have some drawbacks, including being hazardous, easily breakable, difficult to dispose off after use, having a short working life span, and the probability of gas leakage due to the high thermal pressure on the glass. When high- and medium-pressure UV lamps, which operate at a high temperature ranging from 600 to 900 °C, are used, the risks are increased to solve these drawbacks, a lot of research has been done. At the laboratory level for the future applications of photo-Fenton techniques, sunlight irradiation is being used as an alternative for Mercury lamps for this task. The solar generative photo-Fenton processes have shown good efficiency with a significant level of mineralization and up to 90% performance in a short reaction span.

The pH of a solution is yet another significant factor affecting the photo-Fenton method efficiency, as it has a significant impact on the catalysts complex formulation or leaching. The photo-Fenton method was found to be a relatively efficient method for degrading various synthetic dyes. The photo-Fenton method, for example, was perhaps the most effective among several advanced oxidation methods used to degrade RB-19 dye, removing 94.5% dissolved organic carbon and 99.4% total color [30].

4.2 Electro-Fenton Process

Electrochemical technology has received a lot of attention in recent decades for removing contaminants from wastewater. There are numerous research papers available that provide a comprehensive description of the electro-Fenton method for synthetic dye degradation in wastewater. Electrochemistry has many notable benefits, including energy consumption, versatility, and environmental effectiveness due to clean electron and mainstream reagents. Consequently, by combining the electrochemistry and the Fenton process, the oxidation performance can significantly increase. The electro-Fenton oxidation process involves electrochemically adding Fe^{2+} or electrochemically reducing Fe^{3+} while simultaneously [30] electron reduction of oxygen on the cathode surface electro generates hydrogen peroxide in acidic solution [43].

$$O_2 + 2H^+ + 2e^- \rightarrow H_2O_2 \tag{15}$$

The major advantage of this indirect electro-oxidation method over the traditional Fenton process to the highest removal rate of organic contaminants because of continuous transformation of Fe^{3+} and Fe^{2+} at the cathode, as shown in Eq. (16) [36].

$$Fe^{3+} + e^- \rightarrow Fe^{2+} \tag{16}$$

In the aqueous medium, Fe^{2+} reacts with H_2O_2 to produce active •OH. The presence of enough Fe^{2+} ions, which efficiently generate hydroxyl radicals, is ensured by continuous Fe^{3+} conversion, resulting in faster removal efficiency of synthetic dyes. A new approach for an electro-Fenton method was recently developed in which RGO (reduced graphene oxide) was electrochemically accumulated over the surface of carbon felt, resulting in improved dye removal, consistency and H_2O_2 production [30].

4.3 *Sono-Fenton Process*

Ultrasonic waves have been used to decontaminate highly polluted wastewater in recent ages. Ultrasonic waves have a frequency of about 20 kHz or higher, which is higher than human hearing ranges upper limit. Ultrasonic energy is used to produce alternating compression and expansion cycles. Acoustic cavitation is the form of microbubbles that occur when ultrasonic waves expand and contract [23]. Later, during the compression wave cycle, these microbubbles grow to a certain extent and then collapse violently, resulting in several hundred of atmospheric pressure and several thousand Kelvins of temperature, which may be a region of 1000 atm and 500 K, respectively [34].

Cavitation or the cold boiling mechanism uses an energy dispensing phenomenon that involves bubble formation and collapse. Organic contaminants are degraded by pyrolytic cleavage or with the production of hydroxyl radical, despite the fact that these extreme conditions are only observable for a shorter time, highly reactive species including hydroxyl and hydrogen radicals are produced under certain vigorous circumstances as explained in Eqs. (17)–(20) [19]. The use of ultrasonic waves within an aqueous medium generates an oxidative atmosphere in this sono-chemical oxidation technique.

$$H_2O +))) \rightarrow {}^{\bullet}OH + H \tag{17}$$

$$O_2 +))) \rightarrow 2\,{}^{\bullet}O \tag{18}$$

$$O + H_2O \rightarrow 2\,{}^{\bullet}OH \tag{19}$$

$$H^{\bullet} + O_2 \rightarrow OH + {}^{\bullet}O \tag{20}$$

Because of the mutual essential oxidation mechanism, the combined effect of the Fenton process and ultrasonic seems to have synergistic effects on the removal of organic contaminants [60]. The mechanism includes H_2O_2 reacting with Fe^{2+} ions to produce active •OH radicals, analogous to the Fenton technique, and the resulting Fe^{3+} ions reacting with H_2O_2 to produce an intermediate iron complex that disintegrates into Fe^{2+} and •OOH during ultrasound circumstances, as explained in Eq. (21).

$$[Fe^{III}(OOH)]^{2-} +))) \rightarrow Fe^{2+} + {}^{\bullet}OOH \tag{21}$$

Fe^{2+} ions then respond to H_2O_2 producing hydroxyl radicals. As a result, the hydroxyl radicals generated by the sono-Fenton method are more concentrated than those produced by ultrasonic waves only. However, the fusion of ultrasonic and the Fenton system ($Fe^{2+/}H_2O_2$) seems advantageous and has been extensively researched in literature.

According to the literature, the sono-Fenton method is a high-performance method in terms of reaction rate and H_2O_2 consumption. Self-formation of oxidant species becomes advantageous in order to minimize the additional expense of H_2O_2 conversely, due to the ultrasonic system's high energy requirements, Sono; Fenton system-based techniques have been limited in their implementation [30].

5 Non-iron Metal Catalysts for Hydroxyl Radical-Based Oxidation

Various studies have indicated that the transition metals apart from iron, such as Ru, Cu, Ag, Mn, and Co, which exist in at least more than an oxidation state, can catalyze the production of hydroxyl radicals through H_2O_2 [3]. The decolorization of blue dye has been reported using colloidal nanoparticles of Ag, Pd, and Au [27]. NiO/Al_2O_3 and many other heterogeneous non-iron catalysts have also been used to ensure fast dye degradation in contaminated water. For catalytic application, heterogeneous catalysts are assumed to be more proficient and environmental friendly. Using H_2O_2 as an oxidant, [30] used a Cu–ethylene diamine complex existed on

Clay montmorillonite K10 as heterogeneous catalyst for the degradation of acid blue 29 (AB 29) dye.

At 40 °C, nearly 88.2% dye removal was observed in 18 min. The researchers also discussed how temperature and reactants' concentration affects dye degradation performance. The efficient Cu facilitates the removal of dye from the solution referred to the accumulation of hydroxyl radicals and peroxo intermediates, which are effective oxidants for the degradation of AB dye. [56] synthesized Pt nanoparticles of various sizes and deposited on mesoporous Co_3O_4 and used to catalyze the oxidative detraction of methylene blue MB dye using H_2O_2 as an oxidant [56]. At room temperature, extremely effective MB degradation was possible. They reported that different factors, such as dye concentration, H_2O_2 concentration, and temperature effect the dye degradation efficiency. An increase in H_2O_2 concentration and temperature led to a significant increase in catalytic activity, whereas raising the initial concentration of the dye in the reactant. The stream led to a decline in activity. Another group reported polyoxolybdate catalyst supported by MgAL-LDH for the removal of Rhodamine B (RB) and methylene blue (MB) dye separately [5]. The catalyst performed better in the existence of H_2O_2 as an oxidant, achieving nearly 100% depredation of RB and MB in reaction times of 80 and 60 min, respectively, at atmospheric conditions. The efficiency of catalytic degradation of both dyes increased dramatically as the concentration of H_2O_2 increased.

Zeolites have been identified as appropriate support for catalyst growth among several supports. Alekhina et al. synthesized and used Co and Ag exchange Y-type zeolites for the catalytic depredation of carnosine as an example of azo dyes [3]. The highest oxidative depredation of dye was observed using H_2O_2, precisely for CoNaY catalyst in a moderately alkaline medium. In another paper, Alekhina et al. pursuing their research on metal iron-exchange zeolites contrasted the catalytic performance of Co and Fe ion-exchanged HY and NaY zeolites for decomposition of carmoisine dye at 60 °C. The researchers also investigate the impact of catalyst formulation conditions on reaction efficiency. They discovered that using CoNaY as a catalyst, it is possible to completely decolorize the carmoisine solution in weekly acidic and in alkaline media. In contrast, in a weakly acidic medium, FeHY was somewhat impact full as a catalyst. Based on all of the preceding studies on various catalysts, it can be concluded that the form of metal and support material utilized, as well as the approach used for catalyst preparation, are all important factors in achieving efficient dye decomposition operation [30].

6 Factor Affecting Catalytic Activity

6.1 Effect of pH on Dye Degradation

It is critical to figure out how pH influences the degradation performance of catalytic systems. The pH of the reaction medium affects catalytic oxidation when H_2O_2 is used as an oxidant [14]. Since it manipulates the formation of hydroxyl radicals and the accumulation of ferrous ions, the Fenton process is highly based on pH of the solution. The degradation of dyes via Fenton reagent is best done in the acidic media. Fenton reagent activity is decreased at high pH leading to the development of relatively inactive ferric hydroxide and iron oxohydroxides precipitates [32]. A reaction media with strongly acidic pH is therefore inefficient [57]. The presence of iron complex species $[Fe\,(H_2O)_6]^{2+}$, which interacts more slowly via H_2O_2 than other species [16], are thought to be a possible cause of decreased activity at very lower pH. Another possibility is that the peroxide is dissolved in the vicinity of a large number of H^+ ions, resulting in the formation of the sustainable oxonium ion $[H_3O_2]^+$. H_2O_2 becomes more stable when oxygen ions are present and its response with ferrous ions is reduced [57]. As a result, the Fenton process efficiency decreases at both high and low pH levels. The quality of dye degradation decreases with increasing alkalinity in Fenton-like oxidation using various metals other than Fe [45]. The immediate modification of hydroxyl radicals to their less reactive conjugate base, •O, is thought to be the cause of this drop in production using oxidized Pd-created tabular reactions. At ambient temperature, it is discovered that the catalytic transformation of H_2O_2 to hydroxyl radicals increased by increasing pH. The pH range that was found to be optimal was 6–9. The reduction of H_2O_2 is high when the acidity of the solution was high. The conversion of H_2O_2 to hydroxyl radicals in the Pt-coated tabular reactor, contrariety did not significantly decrease when pH of the solution was reduced. The oxidized Pd layer was expected to be quite sensitive to an anion and proton interaction. Thus, H_2O_2 molecule sensitivity had to be limited, resulting in decomposition inhibition, while H_2O_2 molecule exposure to Pt substrate was allowed to improve its catalytic ability. As a result, the acceptable pH range for the decay of hydrogen peroxide varies depending on the metal [30]. A catalytic method that uses a sulphate radical as an oxidant is more appropriate for dye extraction across a wider pH range than one uses a hydroxyl radicals. The efficiency of degradation is reduced in highly alkaline media. The excessive production of OH, which produces hydroxyl radicals while consuming sulphate radical, is thought to be the cause of this phenomenon. Dye decomposition efficiency reduces due to the poor oxidation potential and non-selectivity of hydroxyl radicals, including a drop in the concentration of sulphate radicals [24]. Hence, pH of a processing medium must be optimized, which is affected not only by the catalyst but also by the kind of oxidation involved throughout the process.

6.2 Effect of Temperature on Dye Degradation

Many research articles claim that increasing the operating temperature will improve the oxidation and degree of catalytic degradation of synthetic dyes [16]. However, there has been relatively little research on the impact of experimental temperature conditions on the efficiency of the catalyst for synthetic dye degradation. The Fenton-based process has mostly been used at room temperature [59]. A significant progress in the degradation of phenol (a major component of almost all synthetic dyes) via Fenton oxidation has been registered at a high temperature, with a depredation efficiency of nearly 80% at 120 degree Celsius. At a high temperature, iron-catalyzed H_2O_2 reduction into radicals is also high. Salem et al. looked at how temperature affects the dye removal efficiency of acid blue 29 dye with a heterogeneous Cu catalyst while holding the dye, catalyst, and H_2O_2 concentrations stable [47]. After 18 min of reaction time, they found a 51.8–88.2% improvement in decolonization efficiency when they increased the operating temperature from 20 to 40 °C. The temperature has also been linked to increased performance in a shorter residence period. While retaining all other factor constants during a survey of catalytic oxidation of synthetic dyes that use a PdO-coated tubular reactor, HPHT-H_2O a reaction medium, and H_2O_2 as an oxidant, discovered that the reaction was highly temperature-dependent. COD removal of 84.0% was achieved at 200 °C. Using a reaction solution of orange 11 that boosted up to 99.0% when the observational temperature was raised to 300 °C at a constant gauge pressure of 10 MPa [29, 31]. Moreover, catalytic degradation of synthetic dyes at comparable higher temperatures provides a way to enhance activity by significantly increasing the mineralization percentage and the oxidation percentage within a specified reaction time. The catalytic processes that use a sulphate radical as an oxidation agent are primarily studied at room temperature without considering the effect of ambient temperature. Consequently, the significance of temperature as a variable in the catalytic degradation of dye cannot be neglected. Hence, a comprehensive investigation of the effect of temperature seems to be essential [30].

6.3 *Effect of Initial Concentration of Dye on Degradation*

In industrial applications, the dye's initial concentration plays a vital role. In general, to achieve effective and absolute extraction of the dyes, a smaller initial concentration is preferred. Since the dye content of effluent released by factories is massive, dilution is needed before proceeding to catalytic treatment, irrespective of the nature of oxidant used [9]. When the amount of acid blue 29 dye was increased from 1×10^{-4} to 2×10^{-5} M in the presence of heterogeneous Cu catalyst while keeping H_2O_2 concentration constant (0.2 $_{M}$), the dye degradation efficiency substantially decreased from 92.8 to 78.2%. It is discovered that increasing the initial amount of synthetic dye reduced the efficiency of degradation using an oxidized Pd-Coated tubular reactor with constant H_2O_2 concentration as an oxidizing agent at high pressure and temperature. The production of hydroxyl radicals on the catalyst surface was low at relatively high dye concentration (as the dye molecules may occupy the catalyst's active sites). The effectiveness of the degradation process reduced due to high concentration of dye molecules and low number of active radicals [30]. It is noticed by other researchers that increasing the initial dye concentration increases the dye removal rate [4]. This observation was linked to an improvement in the chance of the dye molecules colliding with oxidizing species because dye content within the reaction medium also increased. Moreover, the initial quantity of dyes in reactants also affected the amount of the catalyst oxidizing agent and residence time. For instance, the quantity of Fenton's repaints utilized in the process helps to determine the initial dose of synthetic dye [26].

6.4 Effect of Concentration of the Oxidant on Dye Degradation

The overall performance of the catalytic decomposition processes of dyes is highly dependent on the oxidant concentration [18]. With an increase in H_2O_2 concentration in a reaction stream, the efficiency of synthetic dye degradation improved in the Fenton oxidation phase, the concentration of Fe^{2+} and H_2O_2 determines the fairly constant concentration of hydroxyl radical [4]. Tian et al. mentioned that increasing H_2O_2 concentration in the Fenton technique to 125 mg/l increased color and COD removed to 94 and 50.7%, respectively, while further increasing H_2O_2 concentration reduced the removal efficiency [55]. The researchers have reported similar findings. The decrease in dye degradation efficiency with increasing H_2O_2 concentration above the optimized concentration was associated with the innovation of competition for adsorption on the catalyst's surface where the surplus H_2O_2 restricts dye molecule access. Furthermore, as a radical scavenger, unnecessary H_2O_2 in the catalytic oxidation phase is important [30]. Salem et al. analyzed the influence of H_2O_2 concentration on the reaction rate for the degradation of acid blue 29 dye utilizing a Cu-based catalyst while keeping the catalyst number, dye, and temperature constant as research on iron-free catalytic oxidation [59]. They found that increasing the concentration of H_2O_2 from 0.02 to 0.4 M increased degradation efficiency from 26.6 to 84.3% with 15 min. They linked their results to the increased production of peroxointermediate or hydroxyl radical in the reaction media as H_2O_2 concentration rises. Other researchers have also documented the formation of peroxo-radicals in Cu-facilitated oxidation processes [10]. However, since the residual concentration of untreated H_2O_2 leads to COD and is toxic to many species, calibrations for the desired H_2O_2 concentration must be achieved based on synthetic dye concentration to be degraded [26].

6.5 Effect of Reaction Time on Dye Degradation

Irrespective of the nature of the catalyst used, residence/reaction time is a significant factor in catalytic dye degradation (homogeneous a heterogeneous). In general, as the duration or residence period of a reaction in the decomposition rate rises, while all other variables, such as temperature, pH, catalyst, oxidant, and dye concentrations remain constant. The influence of the reaction period on the Fenton method was investigated by [40]. They found that raising the residence period from 30 to 120 min resulted in a 45–69% improvement in COD and the dye removal. Within 4 h of the homogeneous Fenton oxidation process, Karthikeyan et al. observed a linear rise in COD elimination until 2 h, thereafter started to decrease over residence time [40]. This initial linear rise in COD depletion was referred to the chemical oxidation of organic matter in contaminated water with hydroxyl radicals. The authors have confirmed that heterogeneous Fenton oxidation is more efficient than homogeneous Fenton oxidation. A heterogeneous catalyst removed 90% of COD from the textile contaminated water in about 4 h. While a homogeneous catalyst removed 50% in about 6 h. Typically, a long residence period in an hour is required to achieve dramatically enhanced COD and color removal, whereas dye decomposition with the reliability of above 99% COD removal can be acquired in few seconds while using thin HPHT-H_2O as reaction media metal catalyst-coated tubular reactor. Other factors, such as the concentration of reacting species and temperature, influence the residence period. Literature review on the catalytic degradation of synthetic dyes while using a PdO-coated tubular reactor discovered that 89% TOC removal of Remazol Brilliant Blue R has been acquired at 200 °C within reaction period of 3.9 s, which increased to 92 and 99.9% in the period of 3.6 and 3.2s, respectively [31].

7 Limitations of Plasma Degradation

In plasma processing, the same gas pressure, flow rate, power input could not give the same amount of the required reacting species. The buying of expensive plasma instruments and high vacuum pumps are seen as limiting factors of plasma-based degradation processes. Scaling up and a pilot batch method to the continuous method could also show a few technical challenges. For each process and piece of equipment, optimal process constraints should be established. These difficulties, however, are not impossible to overcome. Treating relatively thin layers before altering the bulk may be a downside for some end uses and strength whenever the goal is to leave the bulk unchanged and only relative thin layer treatment is required [48].

8 Conclusions

Dye is one of our environment's most severe contaminants. It might be very dangerous to the planet if these dyes are not eliminated from textile wastewater before accessing the aquatic environment. Due to their recalcitrant behavior and susceptibility towards biodegradation, traditional methods are ineffective in handling industrially contaminated water having a higher percentage of synthetic dyes. Catalytic oxidation is an advanced oxidation process that is both environmental friendly and extremely effective. Many methods are used to achieve the complete mineralization of dyes. When iron-based catalysts are used to remove dyes from polluted water, the Fenton reaction treatment is very efficient. The development of an iron-free catalytic system for the stimulation of H_2O_2 using various other metals has also become a focus of research. Another effective method for catalytic decomposition of synthetic dyes is sulphate radical-based oxidation. The most difficult aspect of degradation of synthetic dyes is optimizing various reaction variables, although many degradation techniques are highly dependent on various factors such as temperature, pH, residence period, initial dye concentration, and oxidant concentration. Based on the nature of the catalyst and the oxidizing agent to be used, there is an ideal value for almost every parameter catalytic activity that can be increased to the maximum extent by adequately manipulating various factor.

References

1. Abinandan S, Shanthakumar S (2015) Challenges and opportunities in application of microalgae (Chlorophyta) for wastewater treatment: a review. Renew Sustain Energy Rev 52:123–132. https://doi.org/10.1016/j.rser.2015.07.086
2. Aleem M, Rashid AA (2016) Characterization and removal of dyeing effluents by adsorption and coagulation methods. J Agric Res 54:97–106
3. Alekhina M, Khabirova K, Kon'kova T, Prosvirin I (2017) Y-Type zeolites for the catalytic oxidative degradation of organic azo dyes in wastewater. Kinet Catal 58:506–512. https://doi.org/10.1134/S0023158417050019
4. Aljuboury D, Palaniandy P, Abdul Aziz H, Feroz S (2017) Treatment of petroleum wastewater by conventional and new technologies-A review. Global NEST J 19:439–452
5. Amini M, Khaksar M, Ellern A, Woo LK (2018) A new nanocluster polyoxomolybdate [Mo36O110 (NO) 4 (H_2O) 14]· 52H_2O: Synthesis, characterization and application in oxidative degradation of common organic dyes. Chin J Chem Eng 26:337–342. https://doi.org/10.1016/j.cjche.2017.03.031
6. Andreozzi R, Caprio V, Insola A, Marotta R (1999) Advanced oxidation processes (AOP) for water purification and recovery. Catal Today 53:51–59. https://doi.org/10.1016/S0920-5861(99)00102-9
7. Arnold SM, Hickey WJ, Harris RF (1995) Degradation of atrazine by Fenton's reagent: condition optimization and product quantification. Environ Sci Technol 29:2083–2089. https://doi.org/10.1021/es00008a030
8. Azizullah A, Khattak MNK, Richter P, Häder DP (2011) Water pollution in Pakistan and its impact on public health—a review. Environ Int 37:479–497. https://doi.org/10.1016/j.envint.2010.10.007
9. Babuponnusami A, Muthukumar K (2014) A review on Fenton and improvements to the Fenton process for wastewater treatment. J Environ Chem Eng 2:557–572. https://doi.org/10.1016/j.jece.2013.10.011
10. Blaney L, Lawler DF, Katz LE (2019) Transformation kinetics of cyclophosphamide and ifosfamide by ozone and hydroxyl radicals using continuous oxidant addition reactors. J Hazard Mater 364:752–761. https://doi.org/10.1016/j.jhazmat.2018.09.075
11. Budikania TS, Afriani K, Widiana I, Saksono N (2019) Decolorization of azo dyes using contact glow discharge electrolysis. J Environ Chem Eng 7:103466. https://doi.org/10.1016/j.jece.2019.103466
12. Buthiyappan A, Aziz ARA, Daud WMAW (2016) Recent advances and prospects of catalytic advanced oxidation process in treating textile effluents. Rev Chem Eng 32:1–47
13. Capocelli M, Joyce E, Lancia A, Mason TJ, Musmarra D, Prisciandaro M (2012) Sonochemical degradation of estradiols: incidence of ultrasonic frequency. Chem Eng J 210:9–17. https://doi.org/10.1016/j.cej.2012.08.084

14. Chen Y, Li N, Zhang Y, Zhang L (2014) Novel low-cost Fenton-like layered Fe-titanate catalyst: Preparation, characterization and application for degradation of organic colorants. J Colloid Interface Sci 422:9–15. https://doi.org/10.1016/j.jcis.2014.01.013
15. Cruz-González K, Torres-López O, García-León A, Guzmán-Mar J, Reyes L, Hernández-Ramírez A, Peralta-Hernández J (2010) Determination of optimum operating parameters for Acid Yellow 36 decolorization by. ChemEng J 160:199–206. https://doi.org/10.1016/j.cej.2010.03.043
16. Daud N, Ahmad M, Hameed B (2010) Decolorization of Acid Red 1 dye solution by Fenton-like process using Fe–Montmorillonite K10 catalyst. Chem Eng J 165:111–116. https://doi.org/10.1016/j.cej.2010.08.072
17. de Brito Benetoli LO, Cadorin BM, Baldissarelli VZ, Geremias R, de Souza IG, Debacher NA (2012) Pyrite-enhanced methylene blue degradation in non-thermal plasma water treatment reactor. J Hazard Mater 237:55–62. https://doi.org/10.1016/j.jhazmat.2012.07.067
18. De la Cruz N, Giménez J, Esplugas S, Grandjean D, De Alencastro L, Pulgarin C (2012) Degradation of 32 emergent contaminants by UV and neutral photo-fenton in domestic wastewater effluent previously treated by activated sludge. Water Res 46:1947–1957. https://doi.org/10.1016/j.watres.2012.01.014
19. Dewil R, Mantzavinos D, Poulios I, Rodrigo MA (2017) New perspectives for advanced oxidation processes. J Environ Manage 195:93–99. https://doi.org/10.1016/j.jenvman.2017.04.010
20. Dojčinović BP, Roglić GM, Obradović BM, Kuraica MM, Kostić MM, Nešić J, Manojlovi DD (2011) Decolorization of reactive textile dyes using water falling film dielectric barrier discharge. J Hazard Mater 192:763–771. https://doi.org/10.1016/j.jhazmat.2011.05.086
21. Fenton HJH (1894) LXXIII.—Oxidation of tartaric acid in presence of iron. J Chem Soc Trans 65:899–910. https://doi.org/10.1039/CT8946500899
22. Gao J, Wang X, Hu Z, Deng H, Hou J, Lu X, Kang J (2003) Plasma degradation of dyes in water with contact glow discharge electrolysis. Water Res 37:267–272. https://doi.org/10.1016/S0043-1354(02)00273-7
23. Gogate PR (2008) Treatment of wastewater streams containing phenolic compounds using hybrid techniques based on cavitation: a review of the current status and the way forward. Ultrason Sonochem 15:1–15. https://doi.org/10.1016/j.ultsonch.2007.04.007
24. Guerra-Rodríguez S, Rodríguez E, Singh DN, Rodríguez-Chueca J (2018) Assessment of sulfate radical-based advanced oxidation processes for water and wastewater treatment: a review. Water 10:1828. https://doi.org/10.3390/w10121828
25. Hmd RFK (2011) Degradation of some textile dyes using biological and physical treatments
26. Huang R, Fang Z, Yan X, Cheng W (2012) Heterogeneous sono-Fenton catalytic degradation of bisphenol A by Fe3O4 magnetic nanoparticles under neutral condition. Chem Eng J 197:242–249. https://doi.org/10.1016/j.cej.2012.05.035
27. Ilunga AK, Meijboom R (2016) Catalytic oxidation of methylene blue by dendrimer encapsulated silver and gold nanoparticles. J Mol Catal A Chem 411:48–60. https://doi.org/10.1016/j.molcata.2015.10.009
28. Jamroz P, Dzimitrowicz A, Pohl P (2018) Decolorization of organic dyes solution by atmospheric pressure glow discharge system working in a liquid flow-through mode. Plasma Processes Polym 15:1700083
29. Javaid R, Kawanami H, Chatterjee M, Ishizaka T, Suzuki A, Suzuki TM (2010) Fabrication of microtubular reactors coated with thin catalytic layer (M= Pd, Pd−Cu, Pt, Rh, Au). Catal Commun 11:1160–1164. https://doi.org/10.1016/j.catcom.2010.05.018
30. Javaid R, Qazi UY (2019) Catalytic oxidation process for the degradation of synthetic dyes: an overview. Int J Environ Res Public Health 16:2066. https://doi.org/10.3390/ijerph16112066
31. Javaid R, Qazi UY, Kawasaki SI (2016) Highly efficient decomposition of Remazol Brilliant Blue R using tubular reactor coated with thin layer of PdO. J Environ Manage 180:551–556. https://doi.org/10.1016/j.jenvman.2016.05.075
32. Kavitha V, Palanivelu K (2005) Destruction of cresols by Fenton oxidation process. Water Res 39:3062–3072. https://doi.org/10.1016/j.watres.2005.05.011

33. Keeley J, Jarvis P, Judd SJ (2014) Coagulant recovery from water treatment residuals: a review of applicable technologies. Crit Rev Environ Sci Technol 44:2675–2719. https://doi.org/10.1080/10643389.2013.829766
34. Kerabchi N, Merouani S, Hamdaoui O (2019) Relationship between liquid depth and the acoustic generation of hydrogen: design aspect for large cavitational reactors with special focus on the role of the wave attenuation. Int J Green Energy 16:423–434. https://doi.org/10.1080/15435075.2019.1577741
35. Khataee A, Kasiri MB (2010) Photocatalytic degradation of organic dyes in the presence of nanostructured titanium dioxide: Influence of the chemical structure of dyes. J Mol Catal A Chem 328:8–26. https://doi.org/10.1016/j.molcata.2010.05.023
36. Kim Dh, Lee D, Monllor-Satoca D, Kim K, Lee W, Choi W (2019) Homogeneous photocatalytic Fe3+/Fe2+ redox cycle for simultaneous Cr (VI) reduction and organic pollutant oxidation: Roles of hydroxyl radical and degradation intermediates. J Hazard Mater 372:121-128. https://doi.org/10.1016/j.jhazmat.2018.03.055
37. Lloyd RV, Hanna PM, Mason RP (1997) The origin of the hydroxyl radical oxygen in the. Free Radical Biol Med 22:885–888. https://doi.org/10.1016/S0891-5849(96)00432-7
38. Macías-Sánchez J, Hinojosa-Reyes L, Guzmán-Mar JL, Peralta-Hernández JM, Hernández-Ramírez A (2011) Performance of the photo-Fenton process in the degradation of a model azo dye mixture. Photochem Photobiol Sci 10:332–337
39. Mishra S, Maiti A (2020) Biological methodologies for treatment of textile wastewater. In: Environmental processes and management. Springer Publisher, pp 77–107
40. Karthikeyan S, Titus A, Gnanamani A, Mandal AB, Sekaran G (2011) Treatment of textile wastewater by homogeneous and heterogeneous Fenton oxidation processes. Desalination 281:438–445
41. Naz M, Shukrullah S, Ghaffar A, Rehman N, Sagir M (2016) A low-frequency dielectric barrier discharge system design for textile treatment. Synth React Inorg Met-Org Nano-Met Chem 46:104–109. https://doi.org/10.1080/15533174.2014.900789
42. Naz M, Shukrullah S, Rehman S, Khan Y, Al-Arainy A, Meer R (2021) Optical characterization of non-thermal plasma jet energy carriers for effective catalytic processing of industrial wastewaters. Sci Rep 11:1–13. https://doi.org/10.1038/s41598-021-82019-4
43. Nidheesh PV, Gandhimathi R, Ramesh ST (2013) Degradation of dyes from aqueous solution by Fenton processes: a review. Environ Sci Pollut Res 20:2099–2132. https://doi.org/10.1007/s11356-012-1385-z
44. Naushad M, Lichtfouse E (2019) Green materials for wastewater treatment. Springer
45. Oladipo AA, Ifebajo AO, Gazi M (2019) Magnetic LDH-based CoO–NiFe2O4 catalyst with enhanced performance and recyclability for efficient decolorization of azo dye via Fenton-like reactions. Appl Catal B 243:243–252. https://doi.org/10.1016/j.apcatb.2018.10.050
46. Saksono N, Nugraha I, Gozan M, Bismo S (2014) Plasma formation energy and hydroxyl production on contact glow discharge electrolysis. Int J Arts Sci 7:71
47. Salem IA, El-Ghamry HA, El-Ghobashy MA (2014) Catalytic decolorization of Acid blue 29 dye by H2O2 and a heterogeneous catalyst. Beni-Suef Univ J Basic Appl Sci 3:186–192. https://doi.org/10.1016/j.bjbas.2014.10.003
48. Sarmadi M (2013) Advantages and disadvantages of plasma treatment of textile materials. In: 21st International symposium on plasma chemistry (ISPC 21) Sunday
49. Sastrawidana ID, Maryam S, Sukarta IN (2012) Perombakan air limbah tekstil menggunakan jamur pendegradasi kayu jenis Polyporus Sp teramobil pada serbuk gergaji kayu. Bumi Lestari J Environ 12
50. Schmidt-Bleker A, Winter J, Bösel A, Reuter S, Weltmann KD (2015) On the plasma chemistry of a cold atmospheric argon plasma jet with shielding gas device. Plasma Sources Sci Technol 25: 015005
51. Sharma SK (2015) Green chemistry for dyes removal from waste water: research trends and applications. John Wiley & Sons
52. Shukrullah S, Bashir W, Altaf NUH, Khan Y, Al-Arainy AA, Sheikh TA (2020) Catalytic and Non-Catalytic Treatment of Industrial Wastewater under the Exposure of Non-Thermal Plasma Jet. Processes 8:667. https://doi.org/10.3390/pr8060667

53. Tamimi M, Qourzal S, Barka N, Assabbane A, Ait-Ichou Y (2008) Methomyl degradation in aqueous solutions by Fenton's reagent and the photo-Fenton system. Sep Purif Technol 61:103–108. https://doi.org/10.1016/j.seppur.2007.09.017
54. Tendero C, Tixier C, Tristant P, Desmaison J, Leprince P (2006) Atmospheric pressure plasmas: a review. Spectrochim Acta, Part B 61:2–30. https://doi.org/10.1016/j.sab.2005.10.003
55. Tian S, Tu Y, Chen D, Chen X, Xiong Y (2011) Degradation of Acid Orange II at neutral pH using Fe_2 (MoO_4) 3 as a heterogeneous Fenton-like catalyst. Chem Eng J 169:31–37. https://doi.org/10.1016/j.cej.2011.02.045
56. Xaba MS, Noh JH, Meijboom R (2019) Catalytic activity of different sizes of Ptn/Co_3O_4 in the oxidative degradation of Methylene Blue with H_2O_2. Appl Surf Sci 467:868–880. https://doi.org/10.1016/j.apsusc.2018.10.259
57. Xu XR, Li XY, Li XZ, Li HB (2009) Degradation of melatonin by UV, UV/H2O2, Fe_2+/H_2O_2 and UV/Fe_2+/H2O2 processes. Sep Purif Technol 68:261–266. https://doi.org/10.1016/j.seppur.2009.05.013
58. Yoon J, Lee Y, Kim S (2001) Investigation of the reaction pathway of OH radicals produced by Fenton oxidation in the conditions of wastewater treatment. Water Sci Technol 44:15–15. https://doi.org/10.2166/wst.2001.0242
59. Zazo JA, Pliego G, Blasco S, Casas JA, Rodriguez JJ (2011) Intensification of the Fenton process by increasing the temperature. Ind Eng Chem Res 50:866–870. https://doi.org/10.1021/ie101963k
60. Zhang JH, Zou HY, Ning XA, Lin MQ, Chen CM, An TC, Sun J (2018) Combined ultrasound with Fenton treatment for the degradation of carcinogenic polycyclic aromatic hydrocarbons in textile dying sludge. Environ Geochem Health 40:1867–1876. https://doi.org/10.1007/s10653-017-9946-1
61. Zhang Y, Zhou M (2019) A critical review of the application of chelating agents to enable Fenton and Fenton-like reactions at high pH values. J Hazard Mater 362:436–450. https://doi.org/10.1016/j.jhazmat.2018.09.035
62. Zhao C, Arroyo-Mora LE, DeCaprio AP, Sharma VK, Dionysiou DD, O'Shea KE (2014) Reductive and oxidative degradation of iopamidol, iodinated X-ray contrast media, by Fe (III)-oxalate under UV and visible light treatment. Water Res 67:144–153. https://doi.org/10.1016/j.watres.2014.09.009

Dyes Sonolysis: An Industrial View of Process Intensification Using Carbon Tetrachloride

Aissa Dehane and Slimane Merouani

Abstract Throughout the last two decades, there has been a lot of interest in using ultrasonic as an alternate advanced oxidation method for the degradation of textile dyes in wastewater. This technique effectively destroys pollutants by reactive •OH radical and/or pyrolysis in various reaction zones. The addition of CCl_4 increased dye sonolytic degradation by tens to hundreds of times (intensification viewpoint). Despite several investigations on the issue, the mechanism by which CCl_4 enhanced the sonolytic removal of textile dyes has not been clearly determined. This chapter will address this topic. To begin, works done on the CCl_4-induced intensification of organic dyes were reviewed, together with their experimental conditions and noteworthy results. Second, all elements that influence the intensifying aspect of CCl_4 were shown. Thirdly, using a newly constructed model of single bubble sonochemistry, the different reactive species generated through CCl_4 pyrolysis were identified. The model's results were then utilized to explain the intensifying action of CCl_4 in the breakdown of aqueous organic pollutants by ultrasound. Finally, several prospects for large-scale use of this new intensification approach were highlighted, as well as some emphasizing innovation requirements. To the best of our knowledge, this is the first review focuses on ultrasound/CCl_4 technique for fast dye removal from wastewater.

Keywords Textile dyes · Sonochemical treatment · Process intensification · Carbon tetrachloride (CCl_4) · Computational Analysis

---

A. Dehane
Department of Process Engineering, Faculty of Engineering, Badji Mokhtar–Annaba University, 23000 Annaba, Algeria

S. Merouani (✉)
Department of Chemical Engineering, Faculty of Process Engineering, Salah Boubnider Constantine 3 University, 25000 Constantine, Algeria
e-mail: s.merouani@yahoo.fr; s.merouani03@gmail.com

S. S. Muthu and A. Khadir (eds.), *Advanced Oxidation Processes in Dye-Containing Wastewater*, Sustainable Textiles: Production, Processing, Manufacturing & Chemistry, https://doi.org/10.1007/978-981-19-0882-8_4

1 Introduction

Nowadays, water shortage or even lack becomes a serious threat facing the humankind due to political, economical, and climatological reasons. Water consumption from manufacturing is anticipated to rise 400% between now and 2050, while home use is expected to increase 130% [62]. By 2035, up to 40% of the world's population would be living in severely water-stressed places, and ecosystem's capacity to supply fresh water sources will be increasingly jeopardized [62]. Furthermore, more than a quarter of the world's population suffers from water-related health and hygiene issues [86]. In addition to groundwater, a wide spectrum of organic chemicals have been identified in industrial and municipal effluent. These pollutants are substances that are either manufactured or naturally occurring. Over 100,000 dyes have been marketed in recent years, with approximately 20% of yearly dye output (i.e., 7×10^5 tones) wasted during the dyeing process and released in wastewater effluents [37, 85]. Organic dyes are utilized in a variety of sectors, including textiles, optical data discs, cosmetics, paper, food, plastics, medicinal goods, solar cells, and so on [33, 98]. Two-third of the total dyes production is used in textile industry and consumes large amounts of water and other recalcitrant chemicals for wet processing of textiles [37].

In general, the release of these pollutants into the environment has the potential to cause significant issues for human health and the environment. As a result, the adoption of alternative and reliable treatment technologies aimed at mineralizing or transforming these pollutants into less hazardous and biodegradable species is an issue of a major concern. Several innovative technologies involving physical, chemical, and biological methods are being developed for an efficient degradation and eliminations of the different contaminants, including dyes [39, 94, 96]. Advanced oxidation processes (AOPs) are very interesting methods for an effective degradation of pollutants due to their potential of mineralization of organic contaminants to CO_2 and H_2O [94]. The majority of AOPs (i.e., H_2O_2/UV, $K_2S_2O_8$/UV, O_3/UV, Fenton's reactions, TiO_2/UV, O_3/ultrasound, O_3/TiO_2/UV, etc.) are based on the in situ generation of hydroxyl radicals, while some other processes are based (e.g., $K_2S_2O_8$/UV, $K_2S_2O_8$/iron, chlorine/UV, chlorine/iron) on sulfate and chlorine radicals [31, 56, 65, 67, 68, 92, 114, 115].

Over the last two decades, there has been a surge of interest in using ultrasound to remove dye pollutants from wastewater [73]. The application of ultrasound in chemistry and processing is simply known as sonochemistry [59]. Acoustic cavitation, defined as the creation, expansion, and rapid collapse of gas–vapor filled bubbles in a liquid, is the driving force behind sonochemistry [109]. At the end of bubble collapse, high temperature and pressure are attained for short period of time, therefore several radicals (e.g., $^{\bullet}OH$, $HO_2^{\bullet}$ and $H^{\bullet}$) and active species (e.g., H_2O_2) are created, in addition, the present solutes in liquid are either directly pyrolysis inside the bubble and/or at liquid–gas interface or indirectly destructed via free radicals [78, 112]. However, sonolysis alone hardly results in complete mineralization of pollutants streams containing complex mixtures of organic and inorganic compounds

[105]. Therefore, many efforts have been devoted to the research of new additives (e.g., salts, O_3, H_2O_2, Fenton reagents, CCl_4, C_6F_{14}, persulfate, periodate, etc. [8, 30, 33, 38, 63, 72, 74, 85] in order to increase the sono-degradation efficiency of the different contaminants. Carbon tetrachloride in combination with ultrasound has been widely used as a promising accelerator for the decomposition of several pollutants [9, 17, 43, 60, 88, 93, 104]. In addition, CCl_4 shows an efficient improvement of the different dyes degradation due especially to its high volatility and scavenging effect for hydrogen atoms [35, 98, 102, 103].

The present chapter aims principally to the discussion of the enhancing mechanism of the different dyes sono-degradation in the presence of CCl_4. Therefore, a detailed discussion of all available experimental studies was conducted upon the CCl_4-induced intensification effect. In the second part, a detailed theoretical mechanism of CCl_4 pyrolysis within bubbles was discussed, where the different species formed through the CCl_4 pyrolysis are analyzed to elucidate the main substances responsible of the intensification role of CCl_4.

2 Ultrasound/CCl_4: Experiential Outcomes

Table 1 is a summary of the major works done for the degradation of dyes using ultrasound/CCl_4 process. Globally, consequences of using Ultrasound/CCl_4 process are practically the same: more effective degradation efficiency in shorter operation time. The degradation mechanism passes through the reactive chlorine species (RCS: $^{\bullet}CCl_3$, $:CCl_2$, $^{\bullet}Cl$, Cl_2, and HOCl) yielded from the pyrolysis of CCl_4 inside acoustic bubbles. The specific contribution of each oxidant in the overall degradation rates of dyes was not reported. Mostly, exponentially decays dyes concentration was shown under the Ultrasound/CCl_4 system, providing a first-order reaction law. On the other hand, the degradation rate was strongly sensitive to the operation parameters. Generally, best degradation performances were obtained at higher levels of delivered power, frequency, solution temperature, and lower solution pH. An optimum CCl_4 concentration was reported in some cases. All of these findings will be addressed in depth in the sections that follow.

Merouani and Hamdaoui [73] have recently published a fascinating review on the degradation of textile dyes by ultrasound. In general, dyes are extremely water soluble substrates with low vapor pressure. They are nonvolatile chemicals that are likely destroyed by hydroxyl radicals at the outside of the collapsing bubble under sonication alone [73]. However, the local reaction zone at which degradation occurred may be moved between the bulk solution and the interfacial area, depending on several experimental parameters [73]. The concentration of ultrasonically produced OH in the bulk solution is in the whole modest. The highly hooted bubbles may create an appropriate environment for producing extra-radicals and oxidizing species via CCl_4 pyrolysis, which could improve the degradation of hydrophilic dyes.

The majority of the sonochemical degradations of dyes in the presence of CCl_4 are performed under a fixed ultrasound frequency (between 20 and 1700 kHz, Table 1).

Table 1 Sample of experimental works focusing on the intensification of dyes sono-degradation by addition of carbon tetrachloride (*Abbreviation*: *NI:* not indicated, P_c: calorimetric power, P_a: acoustic power, I_n: acoustic intensity, *f:* frequency of ultrasound, *V:* solution volume, T_{liq}: liquid temperature)

n°	Dye	Environmental conditions	Significant results	References
1	Methyl orange (MO)	$f = 20$ kHz-pulsed mode, $P_a = 3.2$ W, ($I_n = 11.32$ W/cm^2), V = 5 mL. $T_{liq} = 14 \pm 1$ °C, Gas: air $[CCl_4]_0 = 0$–49.2 mg L^{-1} $[MO]_0 = 5$–20 mg L^{-1} pH ~ 2–8.25	– The dye decolorization was observed to behave as a pseudo-first-order reaction in kinetics under all conditions – Under appropriate conditions, the rate constant was able to be increased by more than 100 times by adding CCl_4 to the MO solution – The elimination of MO decreased with the increase in its concentration or the solution pH – Degradation of MO is mainly due to •Cl radicals and HClO formed out of CCl_4 sonication, and less to •OH radicals	[99]
2	Rhodamine 6G (Rh6G)	$f = 50$ kHz, $P_a = 31$ W V = 2 L, T_{liq} = NI, Gas = NI, $[CCl_4]_0 = 0.5$–1 g L^{-1} $[Rh6G]_0 = 10$ mg L^{-1} pH = 12.5	– Extent of decolorization (%) increased by 12 and 22% with the addition of 0.5 and 1 g/L of CCl_4 – Use of UV + CCl_4 (1 $_g$/L) increased the decolorization extent of Rh6G to 23.5% compared to 12% in absence of CCl_4 (UV + US)	[3]

(continued)

Therefore, the experimental studies investigating the optimal ultrasound frequency for the maximal degradation of dye are scarce. Ghodbane and Hamdaoui [33] investigated the decolorization of acid blue 25 (AB25) at two frequencies, 1700 and 22.5 kHz, in the presence and absence of CCl_4. This investigation revealed that the best decolorization is obtained at 1700 $_{kHz}$ both with and without the addition of CCl_4.

Table 1 (continued)

n°	Dye	Environmental conditions	Significant results	References
3	Methyl orange (MO)	*f* = 50–60 kHz. P_a = 4.18–1.858 W, V = 100 mL, T_{liq} = 24 °C, Gas: air, CCl_4: 0.02_0.2% (v/v) μL, $[MO]_0$ = 50 mg L^{-1}, pH = NI	– Addition of 0.1 and 0.2% of CCl_4 increases the sono-decomposition of methyl orange (MO) by 1.66 and 6.33 times, respectively – At 0.02% CCl_4, a decrease in 33% in MO decomposition is observed	[2]
4	Acid bleu 25 (AB25)	*f* = 1700 and 22.5 kHz P_a = 14 W, V = 100 mL T_{liq} = 20 ± 1 °C, Gas = NI, $[CCl_4]_0$ = 100–798 mg L^{-1}, $[AB25]_0$ = 50 mg L^{-1}, pH = 1–11.8	– Bleaching rate increases with rise of CCl_4 concentration and decrease in liquid temperature – Decolorization is improved in acidic field (pH ~ 3.2–5.7) – Degradation of AB25 goes up with the increase in ultrasound frequency (22.5–1700 kHz), either with or with addition of CCl_4 – In presence of CCl_4, the decomposition of AB25 is mainly due to the chlorine-containing radicals, HOCl and Cl_2	[33]
5	Methyl orange (MO)	f_1 = 45 kHz, f_2 = 200 kHz, P_{c1} = 40 W, P_{c2} = 86 W, V_1 = 1500 ml, V_2 = 60 ml, T_{liq} = 20 ± 2 °C, Gas = air, $[CCl_4]_0$ = 0–250 mg L^{-1}, $[MO]_0$ = 30 μM pH = 2 and 6	– The dye sono-degradation is enhanced in acidic medium (pH = 2) – More enhancement of MO decolorization is observed at high-intensity ultrasound irradiation – The dye degradation increases with the rise of CCl_4 concentration -(0 to 250 mg/L)	[84]

(continued)

Table 1 (continued)

n°	Dye	Environmental conditions	Significant results	References
6	Acid orange 8 (AO8)	f = 300 kHz, I_n = 1.98–0.19 W/cm^2, V = 100 ml, T_{liq} = 20 ± 2 °C Gas = air + bubbling Ar $[CCl_4]_0$ = 0–5200 μM $[AO8]_0$ = 30–100 μM pH = 6.34	– The dye bleaching goes up with the rise of $[CCl_4]_0$, however, the rate of this degradation is amortized for high concentration of CCl_4 – The degradation of AO8 is mainly due to the action of $^{\bullet}OH$ radicals and reactive chlorine species (RCS)	[34]
7	Rhodamine B (RhB)	f = 300 kHz, P_a = 60 W, V = 300 ml T_{liq} = 25 °C, Gas: air $[CCl_4]_0$ = 50–200 mg/L $[RhB]_0$ = 5 mg/L pH = 5.3	– The addition of 50, 100, and 200 mg/L of CCl_4 increases the initial decomposition rate of RhB by 2.4, 4.6, and 21 times, respectively – The bleaching of RhB is due to the action of reactive chlorine species (HOCl, Cl_2, $\bullet CCl_3$,$:CCl_2$, and $\bullet Cl$)	[74]
8	Malachite green (MG)	f = 20 kHz, P_a = 5–15 W V = 1 L, T_{liq} = 25 °C Gas: air $[CCl_4]_0$ = 0–4 ml L^{-1}, or 0–0.4% (v/v) $[MG]_0$ = 27.5 μM pH = NI	– The presence of CCl_4 enhances the sonolytic degradation of malachite green, with the existence of an optimum dose for CCl_4 – The same efficiency was obtained for sonolysis and sonophotocatalyzed processes in presence of CCl_4	[7]

(continued)

In the absence of CCl_4, the increase in wave frequency from 22.5 to 1700 kHz causes the initial decolorization rate to be increased from 0.0418 to 0.1467 mg L^{-1} min^{-1}, respectively. This means that the decomposition rate of AB25 is increased by 3.5 times as the ultrasound frequency goes up from 22.5 to 1700 kHz. In the presence of CCl_4 (399 mg/L), this initial rate of AB25 decomposition increased from 3.65 to 15.6 mg L^{-1} min^{-1} as the frequency augmented from 22.5 to 1700 kHz, respectively. This means that in presence of CCl_4, the increase in ultrasound frequency from 22.5 to 1700 kHz results in an increase in four times of the initial rate of AB25 decolorization. The same observation was obtained by [84], where the rise of wave

Table 1 (continued)

n°	Dye	Environmental conditions	Significant results	References
9	Methyl orange (MO)	$f = 1.6$ MHz, $P_c = 17.1$ W, V = 90 ml $T_{liq} = 20$ °C, Gas: air $[CCl_4]_0 = 100$ mg/L $[MO]_0 = 10$ mg/L pH = 2	– Addition of 100 mg/L CCl_4 doubles the rate constant of methyl orange sono-degradation – Increase in solution volume from 90 to 100 ml reduces the removal efficiency of MO by ~ 15% in the presence of CCl_4	[97]
10	Methylene blue (MB) and methyl orange (MO)	f = 200 kHz $P_c = 16$ W V = 60 ml $T_{liq} = 10$–40 °C Gas = Ar $[CCl4]_0 = 0$–8.6 mM $[MO]_0 = 150$ μM $[MB]_0 = 150$ μM pH = 6,3	– The sono-degradation rate MB increases with the decrease in liquid temperature or with increasing CCl_4 concentration (0–8.6 μM) – For the same $[CCl_4]_0$, the degradation rate of methyl orange (MO) is greater than that of methylene blue (MB) – The impact of CCl_4 on the degradation rate of MO and MB depends on the hydrophobicity of these species as well as their chemical reactivity	[98]
11	Reactive brilliant red K-2BP (BR K-2BP)	$f = 20$ kHz $P_a = 100$–400 W V = 50 ml, $T_{liq} =$ 20–50 °C. Gas: air $[CCl_4]_0 = 0$–1 ml/L, or 0–0.1% (v/v), $[K\text{-}2BP]_0 =$ 10–40 mg/L pH = 2–11	– The dye was enhanced by addition of CCl_4, due to the generation of $^{\bullet}OH$ radicals in addition to free chlorine and chlorine-containing radicals – The bleaching of K-2BP goes up with decreasing liquid temperature, initial solution pH, initial concentration of the dye, and increasing acoustic power	[101]

(continued)

Table 1 (continued)

n°	Dye	Environmental conditions	Significant results	References
12	Of methyl orange (MO)	f = 20 kHz-pulsed mode $P_c = 3.2$ W, V = 5 ml $T_{liq} = 14 \pm 1$ °C Gas: NI $[CCl4]_0 = 0–20$ mg/L $[MO]_0 = 10$ mg/L pH = 2–8	– The dye degradation increased with the increase in CCl_4 concentration – The maximal decomposition of MO is obtained in acidic medium (pH = 2) – The elimination extent goes up gradually with $[MO]_0$ increase in its concentration, and then amortized from 12 mg/L MO	[64]

frequency from 45 to 200 kHz increases the decomposition yield of methyl orange (MO) from 16 (in 10 min) to 54% (in 10 s), respectively in the presence of 150 mg/L CCl_4. Similar trends exerted by the high-frequency ultrasound are obtained from the degradation of other organic contaminants in the absence of CCl_4 [4, 48, 51], with the existence of an optimum frequency for some of them [4, 42, 48].

The optimum frequency for sonochemistry depends on several factors: temperature and pressure inside the collapsing bubbles, number and distribution of bubbles, initial size and lifetime of acoustic cavities, cavitation threshold, dynamics and symmetry (shape) of bubbles, effect of organic additives on bubble temperature, effect of intermediates and products, etc. [52, 84]. In general, the increase in wave frequency reduces the maximum temperature reached at the end of bubble collapse [22, 25, 113]. In addition, the cavitation threshold goes up with frequency, which creates narrower regions of active bubbles in the sonicated field [21, 58]. The synergy of all these factors leads to reduce the chemical effects. On the other hand, the acoustical events (i.e., number of bubbles) are increased due to the shorter acoustic period (i.e., decrease in wavelength at higher frequency). Moreover, the lifetime of bubbles becomes shorter with increasing frequency [87]. This means that more radicals formed within a bubble can escape from the acoustic cavitation to react with any other species in the medium. These factors cause the chemical effects to be increased. The combination of all these factors explains clearly the results obtained previously by *Ghodbane and Hamdaoui* (*best frequency:* 1.7 MHz) (2009) and *Okitsu et al.* (*optimum frequency*: 200 kHz) (2008). According to the available data, it is worth mentioning that there was no determination of the optimum frequency for a maximal degradation of dyes in the presence of CCl_4, conversely to the optimum frequency of sono-production of H_2O_2 in the absence of CCl_4, which is situated between 200 and 600 kHz [5, 27, 47, 50, 66, 76, 87, 88].

2.1 Effect of Acoustic Intensity

Bejarano-Pérez and Suarez-Herrera [7] found that in the presence of CCl_4, the bleaching rate of malachite green (MG) is increased linearly with the rise of ultrasonic power (5–15 W), where the effect of acoustic power is more significant above 8 W. The same behavior was observed during the degradation of other contaminants, e.g., Rhodamine B [6] and ibuprofen [69], in the absence of CCl_4. Additionally, a similar trend was observed by [101] during the bleaching of azo dye K-2BP in the presence of CCl_4, whereas an optimum of 200 W was obtained for a maximal degradation of K-2BP. Therefore, more increase in acoustic power leads to a reduction of K-2BP removal. A similar behavior has been observed in other studies [1, 35, 61, 104]. It should be noted that the applied acoustic amplitude should be accurately controlled for a maximal energy efficacy.

The rationale for the acoustic intensity effect is the same whether CCl_4 is present or not. An increase in the overall sonochemical effects was almost obtained with raising the acoustic power between the cavitation threshold and a value above with the sound wave is attenuated [10, 20, 26, 27, 36, 71, 74, 90, 91, 95, 100]. The favorable effects of high power may be explained by the influence of this variable on the single bubble yielding and the number of bubbles [27, 79]. As the acoustic power increases, so do the cavity expansion and compression ratios enabling to increase the volume of entrapped water vapor and CCl_4 and also the bubble temperature during the last step of the collapse [21, 22, 77, 80]. As a result, at higher applied sonic intensities, a greater amount of free radicals may be generated, resulting in a greater sonochemical impact within and around the bubble. Furthermore, as the sonic intensity is increased, the number of cavitation bubbles rises [11, 27, 75]. Consequently, increasing the intensity might result in better sonochemical outcomes.

2.2 Effect of CCl_4 Concentration

Generally, it has been confirmed that the addition of CCl_4 in the sonicated solution enhances impressively the decolorization of the target dye, whereas the extent of this process is very dependent on the operational conditions and species properties and dosages. Ghodbane and Hamdaoui [33] have observed at 1700 kHz that the addition of 798 mg L^{-1} CCl_4 increases the initial rate of acid blue (AB25) decolorization to 17.32 mg L^{-1} min^{-1} as compared to that obtained when CCl_4 is absent, i.e., 0.146 mg L^{-1} min^{-1}. This means that the presence of CCl_4 augments the rate of AB25 bleaching by 118-fold. Similarly, [99] reported that the sonolytic decolorization of methyl orange (MO) at 20 kHz behaves as a pseudo-first-order reaction in kinetics either in the absence or presence of carbon tetrachloride. In addition, the rate constant of MO decolorization increased linearly with rising of CCl_4 concentration, where the addition of 49.2 mg L^{-1} of CCl_4 causes the apparent rate constant of MO elimination to be increased from 0.004 to 0.635 min^{-1}. This means that the addition of CCl_4 result in a 160-fold increase in the rate constant of MO decolorization. In addition, [84] indicated that at low-intensity ultrasonic irradiation, the degradation rate of MO is enhanced 14 times by the addition of 250 mg L^{-1} of CCl_4, while at high-intensity irradiation, the rate goes up to 100 times by the addition of 150 mg L^{-1} of CCl_4. A similar tendency has been obtained by [74] at 300 kHz, [101] and [64] at 20 kHz for the degradation of Rhodamine B (5 mg L^{-1}), azo dye K-2BP (20 mg L^{-1}) and methyl orange (10 mg L^{-1}), respectively, in the presence of CCl_4.

On the other hand, [2] indicated that the addition of 100 and 200 mg L^{-1} CCl_4 enhances the rate constant of methyl orange (MO) decomposition by 1.7 and 6.3 times, respectively. Uddin and Okitsu [98] reported that the addition of 2.6 and 8.6 mM of CCl_4 increases the rate of sono-degradation of methylene blue (MB) by 1.4 and 1.7 times, respectively. Moreover, this rate increase goes up to 9.6 times for the degradation of methyl orange (MO) in the presence of 2.6 mM CCl_4. Similarly, [3] found that the rate constant of Rhodamine 6G (Rh6G) degradation is increased to 1.82×10^{-3} min^{-1} in the presence of CCl_4 (1 g L^{-1}) compared to the case where only ultrasonic irradiation is used ($k = 5 \times 10^{-4}$ min^{-1}). This means that the addition of CCl_4 at 1 g L^{-1} improves the sono-degradation of Rh6G by 3.6 times. The same trend is found by Gultken et al. [34] through the decomposition of the textile dye C.I. Acid Orange 8 (30 μM) at 300 kHz, where the addition of CCl_4 at 2070 μM increases the apparent bleaching rate constant to 0.245 min^{-1} compared to that obtained in the absence of CCl_4 ($k = 0.0648$ min^{-1}). This means an increase in 3.8 times of the apparent rate constant of AO8 decolorization in the presence of CCl_4. Thangavadivel et al. [97] reported that the presence of 100 mg L^{-1} CCl_4 doubles the sono-degradation rate constant of methyl orange (10 mg L^{-1}).

Observing these results, we can clearly deduce the existence of some disagreement in the degradation extents; this can be attributed to the different adopted operational conditions such as ultrasound frequency and power, volume of solution, geometry of sonicated vessels and solution pH, etc. Moreover, it is worth mentioning that the concentration of CCl_4 should be optimized for a maximal sono-degradation of dye and in the same time to avoid any residual CCl_4 in the final discharge effluent stream. This exigency has been confirmed by Gultken et al. [34] during the degradation of C.I. Acid Orange 8 (30 μM), where it has been observed that for a CCl_4 concentration greater than or equal to 385 mg/L, the decolorization rate constant of AO8 is hardly increased. This behavior was attributed to increasing degree of competition for $^{\bullet}OH$ radicals and other oxidative species via the accumulation of oxidation/decomposition by-products. The same observation was inferred by [14] for the optimization of chloroalkanes concentrations. Bejarano-Pérez and Suarez-Herrera [7] indicated that the increase in CCl_4 amount causes the degradation rate of malachite green (MG) to be rapidly increased, stabilized at about 1 ml L^{-1} CCl_4 (0.1%, v/v), and then drastically decreased for larger volumes of CCl_4. The last step was explained according to three possibilities: (i) Scattering of the ultrasound energy, i.e., and reduction of bubble yield, in presence of large amount of CCl_4, (ii) reduction of the adiabatic index of bubble in the presence of excess of CCl_4 by evaporation and therefore the total yield of cavitation goes down, (iii) scavenging of oxidizing agents and obstructing the reaction between them and MG through the large amount of intermediaries resulting from the decomposition of CCl_4, especially at high concentrations. The same conclusion was inferred by [101] during the degradation of azo dye K-2BP (20 mg L^{-1}) with an optimum of 0.03 ml CCl_4, i.e., 0.003% (v/v).

Even when sonochemical processes is combined to other advanced oxidation process, the addition of CCl_4 to these techniques has showed an interesting improvement for dyes removal. *Banerjee*'s group [3] observed that the addition of 1 g L^{-1} CCl_4 to the sonophotolytic (UV + US) process enhances the extent of Rh6G decolorization to 23.5% compared to 12% in the absence of carbon tetrachloride. Therefore, the kinetic rate constant goes up to 2.1×10^{-3} min^{-1} (UV + US + CCl_4), which is about four times as compared to that obtained using only ultrasonic irradiations ($k = 5 \times 10^{-4}$ min^{-1}). Similarly, [7] reported that the addition of 0.5 ml/L CCl_4 increases the degradation rate of malachite green (27.5 μM) about 2.5 times, no matter if degradation is performed via sonochemical or sonophotocatalytic reaction, compared to the case in the absence of CCl_4. In addition, [7] demonstrated that in presence of CCl_4, the degradation extent of MG is approximately the same for sonochemical and sonophotocatalytic processes, which indicate that the sonochemical reaction is the only cause for the raising on MG degradation rate values. This is because no significant synergistic effect was observed between sonolysis and photocatalysis. In spite of that, *Bejarano-Pérez*'s team [7] indicated that the synergetic effect between sonolysis and photocatalysis in the presence of CCl_4 appears in terms of the elimination of some undesired reaction intermediaries that can be equally or more harmful than the original pollutant. This finding has been corroborated via UV–Vis spectra obtained from sonochemical and sonophotocatalyzed degradations of MG. It should be noticed that the presence of bicarbonate would reduce the enhancing effect of CCl_4 through the consumption of hydroxyl radicals. This result was confirmed according to the work of [101] during the bleaching of azo dye K-2BP in the presence of CCl_4 and SO_4^{2-}, HCO_3^{-}, or NO_3^{-} anions.

2.3 *Effect of Dyes Concentration*

In general, it is observed that the promoting effect of CCl_4 is further enhanced at lower initial dyes concentration. Wang et al. [99] reported that the decrease in methyl orange (MO) concentration from 20 to 5 mg L^{-1} increases the rate constant of its degradation from 0.135 to 1.181 min^{-1} with a factor of 8.8 times. This means that the increase in MO concentration has approximately the same effect of decreasing CCl_4 concentration. In another study, [101] observed the same tendency for the degradation azo dye K-2BP (20–40 mg L^{-1}) in the presence of CCl_4 (0.06%, v/v). The same observation was obtained by [3] for Rhodamine 6G, Gultken et al. [34] for C.I. Acid orange 8 and [6] for Rhodamine B, in the absence of CCl_4. In addition, the negative effect of concentration rising was observed in other techniques of dye decolorization such as photocatalytic oxidation, e.g., [49, 83].

On the other hand, an opposite behavior was observed by [64] during the decolorization of methyl orange, MO (6–16 mg L^{-1}), in the presence of 20 mg/L of CCl_4, where the degradation extent of MO goes up gradually with the increase in MO concentration, whereas, from 12 mg/L MO, the degradation extent of methyl orange is hardly improved. This result is in line with those obtained by Ferkous et al. [28, 29] for naphthol blue black, Méndez-Arriaga et al. [69] for ibuprofen and [16] for phenol, 4-isopropylphenol, and Rhodamine B, in the absence of CCl_4. This behavior was attributed to the increased probability of reactive species attack on pollutants molecules with increasing of their concentrations. The difference observed between *Luo*'s study and the other works, conducted in the presence of CCl_4, discussed in this part is probably due to the different operational conditions such as acoustic intensity and frequency, volume, and geometry of reactor and species dosages, etc.

2.4 Effect of Solution pH

As the solution pH is changed, the interaction of dyes with acoustic bubble is complicated according to the physicochemical properties of the analyzed dye. Therefore, the sono-degradation of dye in the presence or absence of CCl_4 is either accelerated or amortized; depending on the solution pH. Wang et al. [99] and Luo et al. [64] found that the sono-elimination of methyl orange (10 mg L^{-1}) is increased with decreasing of solution pH value (i.e., best degradation is at pH ~ 2). The same observation is made by [84] at 45 kHz, where the decomposition of MO increased by the decrease in solution pH from 6 to 2 in the presence of CCl_4 at low ultrasonic irradiation. This was explained by the prevailing neutral form of methyl orange, that is poorly hydrophilic compared to the dissociated anions form, at acidic pH, which is easily accumulated at the gas–liquid interface, and attacked by $^{\bullet}Cl$ radicals and other oxidizing species such as HOCl. In the pH range from 2 to 11, [101] found that the maximal degradation rate of azo dye K-2BP is obtained at a pH 2. Whereas, *Wang*'s team (2011) attributed the fast decomposition of K-2BP in acidic medium to the plausible reduction of oxidation potential of oxidizing agents (such as Cl_2, HOCl, and •OH radicals) with increasing the solution pH. Moreover, this may also stem from higher pH creating more free radical scavengers, like carbonate and bicarbonate ions, and decreasing •OH amount. In general, the same trend of increased degradation with pH reduction is obtained for other dyes in the absence of CCl_4, e.g., Behnajady et al. "Rhodamine B"[6], Ince et al. "C.I Acid orange 7 and C.I Reactive orange 16" [45], and Hinge et al. "RhB Rhodamine B and Rhodamine 6G" [41]. Conversely, [33] have obtained a relatively different behavior through the degradation of acid blue 25 (AB25) in the presence of CCl_4, where three regimes were observed: (i) a low decolorization rate of AB25 is observed at pH 1, (ii) in the pH range 3.2–9, higher level of AB25 degradation is obtained compared to that of pH 1, and (iii) the decolorization rate of AB25 is drastically decreased as the solution pH is increased to 11.8. These findings have been explained according to the effect of pH upon AB25, where at pH 1, more AB25 is accumulated in the liquid/gas interface owing to the hydrophobic character of the molecule resulting from the protonation of $-SO_3^-$ group, which reduces the decomposition of CCl_4 as well as the degradation of AB25. At pH 3.2–9.3, the decolorization of AB25 is enhanced because of the improvement of the hydrophilicity of AB25, which implies that degradation is a bulk solution event. In addition, the enhanced bleaching of AB25 is probably due to the acidic pH, where chlorine exists under the hypochlorous acid (HOCl, E° = 1.49 V) form rather than hypochlorite (OCl^-, E° = 0.94 V). At pH 11.8, the rapid decrease in AB25 degradation is owing to the improvement of AB25 hydrophilicity and the decrease in chlorine/hypochlorite production as a result to the formation of chlorite and perchlorate which react slowly with AB25.

2.5 *Effect of Liquid Temperature*

[33] reported that the elimination of acid blue 25 decreased with the rise of liquid temperature between 20 and 50 °C in the presence of CCl_4 at 399 mg L^{-1}. This was ascribed to the increase in solvent vapor pressure inside the acoustic cavitation as the liquid temperature is increased; therefore, the collapse intensity is attenuated as well as the decomposition of CCl_4 and the generation of oxidizing agents, which implies the reduction of AB25 decolorization. In addition, the increase in the bulk temperature impedes the formation of bubbles, which means that the number of active bubbles is reduced, this phenomenon has been confirmed theoretically by Merouani and Hamdaoui [70] through the study of size distribution of active bubbles for the production of hydrogen. It should be noted that the same temperature effect observed by [33] is obtained by [98, 101] during the sono-degradation of methylene blue and azo dye K-2BP, respectively, in the presence of CCl_4. However, the negative effect of the bulk liquid temperature cannot be a generalization, because in some cases the increase in the liquid temperature may have a promoting impact on the degradation of the different species [29, 74, 81]. Recently, [13] have indicated the direct connection of the liquid temperature effect to the initial concentration of Toluidine blue dye at high-frequency ultrasound (1.7 MHz), where this dependency would be closely related to the different reaction zones at low and high initial contaminant concentration. Additionally, *Chadi's* study shows an optimum of 50 °C for a maximal decomposition of Toluidine blue (TB) in the absence of CCl_4. The comparison of *Chadi*'s results to those obtained by [33, 98, 101] shows the necessity of more investigations in order to clearly explore the interconnections and interferences existing between the different operational parameters affecting the sonochemical efficacy.

3 Decomposition Mechanism of Dyes in the Presence of CCl_4

The experimental work conducted by [99] showed that the sono-destruction of methyl orange is difficult in the presence of ultrasonic irradiation alone. However, the addition of CCl_4 can greatly enhance the decolorization of the dye by more than 100 times in appropriate conditions. The experimental study of Wang's team [99] indicates that the promoting effect of CCl_4 is mainly due to HOCl, $^{\bullet}Cl$ radicals and then to $^{\bullet}OH$ radicals to a less extent. This was confirmed after the addition of 1 and 10 mM tert-butanol, i.e., $^{\bullet}OH$ scavenger, where the rate constant of MO degradation goes down by 1.4 and 15.64%, respectively, compared to the case where tert-butanol is absent. Moreover, the decrease in solution pH from 6.4 to 4.0 after a complete decolorization of MO confirms the formation of HOCl and HCl. Similarly, [84] suggested the plausible attack of $^{\bullet}Cl$ radical or HOCl on MO molecules in the presence of CCl_4, where the contribution of these chlorine species is dominant over that of $^{\bullet}OH$ radicals. This is because only 10% of MO (30 μM) bleaching is attained in 60 min at 200 kHz in the absence of CCl_4 (2008). Identical behaviors have been indicated by [2, 3] in the presence of CCl_4 for the degradation of Rhodamine 6G (Rh6G) and methyl orange (MO), respectively. However, no more experimental investigations were conducted by [2, 3, 84] in order to determine the real mechanism responsible of sono-degradation of Rh6G and MO in the presence of CCl_4, i.e., either by reactive chlorine species ($^{\bullet}CCl_3$,:CCl_2, HOCl, $^{\bullet}Cl$, etc.) or by reactive oxygen substances. In contrast, Gultken et al. [34] pointed out to the important contribution of $^{\bullet}OH$ radicals in addition to chlorine species, where 85% bleaching of C.I. Acid orange 8 (30 μM) was accomplished in 30 min without CCl_4, while in the presence of CCl_4 (2070 μM), this percentage increases to 99%. Wang et al. [101] drew the same conclusion through the degradation of azo dye K-2BP in the presence of CCl_4 (0.2–0.6 ml L^{-1}). *Wang*'s team [101] found that the concentration of hydroxyterephthalte, i.e., a fluorescent molecule, increased proportionally with the increase in CCl_4 amount in the interval 0.01–0.03 mL, i.e., 0.02–0.06% (v/v). This was a corroboration of the enhanced generation and role of $^{\bullet}OH$ radicals in the presence of CCl_4 through hydrogen atoms scavenger. Furthermore, [101] found that the addition of ethanol at 0–0.4 ml (0.8%, v/v) in the presence of 30 μL (0.06%, v/v) of CCl_4 decreases gradually the degradation rate of K-2BP to a minimum of 38%, indicating that at this degradation level is only oxidized through chlorine-containing radicals and free chlorine.

On the other hand, [33] indicate that the bleaching mechanism of Acid blue 25 in the presence of CCl_4 is mainly due to the action of chlorine-containing radicals (•CCl_3, :CCl_2, and •Cl), HOCl and Cl_2. This was justified by the use of tert-butanol, where the bleaching rate of AB25 decreased by 58, 89, and 92% for the use of 399, 798, and 1570 mg L^{-1} of tert-butanol, respectively. The same degradation mechanism has been suggested by [74] for the decolorization of Rhodamine B at 300 kHz and [98] for the bleaching of methylene blue and methyl orange. In addition, [98] concluded that the reactivity of chlorine radicals and chlorine species is the controlling parameter for the sono-degradation of methylene blue and methyl orange in the presence of CCl4.

It should be noticed that for the majority of discussed studies in this section, a pH drop was registered during the decolorization of the different dyes; this is mainly due to the generation of HOCl and HCl molecules from the decomposition of CCl_4. In addition, this pH drop is accelerated with the increase in CCl_4 concentration or time of sonication. For example, [33] found that the drop of pH is increased with the increase in CCl_4 dosage between 100 and 798 mg L^{-1}, while Gultken et al. [34] indicated the decrease in solution pH with increase in the sonication time from pH 6.05 at 1 min to pH 2.92 at 6 min, in the presence of 2070 μM of CCl_4.

The analyzation of the different results discussed in this section reveals the existing controversies about the real mechanism involved during the sono-degradation of various dyes. These explanations stem probably from the different adopted operational conditions (i.e., frequency, intensity, volume and geometry of reactor, solution pH, dyes, CCl_4 dosages, etc.), dyes nature, physical properties, and the techniques used for the confirmation of the degradation path of dyes. Consequently, a serious work is needed for an accurate determination of the real mechanism involved during dyes sono-decomposition in the presence of CCl_4.

4 Theoretical Section

This section is principally based on the theoretical works of Dehane et al. [23, 24] in which a detailed computational analysis has been conducted in order to understand the efficiency enhancement of the different nonvolatile contaminants sono-degradation using carbon tetrachloride. In this part of chapter, we focus upon the effect of acoustic intensity and CCl_4 concentration on the generation of oxidizing species formed through the pyrolysis of CCl_4 within the acoustic cavitation. In *Dehane*'s research [23, 24], the thermal decomposition of CCl_4 is followed during the bubble oscillation under ultrasonic irradiation, where transfer phenomena, i.e., mass transport and heat exchange, and reactions heat are considered in the bubble dynamics model.

4.1 Chemical Scheme

Bubble chemistry is known to be considerably complicated by the thermal decomposition of CCl_4; as a result, numerous species are produced at the end of bubble collapse. Theoretically, a colossal number of chemical reactions is suspected for the formation of the different elements from CCl_4 pyrolysis. However, considering the small volume of the bubble and its short time of collapse, a limited number of chemical reactions is expected to take place during bubble collapse. The chemical mechanism of Table 2 [23, 24] is based principally on the experimental results of [43], in addition to other works [15, 32, 44, 46, 106], where the principal products

Table 2 Reactions scheme describing CCl_4 pyrolysis in collapsing argon bubble [23, 24]. M is the third Body. Subscript "f" denotes the forward reaction and "r" denotes the reverse reaction. A is in (m^3/mol s) for two body reaction [(m^6/mol^2 s) for a three-body reaction], and E_a is in (KJ/mol) and ΔH in (KJ/mol)

n°	Reaction	A_f	n_f	E_{af}	A_r	n_r	E_{ar}	ΔH
1	$H_2O + M \rightleftharpoons H^\bullet + {}^\bullet OH + M$	1.912×10^7	−1.83	28.35	2.2×10^{10}	−2.0	0.0	508.82
2	${}^\bullet OH + M \rightleftharpoons O + H^\bullet + M$	9.88×10^{11}	−0.74	24.43	4.714×10^6	−1.0	0.0	436.23
3	$O + O + M \rightleftharpoons O_2 + M$	4.515×10^{11}	−0.64	28.44	6.165×10^3	−0.5	0.0	505.4
4	$H^\bullet + O_2 \rightleftharpoons O + {}^\bullet OH$	1.915×10^8	0.0	3.93	5.481×10^5	0.39	-7.01×10^{-2}	69.17
5	$H^\bullet + O_2 + M \rightleftharpoons HO_2^\bullet + M$	1.475	0.6	0.0	3.09×10^6	0.53	11.7	−204.80
6	$O + H_2O \rightleftharpoons {}^\bullet OH + {}^\bullet OH$	2.97	2.02	3.21	1.465×10^{-1}	2.11	-6.94×10^{-1}	72.59
7	$HO_2^\bullet + H^\bullet \rightleftharpoons H_2 + O_2$	1.66×10^7	0.0	1.97×10^{-1}	3.164×10^6	0.35	13.3	−239.67
8	$HO_2^\bullet + H^\bullet \rightleftharpoons {}^\bullet OH + {}^\bullet OH$	7.079×10^7	0.0	7.06×10^{-2}	2.027×10^4	0.72	8.8	−162.26
9	$HO_2^\bullet + O \rightleftharpoons {}^\bullet OH + O_2$	3.25×10^7	0.0	0.0	3.252×10^6	0.33	12.75	−231.85
10	$HO_2^\bullet + {}^\bullet OH \rightleftharpoons H_2O + O_2$	2.89×10^7	0.0	-1.19×10^{-1}	5.861×10^7	0.24	16.53	−304.44
11	$H_2 + M \rightleftharpoons H^\bullet + H^\bullet + M$	4.577×10^{13}	−1.4	24.98	1.146×10^8	−1.68	1.96×10^{-1}	444.47
12	$O + H_2 \rightleftharpoons H^\bullet + {}^\bullet OH$	3.82×10^6	0.0	1.9	2.667×10^{-2}	2.65	1.17	8.23
13	${}^\bullet OH + H_2 \rightleftharpoons H^\bullet + H_2O$	2.16×10^2	1.52	8.25×10^{-1}	2.298×10^3	1.40	4.38	−64.35
14	$H_2O_2 + O_2 \rightleftharpoons HO_2^\bullet + HO_2^\bullet$	4.634×10^{10}	−0.35	12.12	4.2×10^8	0.0	2.87	175.35
15	$H_2O_2 + M \rightleftharpoons {}^\bullet OH + {}^\bullet OH + M$	2.951×10^8	0.0	11.59	1.0×10^2	−0.37	0.0	217.89
16	$H_2O_2 + H^\bullet \rightleftharpoons H_2O + {}^\bullet OH$	2.410×10^7	0.0	9.5×10^{-1}	1.269×10^2	1.31	17.08	−290.93
17	$H_2O_2 + H^\bullet \rightleftharpoons H_2 + HO_2^\bullet$	6.025×10^7	0.0	1.9	1.041×10^5	0.70	5.74	−64.32
18	$H_2O_2 + O \rightleftharpoons {}^\bullet OH + HO_2^\bullet$	9.550	2.0	9.5×10^{-1}	8.66×10^{-3}	2.68	4.45	−56.08
19	$H_2O_2 + {}^\bullet OH \rightleftharpoons H_2O + HO_2^\bullet$	1.0×10^6	0.0	0.0	1.838×10^4	0.59	7.4	−128.67
20	$CCl_4 \rightleftharpoons {}^\bullet CCl_3 + {}^\bullet Cl$	3.236×10^{22}	−2.29	288.9	3.388×10^{10}	−1.29	−5.9	288.245

(continued)

Table 2 (continued)

n°	Reaction	A_f	n_f	E_{af}	A_r	n_r	E_{ar}	ΔH
21	$^{\bullet}CCl_3 \rightleftharpoons :CCl_2 + {}^{\bullet}Cl$	1.09×10^9	0.0	228.1	1.0×10^4	0.0	0.0	281.874
22	$2^{\bullet}CCl_3 \rightleftharpoons C_2Cl_6$	4.55×10^6	−1.6	1.5	3.83×10^{26}	−4.5	288	−304.4
23	$2^{\bullet}Cl + M \rightleftharpoons Cl_2 + M$	1.0	0.23	−45.9	3.981×10^7	0.0	198.5	−242.604
24	$C_2Cl_6 + {}^{\bullet}CCl_3 \rightleftharpoons CCl_4 + C_2Cl_5$	7.94×10^5	0	59.9	2.95×10^2	0.8	39.5	34.167
25	$C2Cl5 + Cl_2 \rightleftharpoons C_2Cl_6 + {}^{\bullet}Cl$	2.04×10^5	0.0	9.9	3.51×10^{11}	−1.1	64.7	−79.81
26	$C_2Cl_5 + {}^{\bullet}Cl \rightleftharpoons C2Cl_4 + Cl_2$	2.45×10^7	0.0	0.0	6.37×10^9	−0.4	195	−184.502
27	$\bullet\ Cl + H_2 \rightleftharpoons HCl + H^{\bullet}$	94.56	1.72	12.837	55.99	1.72	9.09	4.4
28	$CCl_4 + H^{\bullet} \rightleftharpoons HCl + {}^{\bullet}CCl_3$	172	1.8	17.14	1.67×10^{-7}	3.72	155.36	−143.4
29	$^{\bullet}Cl + H2O \rightleftharpoons {}^{\bullet}OH + HCl$	1.17×10^2	1.67	63.84	4.12×10^{-1}	2.12	−5.37	65.514
30	$2C_2Cl_5 \rightleftharpoons C_2Cl_6 + C_2Cl_4$	8.13×10^5	0.0	59.03	3.023×10^2	0.76	38.63	−264.31
31	$Cl_2 + H_2O \rightleftharpoons HOCl + HCl$	64.56	0.72	10.8	43.1	1.03	8.7	73.775

of CCl_4 sono-pyrolysis under argon atmosphere are the reactive chlorine species, C_2Cl_4, C_2Cl_6, and HOCl. Water vapor dissociation and it related reactions are all included in this mechanism. Overall, 31 reversible chemical reactions, involving H_2O, $H^•$, O, $^•OH$, $HO_2^•$, H_2O_2, H_2, O_2, CCl_4, $^•CCl_3$, $:CCl_2$, $^•Cl$, Cl_2, C_2Cl_4, C_2Cl_5, C_2Cl_6, HCl, and HOCl are considered. The scheme in Table 2 includes 12 reversible reactions for Cl-containing species, which are the principal reactions reported in literature [1, 9, 17, 19, 43, 44, 46, 60, 88, 89, 93, 104, 108]. The scheme in Table 2 has been partially validated from shock-tube and reactor-type experiments [55, 82, 117] as well as RF thermal plasma reactor [53, 54], hydrogen flame studies [18], and discharge flow/resonance fluorescence technique [12].

Firstly, the generation of Cl-species starts with the thermal decomposition of CCl_4 molecules ($CCl_4 \rightleftharpoons {}^•CCl_3 + {}^•Cl$, reaction 20), which is more rapid as it is compared to the pyrolysis of water molecules (H–OH bond energy is 497 kJ/mol, Cl–CCl_3 is 288 $_{kJ}$/mol). In parallel, CCl_4 molecules react with hydrogen atoms and prevent their recombination with hydroxyl radicals ($CCl_4 + H^• \rightleftharpoons HCl + {}^•CCl_3$, reaction 28) [102, 116]. On the other hand, no chemical reaction is expected between CCl_4 and hydroxl aradicals, this was confrmed experimentally by [102] using terephthalic acid, in which formation of hydroxyterephthalate goes up with CCl_4 added. Additionally, the inhibition of levofloxacin [35] and p-nitrophenol [102] degradation with addition of t-bunatol and ethanol, respectively, confirms the promoting enhancing of CCl_4 towards the fotmation of •OH radicals. On another hand, hexachloroethane (C_2Cl_6) is formed though the self-reaction of trichloromethyl radicals ($2{}^•CCl_3 \rightleftharpoons C_2Cl_6$, Reaction 22) [43, 44]. Furtheremore, tetrachloroethylene is formed via hexachloroethane cleavage (Reactions 24–26 and 30) rather than from dichlorocarbene dimerisation, which is not promoted. Dichlorocarbene ($:CCl_2$) may be formed through the decomposition of trichloromethyl radicals ($•CCl_3 \rightleftharpoons :CCl_2 + {}^•Cl$, Reaction 21) as the sono-formation of$:CCl_2$ is detected experimentally under an argon atmosphere [43]. Molecular chlorine could be formed through the recombination of chlorine radicals (reaction 23 of Table 2) [32]. Hypochlorous and hydrochloric acids are generated according to reactions 27, 29, and 31 [15, 32].

4.2 *Bubble Activity in the Presence of CCl_4*

The effect of CCl_4 mole fraction, or its concentration in the bulk, upon the total yield of the bubble, the bubble temperature and the maximal conversion of CCl_4 has been analyzed over an acoustic intensity range from 0.7 to 1.5 W/cm^2, and under an ultrasonic frequency of 355 kHz [23]. Firstly, it has been found that the increase inin acoustic intensity affects positively the total production of the bubble, the peak temperature, and the maximal decomposition of CCl_4 over the whole range of CCl_4 mole fraction (0–0.1), as revealed in Figs. 1, 2 and 3. These results are in good concordance with several theoretical [27, 80, 113] and experimental [7, 40, 50, 107] works. Secondly, for each acoustic intensity, the total production of the bubble goes up with increasing CCl_4 concentration in the bulk liquid, this was observed

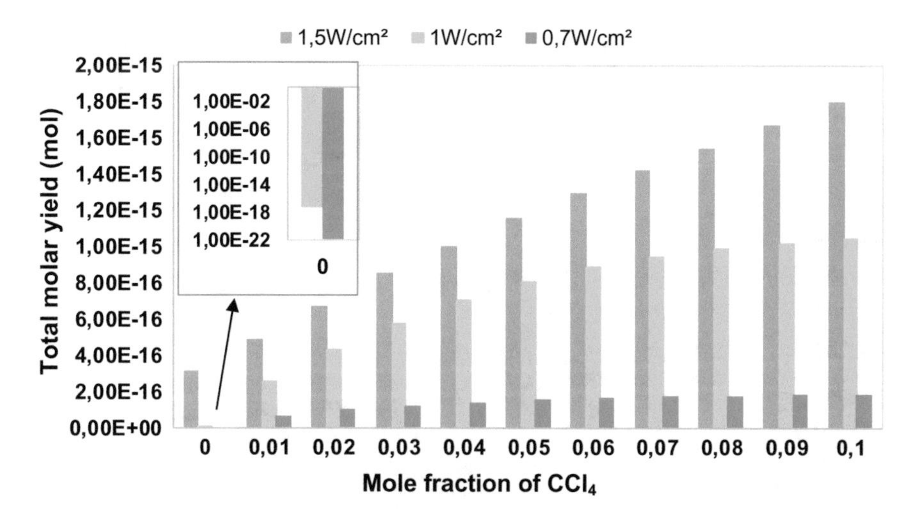

Fig. 1 Total production of all species created inside the bubble at the end of the first bubble collapse versus initial CCl_4 mole fraction inside the bubble (frequency: 355 kHz). Modified from [23]

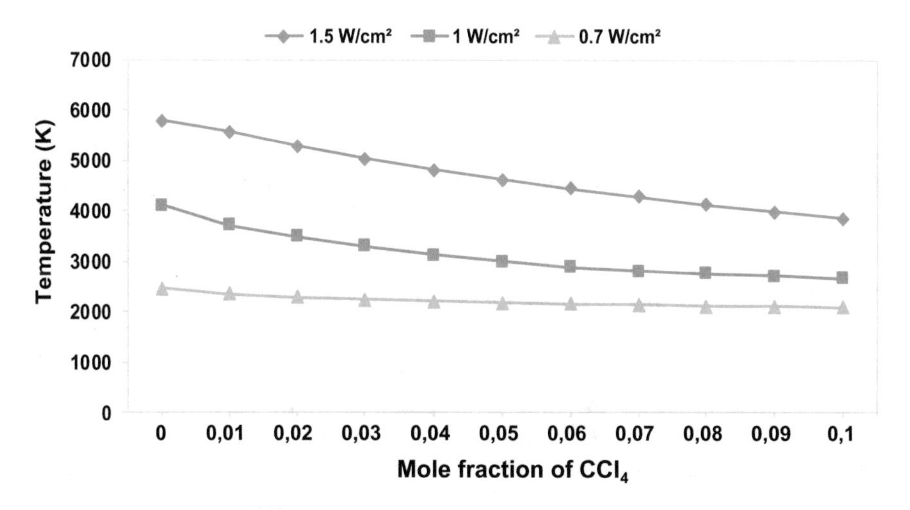

Fig. 2 Maximum bubble temperature versus CCl_4 mole fraction of carbon and acoustic intensity (frequency: 355 kHz). Modified from [23]

in spite of the decrease in the bubble temperature with the same increase in CCl_4 mole fraction, as in Figs. 1 and 2. Similar trends have been observed in pure water through many theoretical cases [81, 110, 111]. Thirdly, it has been observed that the effect of CCl_4 mole fraction inside the bubble on its maximal conversion is acoustic intensity-dependent, where the gradual increase in CCl_4 mole fraction causes its maximal decomposition extent to be decreased depending on the applied intensity, as in Fig. 3. These results are in good agreement with those found by [43, 57, 107].

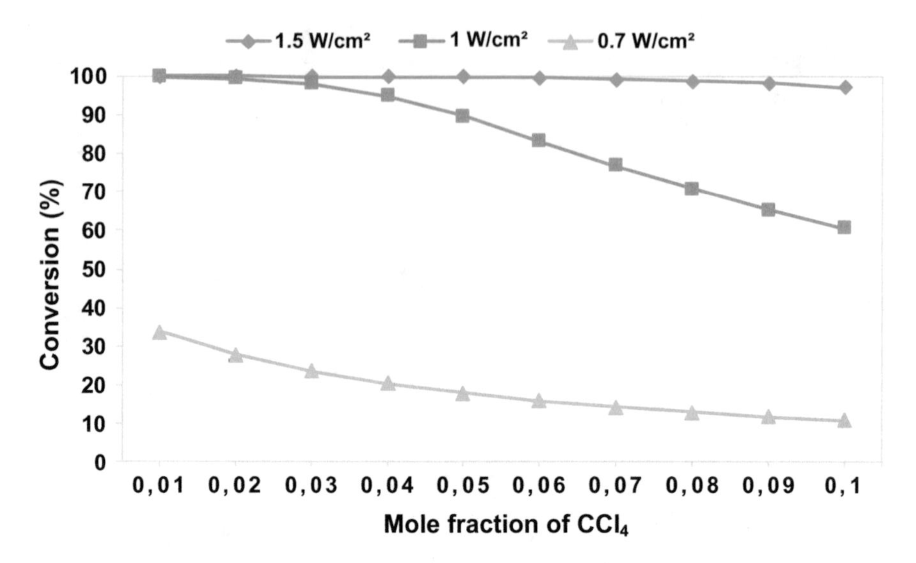

Fig. 3 Conversion of carbon tetrachloride versus initial CCl_4 mole fraction (frequency: 355 kHz). Modified from [23]

It is worth mentioning that the decrease in the bubble temperature as well as the maximal degradation of CCl_4 with increase in CCl_4 mole fraction inside the bubble are owing to the increase in bubble heat capacity with increasing CCl_4 mole fraction in addition to the highly endothermal dissociation of CCl_4 and H_2O molecules.

4.3 Decomposition Products of CCl_4

This section aims to respond to the discussed controversies about the mechanism of efficiency enhancement of the different dyes elimination by using carbon tetrachloride (Sect. 3). Therefore, in this part, we try to expose the important points obtained in Dehane et al. [23, 24] in order to reveal the main oxidants responsible of dyes sono-degradation via CCl_4 pyrolysis inside an argon bubble. The evolution of the molar yield of the important species created during the CCl_4 pyrolysis within the acoustic bubble is given in Figs. 4 and 5, for various CCl_4 concentration in bulk liquid at ambient temperature (20 °C) and pressure (1 atm). Firstly, according to Figs. 4 and 5a and b, it can be seen that is in spite of the increase in acoustic intensity from 1 (T_{max} ~ 4000–2700 K) to 1.5 W cm^{-2} (T_{max} ~ 5900–3900 K), a slight difference is observed for the overall production trend of the different species. Secondly, it has been obtained that for an acoustic intensity of 1.5 W cm^{-2} and CCl_4 mole fraction lower than ~ 0.7%, i.e., corresponding to $\leq 4 \times 10^{-4}$ M CCl_4, molar yields of reactive chlorine species, i.e., $\bullet CCl_3$, $:CCl_2$, $\bullet Cl$, Cl_2, and HOCl, and $\bullet OH$ radicals are approximately in the same range (Fig. 4a), whereas, for CCl_4 mole fraction greater than ~

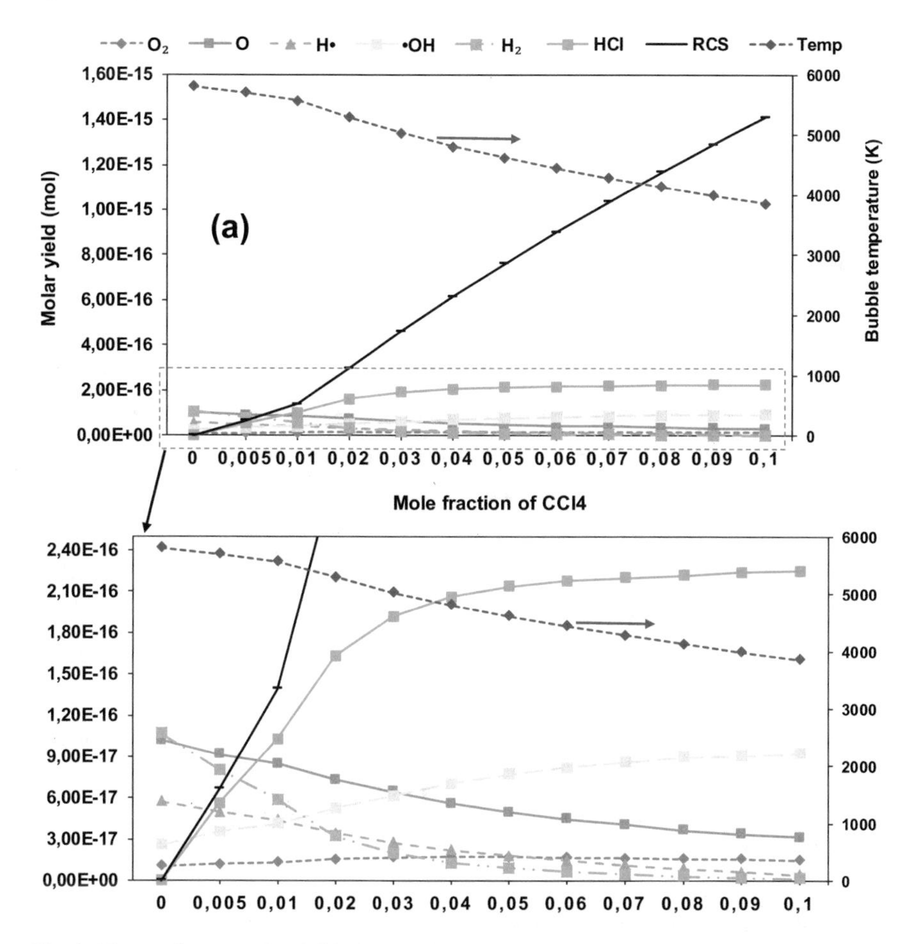

Fig. 4 The maximum molar yield of the different species and the maximum bubble temperature all versus initial CCl_4 mole fraction within the bubble and acoustic (frequency: 355 kHz: Intensity: 1 W/cm^2). **a** The production of O_2, O, H$^•$, $^•$OH, H_2, HCl and total yield of reaction chlorine species (RCS: $^•CCl_3$, :CCl_2, •Cl, Cl_2 and HOCl) and **b** The yield of •CCl_3, :CCl_2, •Cl, Cl_2, HOCl. Modified from [23]

0.7%, the production of reactive chlorine species becomes dominant. On the other hand, for an acoustic intensity of 1 W cm^{-2}, the produced amount of reactive chlorine species is dominant no matter the initial mole fraction of CCl_4 inside bubble, Fig. 5a. Thirdly, it was found that the most important reactive chlorine species are •CCl_3 and •Cl radicals and then:CCl_2, HOCl, and Cl_2 to a less extent, as shown in Figs. 4 and 5b. Fourthly, as it can be seen in Figs. 4 and 5a–b, the important quantities of HCl and HOCl are the main reasons of pH drop of the sonicated medium in the presence of CCl_4, whereas the amount of HCl is more important than that of hypochlorous acid. Moreover, if the reaction of $^•$Cl radicals with water molecules in aqueous phase ($^•Cl + H_2O \rightleftharpoons HCl + {}^•OH$) is taken into account [15] more HCl is produced outside

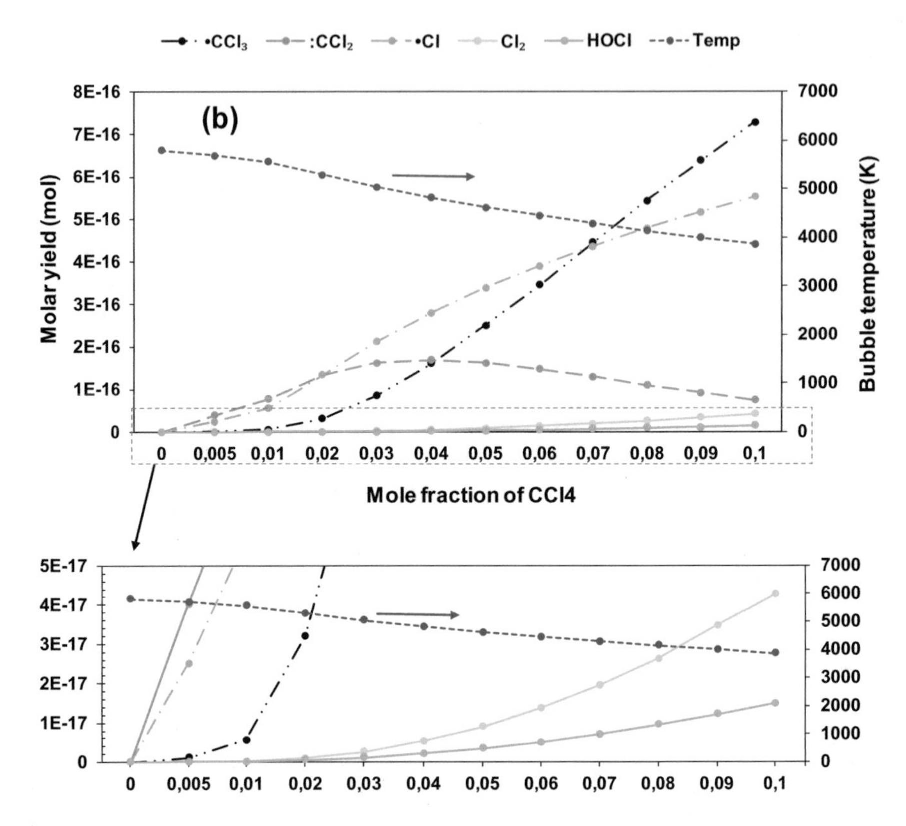

Fig. 4 (continued)

the bubble in addition to $^{\bullet}$OH radicals. Therefore, more decrease in solution pH is observed. Finally, it can concluded that the main species responsible of the elimination enhancement of dyes are chlorine-containing radicals ($^{\bullet}CCl_3$, :CCl_2, and $^{\bullet}$Cl), HOCl, and $^{\bullet}$OH radicals.

5 Conclusion and Future Perspectives

In this chapter, a detailed literature analysis is conducted about the different operational parameters (i.e., frequency, intensity, solution pH, CCl_4 and dyes dosages, liquid temperature) affecting dyes degradation in the presence of CCl_4. In addition, the different enhancing mechanisms of dyes decomposition via CCl_4 are discussed thoroughly, where the different controversies are highlighted. First, it has been found that the increase in ultrasound frequency, acoustic intensity and the initial concentration of CCl_4 clearly improves the degradation rate of dyes, whereas this increase

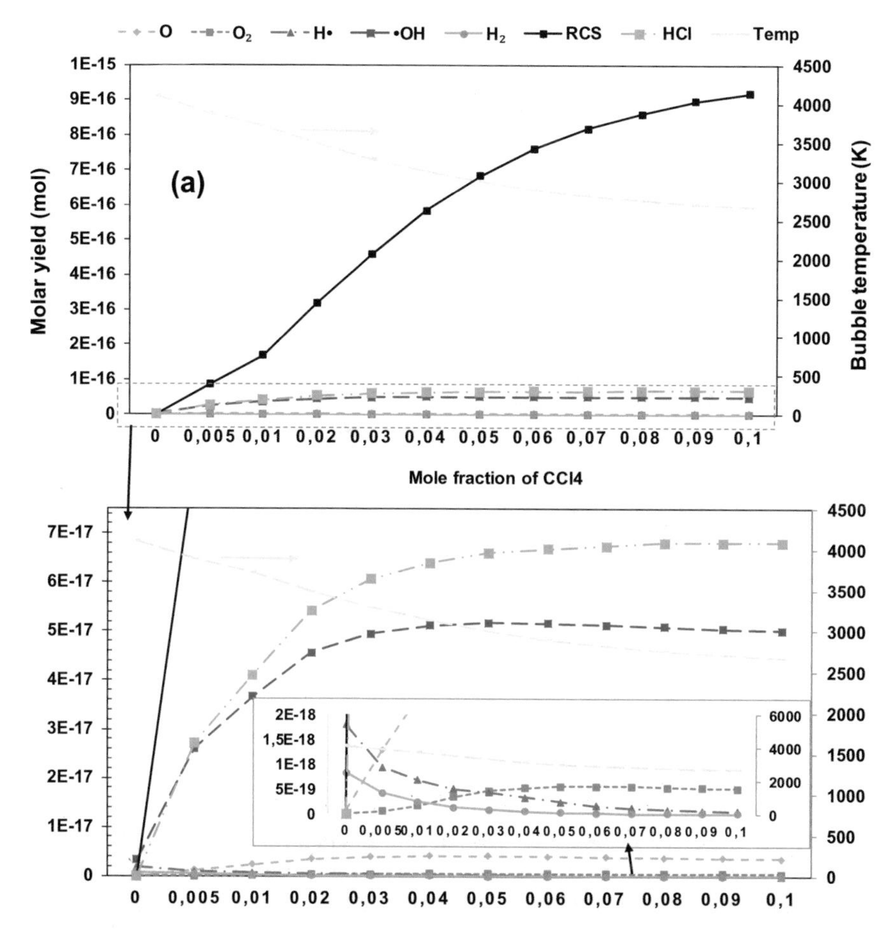

Fig. 5 The maximum molar yield of the different species and the maximum bubble temperature all versus initial CCl_4 mole fraction within the bubble (frequency: 355 kHz: Intensity: 1 W/cm^2). **a** The production of O_2, O, $H^•$, $^•OH$, H_2, HCl and total yield of reaction chlorine species (RCS: $^•CCl_3$, $:CCl_2$, •Cl, Cl_2 and HOCl) and **b** The yield of •CCl_3, $:CCl_2$, $^•Cl$, Cl_2, HOCl. Modified from [23]

in frequency, intensity, and $[CCl_4]_0$ should carefully controlled to avoid the negative effect of each of these operational conditions. Additionally, it was observed that even when sonochemical technique is combined to the different advanced oxidation processes, such as photolytic and photocatalytic, the addition of CCl_4 enhances notably the efficiency of these hybrid methods. On the other hand, the increase in the initial dye concentration reduces its degradation rate, in the presence of CCl_4, even when some improvement is observed through the increase in its initial concentration. Similarly, according to the available data in literature, it was found that the rise of liquid temperature decreases the bleaching rate of dyes. Moreover, the solution pH is considered as an important parameter for a better control of dye elimination, for the

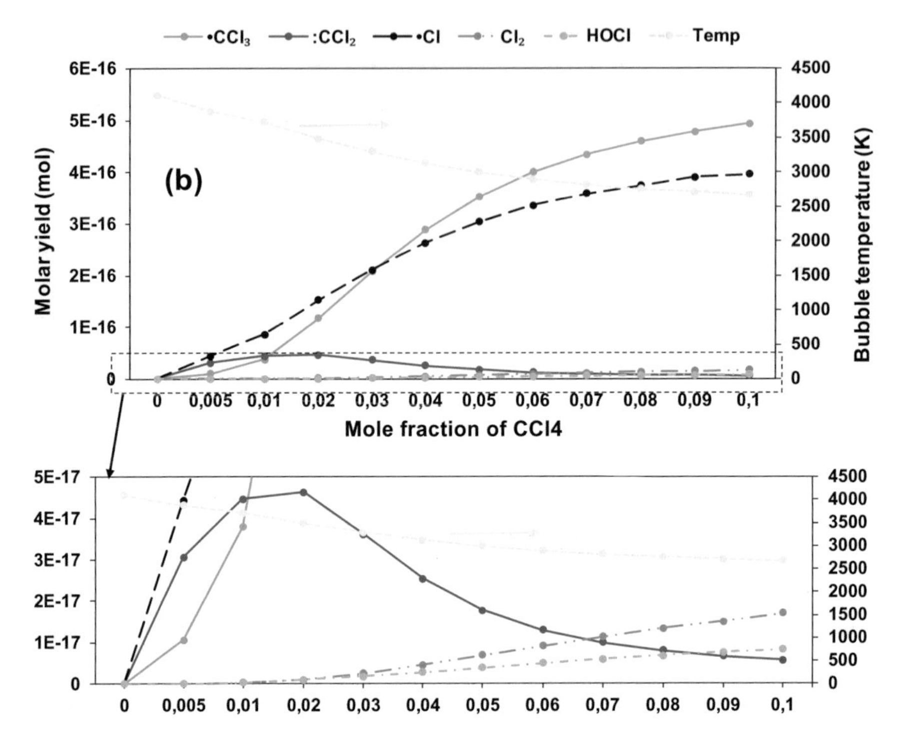

Fig. 5 (continued)

majority of studies the decrease in solution pH affects positively the decolorization rate of dyes. However, this is not always the case where for some dyes degradation at neutral pH of solution is preferred. Secondly, the detailed analysis of the available mechanisms in literature of CCl_4 enhancing effect towards the elimination of dyes reveals the existing debates about the principal species, i.e., reactive chlorine species or hydroxyl radicals, responsible of dyes elimination.

In the second section of this chapter, theoretical results for CCl_4 pyrolysis was shown in order to determine the real mechanism of dyes decolorization in presence of CCl_4. The most important conclusion inferred from the study is that the main path of dyes elimination in presence of CCl_4 is their reaction with chlorine-containing radicals ($^{\bullet}CCl_3$, $:CCl_2$, $^{\bullet}Cl$), Cl_2, HOCl, and $^{\bullet}OH$ radicals.

Finally, from a point of view of practical application of CCl_4 as an enhancer for the sonochemical degradation of dyes, some points need to be investigated:

- Geometry of reactors and sonicated volume, in addition to the type and distribution of transducers within the irradiated solution.
- Optimization of the different operational conditions, such as ultrasound frequency, acoustic intensity, CCl_4 and dye dosages, liquid temperature, solution pH, etc., with respect to the irradiated volume and reactor design.

- Ultrasonic irradiation mode and the possibility of application of dyes sono-degradation under a continuous flow.
- The upscaling of sonolytic process to plant size.

References

1. Alippi A, Cataldo F, Galbato A (1992) Ultrasound cavitation in sonochemistry: decomposition of carbon tetrachloride in aqueous solutions of potassium iodide. Ultrasonics 30:148–151. https://doi.org/10.1016/0041-624X(92)90064-S
2. Ara J, Ashifuzzaman M, Hossain MJ, Razzak SMA, Monira S (2019) Ultrasound assisted sonochemical decomposition of methyl orange in the presence of H_2O_2 and CCl_4. Asian J Appl Chem Res 4:1–10. https://doi.org/10.9734/AJACR/2019/v4i230107
3. Banerjee BS, Khode AV, Patil AP, Mohod AV, Gogate PR (2004) Sonochemical decolorization of wastewaters containing Rhodamine 6G using ultrasonic bath at an operating capacity of 2 L. Desalin Water Treat 52:1378–1387. https://doi.org/10.1080/19443994.2013.786656
4. Beckett MA, Hua I (2000) Elucidation of the 1,4-dioxane decomposition pathway at discrete ultrasonic frequencies. Environ Sci Technol 34:3944–3953. https://doi.org/10.1021/es000928r
5. Beckett MA, Hua I (2001) Impact of ultrasonic frequency on aqueous sonoluminescence and sonochemistry. J Phys Chem A 105:3796–3802. https://doi.org/10.1021/jp003226x
6. Behnajady MA, Modirshahla N, Tabrizi SB, Molanee S (2008) Ultrasonic degradation of Rhodamine B in aqueous solution: Influence of operational parameters. J Hazard Mater 152:381–386. https://doi.org/10.1016/j.jhazmat.2007.07.019
7. Bejarano-Pérez NJ, Suarez-Herrera MF (2008) Sonochemical and sonophotocatalytic degradation of malachite green: The effect of carbon tetrachloride on reaction rates. Ultrason Sonochem 15:612–617. https://doi.org/10.1016/j.ultsonch.2007.09.009
8. Bekkouche S, Bouhelassa M, Ben A, Baup S, Gondrexon N, Pétrier C, Merouani S, Hamdaoui O (2017) Synergy between solar photocatalysis and high frequency sonolysis toward the degradation of organic pollutants in aqueous phase–case of phenol. Desalin Water Treat 20146:1–8. https://doi.org/10.5004/dwt.2017.20146
9. Bhatnagar A, Cheung HM (1994) Sonochemical destruction of chlorinated C_1 and C_2 volatile organic compounds in dilute aqueous solution. Environ Sci Technol 28:1481–1486. https://doi.org/10.1021/es00057a016
10. Boutamine Z, Merouani S, Hamdaoui O (2017) Sonochemical degradation of Basic Red 29 in aqueous media. Turkish J Chem 41:99–115. https://doi.org/10.3906/kim-1603-82
11. Brotchie A, Grieser F, Ashokkumar M (2009) Effect of power and frequency on bubble-size distributions in acoustic cavitation. Phys Rev Lett 102(084302):1–4. https://doi.org/10.1103/PhysRevLett.102.084302
12. Bryukov MG, Slagle IR, Knyazev VD (2001) Kinetics of reactions of H atoms with methane and chlorinated methanes. J Phys Chem A 105:3107–3122. https://doi.org/10.1021/jp0023359
13. Chadi NE, Merouani S, Hamdaoui O, Bouhelassa M (2018) New aspect of the effect of liquid temperature on sonochemical degradation of nonvolatile organic pollutants in aqueous media. Sep Purif Technol 200:68–74. https://doi.org/10.1016/j.seppur.2018.01.047
14. Chakinala AG, Gogate PR, Chand R, Bremner DH, Molina R, Burgess AE (2008) Intensification of oxidation capacity using chloroalkanes as additives in hydrodynamic and acoustic cavitation reactors. Ultrason Sonochem 15:164–170. https://doi.org/10.1016/j.ultsonch.2007.02.008
15. Chendke PK, Fogler HS (1983) Sonoluminescence and sonochemical reactions of aqueous carbon tetrachloride solutions. J Phys Chem 318:1362–1369. https://doi.org/10.1021/j100231a019

16. Chiha M, Merouani S, Hamdaoui O, Baup S, Gondrexon N, Pétrier C (2010) Modeling of ultrasonic degradation of non-volatile organic compounds by Langmuir-type kinetics. Ultrason Sonochem 17:773–782. https://doi.org/10.1016/j.ultsonch.2010.03.007
17. Colussi AJ, Hung H-M, Hoffmann MR (1999) Sonochemical degradation rates of volatile solutes. J Phys Chem A 103:2696–2699. https://doi.org/10.1021/jp984272o
18. Conaire MÓ, Curran HJ, Simmie JM, Pitz WJ, Westbrook CK (2004) A comprehensive modeling study of hydrogen oxidation. Int J Chem Kinet 36:603–622. https://doi.org/10.1002/kin.20036
19. Corzo BA, Suárez-Herrera MF (2018) Effect of carbon tetrachloride on the luminol sonochemiluminescence reaction kinetics during multibubble cavitation. Ultrason Sonochem 48:281–286. https://doi.org/10.1016/j.ultsonch.2018.06.005
20. Dalhatou S, Pétrier C, Laminsi S, Baup S (2015) Sonochemical removal of naphthol blue black azo dye: influence of parameters and effect of mineral ions. Int J Environ Sci Technol 12:35–44. https://doi.org/10.1007/s13762-013-0432-8
21. Dehane A, Merouani S, Hamdaoui O, Alghyamah A (2021a) A complete analysis of the effects of transfer phenomenons and reaction heats on sono-hydrogen production from reacting bubbles: Impact of ambient bubble size. Int J Hydrogen Energy 46:18767–18779. https://doi.org/10.1016/j.ijhydene.2021.03.069
22. Dehane A, Merouani S, Hamdaoui O, Alghyamah A (2021b) A comprehensive numerical analysis of heat and mass transfer phenomenons during cavitation sono-process. Ultrason Sonochem 73:105498. https://doi.org/10.1016/j.ultsonch.2021.105498
23. Dehane A, Merouani S, Hamdaoui O (2021c) Carbon tetrachloride (CCl_4) sonochemistry: A comprehensive mechanistic and kinetics analysis elucidating how CCl_4 pyrolysis improves the sonolytic degradation of nonvolatile organic contaminants. Sep Purif Technol 275:118614. https://doi.org/10.1016/j.seppur.2021.118614
24. Dehane A, Merouani S, Hamdaoui O (2021d) Effect of carbon tetrachloride (CCl_4) sonochemistry on the size of active bubbles for the production of reactive oxygen and chlorine species in acoustic cavitation field. Chem Eng J 426:130251. https://doi.org/10.1016/j.cej.2021.130251
25. Dehane A, Merouani S, Hamdaoui O (2021e) Theoretical investigation of the effect of ambient pressure on bubble sonochemistry: Special focus on hydrogen and reactive radicals production. Chem Phys 547:111171. https://doi.org/10.1016/j.chemphys.2021.111171
26. Fassi S, Petrier C (2016) Effect of potassium monopersulfate (oxone) and operating parameters on sonochemical degradation of cationic dye in an aqueous solution. Ultrason Sonochem 32:343–347. https://doi.org/10.1016/j.ultsonch.2016.03.032
27. Ferkous H, Merouani S, Hamdaoui O, Rezgui Y, Guemini M (2015a) Comprehensive experimental and numerical investigations of the effect of frequency and acoustic intensity on the sonolytic degradation of naphthol blue black in water. Ultrason Sonochem 26:30–39. https://doi.org/10.1016/j.ultsonch.2015.02.004
28. Ferkous H, Hamdaoui O, Merouani S (2015b) Sonochemical degradation of naphthol blue black in water: effect of operating parameters. Ultrason Sonochem 26:40–47. https://doi.org/10.1016/j.ultsonch.2015.03.013
29. Ferkous H, Merouani S, Hamdaoui O (2016) Sonolytic degradation of naphtol blue black at 1700 kHz: effects of salts, complex matrices and persulfate. J Water Process Eng 9:67–77. https://doi.org/10.1016/j.str.2014.12.012
30. Ferkous H, Merouani S, Hamdaoui O, Pétrier C (2017) Persulfate-enhanced sonochemical degradation of naphthol blue black in water: Evidence of sulfate radical formation. Ultrason Sonochem 34:580–587. https://doi.org/10.1016/j.ultsonch.2016.06.027
31. Fernández-castro P, Vallejo M, Fresnedo M, Román S, Ortiz I (2015) Insight on the fundamentals of advanced oxidation processes . role and review of the determination methods of reactive oxygen species. J Chem Technol Biotechnol 796–820. https://doi.org/10.1002/jctb.4634
32. Francony A, Pétrier C (1996) Sonochemical degradation of carbon tetrachloride in aqueous solution at two frequencies: 20 kHz and 500 kHz. Ultrason Sonochem 3:S77–S82. https://doi.org/10.1016/1350-1477(96)00010-1

33. Ghodbane H, Hamdaoui O (2009) Intensification of sonochemical decolorization of anthraquinonic dye Acid Blue 25 using carbon tetrachloride. Ultrason Sonochem 16:455–461. https://doi.org/10.1016/j.ultsonch.2008.12.005
34. Gültekin I, Tezcanli-güyer G, Ince NH (2009) Sonochemical decay of C.I. Acid Orange 8: Effects of CCl4 and t-butyl alcohol. Ultrason Sonochem 16:577–581. https://doi.org/10.1016/j.ultsonch.2008.12.007
35. Guo W, Shi Y, Wang H, Yang H, Zhang G (2010) Intensification of sonochemical degradation of antibiotics levofloxacin using carbon tetrachloride. Ultrason Sonochem 17:680–684. https://doi.org/10.1016/j.ultsonch.2010.01.004
36. Guzman-Duque F, Pétrier C, Pulgarin C, Penuuela G, Torres-Palma RA (2011) Effects of sonochemical parameters and inorganic ions during the sonochemical degradation of crystal violet in water. Ultrason Sonochem 18:440–446. https://doi.org/10.1016/j.ultsonch.2010.07.019
37. Hai FI, Yamamoto K, Fukushi K (2007) Hybrid treatment systems for dye wastewater. Crit Rev Environ Sci Technol 37:315–377. https://doi.org/10.1080/10643380601174723
38. Hamdaoui O, Merouani S (2017) Improvement of sonochemical degradation of brilliant blue R in water using periodate ions: Implication of iodine radicals in the oxidation process. Ultrason Sonochem 37:344–350. https://doi.org/10.1016/j.ultsonch.2017.01.025
39. Hendricks D (2011) Fundamentals of water treatment unit processes: Physical, chemical, and biological. IWA Publishing, London, UK
40. Henglein A (1995) Chemical effects of continuous and pulsed ultrasound in aqueous solutions. Ultrason Sonochem 2:115–121. https://doi.org/10.1016/1350-4177(95)00022-X
41. Hinge SP, Orpe MS, Sathe K V, Tikhe GD, Pandey NS, Bawankar KN, Bagal MV, Mohod V, Gogate PR (2016) Combined removal of Rhodamine B and Rhodamine 6G from wastewater using novel treatment approaches based on ultrasonic and ultraviolet irradiations. 3994:0–13. https://doi.org/10.1080/19443994.2016.1143404
42. Hu Y, Zhang Z, Yang C (2008) Measurement of hydroxyl radical production in ultrasonic aqueous solutions by a novel chemiluminescence method. Ultrason Sonochem 15:665–672. https://doi.org/10.1016/j.ultsonch.2008.01.001
43. Hua I, Hoffmann MR (1996) Kinetics and mechanism of the sonolytic degradation of CCl4: intermediates and byproducts. Environ Sci Technol 30:864–871. https://doi.org/10.1021/es9502942
44. Hung HM, Hoffmann MR (1999) Kinetics and mechanism of the sonolytic degradation of chlorinated hydrocarbons: Frequency effects. J Phys Chem A 103:2734–2739. https://doi.org/10.1021/jp9845930
45. Ince NH, Tezcanli-Güyer G (2004) Impacts of pH and molecular structure on ultrasonic degradation of azo dyes. Ultrasonics 42:591–596. https://doi.org/10.1016/j.ultras.2004.01.097
46. Jennings BH, Townsend SN (1961) The sonochemical reactions of carbon tetrachloride and chloroform in aqueous suspension in an inert atmosphere. J Phys Chem 65:1574–1579. https://doi.org/10.1021/j100905a025
47. Jiang Y, Petrier C, Waite TD (2006) Sonolysis of 4-chlorophenol in aqueous solution: effects of substrate concentration, aqueous temperature and ultrasonic frequency. Ultrason Sonochem 13:415–422. https://doi.org/10.1016/j.ultsonch.2005.07.003
48. Kang J-W, Hung H-M, Lin A, Hoffmann MR (1999) Sonolytic destruction of methyl tert-Butyl ether by ultrasonic irradiation: The role of O_3, H_2O_2, frequency, and power density. Environ Sci Technol 33:3199–3205. https://doi.org/10.1021/es9810383
49. Kansal SK, Singh M, Sud D (2007) Studies on photodegradation of two commercial dyes in aqueous phase using different photocatalysts. J Hazard Mater 141:581–590. https://doi.org/10.1016/j.jhazmat.2006.07.035
50. Kanthale P, Ashokkumar M, Grieser F (2008) Sonoluminescence, sonochemistry (H_2O_2 yield) and bubble dynamics: Frequency and power effects. Ultrason Sonochem 15:143–150. https://doi.org/10.1016/j.ultsonch.2007.03.003
51. Kobayashi D, Honma C, Suzuki A, Takahashi T, Matsumoto H, Kuroda C, Otake K, Shono A (2012) Comparison of ultrasonic degradation rates constants of methylene blue at 22.8 kHz,

127 kHz, and 490 kHz. Ultrason-Sonochem 19:745–749. https://doi.org/10.1016/j.ultsonch.2012.01.004
52. Koda S, Kimura T, Kondo T, Mitome H (2003) A standard method to calibrate sonochemical efficiency of an individual reaction system. Ulrason Sonochem 10:149–156. https://doi.org/10.1016/S1350-4177(03)00084-1
53. Kovács T, Turányi T (2006) Modelling of carbon tetrachloride decomposition in oxidative RF thermal plasma. Plasma Chem Plasma Process 26:293–318. https://doi.org/10.1007/s11090-006-9003-9
54. Kovács T, Turányi T, Föglein K, Szépvölgyi J (2005) Kinetic modeling of the decomposition of carbon tetrachloride in thermal plasma. Plasma Chem Plasma Process 25:109–119. https://doi.org/10.1007/s11090-004-8837-2
55. Kumaran SS, Lim KP, Michael JV (1994) Thermal rate constants for the $Cl+H_2$ and $Cl+D_2$ reactions between 296 and 3000 K. Plasma Chem Plasma Process 110:9487. https://doi.org/10.1063/1.468486
56. DE Laat J, Stefan M (2017) UV/chlorine process. In: Stefan MI (ed) Advanced oxidation processes for water treatment. IWA Publishing, London, UK, pp 383–428
57. Lee M, Oh J (2010) Sonolysis of trichloroethylene and carbon tetrachloride in aqueous solution. Ultrason Sonochem 17:207–212. https://doi.org/10.1016/j.ultsonch.2009.06.018
58. Leighton TG (2007) What is ultrasound? Prog Biophys Mol Biol 93:3–83. https://doi.org/10.1016/j.pbiomolbio.2006.07.026
59. Leong T, Ashokkumar M, Sandra K (2011) The fundamentals of power ultrasound - A review. Acoust Aust 39:54–63
60. Lim M, Son Y, Khim J (2011) Frequency effects on the sonochemical degradation of chlorinated compounds. Ultrason Sonochem 18:460–465. https://doi.org/10.1016/j.ultsonch.2010.07.021
61. Lim MH, Kim SH, Kim YU, Khim J (2007) Sonolysis of chlorinated compounds in aqueous solution. Ultrason Sonochem 14:93–98. https://doi.org/10.1016/j.ultsonch.2006.03.003
62. Guppy L, Anderson K (2017) Global water crisis: the Facts, United Nations University Institute for Water, Environment and Health. Hamilton, Canada
63. Lu X, Qiu W, Peng J, Xu H, Wang D, Cao Y, Zhang W, Ma J (2021) A Review on additives-assisted ultrasound for organic pollutants degradation. J Hazard Mater 403:123915. https://doi.org/10.1016/j.jhazmat.2020.123915
64. Luo W, Chen Z, Zhu L, Chen F, Wang L, Tang H (2007) A sensitive spectrophotometric method for determination of carbon tetrachloride with the aid of ultrasonic decolorization of methyl orange. Anal Chim Acta 588:117–122. https://doi.org/10.1016/j.aca.2007.01.077
65. Mantzavinos D, Kassinos D, Parsons SA (2009) Applications of advanced oxidation processes in wastewater treatment. Water Res 43:3901. https://doi.org/10.1016/j.watres.2009.08.024
66. Mark G, Tauber A, Laupert R, Schuchmann H-P, Schulz D, a AM, Sonntag C von, (1998) OH-radical formation by ultrasound in aqueous solution-Part II: Terephthalate and Fricke dosimetry and the influence of various conditions on the sonolytic yield. Ultrason Sonochem 5:41–52. https://doi.org/10.1016/S1350-4177(98)00012-1
67. Meghlaoui FZ, Merouani S, Hamdaoui O, Bouhelassa M, Ashokkumar M (2019) Rapid catalytic degradation of refractory textile dyes in Fe (II)/chlorine system at near neutral pH: Radical mechanism involving chlorine radical anion ($Cl_2^{\bullet -}$)-mediated transformation pathways and impact of environmental matrices. Sep Purif Technol 227:115685. https://doi.org/10.1016/j.seppur.2019.115685
68. Meghlaoui FZ, Merouani S, Hamdaoui O, Alghyamah A, Bouhelassa M, Ashokkumar M (2020) Fe(III)-catalyzed degradation of persistent textile dyes by chlorine at slightly acidic conditions: the crucial role of Cl2•– radical in the degradation process and impacts of mineral and organic competitors. Asia-Pacific J Chem Eng 1–12. https://doi.org/10.1002/apj.2553
69. Méndez-Arriaga F, Torres-Palma RAA, Pétrier C, Esplugas S, Gimenez J, Pulgarin C (2008) Ultrasonic treatment of water contaminated with ibuprofen. Water Res 42:4243–4248. https://doi.org/10.1016/j.watres.2008.05.033

70. Merouani S, Hamdaoui O (2016) The size of active bubbles for the production of hydrogen in sonochemical reaction field. Ultrason Sonochem 32:320–327. https://doi.org/10.1016/j.ultsonch.2016.03.026
71. Merouani S, Hamdaoui O (2017) Computational and experimental sonochemistry. Process Eng J 1:10–18
72. Merouani S, Hamdaoui O (2019) Sonolytic ozonation for water treatment: efficiency, recent developments, and challenges. Curr Opin Green Sustain Chem 18:98–108. https://doi.org/10.1016/j.cogsc.2019.03.003
73. Merouani S, Hamdaoui O (2021) Sonochemical treatment of textile wastewater. In: Inamuddin MP, Asiri A (eds) Water pollution and remediation: photocatalysis. Springer-Nature Switzerland
74. Merouani S, Hamdaoui O, Saoudi F, Chiha M (2010) Sonochemical degradation of Rhodamine B in aqueous phase: Effects of additives. Chem Eng J 158:550–557. https://doi.org/10.1016/j.cej.2010.01.048
75. Merouani S, Hamdaoui O, Rezgui Y, Guemini M (2013) Effects of ultrasound frequency and acoustic amplitude on the size of sonochemically active bubbles-theoretical study. Ultrason Sonochem 20:815–819. https://doi.org/10.1016/j.ultsonch.2012.10.015
76. Merouani S, Ferkous H, Hamdaoui O, Rezgui Y, Guemini M (2014a) A method for predicting the number of active bubbles in sonochemical reactors. Ultrason Sonochem 22:51–58. https://doi.org/10.1016/j.ultsonch.2014.07.015
77. Merouani S, Hamdaoui O, Rezgui Y, Guemini M (2014b) Energy analysis during acoustic bubble oscillations: Relationship between bubble energy and sonochemical parameters. Ultrasonics 54:227–232. https://doi.org/10.1016/j.ultras.2013.04.014
78. Merouani S, Hamdaoui O, Rezgui Y, Guemini M (2014c) Sensitivity of free radicals production in acoustically driven bubble to the ultrasonic frequency and nature of dissolved gases. Ultrason Sonochem 22:41–50. https://doi.org/10.1016/j.ultsonch.2014.07.011
79. Merouani S, Hamdaoui O, Rezgui Y, Guemini M (2015) Computer simulation of chemical reactions occurring in collapsing acoustical bubble: Dependence of free radicals production on operational conditions. Res Chem Intermed 41:881–897. https://doi.org/10.1007/s11164-013-1240-y
80. Merouani S, Hamdaoui O, Rezgui Y, Guemini M (2016a) Computational engineering study of hydrogen production via ultrasonic cavitation in water. Int J Hydrogen Energy 41:832–844. https://doi.org/10.1016/j.ijhydene.2015.11.058
81. Merouani S, Hamdaoui O, Boutamine Z, Rezgui Y, Guemini M (2016b) Experimental and numerical investigation of the effect of liquid temperature on the sonolytic degradation of some organic dyes in water. Ultrason Sonochem 28:382–392. https://doi.org/10.1016/j.ultsonch.2015.08.015
82. Mueller MA, Kim TJ, Yetter RA, Dryer FL (1999) Flow reactor studies and kinetic modeling of the H2/O2 reaction. Int J Chem Kinet 31:113–125. https://doi.org/10.1002/(SICI)1097-4601(1999)31:2%3c113::AID-KIN5%3e3.0.CO;2-0
83. Neppolian B, Choi HC, Sakthivel S, Arabindoo B, Murugesan V (2002) Solar/UV-induced photocatalytic degradation of three commercial textile dyes. J Hazard Mater 89:303–317. https://doi.org/10.1016/S0304-3894(01)00329-6
84. Okitsu K, Kawasaki K, Nanzai B, Takenaka N, Bandow H (2008) Effect of carbon tetrachloride on sonochemical decomposition of methyl orange in water. Chemosphere 71:36–42. https://doi.org/10.1016/j.chemosphere.2007.10.056
85. Pang YL, Abdullah AZ, Bhatia S (2011) Review on sonochemical methods in the presence of catalysts and chemical additives for treatment of organic pollutants in wastewater. Desalination 277:1–14. https://doi.org/10.1016/j.desal.2011.04.049
86. Pera-titus M, Garc V, Baños MA, Giménez J, Esplugas S (2004) Degradation of chlorophenols by means of advanced oxidation processes: a general review. Appl Catal B Environ 47:219–256. https://doi.org/10.1016/j.apcatb.2003.09.010
87. Petrier C, Jeunet A, Luche J, Reverdyt G (1992) Unexpected frequency effects on the rate of oxidative processes induced by ultrasound. J Am Chem Soc Soc 114:3148–3150. https://doi.org/10.1021/ja00034a077

88. Pétrier C, Francony A (1997) Ultrasonic waste-water treatment: incidence of ultrasonic frequency on the rate of phenol and carbon tetrachloride degradation. Ultrason Sonochem 4:295–300. https://doi.org/10.1016/S1350-4177(97)00036-9
89. Rajan R, Kumar R, Gandhi K (1998) Modeling of sonochemical decomposition of CCl_4 in aqueous solutions. EnvironSciTechnol 32:1128–1133. https://doi.org/10.1021/es970272a
90. Rayaroth MP, Aravind UK, Aravindakumar CT (2015) Sonochemical degradation of Coomassie Brilliant Blue: Effect of frequency, power density, pH and various additives. Chemosphere 119:848–855. https://doi.org/10.1016/j.chemosphere.2014.08.037
91. Rehorek A, Tauber M, Gübitz G (2004) Application of power ultrasound for azo dye degradation. Ultrason Sonochem 11:177–182. https://doi.org/10.1016/j.ultsonch.2004.01.030
92. Remucal CK, Manley D (2016) Emerging investigators series: The efficacy of chlorine photolysis as an advanced oxidation process for drinking water treatment. Environ Sci Water Res Technol 2:565–579. https://doi.org/10.1039/c6ew00029k
93. Spurlock LA, Reifsneider SB (1970) Chemistry of ultrasound. I. Reconsideration of first principles and the applications to a dialkyl sulfide. J Am Chem Soc 92:6112–6117. https://doi.org/10.1021/ja00724a003
94. Stefan MI (2017) Advanced oxidation processes for water treatment: fundamentals and applications. IWA Publishing, London, UK
95. Taamallah A, Merouani S, Hamdaoui O (2016) Sonochemical degradation of basic fuchsin in water. Desalin Water Treat 57:27314–27330. https://doi.org/10.1080/19443994.2016.1168320
96. Tarr MA (2003) Chemical degradation methods for wastes and pollutants: environmental and industrial applications. Marcel Dekker Inc., New York
97. Thangavadivel K, Owens G, Okitsu K (2013) Removal of methyl orange from aqueous solution using a 1.6 MHz ultrasonic atomiser. RSC Adv 3:23370–23376. https://doi.org/10.1039/c3ra44343d
98. Uddin H, Okitsu K (2016) Effect of CCl4 or C6F14 on sonochemical degradation of dyes and phenolic compounds in an aqueous solution. J Water Process Eng 12:66–71. https://doi.org/10.1016/j.jwpe.2016.05.001
99. Wang L, Zhu L, Luo W, Wu Y, Tang H (2007) Drastically enhanced ultrasonic decolorization of methyl orange by adding CCl4. Ultrason Sonochem 14:253–258. https://doi.org/10.1016/j.ultsonch.2006.05.004
100. Wang X, Yao Z, Wang J, Guo W, Li G (2008) Degradation of reactive brilliant red in aqueous solution by ultrasonic cavitation. Ultrason Sonochem 15:43–48. https://doi.org/10.1016/j.ultsonch.2007.01.008
101. Wang X, Wei Y, Wang C, Guo W, Wang J, Jiang J (2011) Ultrasonic degradation of reactive brilliant red K-2BP in water with CCl4 enhancement: Performance optimization and degradation mechanism. Sep Purif Technol 81:69–76. https://doi.org/10.1016/j.seppur.2011.07.003
102. Wang X, Wei Y, Wang J, Guo W, Wang C (2012) The kinetics and mechanism of ultrasonic degradation of p-nitrophenol in aqueous solution with CCl_4 enhancement. Ultrason Sonochem 19:32–37. https://doi.org/10.1016/j.ultsonch.2010.12.005
103. Wei H, Shi J, Yang X, Wang J, Li K, He Q (2018) CCl4-Enhanced ultrasonic irradiation for ciprofloxacin degradation and antibiotic activity. Water Environ Reasearch 90:579–588. https://doi.org/10.2175/106143017X15131012153077
104. Weissler A, Cooper HW, Snyder S (1950) Chemical effect of ultrasonic waves: Oxidation of potassium iodide solution by carbon tetrachloride. J Am Chem Soc 72:1769–1775. https://doi.org/10.1021/ja01160a102
105. Wood RJ, Lee J, Bussemaker MJ (2017) A parametric review of sonochemistry: Control and augmentation of sonochemical activity in aqueous solutions. Ultrason Sonochem 38:351–370. https://doi.org/10.1016/j.ultsonch.2017.03.030
106. Wu C, Liu X, Fan J, Wang L (2001) Ultrasonic destruction of chloroform and carbon tetrachloride in aqueous solution. J Environ Sci Heal Part A 36:947–955. https://doi.org/10.1081/ESE-100104123

107. Wu JM, Huang HS, Livengood CD (1992) Ultrasonic destruction of chlorinated compounds in aqueous solution. Environ Prog Sustain Energy 11:195–201. https://doi.org/10.1002/ep.670110313
108. Wu ZL, Ondruschka B, Bräutigam P (2007) Degradation of chlorocarbons driven by hydrodynamic cavitation. Chem Eng Technol 30:642–648. https://doi.org/10.1002/ceat.200600288
109. Yasui K (2011) Fundamentals of acoustic cavitation and sonochemistry. In: Ashokkumar M (ed) Pankaj. Theoretical and experimental sonochemistry involving inorganic systems. Springer ScienceþBusiness Media, New York, pp 1–29
110. Yasui K, Tuziuti T, Iida Y, Mitome H (2003) Theoretical study of the ambient-pressure dependence of sonochemical reactions. J Chem Phys 119:346. https://doi.org/10.1063/1.1576375
111. Yasui K, Tuziuti T, Iida Y (2004) Optimum bubble temperature for the sonochemical production of oxidants. Ultrasonics 42:579–584. https://doi.org/10.1016/j.ultras.2003.12.005
112. Yasui K, Tuziuti T, Sivakumar M, Iida Y (2005) Theoretical study of single-bubble sonochemistry. J Chem Phys 122:224706. https://doi.org/10.1063/1.1925607
113. Yasui K, Tuziuti T, Lee J, Kozuka T, Towata A, Iida Y (2008) The range of ambient radius for an active bubble in sonoluminescence and sonochemical reactions. J Chem Phys 128:184705. https://doi.org/10.1063/1.2919119
114. Zeng L, McKinley JW (2006) Degradation of pentachlorophenol in aqueous solution by audible-frequency sonolytic ozonation. J Hazard Mater 135:218–225. https://doi.org/10.1016/j.jhazmat.2005.11.051
115. Zhang BT, Zhang Y, Teng Y, Fan M (2015) Sulfate radical and its application in decontamination technologies. Crit Rev Environ Sci Technol 45:1756–1800. https://doi.org/10.1080/10643389.2014.970681
116. Zheng W, Maurin M, Tarr MA (2005) Enhancement of sonochemical degradation of phenol using hydrogen atom scavengers. Ultrason Sonochem 12:313–317. https://doi.org/10.1016/j.ultsonch.2003.12.007
117. Zhirtsova IV, Zaslonko IS, Karasevich YK, Wagner HG (2000) Kinetics of soot formation in tetrachloromethane pyrolysis. Kinet Catal 41:366–376. https://doi.org/10.1007/BF02755374

UV/H_2O_2 Processes for Dye Removal

Ashish Unnarkat, Swapnil Dharaskar, and Meghan Kotak

Abstract Dyes are an inseparable part of coloring industries and the applications cover the width but finds primary usage in the textile industries. The dyes as pollutant in the water pose serious threats to humans along with plant and animal life. Advanced Oxidation Processes (AOPs) pave its way as a novel approach for the remediation of dye-polluted wastewater. AOPs result in almost complete decolorization and degradation of the dyes. Chapter has provided a detailed review on a specific AOP, and UV/H_2O_2 system and its implications for the degradation of different dyes. The chapter covered the effect of different parameters like pH, H_2O_2 dosage, contact time, initial concentration of dye, and presence of co-ions in the solution on the performance of photocatalytic degradation.

Keywords Microwave · Dye removal · Parameters · Photocatalytic · Degradation

1 Introduction

Dyes are integral part of the textile industries alongside it finds in usage food, pharmaceuticals, cosmetics, biomedical, and electronics. These color-causing dyes are highly toxic, carcinogenic, and have serious consequences on human and environmental health. Color of dyes is due to conjugated system and chromophore groups, mainly dyes are made up of skeleton, chromophore, auxochrome (enhance the color), and soluble part (optional) [1, 2]. The color of a dye and the intended color for the substrate to be dyed are determined by the chromophore (the molecule responsible for

A. Unnarkat · S. Dharaskar (✉)
Department of Chemical Engineering, School of Technology, Pandit Deendayal Energy University, Raisan, Gandhinagar, Gujarat 382426, India
e-mail: swapnildharaskar11@gmail.com; swapnil.dharaskar@sot.pdpu.ac.in

A. Unnarkat
e-mail: ashish.unnarkat@sot.pdpu.ac.in

M. Kotak
Department of Chemical Engineering, Dwarkadas J Sangvi College of Engineering, Mumbai, Maharashtra, India

S. S. Muthu and A. Khadir (eds.), *Advanced Oxidation Processes in Dye-Containing Wastewater*, Sustainable Textiles: Production, Processing, Manufacturing & Chemistry, https://doi.org/10.1007/978-981-19-0882-8_5

color) of its chemical composition. These molecules, like some colorants themselves, absorb certain wavelengths of light while reflecting others, causing the perception of color.

These dyes can be classified on the bases of application and on the basis of structure of dyes.

Dyes are grouped according to various principles, as organic/inorganic, natural/synthetic, and based on chemicals in the dyeing method. The major classes of dyes according to their dyeing process in textile finishing are acid dyes, basic dyes, direct dyes, reactive dyes, disperse dyes, mordant dyes, solvent dyes, sulfur dyes, and vat dyes [3]. Most of these dyes are manufactured through different stages involving nitration, reduction, halogenation, amination, sulfonation, diazotization, and oxidation using benzene, toluene, xylene, naphthalene, and anthracene, as raw materials. The effluents from these industries vary widely in composition, contain organic and inorganic compounds, and have high level of color and COD. Colored textile effluents contain persistent dyestuffs, many of which are reported to be toxic and carcinogenic. These dyes are very stable in the environment and persist under oxidation and reduction conditions, light exposure, and biodegradation. Many processes are known to treat the dye-polluted effluent; these methods are either physiochemical, biological, chemical, or electrochemical. Among these processes, Advanced Oxidation Processes (AOPs)—chemical processes—are novel and relatively superior in performance. Most commonly used AOPs utilizes UV irradiation, ultrasonic, gamma irradiation, H_2O_2, Fenton's reagent (or source of Fe^{2+}), ozone either alone or in combination for the treatment. AOPs result in almost complete decolorization and degradation of the dyes [4–6]. Moreover, 95% of colored compounds are now removed by AOPs. The efficiency of most commonly used processes such as biological processes are time-consuming. Chemical process AOPs destroy the color and COD causing compounds almost completely and fast. Typically, with AOPs complete mineralization and oxidation to inert and less concerning products such as H_2O, CO_2, and other simpler products [3–9]. UV/H_2O_2 system is advantageous and hence widely accepted; the advantages are (i) there is no sludge formation during the treatment, (ii) easy to perform under ambient conditions, and (iii) evolved oxygen in the process can be used for biological decay process [4].

2 Advanced Oxidation Process Using UV/H_2O_2 System

In a typical photocatalytic dye degradation mechanism, following six reactions take place. When the material is irradiated with the UV, photons with higher energy than bandgap of catalytic material are absorbed and the electron from the valance band jumps to the conduction band. The process leaves behind a hole in the valence band and an electron in excited state in the conduction band that is now available on the surface of catalyst. These electron–hole pair now acts as the base for generating radicals, hydroxyl and superoxide, through the reaction with oxidants and adsorbed water molecules. The degradation happens on the surface of the catalyst through adsorption and combination of dye and the radicals present on surface. These two radicals attack the dye molecules and disturb the basic structure and chromophore responsible for color to generate decolored intermediates that are further disintegrated into smaller molecules and finally leading to H_2O and CO_2 [3–6, 10]. Refer to the following scheme for visualizing the chemical reaction steps:

$$\text{Photocatalyst} + h\nu \rightarrow h^+_{VB} + e^-_{CB} \quad (1)$$

$$h^+_{VB} + e^-_{CB} \rightarrow \text{Energy (heat)} \quad (2)$$

$$H_2O + h^+_{VB} \rightarrow \bullet OH \text{ (hydroxyl radical)} + H^+ \quad (3)$$

$$O_2 + e^-_{CB} \rightarrow \bullet O^{2-} \text{ (superoxide radical)} \quad (4)$$

$$\bullet OH + \text{pollutant} \rightarrow \text{Intermediates} \rightarrow H_2O + CO_2 \quad (5)$$

$$\bullet O^{2-} + \text{pollutant} \rightarrow \text{Intermediates} \rightarrow H_2O + CO_2 \quad (6)$$

Current section presents one of the prominent AOPs in the treatment of dye and allied waste water consisting of UV irradiation assisted with hydrogen peroxide as oxidant. This system is either used alone or in combination of photocatalysts/metal oxides for promoting the photocatalytic degradation. The section will also describe the effect of different parameters like pH, H_2O_2 dosage, contact time, initial concentration of dye, and presence of co-ions in the solution. Table 1 provides the detailed comparison of the performance of catalysts for degradation of different dyes in the UV/H_2O_2 system which is presented toward the end of the section.

UV/H_2O_2 system is the most commonly employed AOPs for the dye degradation. A wide variety of synthetic industrial dyes like rhodamine B, methyl orange, methylene blue, direct yellow, reactive red, alizarin violet, and malachite green are among those that are tested for degradation using the system whether in presence or absence of catalyst.

3 Non-catalytic UV/H_2O_2 System

Rhodamine B (RhB) is among the most tested dye by researchers due to its manifold applications. Alhamdei et al. presented the degradation of Rhodamine B having H_2O_2 as oxidant using UV light. Parametric study showing effect of dye concentration, pH, H_2O_2 dosage, and irradiation time were also reported. It is observed that the dye decoloration improves till the optimum concentration of H_2O_2 (1.67 mM) after which it starts decreasing. The product of degradation was analyzed as the low molecular

Table 1 Performance of catalytic/H_2O_2/UV-based process for degradation of different dyes

S/N	Target dye	Catalyst	Optimal experimental conditions					Performance efficiency Decolorization/degradation			Reference
			Time	pH	Catalyst dosage	Dye dosage	H_2O_2 dosage	Light source	UV or H_2O_2	UV + H_2O_2	
1	Reactive orange 4	NA	150 min	3	30% w/w	5×10^{-4} M	10 mmol	UV	3	59.85	[11]
2	Rhodamine B	NA	NA	7	NA	10 μM	1.67 mM	UV	NA	73%	[12]
3	Direct yellow 106	NA	NA	9.15	NA	20 mg/L	600 mg/L	PL	90%	NA	[13]
4	Reactive orange 107	Fe_3O_4	25 min	5	0.8 g/L	50 mg/L	10 mM	UV	12.1%	99.20%	[14]
5	Methylene blue	1D-Cu	90 min	NA	40 mg	40 mg/L	0.1 mL	UV	NA	84.90%	[15]
6	Indole-3-acetic acid	1D-Cu	4 h	NA	NA	40 mg/L	NA	UV	NA	60.40%	[15]
7	Acridine orange	Fe^{3+}	160 min	2.9	8×10^{-5} M	4×10^{-5} M	0.07 M	Visible	NA	84.50%	[16]
8	Alizarin violet 3B	Fe^{3+}	160 min	2.9	2×10^{-5} M	2×10^{-5} M	0.07 M	Visible	NA	87.50%	[16]
9	Indigo carmin	WO_3	120 min	4	1 g	100 mL	1 mL	UV	17.5%	90%	[17]
10	Reactive red 147	TiO_2	60 min	3.4	0.6 g	50 ppm	0.9 mL	UV	48%	92%	[18]

(continued)

Table 1 (continued)

S/N	Target dye	Catalyst	Optimal experimental conditions					Performance efficiency Decolorization/degradation			Reference
			Time	pH	Catalyst dosage	Dye dosage	H_2O_2 dosage	Light source	UV or H_2O_2	UV + H_2O_2	
11	Disperse red-60	ZnO	60 min	9	0.6 g/L	25–150 mg/L	0.9 mL/L	UV	68.7%	97%	[19]
12	Solophenyl orange Solophenyl blue Solophenyl scarlet Solophenyl yellow Solophenyl black Navy blue	TiO_2–P25	240 min	3–11	0.5 g/L	50 mg/L	50 mmol/L	UV	NA	100%	[20]
13	Rhodamine B	Carbon membrane	8 h	<8	NA	100 ppm	NA	NA	99.78%	NA	[21]
14	4-Chlorophenol (4-CP)		60 min	3	3.89 mM	0.778 mM	156 mM	NA	<10%	100%	[22]
15	Direct yellow 12	TiO_2, Fe_2O_3, ZnO, SnO_2	130 min	6.8	2 g/L	10 ppm	0.988 M	UV	NIL	85.5–98.5%	[23]
16	Reactive blue 181	NA	20 min	3	NA	500 mg/L	500 mg/l	UV	43.95%	99%	[24]
17	Reactive orange 113	Fe^{2+}	2 h	3	5-10 mg/dm^3	50 mg/dm^3	0.4 mg/mg	NA	100%	NA	[25]
18	Rhodamine B	Fe^{2+}	15 min	7	50 μM	2.5 μM	50 μM	UV	70%	100%	[26]
19	Procion H-exl dyes	TiO_2	8 min	3	1 g/L	75 mg/L	0.5% w/w	UV	NA	100%	[27]

(continued)

Table 1 (continued)

S/N	Target dye	Catalyst	Optimal experimental conditions					Performance efficiency Decolorization/degradation			Reference
			Time	pH	Catalyst dosage	Dye dosage	H_2O_2 dosage	Light source	UV or H_2O_2	UV + H_2O_2	
20	Mordant red 73	Fe^{2+}	3 h	3	0.056–0.11 g/L	5×10^{-2} mM	2.5 mM	UV	NA	85%	[28]
21	Ponceau 4R	Cobalt (II)/$NaHCO_3$	2 h	8.3	11.16 μM	20 mg/L	1.5–4.5 stoichiometric dosage	NA	NA	95%	[29]
22	Reactive blue 19	g-MnO_2/MWCNT	180 min	2–10	0–0.05 g	50–500 mg/L	0.076–0.333 M	NA	NA	78.60%	[30]
23	Blue sulfur dye	NA	3 h	7.29	NA	70 ppm	3–10 cm^3	UV	90.69%	NA	[31]
24	Methylene blue (MB) ethyl violet (EV)	AgNPs	2–80 min	4.0–9.0	0.03 and 0.05 mg/mL	50 mg/L	0.1–0.7 mL	UV	NA	70 and 57%	[32]
25	Methyl orange	Flower-like Rutile TiO_2	90 min	NA	NA	43 m^2/g	0.5–0.1 mM	UV	83.60%	NA	[33]
26	Remazol black-B	NA	30 min	7	NA	40–60 mg/L	770 mg/L	UV	82%	NA	[34]
27	Rhodamine B	HNO_3	70 min	6.5–12	0.01 mol dm^{-3}	6.26×10^{-6} mol dm^{-3}	NA	UV	NA	96%	[35]
28	Acid blue 74		200 min	5.5	25 mM	10^{-4} M	0.035 M	UV	NA	86%	[36]
29	Rhodamine B	$CoMoS_2$	60 min	3.2–10.7	0.4 g/l	10 mg/L	10 mmol/L	NA	NA	98	[37]
30	Basic blue 9 Acid red 183	Co^{2+}/PMS	5 min/30 min	7	0.13 mM/0.40 mM	7 mg/L, 160 mg/l	8×10^{-2} M	NA	NA	95%	[38]

(continued)

Table 1 (continued)

S/N	Target dye	Catalyst	Optimal experimental conditions					Performance efficiency Decolorization/degradation			Reference
			Time	pH	Catalyst dosage	Dye dosage	H_2O_2 dosage	Light source	UV or H_2O_2	UV + H_2O_2	
31	Reactive brilliant blue acid scarlet Reactive brilliant red	Fe^{2+}	30 min	3–4	20 mg/L	50 mg/L	100 mg/L	UV	NA	COD = 80% and TOC = 65%	[39]
32	Magenta flexographic	nZVI	60 min	2	60 mg/L	180 mg/L	11 mM	NA	91.95	NA	[40]
33	Bisphenol A (BPA)	Fe^{2+}	60 min	2.8–3.0				UV	NA	80–96%	[5]
34	Acridine orange Alizarin violet 3B	Fe^{3+}	160 min	2.9	$8 * 10^{-5}$ M and $2 * 10^{-4}$ M	$4 * 10^{-5}$ M and $2 * 10^{-4}$ M	0:07 M	NA	84.50%	NA	[16]
35	Rhodamine B	ZnO nanopowder	8 min	12	200 mg/100 ml	100 ml of 10 ppm	NA	UV	NA	>95%	[41]
36	Malachite green	$MgFe_2O_4$	50 s	7	0.04 g	10 mg/L	0.1 ml	UV		100%	[42]
37	Rhodamine B	Fe_3O_4	15 min	5	0.5 g/L	50 mL	40 mmol/L	NA	NA	95%	[43]
38	Methylene blue	ZnO	150 min	NA	0.25 mg	1.2 mg/L	20 ml	NA	NA	91%	[44]
39	Tetracycline (TTC)	Pyrite (Fe)	30 min	4.1	1 g/L	50 mg/L	5 mmol/L	NA	NA	85%	[45]
40	Reactive green 19	NA	20 min	6.5	NA	$6.3 * 10^{-5}$ M	2.5 and 120 mM	UV	100%	NA	[46]

weight aliphatic alcohols and acids. The study reported different sodium salts as co-existing chemicals have an inhibiting effect on the photodegradation performance. Primarily the hydroxyl radical generated from H_2O_2 on irradiation with UV light is seen to be responsible for the dye degradation [12]. In another study, Rhodamine B dye is degraded using hydroxyl and sulfate radicals in UV/H_2O_2 and UV/persulfate-based advanced oxidation processes. Low pH range was found to be favorable for both the systems. The Rhodamine B dye upon degradation gave rise to multiple low molecular weight carboxylic acids, namely; formic acid, acetic acid, and oxalic acid alone with some other inorganic components. Degradation corresponds to the destruction of aromatic benzene rings and the chromophore structures and the N-de-ethylation therein by the radicals.

Profiles from the degradation of RhB dye and products therein are presented in Fig. 1. At first formation and then decay curve were observed for lactic acid (i.e., $C_3H_5O_3^-$) in both UV/H_2O_2 and UV/PS processes. The amount of all other small carboxylic acids showed an increasing formation trend as a function of reaction time in both processes. Specifically, $C_3H_5O_3^-$ was one of the intermediates formed before RhB was mineralized into $HCOO^-$ and CH_3COO^- (smaller molecular weight). The subsequent decomposition of these intermediates upon further attack by radicals leads to a significant accumulation of $HCOO^-$ and CH_3COO^- in both UV/H_2O_2 and UV/PS processes. Meanwhile, a gradual accumulation of $NH4^+$ ions was recorded in both systems due to the oxidation of the N-de-ethylation. The degradation efficiencies of the RhB were almost 100% and the TOC removal was 50% and 60% for both UV/H_2O_2 and UV/PS processes, respectively [26].

Muruganandham et al. used H_2O_2/UV system for the degradation and decolorization of the reactive orange 4 (RO4), a reactive azo dye. The higher concentration of dye lowers the removal rate thorough photo-oxidation. It has been observed that instead of UV-H_2O_2 system, solar-H_2O_2 also works effectively for the photo-oxidation of the reported dye. Degradation followed the reaction of hydroxyl radicals generated upon photolysis of hydrogen peroxide with the dye to give less harmful

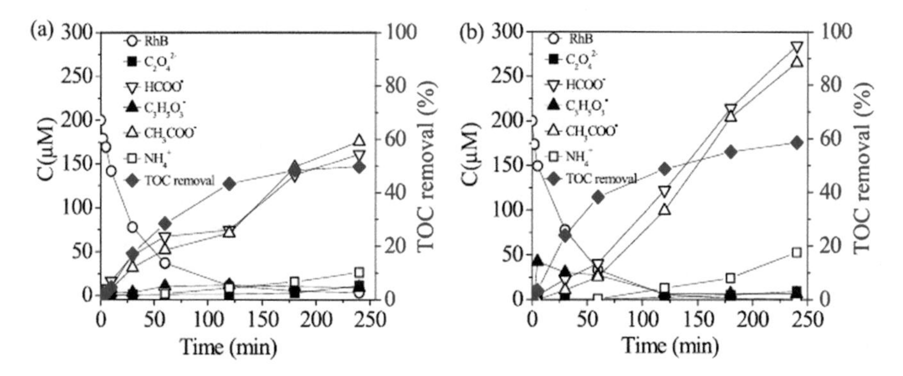

Fig. 1 Degradation and transformation of RhB and the subsequent formation of the small molecular acids and $NH4^+$ in **a** UV/H_2O_2 system and **b** UV/persulfate system [26]

products. Alone H_2O_2 did not find encouraging results as it showed 8.55% efficiency for decolorization and only 3.0% for degradation. The performance has significantly changed with the combined action of UV and H_2O_2 causing 88.68% decolorization and 59.85% degradation for the RO4 dye. Acidic pH is said the preferred for the purpose as the hydroperoxyl anion generated in the alkaline range reduces the removal rate. Higher concentration of peroxide and higher UV power are encouraged to enhance the performance [11].

The investigation by Ince et al. focused on the degradation of commonly used azo dye, Remazol Black-B. In addition, toxicity reduction was also checked for the dye during the treatment by a UV/H_2O_2 process. In a similar trend reported by multiple researchers, the dye was completely decolorized and converted to colorless intermediates but only 44% of the organic carbon was mineralized. SO decolorization efficiency was 100% while degradation was 44%. The influence of the initial H_2O_2 concentration on the rate of dye removal was found to be hindering beyond the operative H_2O_2 dose [34].

The efficiency of the oxidative degradation or mineralization of any organic dye can be estimated from the total organic carbon (TOC) of the wastewater. Any reduction in the TOC value or an increase in the inorganic carbon (IC) value implies to the degradation leading to the formation of an intermediate by ring opening and/or forming smaller molecules [47].

The blue sulfur dye solution as synthetic dye mimicking the textile wastewater system was treated with H_2O_2/UV system and the parameters were optimized. Decolorization efficiency was 100% while the TOC suggested the degradation of 85% which is attributed to the destruction of the chromophore of the dye into smaller fragment intermediates that are no longer visible and are not captured [31]. The azo dye Reactive Green 19 (RG19) is extensively used for dyeing textiles due to its cost-effectiveness and excellent fastness properties. Treatment of wastewaters containing this dye by an advanced oxidation process using UV radiation and hydrogen peroxide was investigated. Total organic carbon (TOC) measurements showed that mineralization was slower than decolorization. Decolorization was complete in 20 min while over 63% of TOC was removed in 90 min. The experimental data presented inferred that neither H_2O_2 nor UV alone had appreciable effects on the dye removal but shown significant effect when applied simultaneously [46]. Basturk et al. presented a study on decolorization of Reactive Blue 181 (RB181), an anthraquinone dye, by UV/H_2O_2 process. Hydroxyl ions and peroxy radicals are said to be responsible for the decolorization. The system effectively decolorized 99% of the dye in the quick span of 20 min under the acidic pH (2.0) condition. The system was not effective in the mineralization which is reflected in the low COD and TOC values [24].

Lei et al. studied 4-chlorophenol (4-CP), a dye intermediate and its degradation using H_2O_2 and phosphomolybic acid (PMA) system. 4-CP gets attached to PMA, thereby leading to ring opening or dichlorination, and the generation of multiple chemical species, such as phenol, benzoquinone, cyclohexanone, 4-chlorohexa-2,4-dien-1-ol, and 3-chlorohexa-2,4-diene-1,6-diol. These species were further mineralized under the effects of free radicals. The results indicated that 4-CP was fully converted, 19.81% to other species, and 44.52% to CO2 and H_2O [22]. Sharma et al. presented removal and mineralization of Bisphenol A (BPA) using inorganic hydrogen peroxide and sodium persulfate as oxidants under UV irradiation at a standard temperature and pH conditions. The system gave 95% BPA removal in 6 h of time for UV/$S_2O_8^{2-}$ system while it reached 85% for UV/H_2O_2 system. The TOC removal for the BPA aqueous solution was found to be 40% and 33% after 240 min of oxidation for UV/$S_2O_8^{-2}$ and UV/H_2O_2 systems, respectively [47].

In a novel approach, Navarro et al. reported the usage of pulsed light (PL) that emits high-intensity light in the broad spectrum of ultraviolet to infrared as the source of irradiation. These lights are common for microbial remediation in food processing industry. Pulsed light was used in case for the degradation of azo dye, direct yellow 106. The process was very fast and could be completed in a quick time of 30 s with 90% decoloration efficiency. The system gives an avenue to be explored for rapid decoloration in waste treatment technologies. It is important to note that neither H_2O_2 nor PL alone were sufficient and were ineffective in the decoloration process. The synergy of the two together only led to the typical radical generation and degrades the dye [13].

Degradation is a complex process and predictions about the intermediates are difficult. Considering the avenues in the field an agent-based model (ABM) for the prediction of degradation of organic compounds is presented wherein the chemical entities of individual molecular species are simulated for the movement and reactions of these entities over time in a defined space. The profiles so obtained for the test case of acetone in the UV/H_2O_2 system are compared with the experimental findings and are found to be consistent. The ABM-based modeling approach produced results that directed the understanding to the micro-level dynamics and underlying molecular interactions happening that led to the by-product formation [48].

Overall it is understood in the non-catalytic UV/H_2O_2 system that by appropriately tuning the process conditions one can effectively degrade the dyes by the UV/H_2O_2 treatment. Oxidant has its role to play for the generation of the radicals and the irradiation promotes the process. It is only the synergy of the two that can easily degrade many complex dyes.

4 Catalytic UV/H_2O_2 System—Iron-Based System

The section will focus on the UV/H_2O_2 system in combination with catalysts/photocatalysts/metal oxides for promoting the catalytic degradation of different dyes. Fenton-like systems are most common wherein forms of iron-based catalyst are used in the process. Reactive Orange 113, an azo derivative belonging to category of synthetic dyes, is degraded using the H_2O_2 Fenton-type system and ozonization process. Fenton system was able to reduce the COD by 40% and TOC by 31%, and the values indicate the degree of degradation caused is not complete. Ozonization system worked better with COD reduction by 44% and TOC by 36%. Decolorization of dye was complete in both the cases. In the typical degradation, at first, the azo bonds lose unsaturated character that is being transformed into hydroxyl structure.

On further oxidation it leads to bond splitting and nitro group formation and mono- or multi-substituted benzene and naphthalene compounds are produced. Continuing the oxidation gives rise to the opening of an aromatic ring to the formation of alcohols, aldehydes, and carboxylic acids. Finally, steps lead to mineralization to give CO_2 and H_2O [25].

Decolorization of the Mordant Red 73 (MR73) azo dye in water was investigated in laboratory-scale experiments using UV/H_2O_2 and photo-Fenton treatments by Elmorsi et al. The photo-Fenton treatment was highly efficient and resulted in 99% decolorization of the dye in 15 min. Fe powder was used as a source of Fe^{2+} ions in the photo-Fenton process. Degradation of dye was estimated from the chemical oxygen demand (COD) values. For the irradiation time of 3 h, H_2O_2/UV process was able to mineralize 65% and photo-Fenton treatment mineralized 85% of the dye [28]. Dong reported the kinetics of photocatalytic degradation of Reactive Brilliant blue KN-R, Acid Scarlet GR, and Reactive Brilliant red X-3B synthetic dyes by UV–vis/H_2O_2/Ferrioxalate complexes were studied. The COD and TOC removal for all the dyes were more than 80% and 65%, respectively, with UV irradiation, while in absence these values were reduced to 60% and 45%, respectively. Effect of UV radiation was significant in the process [39].

Kesi et al. study reveals the potential of nano-zero-valent iron (nZVI) particles as an effective catalyst in the Fenton-like systems. Magenta flexographic dye solution was used for testing the ability of nZVI for the degradation process. The functional nZVI was synthesized using the extract of oak leaves (OAK-nZVI). Definitive screening design was used in the study to understand the influence of four operating variables in decolorization process: initial dye concentration, catalyst dosage, pH value, and hydrogen peroxide concentration. Based on the optimization of parameters, a maximum yield of 91.95% was reported for following conditions: initial dye concentration of 180 mg/L, OAK-nZVI dosage of 60 mg/L, H_2O_2 concentration of 11 mM, and pH value 2 [40].

Synergetic effect of combination of metals is insightful for the process performance. The spinel $MgFe_2O_4$ was synthesized by solid-state reaction method and was applied as a catalyst for the degradation of Malachite Green dye. The catalyst showed 100% degradation of the dye in just 50 $_s$ in the presence of H_2O_2. The performance of the spinel in the process should be further explored for other dyes as well. It is one of the quickest degradations reported for any dye. The high rate of the degradation of the dye was due to Fenton mechanism involving hydroxyl and perhydroxyl radicals [42].

Tetracycline (TTC) is one of the extensively used drugs which accounts for 60% of the all antibiotics applied in the animal's therapeutics. Degradation of tetracycline (TTC) was studied with a pyrite/H_2O_2 process. Pyrite is obtained from the mine waste and mimics the Fenton-like system with H_2O_2. Pyrite used in the study is confirmed to be a mesoporous powder having pure FeS_2. The degradation of TTC in the presence of pyrite/H_2O_2 process was about 85% which is obtained in the acidic pH range and time span on 1 h. Some studies have reported that the oxidation of pyrite in aqueous solutions with dissolved oxygen generates H_2O_2 and hydroxyl radicals. This in situ generated peroxide along with additional H_2O_2 accelerates the degradation process.

Multiple mechanisms were proposed for the degradation; first being generation of hydroxyl radicals from the surface defects, second with pyrite-Fenton reaction, and third being reaction of O_2 as initiator with the pyrite surface. Scavenger's presence has affected the performance; phosphates and carbonates were able to hamper the degradation drastically reducing the degradation to 24%. Chlorides, sulfates, and nitrate could not induce much effect and degradation was in the range of 80–85% [45].

Jaafarzadeh et al. in his study investigated the removal of azo dye Reactive Orange 107 (RO107) by ultrasound assisted with magnetic Fe_3O_4 particles and H_2O_2 making it a sono-Fenton-like degradation process. The system when used of the real-time textile wastewater was able to reduce the COD from 2360 ppm to less than 500 ppm in 3 h of time. The study was intended to overcome the drawbacks of Fenton process that are lower rate of decomposition of peroxide, lower recyclability of catalyst, chemical sludge disposal, and high treatment cost. Ultrasound waves as energy source generate the hydroxyl and peroxy radicals in presence of ferric ions which helps in the degradation of dye. RO107 was 100% removed in 1 h time, but the degradation of TOC reached 87% even after 3 h. The process was able to reduce the components in actual textile wastewater and dye molecules to aliphatic acid, fatty acid, and aldehyde, such as acetic acid, myristic acid, and hexanal, respectively. The products of oxidation are significantly reduced after the treatment [14].

The study by Xie et al. targeted two ionic dyes: cationic Acridine Orange monohydrochloride (AO) and anionic Alizarin Violet 3B (AV) in a homogeneous Fenton degradation system under visible light. The electron spin trapping technique was used to identify the active radicals generated in the process by using the in situ laser irradiation. EPR spectra confirmed the presence of hydroxyl ions. The degradation process was initiated from the hydroxylation of the dye molecules which reacted with the hydroxyl ions generated from the decomposition of peroxide by the Fe source. Hydroxylation of the dye led to the ring opening and the intermediate products of degradation like N,N-dimethyl acetamide and N,N-dimethyl formamide, finally giving CO_2 and some mineralized products [16].

In a novel approach, sono-enhanced degradation of a dye pollutant Rhodamine B (RhB) was investigated by using H_2O_2 as a green oxidant and Fe_3O_4 magnetic nanoparticles as a peroxidase mimetic. It was found that Fe_3O_4 nanoparticles could catalyze the break of H_2O_2 to remove RhB in a wide pH range from 3.0 to 9.0 and its peroxidase-like activity was significantly enhanced by the ultrasound irradiation. As a tentative explanation, the observed significant synergistic effects were attributed to the positive interaction between cavitation effect accelerating the catalytic breakdown of H_2O_2 over Fe_3O_4 nanoparticles, and the function of Fe_3O_4 nanoparticles providing more nucleation sites for the cavitation inception. Because the breakage of H_2O_2 into radicals requires an energy as high as 213.8 kJ/mol, the dye pollutant RhB is hardly degraded by using H_2O_2 alone [43].

5 Photocatalytic UV/H_2O_2 System—TiO_2-Based System

TiO_2 as the photocatalyst has been widely explored for many reactions so does in the dye degradation. Arshad et al. reported photocatalytic degradation of reactive red (RR-147) dye using the combination of UV irradiation, hydrogen peroxide, and TiO_2. Under optimal conditions, 57% COD reduction was noted. The system exhibited maximum degradation efficiency of 92% for the optimum condition of 0.6 g photocatalyst TiO_2 and 50 ppm concentration of dye [18].

Morphological changes always present exciting results in the process. Kőrösia et al. synthesized hierarchically assembled flower-like rutile TiO_2 nanostructures (Fig. 2) with change in the hydrothermal temperature. The flower-like structures looked as a bundle of closely-packed rod-like structures. Overall morphology seems to be like the cauliflower, the size of the particle changes before and after the hydrothermal treatment. The obtained flower-like photocatalysts were tested in the degradation of methyl orange. All the photocatalysts shown an improved rate of degradation compared to the system devoid of H_2O_2. Highest degradation of 83.6% and 82.0% was observed with FLH-R-TiO_2/200 and FLH-R-TiO_2/250, respectively [33].

Soraya et al. presented an optimization and experimental study for the photocatalytic degradation of six commercial dyes, namely, Solophenyl orange TGL,

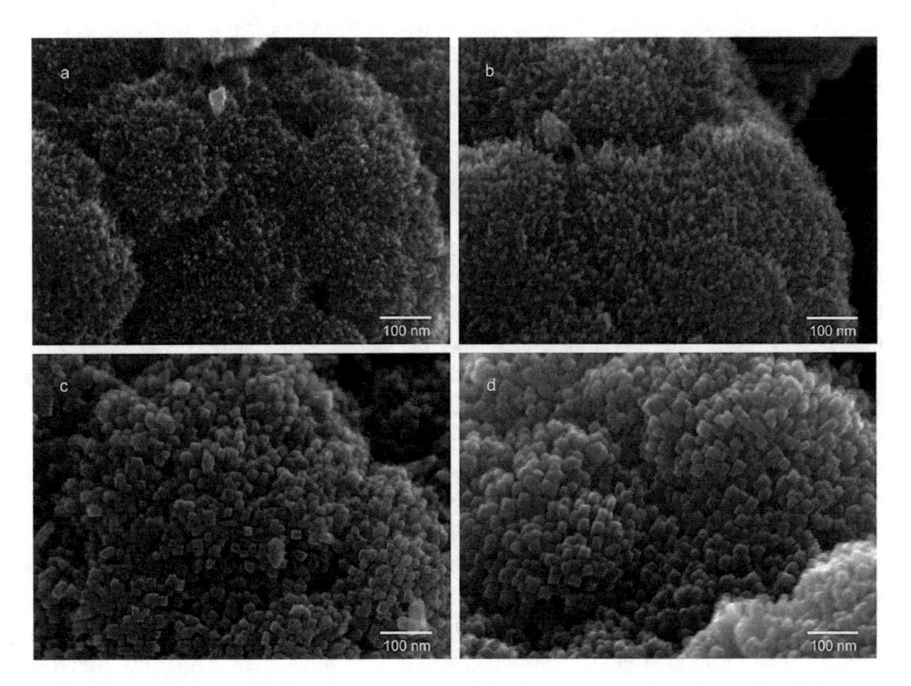

Fig. 2 High-resolution SEM images of **a** FLH-R-TiO_2/AP, **b** FLH-R-TiO_2/150, **c** FLH-R-TiO_2/200, and **d** FLH-R-TiO_2/250 [33]

Solophenyl blue 71, Solophenyl scarlet BNLE, Solophenyl yellow ARL, Solophenyl black FR, and Navy Blue 98. Optimum process parameters were reported based on the statistical approach response surface methodology based on a full factorial experimental design. Two parameters TiO_2 loading and H_2O_2 dosage were optimized. The optimum conditions for the experiment are reported as 0.5 g TiO_2/L and 50 mmol H_2O_2/L, respectively. Combination of UV/H_2O_2 system with photocatalyst was able to complete decolorization in 6 h of irradiation time for the tested azo dyes [20]. Abbasi reported the sono-catalytic process for the degradation of basic Blue 41 dye. H_2O_2 remains as the source for radicals while TiO_2 adds to catalyze the decomposition and further generation of radicals but these were incomplete in the performance without the sonication source. The results indicate the synergetic effect of nano-TiO_2, H_2O_2, and sonication together only yield good dye removal efficiency (~89.5%). Ultrasound irradiation was described as a possible generator of highly active radicals needed for the degradation process [49].

6 Photocatalytic UV/H_2O_2 System—ZnO-Based System

Jamil et al. reported the combination of ZnO with UV- and H_2O_2-based advanced oxidation process for the degradation of Disperse Red-60 (DR-60). The maximum degradation of 97% was achieved at optimum conditions of H_2O_2 (0.9 mL/L), ZnO (0.6 g/L) at pH 9.0 in 60 min irradiation time. Significant reduction in the COD (79%) and BOD (60%) level was observed while DO have seen a rise to 85.6% [19]. Nagaraja et al. presented usage of ZnO nanopowder was for the photocatalytic degradation of Rhodamine B (RB) comparing solar and UV irradiation. The crystallite size was in the range 12–50 nm. The results indicated that the decolorization of RB occurs at a faster rate under solar light in comparison to that of UV light. 99% of decolorization was observed in 8 min of under solar light, while it took 80 min in the presence of UV irradiation to get the same results. Similarly, the chemical oxygen demand (COD) reduction also takes place at a faster rate under solar light as compared to that of UV light. The crystallite size was the nanopowder which was altered by changing the calcination temperature from 600 to 1000 °C. The larger particle size has the larger diffusion path length and hence bulk recombination of charge carriers prevails over the interfacial charge transfer process and decreases the photocatalytic activity. For smaller particles, it happens vice versa [41]. ZnO nanostructures with two different morphologies were synthesized from the two different initial sources of zinc, zinc powder and zinc sheets, in study by Mohammadzadeh et al. The tetrapod- and multipod-structured ZnO particles are shown in Figs. 3 and 4. The bandgap energy of ZnO nanostructures was slightly different from the sheet (3.20 eV) and powder (3.21 eV) Zn sources. The photodegradation efficiency for multipod was 93% and for tetrapod structure was 91%, and these values were further improved to 97% and 94%, respectively, with the addition of peroxide. The change was attributed to the decreased recombination of electron–hole pairs on the surface of photocatalyst by the addition of peroxide [44].

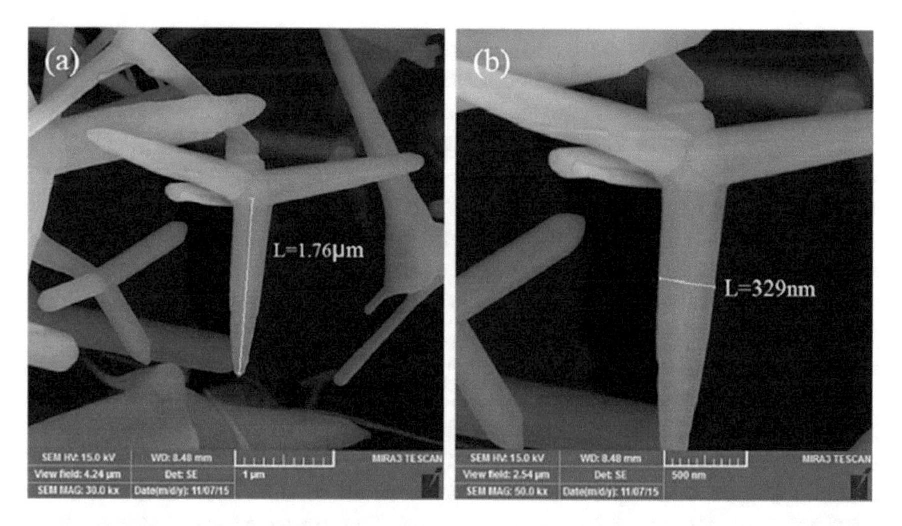

Fig. 3 FESEM micrographs with two different magnifications of tetrapod structures synthesized from the Zn powder at 957 °C [44]

Fig. 4 FESEM micrographs of ZnO nanostructures synthesized from the bulk Zn (Zinc sheets) at 957 °C under air atmosphere [44]

7 Photocatalytic UV/H_2O_2 System—Cobalt-Based System

The article by Macías-Quiroga et al. presented the response surface methodology (RSM) for the optimization of decolorization process for an azo dye, ponceau 4R in the cobalt-based catalytic system assisted with bicarbonate and peroxide. Molar ratio of H_2O_2 to bicarbonate, dosage of H_2O_2, and concentration of catalyst are the three experimental variables that were optimized using the Central Composite Design (CCD) approach for the decolorization of Ponceau 4R system. The optimal conditions were found as H_2O_2 dosage—4.73 times the stoichiometric dosage, molar ratio of H_2O_2 to bicarbonate—1.7, and the concentration for catalyst—11.16 μM to achieve 96.31% decolorization efficiency [29].

Ling et al. report cobalt-based catalytic degradation of basic blue 9 and acid red 183 using H_2O_2 and peroxymonosulfate (PMS). Active centers of cobalt metal were responsible for producing the hydroxyl and sulfate radicals from oxidants. Performance of sulfate system was better compared to the peroxide system. The ratio of metal to oxidant was optimized for the two systems. The optimum value was found to be 6 and 3 for Co^{2+}/H_2O_2 and Co^{2+}/PMS, respectively [38].

Novel catalyst is reported in the study by Han et al. The nanocatalyst consists of reduced graphene oxide nanosheets over which $CoMoS_2$ nanosphere is embedded. The nanocatalyst was said to be 21 times highly reactive compared to the conventional Fenton catalysts. The catalyst has remarkable reactivity for in situ generation of H_2O_2 and decomposition to radicals at different active centers, for the efficient degradation of the pollutants. Rhodamine B (RhB, 96%), methylene blue (MB, >98%), acid orange 7 (AO7, 99%), and methyl orange (MO, >98%) were tested for the catalytic degradation in presence and absence of additional peroxide. Mo-S-C bonding bridges in catalyst lead to activation of the π electrons and their transfer from graphene nanosheets to the metal centers. The formed Mo-O-Co further leads to a distribution of orientations of the electrons around the metal centers due to the different electronegativities of Mo and Co. During the reaction, the dissolved O_2 is efficiently reduced to peroxy radical or superoxide radical around the electron-rich Mo center, and these radicals are further reduced to H_2O_2 around the Co center. The generated H_2O_2 is finally reduced to hydroxyl radical that is responsible for degrading dyes. The mechanism is represented in Fig. 5 [37].

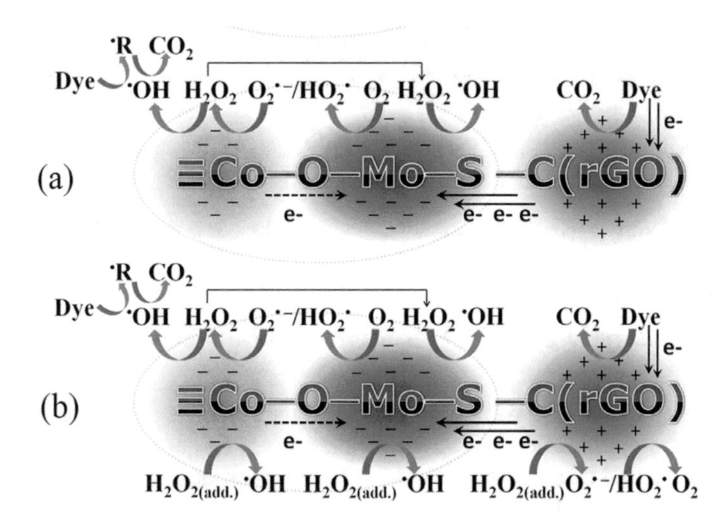

Fig. 5 Schematic for the novel Fenton-like reaction process in CMS-rGO NSs systems under different conditions. **a** Catalyst + H_2O/O_2 + Dye and **b** Catalyst + H_2O/O_2 + H_2O_2 + dye [37]

8 Catalytic UV/H_2O_2 System—Miscellaneous

Micropowders of copper one-dimensional complex as the catalyst was reported by Wang et al. for the photodegradation of methylene blue (MB) and indole-3-acetic acid (IAA) under UV irradiation assisted with H_2O_2.

Overall degradation efficiency for MB was 84.9% and that for IAA was 60.4%. The experimental results in Fig. 6 show that the degradation is primarily influenced by the presence of the complex micropowder. Presence of peroxide with UV irradiation too does not lead to significant degradation. It is only when the catalyst was added along with UV/H_2O_2 system that degradation was significantly high and reached 84.9% in 90 min time. Similar observations were made for the IAA where the degradation was drastically improved on the addition of catalyst. For IAA degradation was close to 12% in absence of catalyst which improved to 60.4% with the addition of catalyst. The possible mechanism of degradation in the presence of copper complex is presented in Fig. 7. In a typical mechanism, with UV irradiation, the electron transfers from the highest occupied molecular orbital to the lowest unoccupied molecular orbital. This transfer of electrons led to the holes behind. The electron transfer process generates the hydroxyl ions that are the strong oxidant responsible for the cleavage of the dye molecules [15].

Eroi et al. studied the effect of different sizes and morphologies tungsten oxide crystals as a heterogeneous Fenton-like catalyst for the degradation Indigo Carmine (IC) in the presence of H_2O_2 as oxidant. Indigo carmin release in the environment poses a serious threat to the human health-inducing cardiovascular and respiratory issues. Catalysts with different shapes were evolved with the changes in the calcination temperature. For non-calcined sample (NCT), it was more of irregular shape while the particles turned into broad rod-like structures with the rise in temperature to 600 °C (CT6). The material is confirmed to be hydrated tungsten oxide when not calcined, hexagonal phase and pure monoclinic crystalline tungsten oxide when calcined to 200 (CT2) and 600 °C, respectively. Different form of tungsten oxide showed variation in the performance. For NCT/H_2O_2, rate of removal was 45%,

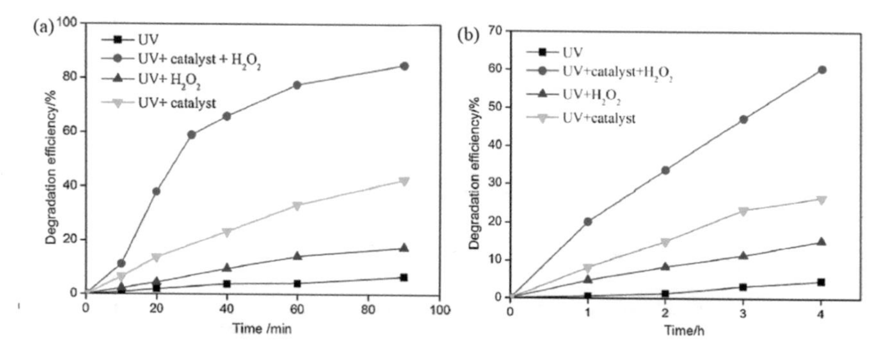

Fig. 6 Degradation performance of **a** methylene blue (MB) and **b** Indole-3-acetic acid (IAA) for different processes [15]

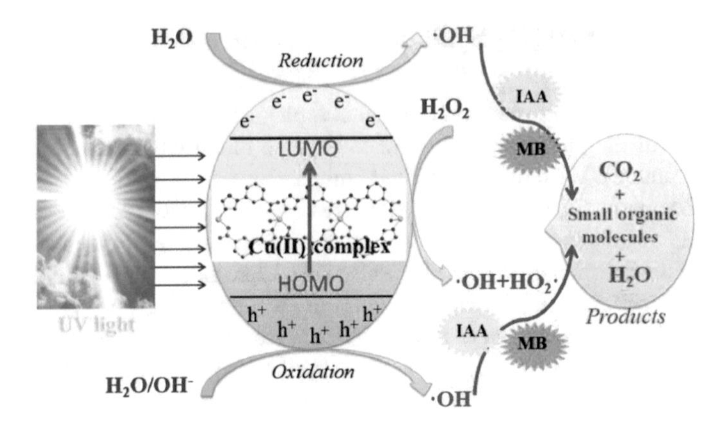

Fig. 7 Mechanism for the degradation of MB and IAA in the presence of 1D-Cu complex and H_2O_2 [15]

CT2/ H_2O_2 had 70%, and CT6/H_2O_2 showed 90% for the dye degradation (Fig. 8) [17].

The catalytic oxidation of an anionic Reactive Blue 19 (RB19) dye solution using a novel synthesized γ-MnO_2/MWCNT nanocomposite catalyst with H_2O_2 was reported. Morphological investigation confirmed the core–shell-type nanostructure formed. The birnessite-type manganese oxide is uniformly grown over as nanoflakes growth on the surface of the activated MWCNT. The schematic of the same is shown in Fig. 9. The developed catalyst 0.5MnO_2/MWCNT with H_2O_2 was able to completely decolorize the RB19 dye solution at close to neutral pH range. Lower

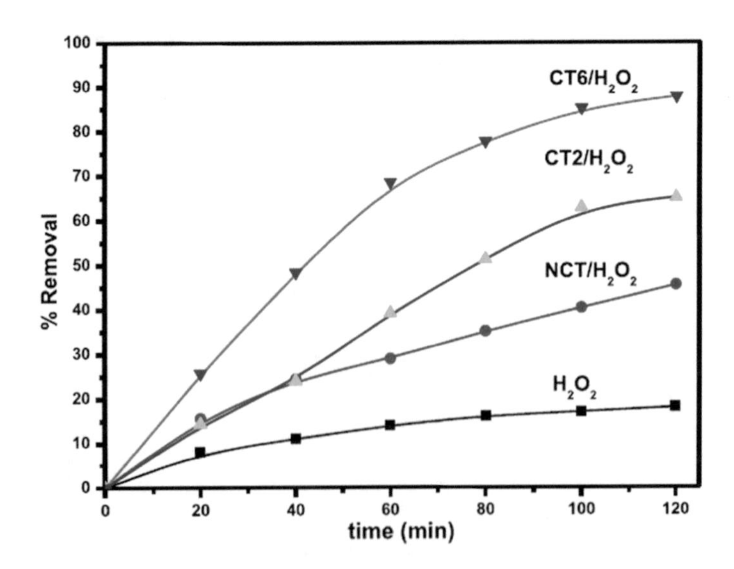

Fig. 8 Removal rate of indigo carmine with NCT/H_2O_2, CT2/H_2O_2, and CT6/H_2O_2 [17]

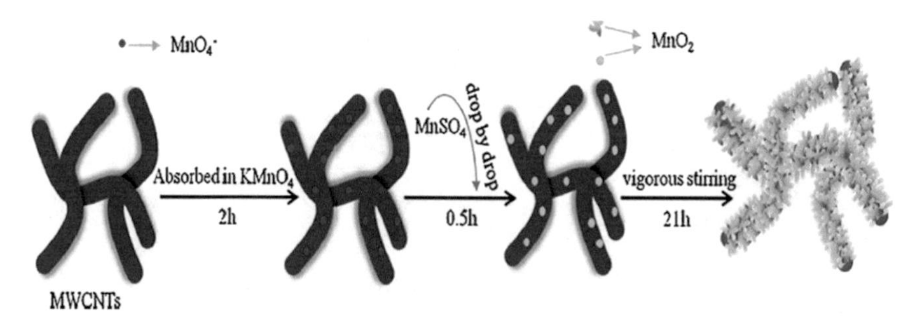

Fig. 9 Schematic diagram for the synthesis of MnO_2/MWCNT nanocomposite catalysts [30]

concentration of dye, higher dosage of H_2O_2, and acidic pH range are obtained as preferred range of operation [30].

Silver nanoparticles were synthesized using an aqueous extract of waste tea and employed in the degradation of two dyes, methylene blue (MB) and ethyl violet (EV), with and without hydrogen peroxide. Silver nanoparticles are well known for their anti-microbial activity. In the present case, this was taken as shuttle for the electron transfer between the tea extract and the organic dye. It is also responsible for the decomposition and allied reaction with peroxide to give radicals. The combined catalytic and the radicals affect the degradation of dyes [32].

The study by Sohrabi et al. was very indicative of performance of different catalysts. In case, the degradation of diazo Direct Yellow 12 dye (Chrysophenine G) was studied by using four different four photocatalysts, namely, TiO_2, ZnO, Fe_2O_3, and SnO_2 under UV irradiation. Dye solution was tested for degradation and decolorization with UV and TiO_2 alone and in combination. The results depict no significant changes in the degradation efficiency when UV or TiO_2 were used standalone. However, the combination of UV/TiO_2 system yields about 85.83% degradation of the solution as depicted in Fig. 10. UV/H_2O_2/TiO_2 system shows a comparable rate to UV/TiO_2 system and was able to give complete degradation of the dye at the system conditions (Fig. 11) [23].

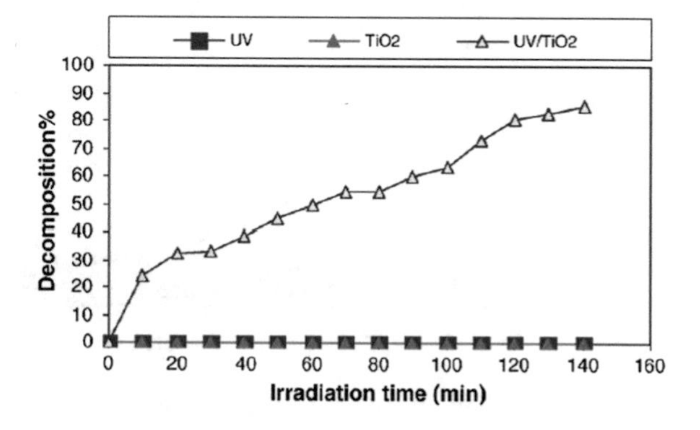

Fig. 10 Change in the decomposition of aqueous solution of direct yellow 12 dye as a function of irradiation time [23]

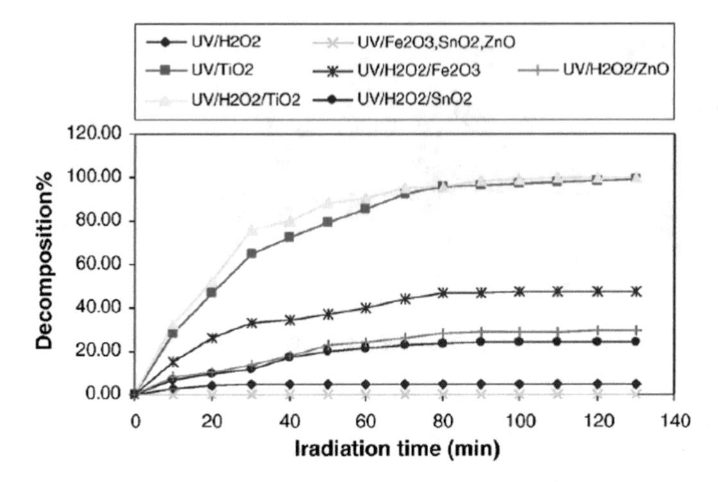

Fig. 11 Comparison of degradation rates for the decomposition of direct yellow 12 dye in the UV/H_2O_2 systems [23]

In similar comparative study, decolorization and degradation of three different commercial reactive azo dyes, namely, Procion Navy H-exl, Procion Crimson H-exl, and Procion Yellow H-exl with combination of UV, H_2O_2, TiO_2, and Fenton process. System parameters dictate the performance of process for decolorization and degradation. Fenton (H_2O_2/Fenton) and photo-Fenton (UV/H_2O_2/Fenton) processes were investigated. The Fenton processes show the highest rate for decolorization, reaching 96% in span of 5 min. TiO_2/UV/H_2O_2 process though slower compared to Fenton process was able to achieve 100% decolorization in 8 min. The rate of decolorization improves in the presence of UV irradiation (Fig. 12).

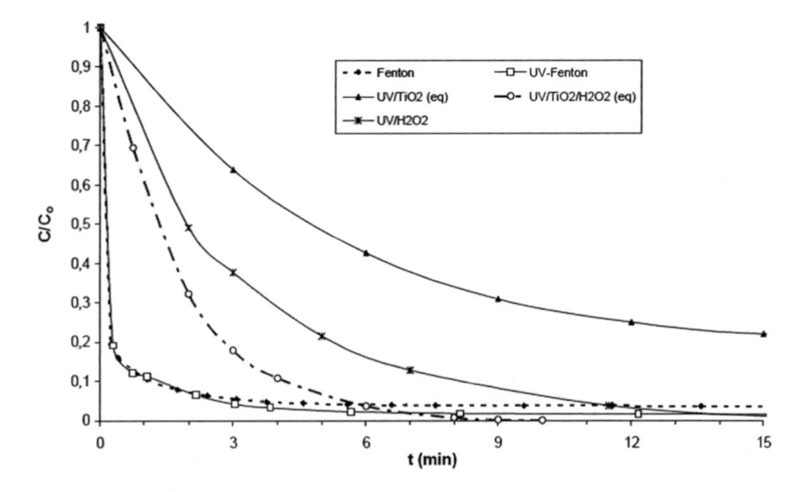

Fig. 12 Effect of different AOPs at the optimum conditions on dye degradation (C/C_o) [27]

The five different AOPs were compared for the degradation of dyes. Degradation rates also differ a lot among the five processes. The fastest mineralization is obtained with H_2O_2/UV process. The degradation rate wasn't much altered in either of the process. It is observed that with lower concentration of H_2O_2 (0.1% w/w) the mineralization was incomplete due to exhaustion of hydroxyl radical while with the higher concentration of H_2O_2 (1.0% w/w) complete degradation of all the organic fragments was possible to the final products [27].

9 Parameters Affecting Photocatalytic Dye Degradation

- **pH of Solution**

pH plays a substantial role in the degradation of H_2O_2/UV process with or without catalyst. The reported studies concluded to be more effective in an acid pH range when it comes to degradation or decolorization. Khan et al. have studied the degradation of model dye rhodamine B in the presence of different acids (nitric, sulphuric, hydrochloric, citric, and acetic acid), bases (sodium hydroxide and sodium carbonate), and both type of ionic surfactants (CTAB, SDS, and Tween 80). It was reported that HNO_3 was able to degrade the target dye completely in the quick time of 10 min and acted as the strong promoter. Citric acid, sulfuric acid, hydrochloric acid, and acetic acid degraded RhB to 96%, 88%, 78%, and 58%, respectively. The base components were able to degrade maximum of 38%. The addition of these components was effective only with UV irradiation. Surfactants were seen to influence the degradation process to an extent. However above the post-micellar concentration can slow the degradation by multiple interactions [35]. The decrease in degradation efficiency at higher pH (alkaline range) can be attributed to the fact that a part of H_2O_2 decomposes in the presence forming dioxygen and water rather than producing hydroxyl radicals under UV radiation. Without having substantial hydroxyl radicals the overall performance reduces. A detailed review on the AOPs at basic pH conditions gives a holistic view on the topic [7]. Even the co-presence of scavenging components like carbonates, phosphates, sulfate, and chlorides has the role to play in scavenging the hydroxyl radicals.

- **H_2O_2 Dosage**

Dose of H_2O_2 also plays a vital role in degradation process. The degradation efficiency increases with the dosage of H_2O_2 up to a certain optimum value after which the efficiency goes down [23, 26, 33, 46]. High level of H_2O_2 acts as a scavenger for hydroxyl radicals [50]. At high concentration, H_2O_2 starts to compete with the dye for reaction with hydroxyl radicals since HO_2• is less reactive than OH• radical. In addition, the OH• radicals generated at high concentration will readily dimerize to H_2O_2 [51]. Hence, the dose of H_2O_2 needs to be optimized so as to maximize the efficiency.

- **Contact Time**

Contact time is evidently known to affect the degradation process. Different dyes require different contact times for the degradation to get completed. The contact time may range from minutes to hours depending on the system (H_2O_2/UV + H_2O_2/catalytic process). Decolorization of the dyes is fast and further mineralization takes more contact time. Mordant Red 73 (MR73) azo dye 99% decolorization was achieved in just 15 min [28]. For degradation of malachite green dye with $MgFe_2O_4$, 100% degradation of the dye in just 50 s in the presence of H_2O_2 [42] while for Disperse Red-60 (DR-60) dye, the maximum degradation of 97% was achieved at optimum conditions in 60 min irradiation time [19]. The degradation of TTC in the presence of pyrite/H_2O_2 process was about 85% which is obtained in time span of 1 h [45]. Higher reaction time enhances the efficacy for color removal and mineralization process.

- **Initial Concentration of Dye**

Initial concentration of dye has an inverse relation to the efficiency of H_2O_2/UV process. The primary reason being the imbalance caused in the number of dye molecules and that of radicals responsible for degradation, in case hydroxyl and perhydroxyl. The numbers of radicals become insufficient with the increase in dye concentration. With multiple mechanisms working during the process, the competitive effect is not suitable. Moreover, higher concentration of dye will increase the optical density of solution and hence becomes more impermeable to UV radiation. Therefore, the rate of photolysis of hydrogen peroxide which depends directly on the incident intensity also decreases. Overall, with the increase in the initial dye concentration, the generation of hydroxyl radicals decreases and thus the efficiency of degradation decreases.

- **Presence of Co-ions in the Solution**

Aleboyeh et al. presented a detailed study on the effect of dyeing auxiliaries, specifically carbonate, bicarbonate, and chloride anions on UV/H_2O_2-based decolorization and degradation efficiency of Indigoid dye, Acid Blue 74 (AB74). All the three anions are understood to be inhibiting the degradation process; the extent of hindrance caused by the co-existing ion was evaluated in terms of color removal and total organic carbon (TOC). The presence of carbonate, bicarbonate, and chloride anions had no considerable effect on the total decolorization time. The degradation efficiency has shown a change with the different anions. With the presence of bicarbonate anion the degradation efficiency was increased initially, with the formation of active oxygen species that participate in the degradation process while in later part the rate was considerably decreased. The competitive rate for the active oxygen species was higher compared to the consumption of hydroxyl radicals by these anions that led to the improved rate of degradation. Adding the chloride had less decreasing effect compared to the bicarbonate [36]. Similar study wherein the presence of inorganic salts having chlorides, carbonates, phosphates, and sulfates (NaCl, Na_2CO_3,

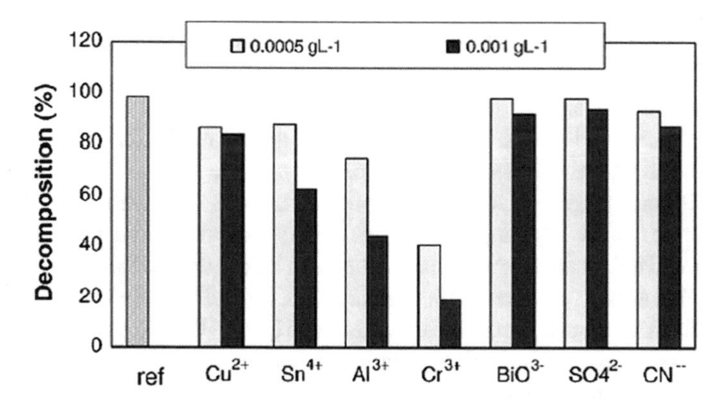

Fig. 13 Comparison of degradation rates for the decomposition of Direct Yellow 12 dye in the presence of cations and anions [23]

$NaHCO_3$, Na_2SO_4, $NaNO_3$, and Na_3PO_4) was investigated during the degradation Procion H-exl dyes. Decolorization efficiency was severely affected by the addition of the inorganic salts. Depending on the salt the decolorization reduced from 4.8% (nitrate) to 100% (phosphate) [27] (Fig. 13).

The occurrence of dissolved inorganic ions in the wastewater is inevitable. The UV/TiO_2 system was tested for the solution having co-presence of dissolved inorganic anions and cations (Cu^{2+}, Al^{3+}, Cr^{3+}, Sn^{4+}, BiO^{3-}, $SO_4{}^{2-}$, and CN^-). The results reflect the reduction in the effectiveness of system. It is noted that the inorganic ions compete with the dye molecules for the active sites on the TiO_2 surface along or at times these ions deactivate the photocatalyst. Overall effect of the co-existing ions is the reduction in the rate which is predominant at higher concentrations [23].

10 Conclusion

Dyes are an inseparable part of coloring industries and the applications cover the width but finds primary usage in the textile industries. The dyes as pollutant in the water pose serious threats to humans along with plant and animal life. Advanced Oxidation Processes (AOPs) pave its way as a novel approach for their mediation of dye-polluted wastewater. AOPs result in almost complete decolorization and degradation of the dyes. Chapter has provided a detailed review on a specific AOP, and UV/H_2O_2 system and its implications for the degradation of different dyes. The chapter covered the effect of different parameters like pH, H_2O_2 dosage, contact time, initial concentration of dye, and presence of co-ions in the solution on the performance of photocatalytic degradation. H_2O_2 and UV alone were ineffective in most cases. It is only the synergy of the two that was able to easily degrade many complex dyes. Over photocatalysts are inevitable for improved rate in the degradation of dyes. Iron (Fenton-like systems), TiO_2, ZnO, cobalt, and copper-based catalysts are mostly studied for the degradation of dyes. pH plays a substantial role in the degradation of H_2O_2/UV process with or without catalyst. The reported studies concluded to be more effective in an acid pH range when it comes to degradation or decolorization. Dosage of H_2O_2 was critical and has to be optimized, the right dosage gives the best results. Dye degradation has clearly shown direct dependence on the contact time. Higher the contact time, higher will be the degradation; however, type of dye and the system parameters dictate the rate of degradation. Initial concentration of dye has an inverse relation to the efficiency of H_2O_2/UV process. The numbers of radicals become insufficient with the increase in dye concentration and the imbalance caused is responsible for the reduced efficiency. The presence of dissolved inorganic ions in the wastewater also results in the reduction of the effectiveness of system. UV/H_2O_2 system with and without catalyst is an impressive AOP for the degradation and decolorization of dyes. The system has plenty advantages over the other AOPs. Correct tuning of the system parameters and the economic optimization of the process must be worked out for its effective application in the dye wastewater remediation.

References

1. Sudha M, Saranya A, Selvakumar G, Sivakumar N (2018) Microbial degradation of azo dyes: a review review article microbial degradation of azo dyes: a review. Int J Curr Microbiol Appl Sci 3:670–690
2. Thakur S, Chauhan MS (2018) Treatment of dye wastewater from textile industry by electrocoagulation and fenton oxidation: a review. 117–129. https://doi.org/10.1007/978-981-10-5795-3_11
3. Anwer H, Mahmood A, Lee J et al (2019) Photocatalysts for degradation of dyes in industrial effluents: opportunities and challenges review article Hassan. Nano Res 12:955–972. https://doi.org/10.1007/s12274-019-2287-0
4. Kalra SS, Mohan S, Sinha A, Singh G (2011) Advanced oxidation processes for treatment of textile and dye wastewater: a review. IPCBEE 4:271–275
5. Oturan MA, Aaron JJ (2014) Advanced oxidation processes in water/wastewater treatment: principles and applications. A review. Crit Rev Environ Sci Technol 44:2577–2641. https://doi.org/10.1080/10643389.2013.829765
6. Lavanya C, Dhankar R, Chhikara S, Sheoran S (2014) Review article degradation of toxic dyes: a review. Int J Curr Microbiol Appl Sci 3:189–199
7. Boczkaj G, Fernandes A (2017) Wastewater treatment by means of advanced oxidation processes at basic pH conditions: a review. Chem Eng J 320:608–633. https://doi.org/10.1016/j.cej.2017.03.084
8. Scholz DAYM (2019) Textile dye wastewater characteristics and constituents of synthetic effluents: a critical review. Springer, Berlin Heidelberg
9. Rekhate CV, Srivastava JK (2020) Recent advances in ozone-based advanced oxidation processes for treatment of wastewater: a review. Chem Eng J Adv 3:100031. https://doi.org/10.1016/j.ceja.2020.100031
10. Asghar A, Raman AAA, Daud WMAW (2015) Advanced oxidation processes for in-situ production of hydrogen peroxide/hydroxyl radical for textile wastewater treatment: a review. J Clean Prod 87:826–838. https://doi.org/10.1016/j.jclepro.2014.09.010

11. Muruganandham M, Swaminathan M (2004) Photochemical oxidation of reactive azo dye with UV-H_2O_2 process. Dye Pigment 62:269–275. https://doi.org/10.1016/j.dyepig.2003.12.006
12. AlHamedi FH, Rauf MA, Ashraf SS (2009) Degradation studies of Rhodamine B in the presence of UV/H_2O_2. Desalination 239:159–166. https://doi.org/10.1016/j.desal.2008.03.016
13. Navarro P, Gabaldón JA, Gómez-López VM (2017) Degradation of an azo dye by a fast and innovative pulsed light/H_2O_2 advanced oxidation process. Dye Pigment 136:887–892. https://doi.org/10.1016/j.dyepig.2016.09.053
14. Jaafarzadeh N, Takdastan A, Jorfi S et al (2018) The performance study on ultrasonic/Fe_3O_4/H_2O_2 for degradation of azo dye and real textile wastewater treatment. J Mol Liq 256:462–470. https://doi.org/10.1016/j.molliq.2018.02.047
15. Wang D, Zhao P, Yang J et al (2020) Photocatalytic degradation of organic dye and phytohormone by a Cu(II) complex powder catalyst with added H_2O_2. Colloids Surf A Physicochem Eng Asp 603:125147. https://doi.org/10.1016/j.colsurfa.2020.125147
16. Xie Y, Chen F, He J et al (2000) Photoassisted degradation of dyes in the presence of Fe^{3+} and H_2O_2 under visible irradiation. J Photochem Photobiol A Chem 136:235–240. https://doi.org/10.1016/S1010-6030(00)00341-5
17. Eroi SN, Ello AS, Diabaté D, Ossonon DB (2021) Heterogeneous WO_3/H_2O_2 system for degradation of Indigo Carmin dye from aqueous solution. South African J Chem Eng 37:53–60. https://doi.org/10.1016/j.sajce.2021.03.009
18. Arshad R, Bokhari TH, Javed T et al (2020) Degradation product distribution of reactive red-147 dye treated by UV/H_2O_2/TiO_2 advanced oxidation process. J Mater Res Technol 9:3168–3178. https://doi.org/10.1016/j.jmrt.2020.01.062
19. Jamil A, Bokhari TH, Iqbal M (2019) ZnO/UV/H_2O_2 based advanced oxidation of disperse red dye
20. Palácio SM, Espinoza-Quiñones FR, Módenes AN et al (2012) Optimised photocatalytic degradation of a mixture of azo dyes using a TiO_2/H_2O_2/UV process. Water Sci Technol 65:1392–1398. https://doi.org/10.2166/wst.2012.015
21. Tao P, Xu Y, Song C et al (2017) A novel strategy for the removal of Rhodamine B (RhB) dye from wastewater by coal-based carbon membranes coupled with the electric field. Sep Purif Technol 179:175–183. https://doi.org/10.1016/j.seppur.2017.02.014
22. Lei M, Gao Q, Zhou K et al (2021) Catalytic degradation and mineralization mechanism of 4-chlorophenol oxidized by phosphomolybdic acid/H_2O_2. Sep Purif Technol 257:117933. https://doi.org/10.1016/j.seppur.2020.117933
23. Sohrabi MR, Ghavami M (2010) Comparison of direct yellow 12 dye degradation efficiency using UV/semiconductor and UV/H_2O_2/semiconductor systems. Desalination 252:157–162. https://doi.org/10.1016/j.desal.2009.10.009
24. Basturk E, Karatas M (2015) Decolorization of antraquinone dye reactive blue 181 solution by UV/H_2O_2 process. J Photochem Photobiol A Chem 299:67–72. https://doi.org/10.1016/j.jphotochem.2014.11.003
25. Gutowska A, Kałuzna-Czaplińska J, Jóźwiak WK (2007) Degradation mechanism of reactive orange 113 dye by H_2O_2/Fe^{2+} and ozone in aqueous solution. Dye Pigment 74:41–46. https://doi.org/10.1016/j.dyepig.2006.01.008
26. Ding X, Gutierrez L, Croue JP et al (2020) Hydroxyl and sulfate radical-based oxidation of RhB dye in UV/H_2O_2 and UV/persulfate systems: kinetics, mechanisms, and comparison. Chemosphere 253:126655. https://doi.org/10.1016/j.chemosphere.2020.126655
27. Riga A, Soutsas K, Ntampegliotis K et al (2007) Effect of system parameters and of inorganic salts on the decolorization and degradation of Procion H-exl dyes. Comparison of H_2O_2/UV, Fenton, UV/Fenton, TiO_2/UV and TiO_2/UV/H_2O_2 processes. Desalination 211:72–86. https://doi.org/10.1016/j.desal.2006.04.082
28. Elmorsi TM, Riyad YM, Mohamed ZH, Abd El Bary HMH (2010) Decolorization of mordant red 73 azo dye in water using H_2O_2/UV and photo-Fenton treatment. J Hazard Mater 174:352–358. https://doi.org/10.1016/j.jhazmat.2009.09.057
29. Macías-Quiroga IF, Rojas-Méndez EF, Giraldo-Gómez GI, Sanabria-González NR (2020) Experimental data of a catalytic decolorization of Ponceau 4R dye using the cobalt

(II)/$NaHCO_3$/H_2O_2 system in aqueous solution. Data Br 30. https://doi.org/10.1016/j.dib.2020.105463
30. Fathy NA, El-Shafey SE, El-Shafey OI, Mohamed WS (2013) Oxidative degradation of RB19 dye by a novel γ-MnO_2/MWCNT nanocomposite catalyst with H_2O_2. J Environ Chem Eng 1:858–864. https://doi.org/10.1016/j.jece.2013.07.028
31. Amin H, Amer A, El Fecky A, Ibrahim I (2008) Treatment of textile waste water using H_2O_2/UV system. Physicochem Probl Miner Process 42:17–28
32. Qing W, Chen K, Wang Y et al (2017) Green synthesis of silver nanoparticles by waste tea extract and degradation of organic dye in the absence and presence of H_2O_2. Appl Surf Sci 423:1019–1024. https://doi.org/10.1016/j.apsusc.2017.07.007
33. Korösi L, Prato M, Scarpellini A et al (2016) H_2O_2-assisted photocatalysis on flower-like rutile TiO_2 nanostructures: rapid dye degradation and inactivation of bacteria. Appl Surf Sci 365:171–179. https://doi.org/10.1016/j.apsusc.2015.12.247
34. Ince NH, Stefan MI, Bolton JR (2017) UV/H_2O_2 degradation and toxicity reduction of textile azo dyes: Remazol black-B, a case study. J Adv Oxid Technol 2:442–448. https://doi.org/10.1515/jaots-1997-0312
35. Khan AM, Mehmood A, Sayed M et al (2017) Influence of acids, bases and surfactants on the photocatalytic degradation of a model dye rhodamine B. J Mol Liq 236:395–403. https://doi.org/10.1016/j.molliq.2017.04.063
36. Aleboyeh A, Kasiri MB, Aleboyeh H (2012) Influence of dyeing auxiliaries on AB74 dye degradation by UV/H_2O_2 process. J Environ Manage 113:426–431. https://doi.org/10.1016/j.jenvman.2012.10.008
37. Han M, Lyu L, Huang Y et al (2019) In situ generation and efficient activation of H_2O_2 for pollutant degradation over $CoMoS_2$ nanosphere-embedded rGO nanosheets and its interfacial reaction mechanism. J Colloid Interface Sci 543:214–224. https://doi.org/10.1016/j.jcis.2019.02.062
38. Ling SK, Wang S, Peng Y (2010) Oxidative degradation of dyes in water using Co^{2+}/H_2O_2 and Co^{2+}/peroxymonosulfate. J Hazard Mater 178:385–389. https://doi.org/10.1016/j.jhazmat.2010.01.091
39. Dong X, Ding W, Zhang X, Liang X (2007) Mechanism and kinetics model of degradation of synthetic dyes by UV-vis/H_2O_2/ferrioxalate complexes. Dye Pigment 74:470–476. https://doi.org/10.1016/j.dyepig.2006.03.008
40. Kecić V, Kerkez Đ, Prica M et al (2018) Optimization of azo printing dye removal with oak leaves-nZVI/H_2O_2 system using statistically designed experiment. J Clean Prod 202:65–80. https://doi.org/10.1016/j.jclepro.2018.08.117
41. Nagaraja R, Kottam N, Girija CR, Nagabhushana BM (2012) Photocatalytic degradation of Rhodamine B dye under UV/solar light using ZnO nanopowder synthesized by solution combustion route. Powder Technol 215–216:91–97. https://doi.org/10.1016/j.powtec.2011.09.014
42. Das KC, Dhar SS (2020) Rapid catalytic degradation of malachite green by $MgFe_2O_4$ nanoparticles in presence of H_2O_2. J Alloys Compd 828:154462. https://doi.org/10.1016/j.jallcom.2020.154462
43. Wang N, Zhu L, Wang M et al (2010) Sono-enhanced degradation of dye pollutants with the use of H_2O_2 activated by Fe_3O_4 magnetic nanoparticles as peroxidase mimetic. Ultrason Sonochem 17:78–83. https://doi.org/10.1016/j.ultsonch.2009.06.014
44. Mohammadzadeh A, Khoshghadam-Pireyousefan M, Shokrianfard-Ravasjan B et al (2020) Synergetic photocatalytic effect of high purity ZnO pod shaped nanostructures with H_2O_2 on methylene blue dye degradation. J Alloys Compd 845:156333. https://doi.org/10.1016/j.jallcom.2020.156333
45. Mashayekh-Salehi A, Akbarmojeni K, Roudbari A et al (2021) Use of mine waste for H_2O_2-assisted heterogeneous Fenton-like degradation of tetracycline by natural pyrite nanoparticles: catalyst characterization, degradation mechanism, operational parameters and cytotoxicity assessment. J Clean Prod 291:125235. https://doi.org/10.1016/j.jclepro.2020.125235

46. Zuorro A, Lavecchia R (2014) Evaluation of UV/H_2O_2 advanced oxidation process (AOP) for the degradation of diazo dye reactive green 19 in aqueous solution. Desalin Water Treat 52:1571–1577. https://doi.org/10.1080/19443994.2013.787553
47. Sharma J, Mishra IM, Kumar V (2015) Degradation and mineralization of Bisphenol A (BPA) in aqueous solution using advanced oxidation processes: UV/H_2O_2 and $UV/S_2O8(2-)$ oxidation systems. J Environ Manag 156:266–275. https://doi.org/10.1016/j.jenvman.2015.03.048
48. Zupko R, Kamath D, Coscarelli E et al (2020) Agent-based model to predict the fate of the degradation of organic compounds in the aqueous-phase UV/H_2O_2 advanced oxidation process. Process Saf Environ Prot 136:49–55. https://doi.org/10.1016/j.psep.2020.01.023
49. Abbasi M, Razzaghi-Asl N (2008) Sonochemical degradation of basic blue 41 dye assisted by nanoTiO_2 and H_2O_2. J Hazard Mater 153:942–947. https://doi.org/10.1016/j.jhazmat.2007.09.045
50. Dionysiou DD, Suidan MT, Baudin I, Laîné JM (2004) Effect of hydrogen peroxide on the destruction of organic contaminants-synergism and inhibition in a continuous-mode photocatalytic reactor. Appl Catal B Environ 50:259–269. https://doi.org/10.1016/j.apcatb.2004.01.022
51. Aleboyeh A, Moussa Y, Aleboyeh H (2005) The effect of operational parameters on UV/H_2O_2 decolourisation of acid blue 74. Dye Pigment 66:129–134. https://doi.org/10.1016/j.dyepig.2004.09.008

Ozone-Based Processes in Dye Removal

Qomarudin Helmy, I. Wayan K. Suryawan, and Suprihanto Notodarmojo

Abstract The general practice carried out by the textile industry in treating its wastewater is chemically, physically, biologically or a combination of the three. As a case study that occurred in Indonesia, regulations regarding color parameters were not regulated until 2019. To cope with the new and more stringent regulatory threshold values, many textile industries have to modify or even rebuild their existing wastewater treatment plants (WWTPs) by adapting the latest wastewater treatment technologies. One promising alternative that can be added and/or modified to the existing WWTP is advanced oxidation using ozone as an oxidizing agent. The use of ozone in textile industry wastewater treatment applications has several advantages, including a high oxidation power that requires a relatively short contact time (CT) in the order of minutes to oxidize impurities contained in wastewater. This chapter will discuss a brief history of ozone use in water and wastewater treatment, its chemistry and generation methods, degradation process, mechanisms, and factors affecting dye removal using ozone, also its practical application and integration with the existing WWTP process.

Keywords Textile · Wastewater · Dye · Ozone · Oxidation · Decolourization · Hydroxyl radical · Degradation · Organic compound · Color removal · Color standard

---

Q. Helmy (✉)
Bioscience and Biotechnology Research Center, Institute of Technology Bandung, Ganesa St. 10, Bandung, West Java 40132, Indonesia
e-mail: helmy@tl.itb.ac.id

Q. Helmy · S. Notodarmojo
Water and Wastewater Engineering Research Group, Faculty of Civil and Environmental Engineering, Institute of Technology Bandung, Ganesa St. 10, Bandung, West Java 40132, Indonesia

I. W. K. Suryawan
Environmental Engineering Study Program, University of Pertamina, Teuku N.A. St., Simprug, Kebayoran Lama, Jakarta 12220, Indonesia
e-mail: i.suryawan@universitaspertamina.ac.id

S. S. Muthu and A. Khadir (eds.), *Advanced Oxidation Processes in Dye-Containing Wastewater*, Sustainable Textiles: Production, Processing, Manufacturing & Chemistry, https://doi.org/10.1007/978-981-19-0882-8_6

1 Introduction

Textile manufacturing is one of the industries that consume a lot of water and generates a lot of wastewater. Over 80% of the water consumed by the textile industry will eventually degrade into wastewater that must be treated before being released into the environment. The average amount of water consumed in the wet process of textile manufacturing is 150 m^3/ton of product [40]. Fabric dyeing is one of the textile processes that consumes the most water and produces the most wastewater. This process consumes between 80 and 200 L of water per kilogram of cloth [63, 70, 100] when dyeing cotton cloth conventionally. This raises questions about the availability and necessity of safe drinking water, particularly in industrial areas. The textile industry, in general, relies on deep groundwater as a source of raw water. This activity has the potential to degrade the aquifer and result in land subsidence. Between 2000 and 2012, land surface measurements revealed that several locations in Indonesia's textile industry experienced land subsidence ranging from 8 to 17 cm/year, which was attributed to groundwater extraction by textile industry activities [32].

The textile industry's wastewater contains residual dyes that are hazardous, toxic, mutagenic, and carcinogenic [42]. For instance, consider dyes containing an azo chromophore, which accounts for between 50 and 70% of all dyes used in the textile industry. This azo compound has been shown in numerous studies to cause cell mutations and may be carcinogenic. Another chromophore is phthalocyanine, which contains three to four cyanin groups bonded to Cu, Cr, or Co metals. Because these dyes have a large molecular size, they are extremely difficult to decompose naturally. 10–50% of the dye used in the dyeing process is wasted as wastewater because it cannot be fixed to the fiber. Around 700,000 tons of dye are consumed annually, putting a strain on the textile industry, which is widely regarded as a major polluter [80]. Around 280,000 tons of textile dyes are discharged as textile wastewater effluent each year [41]. Dyes are non-biodegradable due to their complex chemical structure that contains cyclic bonds and high molecular weight. Certain dyes also contain azo compounds that are capable of producing one or more aromatic amine compounds at concentrations greater than 30 ppm [17]. Additionally, certain azo compounds that are carcinogenic are prohibited.

Colored waste is psychologically more frightening to society in addition to its very disturbing aesthetics. The minimum concentration of color that can be seen visually in river flows is around 1–10 mg/L, depending on color, illumination, and the degree of water clarity. In addition to aesthetics, colored substances in water will block the transmission of light through the water so that it can interfere with the photosynthesis process which results in an ecological imbalance. In general, the textile industry in Indonesia uses WWTP with coagulation-flocculation processes, biological processes, or a combination of both processes, with high BOD and COD removal efficiency but not for color removal. Several AOPs, such as ozone, Fenton, photo-Fenton, photocatalytic, and UV-based oxidation have been carried out to overcome this problem. It is reported that the use of ozone has a good ability to remove color, but its low solubility in water and the use of high energy in producing ozone are one of the drawbacks of this method.

Ozone (O_3) is an allotropic form of oxygen (O_2) that consists of identical atoms but is consolidated in a different structure. The distinction is that oxygen has only two oxygen atoms, whereas ozone has three. Ozone has a low molecular weight (MW = 48 g/mol) and is made up of three oxygen atoms arranged in chains. Ozone is a gaseous compound found in the atmosphere that is formed as a result of ultraviolet radiation [43]. For centuries, ozone has been used to disinfect water in drinking water treatment plants and to help remove unpleasant odors and organic/inorganic contaminants [54, 72]. Ozone has been used for a long time in European countries, and its use in the food and beverage industries has recently become more widespread [33]. In 1982, the Food and Drug Administration (FDA) of the United States granted ozone in bottled water safe status. In addition, in 1997, the FDA declared ozone to be Generally Recognized as Safe (GRAS) for use in food processing [30]. In addition, the FDA recognized and approved ozone as an antimicrobial agent as a food additive in 2001 [26, 58].

Ozone is naturally formed in small amounts (0.05 mg/L) in the stratosphere by the action of ultraviolet irradiation on oxygen. Small amounts of ozone are also formed in the troposphere as a byproduct of photochemical reactions involving oxygen, nitrogen, and hydrocarbons emitted by industries, oil-fired engine exhaust, forest fires, and volcanic eruptions. However, the gas produced is extremely unstable and decomposes rapidly in the air [47]. Diatomic oxygen molecules must first be broken down in order to produce ozone. As a result, the produced oxygen free radical freely reacts with other diatomic oxygens to form triatomic ozone molecules. However, breaking the O–O bond requires a significant amount of energy [10, 34]. When ozone is used in industry, it is typically produced at the point of application and in closed systems. To initiate the formation of oxygen free radicals and thus generate ozone, ultraviolet radiation (wavelength of 188 nm) and corona discharge techniques can be used. Corona discharge techniques are commonly used for large-scale applications to generate commercially feasible ozone concentrations [24].

This chapter summarizes ozone's physicochemical properties, the mechanism of ozone formation, and the use of ozone in water and wastewater treatment. Ozone can be produced using a variety of methods, including quiescent discharge, phosphorus contact, photochemical reactions, and electrochemical reactions, all of which involve the reaction of oxygen atoms with oxygen molecules. However, ozone depletion is caused by side reactions to ozone formation, such as thermal decomposition and cooling reactions by reactive species. Ozone's solubility in water is greater than that of oxygen, indicating that it can be used safely in water and wastewater treatment. According to the ozone resonance structure, one oxygen atom in the ozone molecule is electron deficient and has electrophilic properties, while the other oxygen atom has nucleophilic properties.

1.1 History of Ozone Use in Water and Wastewater Treatment

Ozone has extensive research and application history. In 1785, a Dutch chemist named Van Marum was perhaps the first scientist to detect the presence of ozone through the distinct odor of the air in the vicinity of his electrostatic generator when exposed to electric sparks. Another scientist, William Cruickshank, a Scottish chemist discovered the same distinctive gas odor near the anode during water electrolysis in 1801 [51, 64, 84]. M. B. Rubin has written an excellent series of papers on the history of ozone discovery [9, 49, 68, 76–79, 98, 99]. The history of ozone application and regulation in water and wastewater is shown in Table 1.

Due to the Corona Virus Outbreak in 2019, treating wastewater with ozone, primarily for disinfection, has been a primary priority. According to a paper published in 2021, the virus has the potential to spread through the raw water and wastewater networks. In places with inadequate sanitation systems, particularly in developing nations, the risk of exposure via the fecal–oral route due to its excretion into the sewer has also been emphasized. Although the virus's infectivity is unknown, the presence of the virus in human feces was verified even one month after the patient had tested negative for COVID-19. The risk is higher in third-world countries where open defecation is common. According to WHO figures from 2010, 1.1 billion people, or 17% of the world's population, still defecate in open spaces. Open defecation is practiced by up to 81% of the population in ten nations around the world, with Indonesia being the second-largest behind India in terms of open defecation [92].

Commercially produced ozone for oxidation reactions is always created as a gas, from the air at concentrations ranging from 1.0 to 2.0 wt%, or from liquid oxygen at concentrations more than 2%. Ozone cannot be kept as gas and transported because it is extremely reactive and has a short half-life. As a result, ozone is always produced on-site and used immediately. When ozone is used as a gas to purify drinking water, it is primarily for its oxidative properties. Because of its high oxidizing potential, ozone can effectively reduce or eliminate color, residual taste, and odor. More importantly, ozone is more effective than other disinfectants at killing dormant bacteria and viruses. Metal contaminants can be oxidized by ozone, for example, iron and manganese can be oxidized into iron (III) and manganese (III) forms, which are easier to remove by filtration. The same procedure is used to free organically bound heavy metals that would otherwise be difficult to remove. When used correctly in the water treatment process, ozone does not produce halogenated chemicals like Trihalomethanes (THMs), which are produced when chlorine is introduced to raw water containing humic components. When employed in the last phases of water treatment, ozone can be used as an oxidant. Ozone is used to purify drinking water in over 2,000 big installations across the world, not to mention in small-scale domestic drinking water systems. Ozone is a disinfectant that may be used to clean municipal and industrial wastewater, as well as a number of complicated and dangerous substances.

Table 1 Ozone's history, uses, and regulation

Year	Achievement	Reference
1985	Martinus Van Marum was probably the first known scientist who observed the presence of the specific odor which will later be named ozone	[51, 76, 84]
1801	William Cruickshank observed the same specific gas odor that formed near the anode during the electrolysis of water	[73]
1840	Christian Friedrich Schonbein, a chemistry professor at the University of Basel, concluded that the odor created by electric sparking was created by an unknown molecule he named ozone, which he derived from the Greek word "Ozein," which means to smell	[68, 76]
1856	Thomas Andrew, a chemistry professor at Queen's University of Belfast, demonstrated that ozone, regardless of its source, is a single body with equal properties and constitution, and is not a compound substance, but oxygen in a transformed or allotropic state	[77, 99]
1857	Werner von Siemens invented the apparatus of an electric discharge ozone generator, and only this invention made industrial applications of ozone possible at that time	[73]
1865	Jacques Louis Soret established the chemical geometry of ozone and the link among oxygen and ozone by observing that three volumes of oxygen make two volumes of ozone	[77]
1870	Lender, a German physician, conducted the first study on the biological effects of using ozone in water treatment and its antibacterial capabilities. This discovery ultimately transformed medical treatment during this time period, more than half a century before penicillin was discovered	[84]
1893	In Oudshoorn, the Netherlands, the first ozone-based drinking water treatment prototype plant was developed. After sedimentation and filtering, ozone was used to purify the Rhine River's water. After seeing this apparatus, a group of French scientists and chemists decided to establish their own facility in Nice, France	Rice et al. (2018) [75]
1900	Tesla Ozone Co. is founded and begins marketing ozone generators for medical applications	[67]
1906	France commissions the installation of its first ozone disinfection unit at their water treatment plant in Nice, where ozone was being used to disinfect nearly 22,500 m^3/day raw water collected from the Vesubie River after being filtered by a slow sand unit	[73]
1909	Ozone is used as a preservative for meat cold storage in Germany	[34]
1914	Popularity in ozone for water treatment starts to decline as a result of numerous investigations that resulted in the development of inexpensive chlorine gas as a disinfectant	[98]

(continued)

Table 1 (continued)

Year	Achievement	Reference
1916	Throughout Europe, around 49 ozone installations were in operation, with 26 of them located in France	[49]
1920	Dr. Edwin Parr, a Swiss dentist, began employing ozone as a component of his sterilization system. Dr. Charles S. Neiswanger, Professor of Medicine at the Chicago Hospital College of Medicine, publishes "Electro Therapeutical Practice," which includes a chapter on "Ozone as a Therapeutic Agent"	[29, 64]
1931	Dr. E. A. Fisch pioneered its use in dentistry through his use of ozonated water during dental treatments	[50]
1936	In France, ozone is used to purify seafood	[34]
1939	Ozone was utilized to inhibit yeast and mold growth during the preservation of fruits	[34]
1942	In the United States of America, ozone is utilized in egg storage areas and cheese storage facilities	[34]
1957	Ozone is used in Germany's drinking water to oxidize humic compounds, undesirable odors and tastes, iron, and manganese	[49]
1964	The French built an ozone plant to boost particulate matter removal as a result of spontaneous flocculation in ozone contact ponds	[49]
1965	The United Kingdom and Ireland begin using ozone to regulate the color of surface water. Switzerland began oxidizing micropollutants such as phenolic substances and certain pesticide residues using ozone	[49]
1970	In France, ozone is used to suppress algae overgrowth	[49]
1982	United States Food and Drug Administration (FDA) has designated ozone disinfection as GRAS (generally regarded as safe) for use in the bottled drinking water business	[34]
1987	After seven years of testing, a 600 million gallon per day ozonation plant was constructed in Los Angeles, California, USA	[49]
1992	Russia reports the use of ozone in salt water baths for burn treatment applications	[67]
1996	Japan's government, Canada's Food Inspection Agency (CFIA), and Australia's government have all approved the use of ozone in direct contact with all kinds of food	[34]
1997	In the United States, ozone is classified as Generally Recognized as Safe/GRAS	[34]
2001	The United States Department of Agriculture has approved the use of ozone as an antimicrobial agent in direct contact with food. It is now permitted to be used on all meat and poultry products	[103]
2004	The International Ozone Association reports that between 1969 and 2004, 894 ozone installation projects were completed, totaling 21,246 kg/h of ozone	[49]
2008	Montreal, Canada, intends to be the world's first metropolis to disinfect all of its waste water treatment plants using ozonation, with an ozone capacity of approximately 1,800 kg/h	[49]

(continued)

Table 1 (continued)

Year	Achievement	Reference
2009	Over 50 water treatment plants in Japan have installed ozonation as an advanced process for the removal of color, odor, and taste, as well as for the control of trihalomethanes (THMs) formation, with an ozone capacity of approximately 800 kg/h (42,000 lb/day). Ozone-treated reclaimed wastewater is being used in excess of 100,000 m^3/day	[36, 88]
2010	The installed ozone capacity to treat drinking water in the United States surpasses 525,000 lb/day. During the period 2005–2010, there were seven ozone-treated wastewater treatment plants, with a flow capacity of 60 MGD and an ozone capacity of 2,000 lb/day	[49]
2020	Coronavirus disease (COVID-19) virus inactivation using ozone as a possible oxidant. Even after final disinfection with chlorine, ozonated nanobubbles were used in the Hospital Wastewater Treatment Plant to remove the lingering SARS-CoV-2 remains	[3, 8, 46, 52, 89, 97]

1.2 Ozone Chemistry

Ozone generation can only be done on-site with an ozone generator that can be run economically and consistently due to the unstable nature of ozone, which spontaneously reverts back to oxygen. Understanding the physical and chemical properties of ozone, particularly its solubility and chemical reactivity to the pollutant chemicals of concern, is critical for its application in water and wastewater treatment. Ozone is made up of three oxygen atoms that are bound together at an angle of 116.8° by the same oxygen–oxygen bonds. It can't form a triangle because of the steric hindrance, which prevents each oxygen atom from making the anticipated two bonds. Instead, each oxygen creates only one link, with the rest of the molecule's negative charge dispersed. Ozone is a dipole molecule, which gives it the traits of being electrophilic and reacting very selectively (Fig. 1).

Although ozone is around 10 times more soluble in water than oxygen, it is only modestly soluble in water [98]. At 20 °C, the solubility of 100% ozone is around 570 mg/L, while at 30 °C, it drops to 400 mg/L (Table 2). Although ozone has a higher solubility than oxygen, chlorine has a solubility in water that is 12 times that of ozone. The presence of sensitizing impurities, such as metal oxides and heavy-metal cations, as well as temperature and pressure, affect the stability [37]. In general, lowering the temperature or increasing the pressure increases ozone solubility in the aqueous phase. Ozone concentrations in water treatment are typically less than 12%, limiting its mass transfer pushing power into the water. This results in a modest ozone concentration in the water, ranging from 0.1 to 1 mg/L. Furthermore, because ozone has a short half-life in water (20 min at 20 °C), if we need a specific amount of ozone to react, we must inject a bigger amount of ozone.

Numerous studies have demonstrated that ozone degrades autonomously during water treatment via a complex mechanism including the formation of hydroxyl free radicals. Although hydroxyl free radicals are one of the most reactive oxidizing

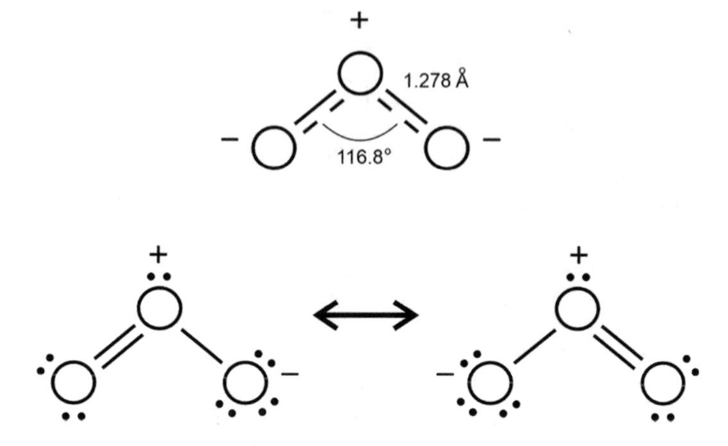

Fig. 1 Schematic of the two ozone resonance structures, which include a triatomic structure with no unpaired electrons and a twisted structural shape, length, and angle generated by three oxygen atoms

Table 2 Ozone and oxygen physical and chemical properties

No	Property	Ozone	Oxygen
1	Molecular weight, g/mol	48	32
2	Density (at 101 kPa), kg/m^3		
	– Gas (0 °C)	2.144	1.429
	– Liquid (−183 °C)	1571	1142
3	Color	Gas: blue-colored Dissolved: purple-blue	Gas: colorless Dissolved: light blue
4	Boiling point (101 kPa), °C	−112	−183
5	Melting point, °C	−192.7	−218.8
6	Solubility in water, mg/L		
	0 °C	1090	14.6
	10 °C	780	11.3
	20 °C	570	9.1
	30 °C	400	7.6
	40 °C	270	6.5
	50 °C	190	5.6
7	Oxydation potential, eV	2.07	1.23
8	Ozone half-life, Gaseous	Dissolved in water	
	1.5 s at 250 °C	8 min at 35 °C	
	3 days at 20 °C	12 min at 30 °C	
	8 days at −25 °C	15 min at 25 °C	
	18 days at −35 °C	20 min at 20 °C	
	3 months at −50 °C	30 min at 15 °C	

Summarized from Kirschner [48], USEPA [93], Rice er al. [74], Rice and Browning [73]

agents in water, with reaction rates of 10^{10}–10^{13} M^{-1} s^{-1}, their half-life is in the microsecond range, and hence their concentrations in water can never exceed 10^{-12} M [93].

$$O_3 + H_2O \rightarrow HO_3 + OH^* \quad (1)$$

$$HO_3 \rightarrow O_3^* + H^+ \quad (2)$$

$$HO_3 \leftrightarrow OH^* + O_2 \quad (3)$$

When ozone is in contact with water, several types of oxidizing agents will be produced which will compete for the substrate. In Eq. (1) more HO_3 will be formed than OH*, even though oxidation using dissolved ozone tends to take place more slowly than oxidation using hydroxyl free radicals. On the other hand, at an acidic pH, oxidation using hydroxyl free radicals tends to be small and substrate oxidation is dominated by dissolved ozone as in Eq. (2). On the other hand, at alkaline pH, UV exposure, and the addition of a catalyst, free hydroxyl will dominate as in Eq. (3). In alkaline conditions, the decomposition of ozone in water can also be described in Eqs. (4)–(10) [37].

$$OH^- + O_3 \rightarrow O_2 + HO_2^- \xleftrightarrow{H^+} H_2O_2 \quad (4)$$

$$HO_2^- + O_3 \rightarrow HO_2^* + O_3^{*-} \quad (5)$$

$$HO_2^* \leftrightarrow H^+ + O_2^{*-} \quad (6)$$

$$O_2^{*-} + O_3 \rightarrow O_2 + O_3^{*-} \quad (7)$$

$$O_3^{*-} + H^+ \rightarrow HO_3^* \quad (8)$$

$$HO_3^* \rightarrow HO^* + O_2 \quad (9)$$

$$HO^* + O_3 \leftrightarrow HO_2^* + O_2 \quad (10)$$

Ozone can oxidize substances in aqueous solution by direct oxidation by molecular ozone or via oxidation by hydroxyl free radicals created during ozone breakdown. Additionally, ozone leaves no residue in the media, making it extremely successful in a variety of applications such as oxidizing agents, disinfectants, color removers, odorants, and tastes. The primary disadvantage of this oxidizing agent is that it must be created on-site, necessitating the installation of production equipment at the point of application, which results in expensive expenditures [35, 62]. To do ozone oxidation, the ozone must be dissolved in water. To get optimal oxidation outcomes, the ozone content in the water must be maintained at a high level. In comparison to other gases, ozone's solubility is difficult to estimate because it is influenced by a variety of parameters such as temperature, pH, and other solutes. The following strategies can be used to improve the solubility of ozone in water:

(a) Increase the ozone concentration in the air
(b) Increase the gas pressure
(c) Lower the liquid's temperature
(d) Increase the pH
(e) Make contact with UV.

Through a multitude of mechanisms, ozone decomposes spontaneously. Although the precise process and related responses have not been determined, numerous researchers have presented several hypotheses [6, 12, 19, 38, 98]. Ozone decomposes to create free radicals (OH*/hydroxy radicals) with a very high oxidation potential of 2.8 V, making them a more powerful oxidizing agent than ozone. As a result, the ozonation process in water is always biphasic, involving ozone (direct oxidation) and OH* (indirect oxidation). Because it is believed that hydroxyl free radicals are one of the intermediate products that react directly with compounds in water, ozone demands are associated with the following:

(a) In the presence of radical scavengers, carbonate or bicarbonate ions (which are typically quantified as alkalinity), hydroxyl radicals will react with carbonate radicals to create carbonate radicals.

$$HCO_3^- + OH^* \rightarrow CO_3^{*-} + H_2O \quad (11)$$

(b) When natural organic compounds (NOM) are present in water, an oxidation process occurs, resulting in the formation of aldehydes, organic acids, and ketoacids. These oxidation by-products are generally more conducive to biological breakdown and are one of the tactics used in wastewater treatment plants to boost the biodegradability of chemicals (BOD/COD ratio). Meanwhile, in suitable conditions, synthesized organic molecules can be oxidized. In the case of mineralization of such compounds, the process should be dominated by hydroxyl free radical oxidation, as in advanced oxidation processes (AOPs), such as O_3/UV, Fenton, and/or O_3/H_2O_2.
(c) When bromide ion is oxidized, bromate ion, hypobromite ion, hypobromous acid, bromamines, and brominated organics are formed.

1.3 Methods of Ozone Production

Probably the first scientist to observe and report the presence of a distinct odor in the air near his electrostatic generator when exposed to electric sparks is a Dutch chemist Martinus van Marum in 1785. When a high voltage, alternating electric discharge is passed into a gas stream containing oxygen, the molecular oxygen is degraded to atomic oxygen. While some of the freed oxygen atoms mix with oxygen to generate ozone, others just recombine to form oxygen. This is one of the most common methods of producing ozone gas. The three most popular means of creating ozone are corona discharge, UV lamp, and cold plasma.

Corona Discharge

Corona discharge in a dry process gas containing oxygen is the most extensively utilized method of generating ozone for water treatment at the moment. Corona develops as a result of rapid ionization between two electrodes generated by a sufficiently strong electric field. When two electrons are exposed to a sufficiently high voltage, the electric field between the two electrodes becomes sufficiently strong to allow the electrons to travel between the electrodes. Electrons clash with unbound molecules due to their mobility. The impact provides sufficient energy for the liberated molecule to release its outer electron. The collision event generates two new electrons, those that collide with the molecule and those that exit the molecule. Because the two electrons are still under the pressure of the electric field, the two electrons will migrate and collide with other free molecules [27]. The corona is created during the ozone production process as a result of the ionization of oxygen. Two electrodes with varying voltages are used, one is high voltage and the other is low voltage. These electrodes are connected to a source of extremely high voltage. A dielectric medium separates the two electrodes, and a tiny discharge gap is given. Figure 2 depicts a schematic of ozone generation by the corona discharge approach.

The presence of a high electric field inside the electrode causes the ionization of air-containing oxygen, which flows in the electrode. The movement of electrons allows the collision of electrons with oxygen molecules and will produce two oxygen atoms (O). Furthermore, these oxygen atoms will naturally collide with the surrounding oxygen molecules to form ozone. The reaction equation for the formation of ozone can be seen in Eqs. (12–13).

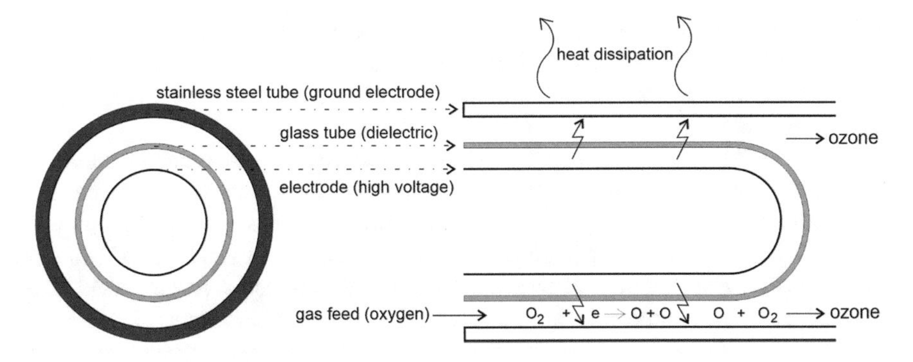

Fig. 2 Electrical discharge method of generating ozone

$$e^- + O_2 \rightarrow 2O + e^- \quad (12)$$

$$O + O_2 \rightarrow O_3 \quad (13)$$

This method is the most widely used method of ozone formation in various industrial activities because it has advantages such as high ozone productivity, does not require complicated maintenance, and is easy to apply. The drawback is that the amount of energy consumed is quite large while the concentration of ozone produced is low, so this technology is considered expensive. Ozone generation by electrical discharge produces heat, where excessive heat can cause decomposition of ozone in the product gas. This makes heat dissipation an important part of the ozone generator unit and must be carried out as quickly and efficiently as possible. Heat can be removed by using heat sinks, water coolers, and/or air coolers.

In a corona discharge, two electrodes, one low voltage (ground) and one high voltage, are separated by a dielectric medium in a tiny discharge gap. When electrons have enough energy to split the oxygen molecules, a certain fraction of these collisions occur, resulting in the formation of ozone molecules from each oxygen atom. The efficiency of corona discharge ozone formation is dependent on the strength of the micro discharges, which is influenced by numerous elements such as gap width, gas pressure, dielectric and metallic electrode characteristics, power supply, and moisture presence. When there is a weak discharge, the majority of the energy is absorbed by the ions, whereas when there is a powerful discharge, almost all of the energy is transferred to the electrons responsible for ozone formation. The optimal value is a trade-off between avoiding energy loss for the ions and obtaining a decent conversion efficiency of oxygen atoms to ozone. If air is used as the feed gas, it must be dry, as wet air produces nitrogen oxides in the ozone generator, which react with the generator to form nitric acid, corroding it and necessitating frequent maintenance. When air is used as the feed gas, 1–3% ozone is produced; when high purity oxygen is used, up to 16% ozone is produced [74].

Ultraviolet Light Lamp

This ozone formation mechanism is comparable to how UV energy from the sun divides O_2 into individual oxygen atoms. When ultraviolet light with a wavelength of 254 nm strikes an oxygen atom, it converts it to ozone. Ozone is formed when an oxygen molecule breaks into two oxygen atoms (2O). These two oxygen atoms join with another oxygen molecule (O_2) to form Ozone (O_3). Although ultraviolet light is produced naturally by the sun's beams, this process is thought to be less efficient than corona discharge. For decades, ultraviolet lamps have been employed to generate ozone. This lamp generates ultraviolet light at a wavelength of 185 nm (nm). The electromagnetic spectrum is used to quantify light, and its increments are referred to as nanometers. When exposed to ultraviolet light, oxygen molecules in their ground state absorb energy and dissociate to varying degrees depending on the precise energy and wavelength of the absorbed light. Following that, the oxygen atom combines with other oxygen molecules to generate ozone (Fig. 3). Due to the fact that

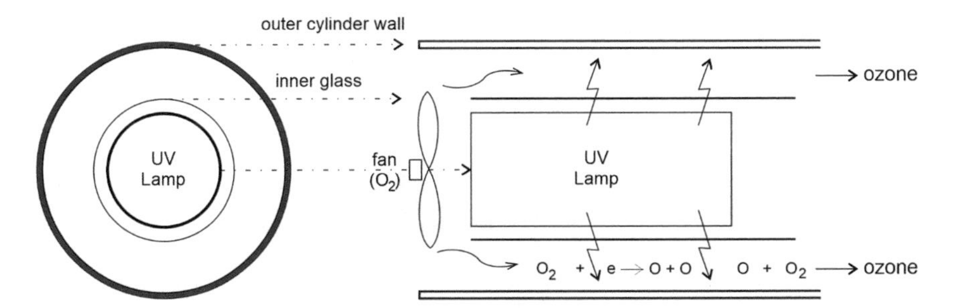

Fig. 3 Ozone production by photochemical means employing a tubular ultraviolet lamp with a cylindrical outer container wall

current technologies use mercury-based UV emission lamps, the 254 nm wavelength is transmitted simultaneously with the 185 nm wavelength, and ozone photolysis occurs concurrently with its formation. Additionally, the relative emission intensity is 5–10 times greater at 254 nm than at 185 nm. UV light has several advantages over corona discharges. It is less expensive, easier to install and operate, and its ozone emission is less impacted by moisture.

Cold Plasma

Plasma is the fourth phase element found in nature, following the solid, liquid, and gas phases. In contrast to the typical gas phase, plasma contains gas in which the components of the atomic nucleus (ions) and electrons have been separated and exhibit reactive properties as a result of the energy received. Plasma can develop spontaneously, as it does in the sun and the elements of stars in space. Plasma can also be generated by injecting a large amount of energy into a gas medium, causing the gas to undergo a dissociation and ionization process. Both reactions, depending on the amount of energy transferred, change neutral gas into highly reactive negatively and positively charged particles or ions, either partially or entirely. Plasma can be classified into two types based on its temperature: high-temperature plasma (thermal/equilibrium plasma) and low-temperature plasma (cold plasma/non-equilibrium plasma) [45]. It is similar to a corona discharge tube in construction, except that the anode and cathode of this cold plasma are contained within a glass rod filled with noble gases. The voltage jumps between the anode and cathode rods in this configuration, generating an electrostatic or "plasma" field. The advantage of the cold plasma system is that no heat release is delivered to the gas as it passes through the electrostatic field. Due to this design feature, cold plasma technology has an extremely extended service life. Plasma is a gas that has been partially or totally ionized by an incandescent discharge [44]. This phenomenon can occur when the voltage difference between the two electrodes is extremely large [61]. When gas ionization occurs, free electrons smash with gas ions to form radical active species (OH^*, O^*, and H^*), molecules (H_2O_2 and O_3), and ultraviolet light [44, 56, 102] (Fig. 4).

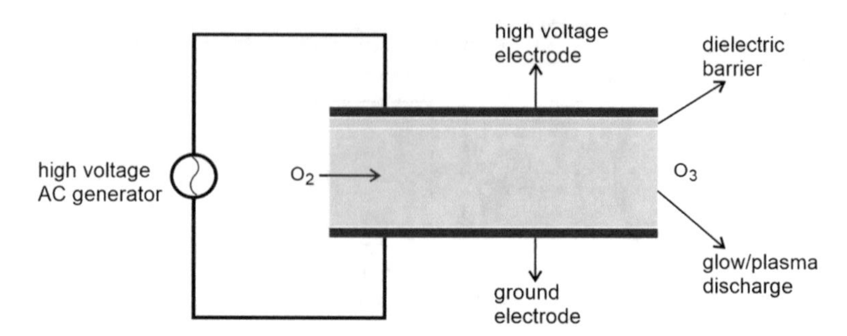

Fig. 4 Schematic of a cold plasma dielectric ozone generator

2 Ozone for Dye Removal

2.1 *Process and Mechanisms*

Ozone can be produced naturally or artificially. Ozone is formed naturally when ultraviolet (UV) rays from sunlight breakdown oxygen gas in free air. The oxygen molecule disintegrates into two oxygen atoms (O*), which naturally collide with the surrounding oxygen gas molecules to generate ozone (O_3). Ozone can be synthesized artificially in a variety of ways, including collisions and light absorption. Ozone can be created via collision in a variety of ways, including Dielectric Barrier Discharge Plasma (DBDP), Corona Discharge, and electrolysis. Meanwhile, ozone can be created via light absorption such as UV radiation. Ozone's concentration will degrade during the ozone process as a result of reactions between radicals and non-radical molecules. Ozonation generates oxidizing hydroxyl radicals, which are capable of decomposing organic contaminants in wastewater. The potential for radical oxidation is extremely high, which indicates that pollutants may exhibit reactivity. Additionally, reactive oxygen species (ROS) such as HO•, O_2• –/HO_2•, and H_2O_2 are generated and contribute to redox reactions that enable pollutants to be transformed [53]. Organic contaminants undergo radical reactions in response to substances in the waste. Organic components can stimulate the development of unstable radicals that are easily oxidized to water, carbon dioxide, and acids [53].

The majority of pollutants in textile wastewater are dyestuffs, particularly synthetic colors. Synthetic dyes are electron-delocalized compounds that contain two groups, chromophore and auxochrome. While chromophores act as electron acceptors, auxochromes act as electron donors, regulating their solubility and color. The azo group (–N=N–), the carbonyl group (–C=O), the ethylene group (–C=C–), and the nitro group (–NO_2) are all key chromophore groups that can produce color. While many auxochrome groups such as –NH_2, –COOH, –SO_3H, and –OH are polar and thus dissolve in water [69]. There are numerous synthetic dyes available today, and their application is determined by the type of fiber to be dyed, the required

Table 3 Type, application, and characteristics of dyes

No	Type of dyes	Application	Characteristics
1	Direct dyes	Viscose, Cotton	Direct dye having affinity to cellulose fibers, dyeing is carried out directly in solution with suitable additives
2	Mordant dyes	Cotton, Wool, Silk	Due to the weak link formed by mordant dyes with the fiber, the dyeing procedure is typically carried out by adding chromium to the dye to form a metal complex
3	Reactive dyes	Viscose, Nylon, Cotton, Wool, Silk	Reactive dyes have a reactive group that can form strong covalent bonds with cellulose, protein, polyamide and polyester fibers, which can be carried out at low and high temperatures
4	Acid dyes	Nylon, Wool, Silk	Acid dyes have a strong bond with protein and polyamide fibers, dyeing is carried out under acidic conditions and directly added to the fiber
5	Basic dyes	Jute, Acrylic, Paper	Basic/cationic dyes have a strong affinity with protein fibers, dyeing is carried out under alkaline conditions and directly added to the fiber
6	Disperse dyes	Nylon, Acrylic, Polyester, Acetate and Tri-acetate fiber	Disperse dyes were first created to color secondary cellulose acetate fibers, which are relatively insoluble in water. The dye is prepared for staining by grinding it into small particles in the addition of a dispersing agent
7	Sulfur dye	Viscose, Linen, Cotton	Sulfur dyes have a strong bond with cellulose fibers, the side group contains sulfur which is able to bind strongly to the fiber
8	Vat dyes	Viscose, Rayon, Linen, Cotton, Wool, Silk	Vat dyes are derived from synthetic indigo. Contain a complex polycyclic molecule based on the quinone structure that is insoluble in water and is well-known for requiring a reducing agent to be absorbed
9	Azoic dyes	Viscose, Cotton	Azoic dyes are those that have at least one azo group (–N=N–) bonded to one or, more commonly, two aromatic rings. Due to the fact that azoic dyes are insoluble in water, their water fastness is good. Azoic dyes exhibit excellent coloring capabilities, ranging from red to yellow

color resistance, and other technical and economic criteria. Table 3 summarizes the classification of textile dyes according to the method of dyeing.

The dyestuff classification shown in Fig. 5 can help choose the treatment approach for use. Because physical separation methods for dissolved dyes are rather challenging, chemical, AOPs, or biological procedures are typically used. In comparison to insoluble dyes, absorption or physical approaches are highly effective. Textile

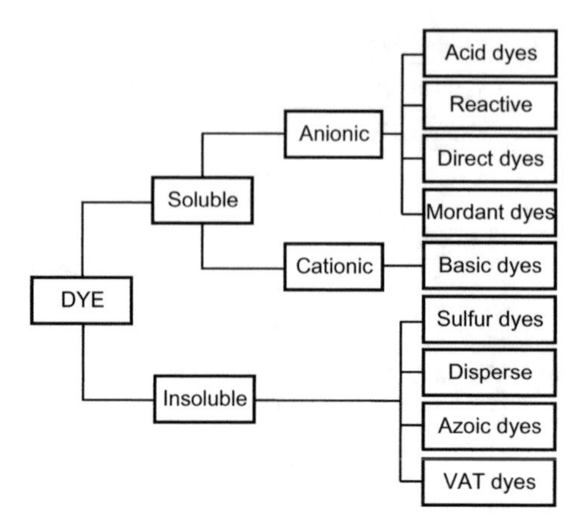

Fig. 5 Textile dyes classifications

manufacturing is one of the industries that generate wastewater with a high concentration of non-biodegradable organic matter. In general, biological wastewater treatment technologies such as activated sludge processes, aerated lagoons, moving bad biofilm reactors, anaerobic–aerobic biofilters, or trickling filters are used to minimize degradable organic contaminants. Meanwhile, wastewater containing contaminants, such as long-chain colors frequently used in the textile industry, contains azo compounds, which normal biological processes cannot rapidly degrade. The ozonation method is effective at degrading specific colors. The previous study has established that three types of dyeing agents, reactive black 5, reactive red 239, CI Reactive Blue 19, and Reactive Orange 16, require many stages of ozone degradation (Figs. 6, 7, 8 and 9). CH_3COOH, H_2C_2O, HCOOH, and CO_2 are used to degrade reactive black 5 dyes [101]. Meanwhile, red 239 reacts to form CH_3COOH and CO_2 [22]. CI Reactive

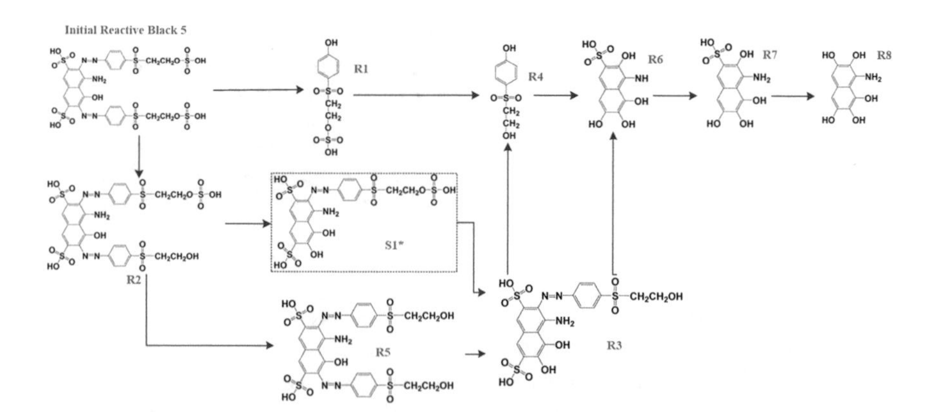

Fig. 6 The ozonation process for the degradation of reactive black 5 dye (adapted from Zheng et al. [101])

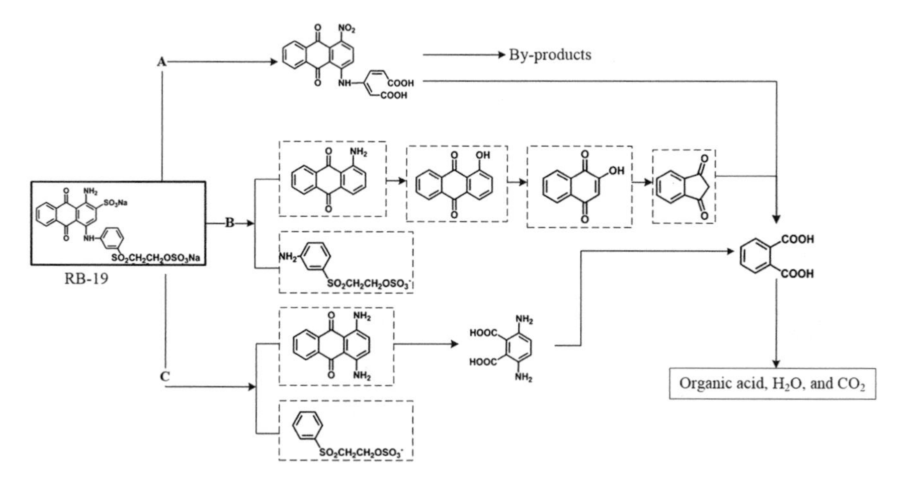

Fig. 7 The ozonation process for the degradation of CI reactive blue 1 dye (adapted from Fanchiang and Tseng [25])

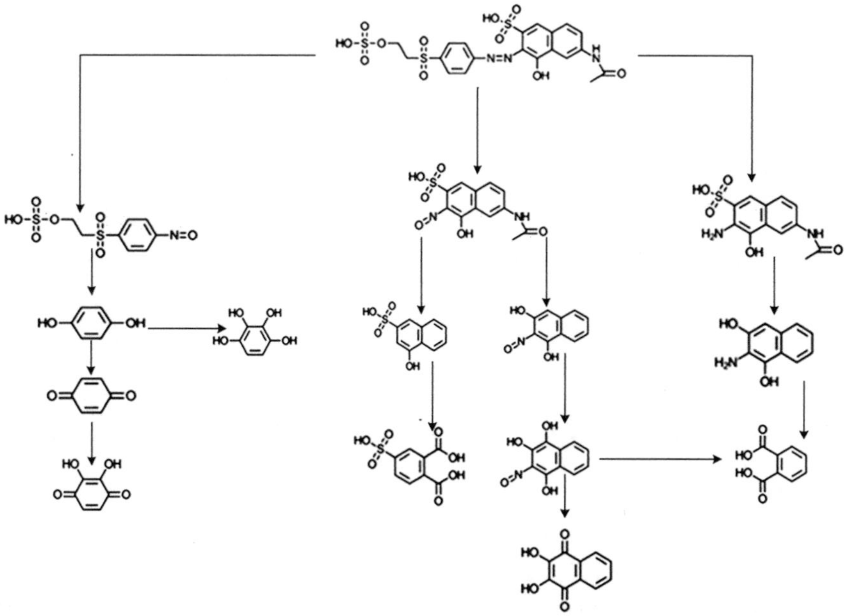

Fig. 8 The ozonation process for the degradation of reactive orange 16 (adapted from Castro et al. [15])

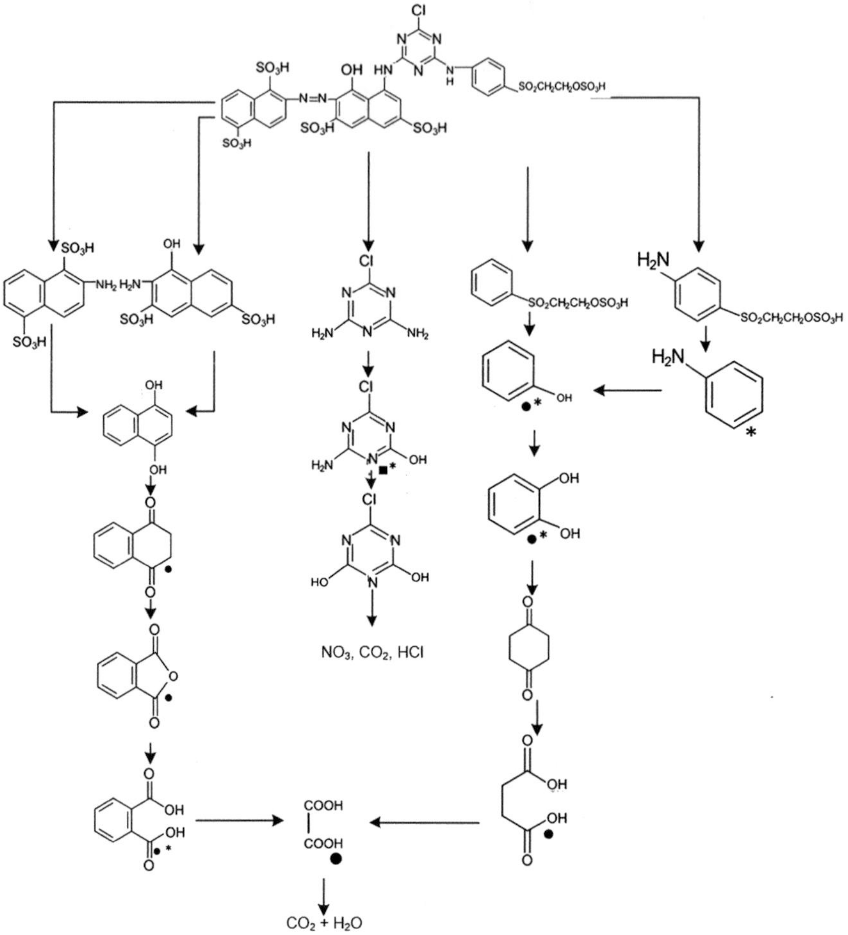

Fig. 9 The ozonation process for the degradation of reactive red 239 dye (adapted from Dias et al. [22])

Blue 19 degrades to CO_2, H_2O, and organic acids when exposed to ozone [25]. In general, the color degradation process using ozone is intended to produce CO_2. It is rather difficult to demonstrate the degrading process in a realistic manner. The biodegradability index (BOD_5/COD) is a regularly used method for defining the process of dye degradation by ozonation.

The increase in the biodegradability index in textile wastewater treatment measures the increased likelihood of dye degradation. The research shows that ozonation as a pre-treatment step could increase the biodegradability index for all types of dyes wastewater or textile wastewater (Table 4). This increase in biodegradability was achieved with moderate COD removal and short ozone times. In addition, the

Table 4 Results of increased biodegradability of dyes by ozonation

No	Dye	Initial biodegradability	Biodegradability after ozone pre-treatment	Dosage	Contact time (min)	Reference
1	Acid red 14	–	±0.45	–	25	[95]
2	C.I. reactive blue 19	0.15	0.33	88.8 mg/min	10	[25]
3	Polypropylene and polyester yarn dyeing industry	0.18	0.32	26 mg/L	75	[91]
4	Reactive black 5	0.056	0.46	2 g/h	270	[23]
5	Reactive black 5	–	0.27	15 mg/L	60	[4]

pre-treatment process is considered feasible for a biodegradability index greater than 0.3.

The degradation of RB5 dye by ozonation involves two major reactions: the detachment of the sulfonate group and the cleavage of the azo link, which happens concurrently with the increase in the OH group generated during the RB5 oxidation process. P2 (when one –O–SO_3H group is removed) and P5 (when both –O–SO_3H groups are removed or degraded) are formed as a result of the sulfonate moiety being released from the parent molecule. Meanwhile, the azo group in the initial RB5 dye molecule will be cleaved to yield S1 and R1 molecules. S1 molecules can also be synthesized as a result of the azo bond being broken in R2. Substance R3 is generated by removing/eliminating sulfonic groups from compound S1 or by breaking azo bonds and subsequent hydroxylation events from compound R5. At R3, the same azo bond cleavage reaction can occur, yielding hydroxylated naphthalene compounds (R6) and alkylsulfonyl phenolic compounds (R4). Then, desulphonation and subsequent hydroxyl addition events on the R6 molecule produce R7, which can be transformed to R8 more quickly. Additionally, R1 is changed to R4 via the loss of a sulfonate group. Finally, additional oxidation of R8 results in the formation of lower molecular weight organic acids such as formic, acetic, and oxalic acids, as well as mineralization to carbon dioxide and water [101].

The probable mechanism for dyestuff degradation via the oxidation process with ozone, as depicted in the above Figure, generally consists of two major mechanisms: the reductive severance of double bonds in dyes, which results in color loss in the visible color absorbance spectrum. Further oxidation of aromatic amine intermediate molecules results in the formation of simpler breakdown products such as aromatic and aliphatic acids that are finally mineralized. The following are possible pathways for dye degradation by ozone oxidation [85]:

1. Azo bond cleavage (–N=N–)
   The mechanism by which azo bonds in dye molecules are reductively broken appears to be the primary reaction that results in the formation of intermediate compounds during the oxidation of azo dyes.
2. Cleavage of carbon–carbon, carbon–nitrogen, and carbon–sulfur bonds
   Hydroxy radicals disrupt the C–C and C–N bonds of the color chromophore groups, as well as the C–S bonds between the aromatic ring and the sulfonate groups. This results in the intermediate product being further degraded.
3. Termination of the naphthalene ring
   The intermediate chemical generated when the adsorbed azo bond is broken undergoes structural modification.
4. Termination of the benzene ring
   Benzene derivatives are degraded further in the presence of an azo dye to form smaller molecular organic acids. Additionally, hydroxy radicals react with COOH molecules to make simpler organic acids such as maleic acid, acetic acid, oxalic acid, and formic acid.
5. Mineralization in the direction of the final product, which is carbon dioxide and water.

2.2 *Factors Affecting Dye Removal*

Pre-treatment with ozone results in an increase in the biological breakdown of organic contaminants. Pre-treatment can enhance the quality of trash and bring it into compliance with existing quality requirements [5, 14]. The primary disadvantage of the ozonation process is its short half-life, which can be extended if the dye is present in acidic circumstances, necessitating the adjustment of the pH of the textile waste [2]. In the ozone operation, the short half-life can be a result of an excessively large diffuser, which results in low ozone solubility. This can be accomplished through the use of porous glass or metal armor, solid catalysts, stirring, contact, and increased retention time provided by massive bubble columns or diffusers [11]. The wastewater from the textile sector is relatively warm, around 70–80 °F. This high temperature is also a hindrance in the textile business, as the wastewater must first be cooled in a cooling tower before further processing. Theoretically, when the temperature rises, the solubility of ozone in water decreases. At 43 °C, the solubility of ozone in the liquid phase drops. In chemical reactions such as the color degradation reaction involving ozone, a temperature increase of ten degrees Celsius doubles the rate of the reaction. Thus, there is an optimal temperature range for ozone solubility and degradation reaction rate. Temperature's effect on ozonization must be considered in light of the repercussions of both products. However, the inclusion of additional elements in the reaction flow, such as homogeneous and heterogeneous catalysts, illumination, initiators, and inhibitors, will affect the temperature dependence in extremely complex modes [55].

Because the ozone breakdown process is accelerated under alkaline settings (pH > 8.5), it is necessary to check the pH of textile waste on a constant basis. Increasing the pH value accelerates the breakdown of ozone in the water. For example, with a pH of 10, ozone in water has a half-life of less than 60 s. Oxidation of organic molecules can occur as a result of an interaction between the ozone molecule and OH radicals [55]. When ozone reacts with hydroxide ions, superoxide anions of O_2 radicals and hydroperoxyl HO radicals are formed [35]. Bicarbonate and carbonate are required for the formation of OH radicals in the system. The OH radical reacts with the carbonate or bicarbonate to form the passive carbonate or bicarbonate radical, which has no further interaction with ozone or organic molecules. The reaction rate of OH radicals is typically 106–109 times quicker than the reaction rate of ozone molecules. Protons H^+ combine with O_3 in an acidic environment to generate O_2 and H_2O, preventing O_3 from reacting directly with pollutants as specified in the equation [11].

The majority of textile dyes are composed of long-chain chemical molecules. Dyestuff is a complicated substance that can be kept within a molecular network. Because the dye is a mixture of near-identical organic compounds, it must have chromogen as the color carrier and Auxochrome as the binder between the color and the fiber. The ozonation mechanism for color removal is quite complex, and its interpretation is difficult to discern based on the makeup of the intermediates and final product. The reaction rate constant is derived using a simple mathematical model, as it is the rate constant for the ozone reaction with dyes. Variation in the amount of ozone produced by the reactor is critical for researching dye breakdown dynamics and ozonization kinetics. For O_3/H_2O_2 and the $O_3/H_2O_2/UV$ processes, the first pseudo-order constant for dye degradation has a high reliance on the hydrogen peroxide concentration [7]. The ozonation method can achieve the optimal condition depending on the type of dye, initial concentration, ozone dose applied, and detention time. The results of the exclusion of various types of research are shown in Table 5. Generally, color removal cannot reach 100%, and some studies indicate that roughly 70% of color is removed after 10–14 min of contact time.

3 Practical Application

3.1 Ozone System Configuration

System for generating and applying ozone to water and/or wastewater typically consist of five components:

(a) Electrical power generation,
(b) Ozone generation method,
(c) Feed gas preparation,
(d) Contacting of liquid with ozone, and
(e) Excess ozone destruction method.

Table 5 References to previous research regarding color removal efficiency with ozonation process

No	Dye	Initial dye concentration (mg/L)	Ozone dose	Dye removal (%)	Detention time (min)	Source
1	Acid red 14	1500	5 g/h	93	25	[22]
2	C.I. reactive black 5 (RB5)	200	3.2 g/h	70.21	10	[21]
3	Direct red 28	1500	5 g/h	92	25	[101]
4	Methylene blue	10	46.32 mg/L	73.01	13	[31]
5	Methylene orange	10	0.83 mg/s	100	15–30	[31]
6	Methylene blue	1500	5 g/h	93.5	25	[82]
7	Reactive black 5	1500	5 g/h	94	25	[95]
8	Reactive red 239	50	40 and 20 mg/L	More than 95	4–12	[95]
9	Reactive red 239	50	20 mg/L	100	20	[95]
10	Reactive red 239	500	16.6 mg/min	90	90	[60]
11	Reactive red X-3B	100	0.66 L/h	92	6	[1]
12	Reactive yellow 176	500	16.6 mg/min	90	90	[96]
13	Reactive black 5	100	40.88 mg/min	96.9	60–300	[86]
14	Acid red 14	1500	–	93	25	[94]

The five fundamental components of an ozonation system, each of which must be considered while developing ozone to ensure both effectiveness and safety. Instrumentation and control systems can be incorporated to guarantee the whole ozonation system operates effectively and safely. Electrical energy generation is inextricably linked to the process of ozone creation, which is often classified according to the frequency of the power provided to the ozone generator. Today, the market offers low-frequency (50 or 60 Hz) ozone generators, medium frequency (60–1000 Hz) ozone generators, and high-frequency (>1000 Hz) ozone generators. Low-frequency generators generated less heat than medium- and high-frequency generators, yet they effectively created more ozone than low-frequency generators. Table 6 summarizes the comparison of the three generator types.

Ozone can be produced on-site using corona discharge or ultraviolet (UV) radiation. When dry air is fed to an ozone generator, UV radiation produces low concentrations of ozone (less than 0.1 wt%), whereas corona discharge produces ozone concentrations ranging from 0.5 to 2.5 wt%. When pure oxygen is used as the feed gas, ozone

Table 6 Ozone generators with varying frequencies of operation (low, medium, and high)

Characteristics	Low frequency	Medium frequency	High frequency
Hertz (Hz)	50–60	60–1000	More than 1000
Level of electronics sophistication	Low	High	High
Turndown ratio	5:1	10:1	10:1
Required cooling water (Liters/kg ozone generated)	4.1–8.2	4.1–12.5	2.1–8.2
Optimum cooling water differential	8° to 10 °F	5° to 8 °F	5° to 8 °F
Typical application range	<225 kg/day	To 900 kg/day	To 900 kg/day
Optimum O_3 production	60–75%	90–95%	90–95%
Operating concentrations			
– Air as feed gas – Oxygen as feed gas	0.5–1.5 wt% 2.0–5.0 wt%	1.0–2.5 wt% 2.0–12 wt%	1.0–2.5 wt% 2.0–12 wt%
Power requirements for air feed system, (kW h/kg O_3)	11–15	11–15	11–15
Power required, (kW h/kg O_3)			
– Air as feed gas – Oxygen as feed gas	17–26 8–13	17–26 8–13	17–26 8–13

Adapted from USEPA [93], with modifications

concentrations of 2–5 wt% are produced, which is typical for low output generators, but can reach 10–12 wt% in general for large-scale applications by ozone generator manufacturers. Oxygen concentrators are frequently used in place of air drying units to supply oxygen-enriched air to ozone generators, resulting in higher output and gas-phase ozone concentrations while avoiding the need for on-site oxygen production or storage facilities. In all cases, ozone is only partially soluble in water and must come into contact with water and/or wastewater in order to maximize ozone transfer to a solution. Many different types of ozone contactors have been developed for this purpose; all of them are effective for the intended water treatment purpose. The contact system design becomes more critical as ozone concentrations increase due to the lower ozone gas-to-liquid ratio. The simplified ozone system configuration is shown in Fig. 10. Furthermore, using pure oxygen as a feed gas can cause oxygen saturation in the treated water, causing operational issues in following the treatment process as well as aesthetic issues in the distribution system [71]. Table 7 summarized comparison of the benefits and drawbacks of using either air or oxygen as a feed system.

The air supply system for an ozone generator is rather complicated, as the air must be adequately conditioned to avoid the generator from being damaged. The air must be pure and dry (dew point −80 to −40 °F) and free of contaminants. A typical air preparation system includes an air compressor, a filter, a drier, and a pressure regulator. To do this, an air dryer device is required to significantly reduce or even eliminate the moisture and pollutants in the free air. Compressed air dryers are frequently used in industry, as the main unit consists of an air compressor and a compressed air

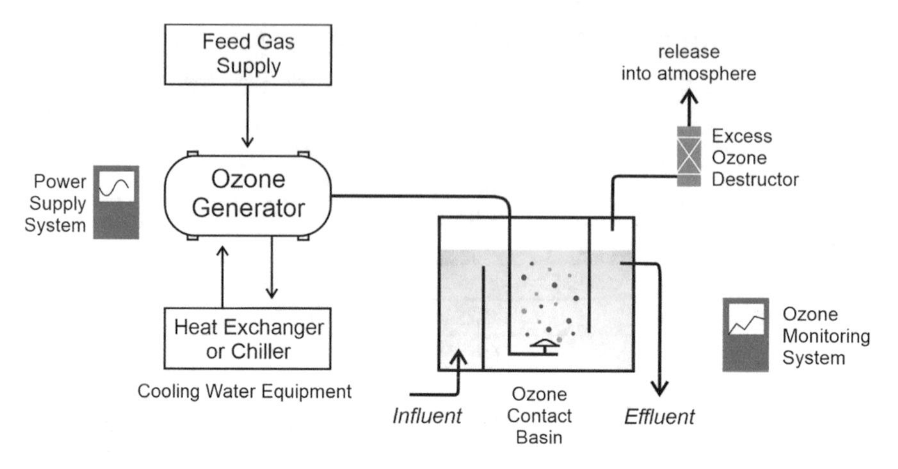

Fig. 10 Simplified typical ozone system configuration

Table 7 Comparison of air and oxygen feed system in ozone generation

Feed gas	Advantages	Disadvantages
Air	Proven technology, simple and uncomplicated mechanical equipment, suitable both for small and large-scale installation systems	Low efficiency, requiring more energy per unit of ozone volume produced, as well as gas handling and adjusting requirements
High purity oxygen	Produce higher ozone concentration for approximately double for the same generator that uses air as the feed gas, suitable both for small and large-scale installation systems	Required oxygen concentrators, oxygen resistant material, and safety concerns
Liquid oxygen (LOX)	Simple to operate and maintain, requires minimal equipment, can store excess oxygen to meet peak demand, and is well-suited for small to medium-sized installation systems	Increased expense of liquid oxygen, on-site storage, which raises safety concerns, and the possibility of LOX loss in storage while not in use
Generation of cryogenic O_2	Equipment is similar to air preparation systems in that it can store excess oxygen to meet peak demand and is capable of being installed on a big scale	Higher capital cost, extensive gas handling equipment that is more complex than the LOX system, and sophisticated in its operation and maintenance

Adapted from USEPA [93], with modifications

storage tank. The compressed air is dried through a process that lowers the dew point temperature, allowing the water vapor contained in the air to condense, resulting in comparatively dry air output from the air dryer. One sort of air dryer that is frequently accessible in the market is the regenerative desiccant air dryer, which removes moisture via the adsorption principle. Desiccant dryers are typically equipped with dual towers/tubes that contain moisture-absorbing materials such as silica gel, activated alumina, zeolite, or other water-absorbing compounds. Water vapor is removed from the drier using an external heat source or by sending a fraction of dry air (between 10 and 30%) at lower pressure via a saturated tower/tube. Additionally, refrigeration dryers and membrane air dryers are available on the market.

Compressed air typically contains water in the form of liquids and vapors that are influenced by the surrounding ambient air conditions. The combination of a 100-cfm compressor and a dryer-cooler operating for 4000 h under typical climatic circumstances produces roughly 8300 L of liquid condensate per year, which increases dramatically with larger compressors or in more humid areas. Water of any kind must be eliminated from the system in order for it to operate effectively and efficiently. That is why dryers are critical for the production of clean, dry air.

Air drying procedures might range from capturing condensed water and avoiding additional moisture condensation to completely removing the water present. The more water eliminated, the more expensive it is. However, if excessive water is allowed to remain in the compressed air supply, future costs will increase due to downtime, increased maintenance, corrosion, product failure, and premature equipment failure. Figure 11 illustrates a typical basic air preparation system with numerous possible unit actions. Aftercoolers bring compressed air's temperature and moisture content down. Separators for bulk liquids are used to separate liquid that has condensed in the distribution system. Particulate filters remove pollutants down to

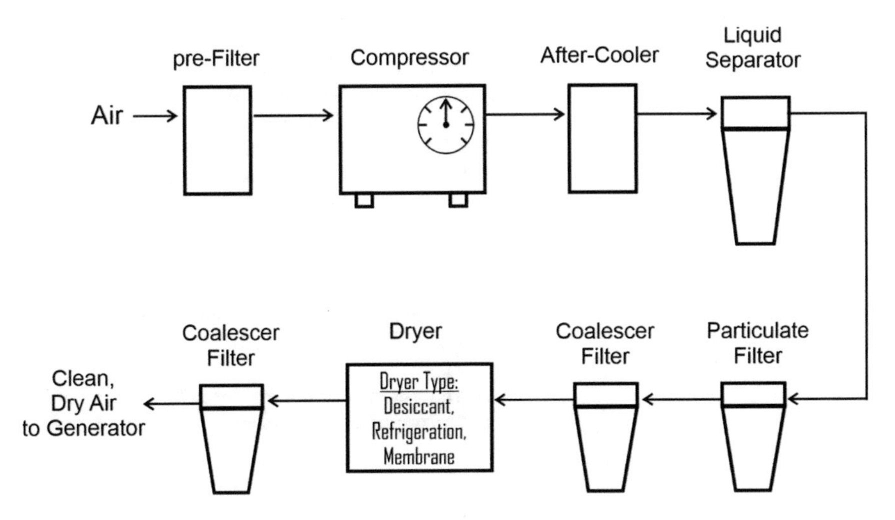

Fig. 11 Simplified typical air preparation system

the size of 1–5 microns and separate bulk liquid from the air stream. Coalescing filters remove liquid aerosols and particulate matter down to a size of 0.01 micron [13].

Ozone contact systems may employ pressurized gas-to-liquid mass transfer methods, ceramic or stainless steel fine bubble diffusers, static mixers, or venturi injectors to combine ozone gas with the treated water and/or wastewater (Table 8, Fig. 12). Small in-line injectors and pressure reaction vessels on a big scale replace massive concrete, cost-effective 20-foot deep bubble diffuser tanks. Once dissolved in water, ozone is ready to operate on water pollutants, disinfecting and/or oxidizing the target pollutant. The final component of ozone treatment is a device for destroying extra ozone that is always present in the off-gas contactor when corona discharges occur. Without a competent off-gas ozone-destroying device, this surplus ozone will create a hazardous oxidizing environment for people and nearby objects. Thermally (370 °C), thermally catalytically, or by flowing through the catalyst medium alone, the ozone off-gas contactor can be easily destroyed.

Table 8 Type of ozone contact system and apparatus

Diffuser type	Characteristics
Stainless steel fine bubble	• Material: SS304, SS316 • Size pore: 2.5–160 μm • Airflow per band: 15–63 Nm^3/h • Withstand pressure: 0.5–3 Mpa • Temperature resistance: 600–1000 °C
Ceramic fine bubble	• Airflow per unit: 1.2–4 Nm^3/h • Effective surface area per unit: 0.5–1 m^2 • Standard gas transfer rate: 0.2–0.6 kg O_2/h. unit
Static mixer (teflon/fiberglass/resin)	A static mixer based on the turbulent mixing concept, the mixing effect is accomplished primarily through internal rotation and shearing of the fluid layers at the point where the rotation direction is reversed. When shearing rates are required to be low, a pitch between two rotations can be specified. This pitch retards the rotation process, resulting in an enhanced mixing effect. Common advantages include the absence of moving components and mechanical seals, low to no maintenance, no leakage, predictable homogeneity, low energy consumption, ease of scaling up, and in-line processing
Jet mixer/aerator	With the use of a low shear rotor–stator mixing head, the Jet Mixer can effectively and uniformly generate a unique tank flow pattern Common advantages: eliminates vortexing-stratification and dead mixing area. Standard gas transfer rate: 1.5–40 kg O_2/h

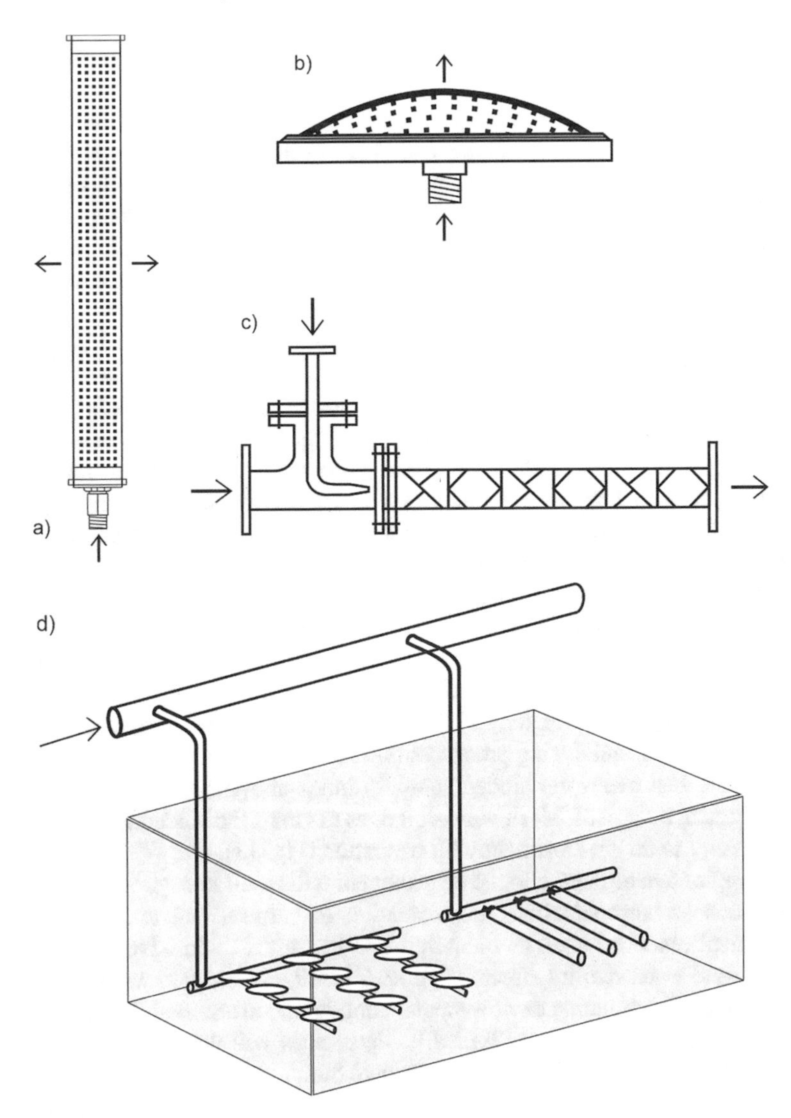

Fig. 12 Schematic of various ozone diffusers. **a** Stainless steel fine bubble, **b** ceramic fine bubble, **c** static mixer, and **d** unit placement perspective

3.2 Integration with Existing WWTP Process

The textile industry usually consists of 4 processes that produce wastewater: desizing, bleaching, dyeing, and mercerization. Based on [39], the effluent of the bleaching and dyeing process needs to be given more attention related to the biodegradability value. At first, the raw material is subjected to a singeing process to burn the hairs on the fabric surface, and then a desizing process is carried out to remove starch.

Starch must be removed from the fabric to not interfere with the following process because it will block the absorption of the substances used in the process. The starch removal process aims to convert water-insoluble starch into water-soluble glucose and maltose compounds. Hydrolysis can occur in hot water, acidic solutions, and alkaline solutions. An oxidizing agent can be used to degrade starch that cannot be hydrolyzed but is easily oxidized. Additionally, it can utilize enzymes that act as catalysts in the conversion of starch to water-soluble sugars [57]. Typically, this procedure uses wastewater that is highly biodegradable for biological wastewater treatment [39].

In the bleaching process, auxiliary chemicals typically employed are sodium hypochlorite, sodium silicate, hydrogen peroxide, and organic stabilizers such as enzymes as bleaching agents. The usage of chloride or peroxides results in inhibition issues, which result in an acidic pH. The dyeing process is the process of giving color evenly to the fabric. Before the dyeing process is carried out, several steps are carried out, namely carrying out removing dirt which consists of desizing to remove starch and then scouring to remove the dirt stuck to the fabric, and then burning the hair and then washing it. After washing, the preparation stage is carried out, namely the selection of the dye, which is then carried out by dissolving and entering the textile materials into the solution [57]. The coloring process requires a substantial volume of water compared to other processes [81]. Not only during immersion, but rinsing also involves a lot of water. In the staining process, many chemicals such as metals, salts, surfactants, organic matter, sulfides are added to the fiber. Wastewater resulting from the bleaching and dyeing processes must be re-analyzed to see the biodegradability value. If it meets the biodegradability index above 0.3, it can be continued with biological treatment. Meanwhile, if it does not meet the biodegradability index, it is necessary to do pre-treatment with ozonation (Fig. 13).

Disposal of dye effluent into the environment will result in environmental disturbances such as eutrophication, the formation of harmful and toxic by-products via chemical reactions such as oxidation and hydrolysis, and adverse ecological, aesthetic, and water quality effects [59]. As a result, color-coded wastewater treatment is critical for reducing environmental contamination and enabling the practical implementation of technology. Typically, these hues will not degrade aerobically and cannot be rapidly handled using traditional biological treatment techniques [20, 28]. The reason that the color is non-biodegradable waste is that the environment lacks the enzymes required for color decomposition. Colored wastewater can be treated using physical and chemical processes such as adsorption, coagulation, sedimentation, and oxidation. However, implementation of this approach is complicated, expensive, and results in incomplete conversion to inorganic chemicals [20]. Because reactive colors containing poly-aromatic compounds are extremely soluble in water, they make absorption by absorbent agents more challenging. Activated sludge has been employed as a sorbent to remove reactive colors in a number of experiments, with the absorption capacity varying according to environmental conditions, pH, type, and color concentration. The most significant constraints are thought to be high operational expenditures, such as restoring the sorbent or processing the resulting solid waste.

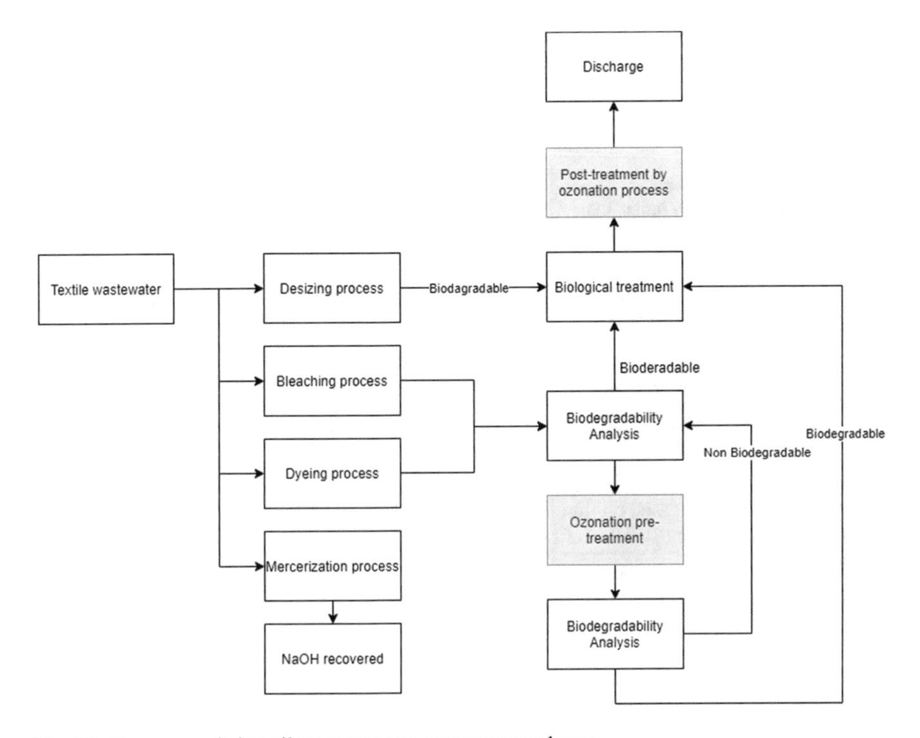

Fig. 13 Recommended textile wastewater management scheme

The results of previous studies showed color removal with ozone pre-treatment was higher than without pre-treatment (Table 9, Fig. 14). The finding confirms that color removal by biological treatment alone is not effective enough compared to biological treatment with ozonation pre-treatment regardless of the type of biological treatment used.

The process by which ozone removes color is either direct, in which the ozone molecule O_3 oxidizes dyes, or indirect, in which ozone decomposes into OH radicals with a greater oxidation potential of 2.8 eV than ozone's 2.07 eV. The direct method happens most frequently when the pH of the solution is between acidic and neutral. Meanwhile, the indirect reaction happens at an alkaline pH because ozone decomposes more rapidly into hydroxyl radicals (OH radicals) [18, 101].

The ozone reaction with a single chromophore group, such as an azo group or a carbon double bond (C=C) in an aromatic ring, is the first stage of the color degradation process in most dyes [18]. Ozone can cleave the conjugate bonds of azo dyes, converting them to molecules with a lower molecular weight, such as organic acids, so deteriorating the hue. Both ozone and hydroxyl radicals, when dissolved, are capable of breaking aromatic rings and oxidizing inorganic and organic molecules [83].

The initial design of textile WWTPs in Indonesia was not specifically designed to eliminate color because before 2019, color parameters were not regulated by national

Table 9 Biological treatment types integrated with ozone pre-treatment process for wastewater containing dye

Dye	Biological treatment	Color removal with ozone pre-treatment	Color removal without ozone pre-treatment	Reference
Reactive black 5	Upflow anaerobic sludge blanket (UASB)	94%	–	[96]
Reactive orange 16	Moving-bed biofilm reactor (MBBR)	More than 97%	0%	[15]
Reactive orange 16	Anaerobic moving-bed biofilm reactor (MBBR)	61 ± 18%	–	[16]
Real textile wastewater	Anoxic-aerobic activated sludge	76.60%	30%	[87]
Remazol black 5	Moving-bed biofilm reactor (MBBR)	86.74%	68.60%	[65]
Remazol black B	Activated sludge (biomass from a municipal wastewater treatment plant)	91%	59.30%	[90]

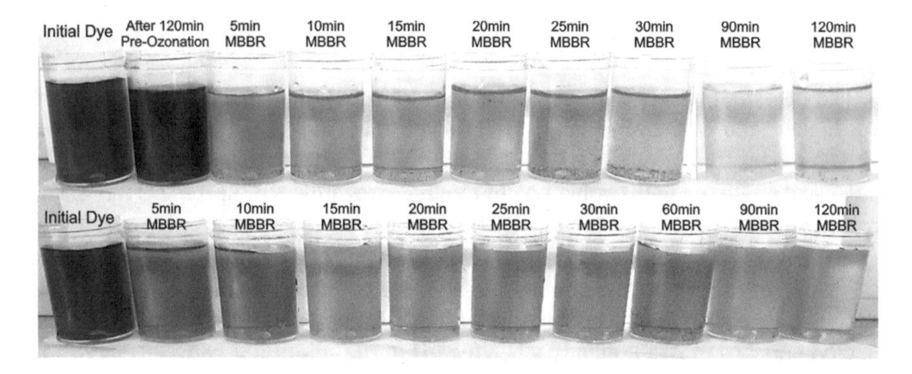

Fig. 14 Visual image of reactive black-5 color gradation with an initial concentration of 50 mg dye/l after 5–120 min degradation processes in the MBBR reactor. Upper image shows the effluent color of the MBBR reactor with pre-treatment using ozonation for 120 min with 3.81 mg O_3/min (color removal eff. 86.74%), and lower image shows the effluent color of the MBBR reactor without pre-treatment (color removal eff. 68.60%). *Source* Modified from Pratiwi [66]

quality standards. Knowing the ineffectiveness of the current WWTP in removing color, the textile industry must modify and improve the WWTP system by adding a decolorization unit combined with the previously existing WWTP unit. Based on the poor performance in decolorization, an appropriate solution is needed to improve the WWTP so that the effluent released complies with quality standards by evaluating the

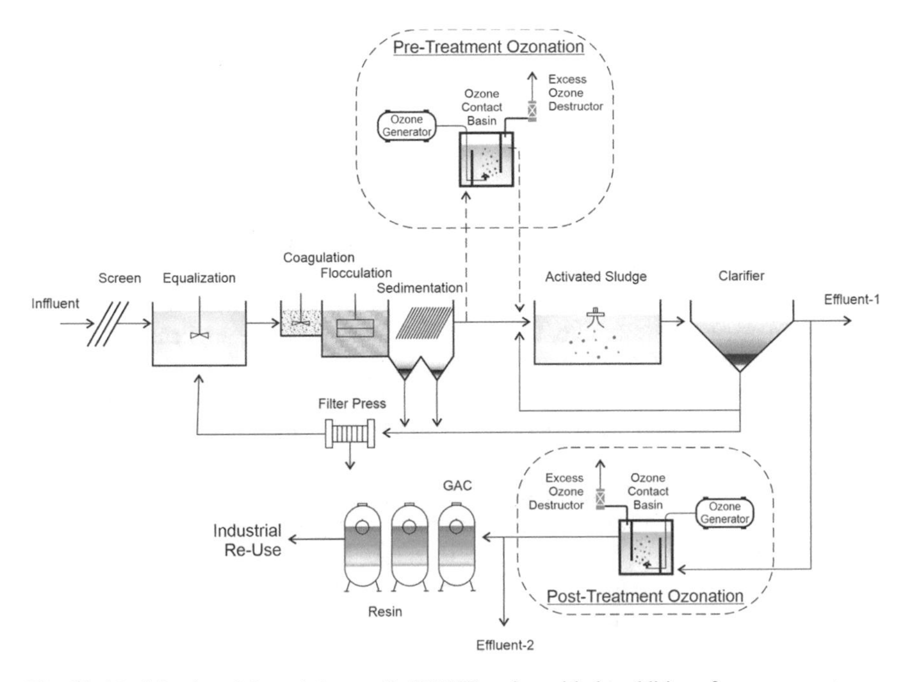

Fig. 15 Modification of the existing textile WWTP option with the addition of an ozone system as pre- and/or post-treatment

existing WWTP unit and planning new alternative designs that are effective and able to provide benefits to the textile industry from both technical and financial aspects. Alternative textile wastewater treatment technology that can be used is the addition of an ozonation unit that can be positioned in front as a pre-treatment of a biological process or post-treatment as a polishing system (Fig. 15).

4 Concluding Remarks

Chemical, physical, biological, or a combination of the three methods can be used to treat textile effluent. Chemical treatment is carried out by coagulation, flocculation, and neutralization. The coagulation and flocculation processes are carried out by adding coagulants and flocculants to stabilize colloidal particles and suspended solids to form flocs that can settle under gravity. The formed flocs are able to absorb the color and then settle as sludge. Physical wastewater treatment can be done by adsorption, filtration, and sedimentation. Adsorption is done by adding adsorbent, activated carbon or the like. Filtration is a procedure that separates solids from liquids by passing them through a filter. Sedimentation is a solid–liquid separation method that involves depositing suspended particles in the presence of gravity in order to separate the two liquids. Biological wastewater treatment is the utilization of microorganism activity to decompose organic materials contained in wastewater. Each of the three processing methods above has advantages and disadvantages. Chemical wastewater treatment will produce large amounts of sludge, thus creating new problems for the handling of the sludge. Although the use of activated carbon results in a high percentage of dye reduction, the cost of activated carbon is relatively high, and thus increases the cost of the equipment used to regenerate the activated carbon. The use of ozone in textile wastewater treatment applications has several advantages including, it has a very large oxidizing power so that it only requires a relatively short contact time (CT) to oxidize the impurities contained in the liquid waste in the order of minutes. Increases the solubility of oxygen in water. In ozone applications, almost no chemicals are needed, except those absolutely necessary in the sedimentation process (coagulant and flocculants aid). Ozone reacts rapidly and is very effective in removing dissolved organic compounds in water (TOC, total organic compounds), hence improving the performance of biological processes. Because ozone decomposes rapidly in water (on the order of 1–15 min), its residual effects are easily removed. On the other hand, some of the disadvantages or impediments of using this ozone technology include the fact that ozone is a hazardous gas (the OSHA TLV is 0.1 ppm) whose toxicity is proportional to its concentration and exposure time. Due to its high energy consumption and relative complexity in comparison to other oxidation processes, ozonation has a high installation cost. It is strongly advised that an ozone destruction catalyst be fitted at the outlet to prevent ozone poisoning and fire hazards. This can result in the formation of undesirable carbonyl compounds (aldehydes and ketones), particularly if the contact duration is too short. Recently, the use of ozone as a disinfectant in air, water, and wastewater treatment has garnered widespread interest globally. The development of disinfection chambers that use ozone as a decontamination agent is gaining momentum, partly as a result of the 2019 COVID-19 pandemic.

References

1. Adelin MA, Gunawan G, Nur M, Haris A, Widodo DS, Suyati L (2020) Ozonation of methylene blue and its fate study using LC-MS/MS. J Phys Conf Ser (IOP Publishing) 1524(1):012079
2. Ahmet B, Ayfer Y, Doris L, Nese N, Antonius K (2003) Ozonation of high strength segregated effluents from a woollen textile dyeing and finishing plant. Dyes Pigments 58:93–98
3. Alimohammadi M, Naderi M (2021) Effectiveness of ozone gas on airborne virus inactivation in enclosed spaces: a review study. Ozone Sci Eng 43(1):21–31. https://doi.org/10.1080/01919512.2020.1822149
4. Alvares ABC, Diaper C, Parson SA (2001) Partial oxidation of hydrolysed and unhydrolysed textile azo dyes by ozone and the effect on biodegradability. Process Saf Environ Prot 79(2):103–108. https://doi.org/10.1205/09575820151095184
5. Baban A, Yediler A, Lienert D, Kemerdere N, Kettrup A (2003) Ozonation of high strength segregated effluents from a woollen textile dyeing and finishing plant. Dyes Pigments 58(2):93–98. https://doi.org/10.1016/S0143-7208(03)00047-0

6. Bailey PS (1972) Organic groupings reactive toward ozone mechanisms in aqueous media. In: Evans FL (ed) Ozone in water and wastewater treatment. Ann Arbor Science, Ann Arbor, Michigan, pp 29–59
7. Bilinska L, Gmurek M, Ledakowicz S (2016) Comparison between industrial and simulated textile wastewater treatment by AOPs – biodegradability, toxicity and cost assessment. Chem Eng J 306:550–559
8. Blanco A, Ojembarrena FB, Clavo B, Negro C (2021) Environ Sci Pollut Res 28:16517–16531
9. Braslavsky S, Rubin MB (2011) The history of ozone Part VIII. Photochemical formation of ozone. Photochem Photobiol Sci 10(10):1515–1520
10. Bocci V (2006) Is it true that ozone is always toxic? The end of a dogma. Toxicol Appl Pharmacol 216(3):493–504
11. Boczkaj G, Fernandes A (2017) Wastewater treatment by means of advanced oxidation processes at basic pH conditions: a review. Chem Eng J 320:608–633. https://doi.org/10.1016/j.cej.2017.03.084
12. Bühler RE, Staehelin J, Hoigné J (1984) Ozone decomposition in water studied by pulse radiolysis. 1. HO_2/O_2^- and HO_3/O_3^- as intermediates. J Phys Chem 88:2560–2564
13. CAGI (2016) Compressed air & gas handbook, 7th edn. Compressed Air and Gas Institute
14. Carmen Z, Daniela S (2012) Textile organic dyes—characteristics, polluting effects and separation/elimination procedures from industrial effluents—a critical overview. In: Puzyn T (ed.) Organic pollutants ten years after the stockholm convention—environmental and analytical update. InTech. ISBN: 978-953-307-917-2
15. Castro FD, Bassin JP, Dezotti M (2017) Treatment of a simulated textile wastewater containing the reactive orange 16 azo dye by a combination of ozonation and moving-bed biofilm reactor: evaluating the performance, toxicity, and oxidation by-products. Environ Sci Pollut Res 24(7):6307–6316
16. Castro FD, Bassin JP, Alves TLM, Sant'Anna GL, Dezotti M (2020) Reactive orange 16 dye degradation in anaerobic and aerobic MBBR coupled with ozonation: addressing pathways and performance. Int J Environ Sci Technol 1–20
17. Cattoor T (2007) European legislation relating to textile dyeing. In: Environmental aspects of textile dyeing. Elsevier Inc., pp 1–29
18. Colindres P, Yee-Madeira RR (2010) Removal of reactive black 5 from aqueous solution by ozone for water reuse in textile dyeing processes. Desalination 258:154–158
19. Criegee R (1975) Mechanismus der ozonolyse (mechanisms of the ozonolysis). Angew Chem 87:765–771
20. Dehghani M, Nasseri S, Mahdavi P, Mahvi A, Naddafi K, Jahed G (2015) Evaluation of acid 4092 dye solution toxicity after UV/ZNO mediated nanophotocatalysis process using daphnia magna bioassay. Colour Sci Technol 5:285–292
21. Dias NC, Alves TL, Azevedo DA, Bassin JP, Dezotti M (2020) Metabolization of by-products formed by ozonation of the azo dye reactive red 239 in moving-bed biofilm reactors in series. Braz J Chem Eng 37(3):495–504
22. Dias NC, Bassin JP, Sant'Anna Jr GL, Dezotti M (2019) Ozonation of the dye reactive red 239 and biodegradation of ozonation products in a moving-bed biofilm reactor: revealing reaction products and degradation pathways. Int Biodeterior Biodegrad 144:104742
23. Dinçer AR (2020) Increasing BOD5/COD ratio of non-biodegradable compound (reactive black 5) with ozone and catalase enzyme combination. SN Appl Sci 2(4):736. https://doi.org/10.1007/s42452-020-2557-y
24. Duguet JP (2004) Basic concepts of industrial engineering for the design of new ozonation processes. Ozone News 32(6):15–19
25. Fanchiang JM, Tseng DH (2009) Degradation of anthraquinone dye CI reactive blue 19 in aqueous solution by ozonation. Chemosphere 77(2):214–221
26. FDA (2001) Secondary direct food additives permitted in food for human consumption. Fed Regist 66(123):33,829–33,830

27. Garniwa I, Sudiarto B, Gaol EHL (2006) Study of partially discharge and flash off corona waves in a simulated cubicle covered with insulating material (Studi gelombang korona peluahan sebagian dan lepas denyar dalam kubikel simulasi dengan dilapisi bahan isolasi). Jurnal Teknologi 1:24–31
28. Girish K (2019) Microbial decolourization of textile dyes and biodegradation of textile industry effluent. In: Kumar S (ed) Advances in biotechnology and bioscience. AkiNik Pub. ISBN 978-93-5335-855-6
29. Giunta R, Coppola A, Luongo C, Sammartino A, Guastafierro S, Grassia A, Giunta L, Mascolo L, Tirelli A, Coppola L (2001) Ozonized autohemotransfusion improves hemorheological parameters and oxygen delivery to tissuesin patients with peripheral occlusive arterial disease. Ann Hematol 80(12):745–748. https://doi.org/10.1007/s002770100377
30. Graham DM (1997) Use of ozone for food processing. Food Technol 51:72–75
31. Güneş Y, Atav R, Namırtı O (2012) Effectiveness of ozone in decolourization of reactive dye effluents depending on the dye chromophore. Text Res J 82(10):994–1000
32. Gumilar I, Abidin HZ, Hutasoit LM, Hakim DM, Sidiq TP, Andreas H (2015) ISEDM 3rd international symposium on earthquake and disaster mitigation land subsidence in bandung basin and its possible caused factors. Proc Earth Planet Sci 12:47–62
33. Guzel-Seydim ZB, Greene AK, Seydim AC (2004) Use of ozone in the food industry. Lebensm-WissUTechnol 37:453–460
34. Goncalves AA (2009) Ozone-an emerging technology for the seafood industry. Braz Arch Biol Technol 52(6):1527–1539
35. Gottschalk C, Libra A, Saupe A (2010) Ozonation of water and wastewater: a practical guide to understanding ozone and its applications. Wiley-VCH, Weinheim
36. Hashimoto T, Nazazawa H, Murakimi T (2009) State of ozonation to municipal wastewater treatment in Japan. In: Proceedings of 19th ozone world congress, Tokyo, Japan, Paper 4-K-1
37. Hoigne J, Bader H (1978) Ozonation of water: kinetics of oxidation of ammonia by ozone and hydroxyl radicals. Swiss Federal Institute for Water Resources and Water Pollution Control. Dubendort, Switzerland
38. Hoigné J, Bader H (1975) Ozonation of water: role of hydroxyl radicals as oxidizing intermediates. Science 190:782–783
39. Holkar C, Jadhav A, Pinjari D, Mahamuni N, Pandit AB (2016) A critical review on textile wastewater treatments: possible approaches. J Environ Manage 182:351–366
40. Hussain T, Wahab A (2018) A critical review of the current water conservation practices in textile wet processing. J Clean Prod (Elsevier Ltd.) 198:806–819
41. Hussain Z, Arslan M, Malik MH, Mohsin M, Iqbal S, Afzal M (2018) Treatment of the textile industry effluent in a pilot-scale vertical flow constructed wetland system augmented with bacterial endophytes. Sci Total Environ 645:966–973. https://doi.org/10.1016/j.scitotenv.2018.07.163
42. Iervolino G, Vaiano V, Palma V (2020) Enhanced azo dye removal in aqueous solution by H_2O_2 assisted non-thermal plasma technology. Environ Technol Innov 19:100969
43. Jakob SJ, Hansen F (2005) New chemical and biochemical hurdles. In: Sun D (ed) Emerging technologies for food technology. Elsivier Ltd., pp 387–418
44. Jiang B, Zheng J, Qiu S, Wu M, Zhang Q, Yan Z, Xue Q (2014) Review on electrical discharge plasma technology for wastewater remediation. Chem Eng J 236:348–368
45. Kasih TP, Nasution J (2016) Development of cold plasma technology for modification of surface characteristics of material without changing the basic properties of the material (Pengembangan Teknologi Plasma Dingin untuk Modifikasi Karakteristik Permukaan Material tanpa Mengubah Sifat Dasar Material). J Res Appl Ind Syst Appl Vol X 3:373–379
46. Kataki S, Chatterjee S, Vairale MG, Sharma S, Dwivedi SK (2021) Resour Conserv Recycl 164:105156. https://doi.org/10.1016/j.resconrec.2020.105156
47. Khadre MA, Yousef AE (2001) Decontamination of a multilarninated aseptic food packaging material and stainless steel by ozone. J. Food Safety 21:1–13
48. Kirschner MJ (2000) Ozone. In: Ullmann's encyclopedia of industrial chemistry. Wiley-VCH Verlag. https://doi.org/10.1002/14356007.a18_349

49. Loeb BL, Thompson CM, Drago J, Takahara H, Baig S (2012) Worldwide ozone capacity for treatment of drinking waterand wastewater: a review. Ozone: Sci Eng 34:64–77
50. Makkar S, Makkar M (2011) Ozone-treating dental infections. Indian J Stomatol 2(4):256–259
51. Manjunath RGS, Singla D, Singh A (2015) Ozone revisited. J Adv Oral Res 6(2):5–9
52. Morrison C, Atkinson A, Zamyadi A, Kibuye F, McKie M, Hogard S, Mollica P, Jasim S, Wert EC (2021) Critical review and research needs of ozone applications related to virus inactivation: potential implications for SARS-CoV-2. Ozone Sci Eng 43:1, 2–20. https://doi.org/10.1080/01919512.2020.1839739
53. Mondal S, Bhagchandani C (2016) Textile wastewater treatment by advanced oxidation processes. J Ad Eng Technol Sci 2:2455–3131
54. Muthukumarappan K, Halaweish F, Naidu AS (2000) Ozone. In Naidu AS (ed) Natural food antimicrobial systems. CRC Press, Boca Raton, FL pp 783–800
55. Munter R (2001) Advanced oxidation processes – current status and prospects. Proc Estonian Acad Sci Chem 50(2):59–80
56. Murugesan PV, Monica VE, Moses JA, Anandharamakrishnan C (2020) Water decontamination using non-thermal plasma: concepts, applications, and prospects. J Environ Chem Eng 8(5):104377
57. Moertinah S (2008) Clean production opportunities in the textile finishing bleaching industry (Peluang-peluang produksi bersih pada industri tekstil finishing bleaching). Diponegoro University, Thesis
58. Nath A, Mukhim K, Swer T, Dutta D, Verma N, Deka BC, Gangwar B (2014) A review on application of ozone in the food processing and packaging. J Food Prod Dev Packag 1:7–21
59. Nakhjirgan P, Dehghani M (2015) The evaluation of the toxicity of reactive red 120 dye by daphina magna bioassay. J Res Environ Health 1:1–9
60. Nashmi OA, Mohammed AA, Abdulrazzaq NN (2020) Investigation of ozone microbubbles for the degradation of methylene orange contaminated wastewater. Iraqi J Chem Pet Eng 21(2):25–35
61. Nur M (2011) Plasma physics and its applications (Fisika Plasma dan Aplikasinya). Diponegoro Univ Press. ISBN: 978-979-097-093-9
62. Parsons S (2004) Advanced oxidation process for water and wastewater treatment. Gray Publishing, Turnbridge Wells, UK
63. Petek J, Glavič P (1996) An integral approach to waste minimization in process industries. Resour Conserv Recycl 17(3):169–188
64. Pulga A (2018) Oxygen-ozone therapy in dentistry: current applications and future prospects. Ozone Therapy 3(3):37–42
65. Pratiwi R, Notodarmojo S, Helmy Q (2018) Decolourization of remazol black-5 textile dyes using moving bed bio-film reactor. IOP Conf Ser Earth Environ Sci (IOP Publishing) 106(1):012089)
66. Pratiwi R (2018) Reactive black 5 (RB 5) dye treatment using moving bed biofilm reactor (MBBR). Thesis, Environmental Engineering Dept, Institute of Technology Bandung
67. Pressman S (2007) The story of ozone. Plasmafire International, Canada
68. Preis S (2008) History of ozone synthesis and use for water treatment. In: Munter R (ed) Encyclopedia of life support systems (EOLSS). EOLSS Publishers, pp 6–192
69. Ramachandran P, Sundharam R, Palaniyappan J, Munusamy AP (2013) Adv Appl Sci Res 4(1):131–145
70. Raja ASM, Arputharaj A, Saxena S, Patil PG (2019) Water requirement and sustainability of textile processing industries. In: Water in textiles and fashion, pp 155–173
71. Rice RG, Overbeck PK, Larson K (1998) Ozone treatment of small water systems. International Ozone Association Pan American Group, Vancouver, British Columbia, Canada.
72. Rice RG (1999) Ozone in the United States of America-State-of-the-art. Ozone Sci Eng 21:99–118
73. Rice RG, Browning ME (1980) Ozone for industrial water and wastewater treatment, a literature survey. USEPA Research and Development, EPA-600/2-80-060

74. Rice RG, Robson CM, Miller GW, Hill AG (1981) Uses of ozone in drinking water treatment. J Am Water Works Assoc 73:44–57
75. Remondino M, Valdenassi L (2018) Different uses of ozone: environmental and corporate sustainability. Literature review and case study. Sustainability 10:4783. https://doi.org/10.3390/su10124783
76. Rubin MB (2001) The history of ozone. The schonbein period, 1839–1868. Bull Hist Chem 26(1):40–56
77. Rubin MB (2002) The history of ozone. II. 1869–1899 (1). Bull Hist Chem 27(2):81–105
78. Rubin MB (2003) The history of ozone. Part III. C.D. Harries and the introduction of ozone into organic chemistry. Helv Chim Acta 86:930–940
79. Rubin MB (2004) The history of ozone. IV. The isolation of pure ozone and determination of its physical properties (1). Bull Hist Chem 29(2):99–106
80. Samsami S, Mohamadi M, Sarrafzadeh MH, Rene ER, Firoozbahr M (2020) Recent advances in the treatment of dye-containing wastewater from textile industries: overview and perspectives. Process Saf Environ Prot 143:138–163
81. Sarayu K, Sandhya S (2012) Current technologies for biological treatment of textile wastewater–a review. Appl Biochem Biotechnol 167:646–661
82. Shen Y, Xu Q, Wei R, Ma J, Wang Y (2017) Mechanism and dynamic study of reactive red X-3B dye degradation by ultrasonic-assisted ozone oxidation process. Ultrason Sonochem 38:681–692
83. Shimizu A, Takuma Y, Kato S, Yamasaki A, Kojima T, Urasaki K, Satokawa S (2013) Degradation kinetics of azo dye by ozonation in water. SJ Fac Sci Tech Seikei University 50:1–4
84. Srikanth A, Sathish M, Harsha AVS (2013) Application of ozone in the treatment of periodontal disease. J Pharm Bioallied Sci 5(Suppl 1):S89–S94. https://doi.org/10.4103/0975-7406.113304
85. Sugiyana D, Soenoko B (2016) Identification of photocatalytic mechanism in the degradation of azo reactive black 5 dye using TiO_2 microparticle catalyst (Identifikasi mekanisme fotokatalitik pada degradasi zat warna azo reactive black 5 menggunakan katalis mikropartikel TiO_2). Arena Tekstil 31(2):115–124. https://doi.org/10.31266/at.v31i2.1939
86. Suryawan IWK, Helmy Q, Notodarmojo S (2018) Textile wastewater treatment: colour and COD removal of reactive black-5 by ozonation. IOP Conf Ser Earth Environ Sci (IOP Publishing) 106(1):012102
87. Suryawan I, Siregar MJ, Prajati G, Afifah AS (2019) Integrated ozone and anoxic-aerobic activated sludge reactor for endek (Balinese textile) wastewater treatment. J Ecol Eng 20(7)
88. Takahara H, Kato Y, Nakayama S, Kobayashi Y, Kudo Y, Tsuno H, Somiya I (2009) Estimation of an appropriate ozonation system in water purification plant. In: Proceedings of 19th ozone world congress, Tokyo, Japan, Paper 17–2
89. Tizaoui C (2020) Ozone: a potential oxidant for COVID-19 virus (SARS-CoV-2). Ozone Sci Eng 42(5):378–385. https://doi.org/10.1080/01919512.2020.1795614
90. Ulson SMDAG, Bonilla KAS, de Souza AAU (2010) Removal of COD and colour from hydrolyzed textile azo dye by combined ozonation and biological treatment. J Hazard Mater 179(1–3):35–42
91. Ulucan-Altuntas K, Ilhan F (2018) Enhancing biodegradability of textile wastewater by ozonation processes: optimization with response surface methodology. Ozone Sci Eng 40(7):1–8. https://doi.org/10.1080/01919512.2018.1474339
92. UNICEF-WHO (2015) Progress on sanitation and drinking water – 2015 update and MDG assessment. WHO Press, Geneva, Switzerland
93. USEPA (1999) Alternative disinfectants and oxidants-guidance manual. Office of Water (4607), EPA 815-R-99-014
94. Venkatesh S, Venkatesh K (2020) Ozonation for degradation of acid red 14: effect of buffer solution. Proc Natl Acad Sci India Sect A 90(2):209–212
95. Venkatesh S, Quaff AR, Pandey ND, Venkatesh K (2015) Impact of ozonation on decolourization and mineralization of azo dyes: biodegradability enhancement, by-products formation, required energy and cost. Ozone Sci Eng 37(5):420–430

96. Venkatesh S, Venkatesh K, Quaff AR (2017) Dye decomposition by combined ozonation and anaerobic treatment: cost effective technology. J Appl Res Technol 15(4):340–345
97. Verinda SB, Yulianto E, Gunawan G, Nur M (2021) Ozonated nanobubbles- a potential hospital waste water treatment during the COVID-19 outbreak in Indonesia to eradicate the persistent SARS-CoV-2 in HWWs. Ann Trop Med Public Health 24(1):197. https://doi.org/10.36295/ASRO.2021.24197
98. Von Sonntag G, von Gunten U (2012) Chemistry of ozone in water and wastewater treatment from basic principles to applications. IWA Publishing
99. Wisniak J (2008) Thomas Andrews. Revista CENIC Ciencias Químicas 39(2):98–108
100. Zheng H, Zhang J, Yan J, Zheng L (2016) An industrial scale multiple supercritical carbon dioxide apparatus and its eco-friendly dyeing production. J CO_2 Util 16:272–281
101. Zheng Q, Dai Y, Han X (2016) Decolourization of azo dye CI reactive black 5 by ozonation in aqueous solution: influencing factors, degradation products, reaction pathway and toxicity assessment. Water Sci Technol 73(7):1500–1510
102. Zhang C, Sun Y, Yu Z, Zhang G, Feng J (2018) Simultaneous removal of Cr(VI) and acid orange 7 from water solution by dielectric barrier discharge plasma. Chemosphere 191:527–536
103. Ziyaina M, Rasco B (2021) Inactivation of microbes by ozone in the food industry. Afr J Food Sci 15(3):113–120. https://doi.org/10.5897/AJFS2020.2074

Ferrite-Based Magnetic Nanoparticle Heterostructures for Removal of Dyes

Bintu Thomas and L. K. Alexander

Abstract Photocatalysis has emerged as one of the technologies for the elimination of environmental contaminants from water. Photocatalysis using magnetic nanoparticles hold a huge potential for water purification applications that necessitates recovery of the catalyst after the photocatalytic process. Water contamination caused by organic dyes is one of the most serious problems aggravating the scarcity of safe and healthy water. In this chapter, we focus on the application of ferrite-based superparamagnetic heterostructures, principally magnetic nanospheres and core@shell nanostructures for the removal of organic dyes from the water system. The performance of nanostructured materials strongly depends on the size, dimensionality and morphologies which are the key factors to their eventual performance and applications. We emphasize the advantages of enhancing the photocatalytic activity by the surface modification of ferrite-based magnetic nanoparticles. Among the several strategies to increase the photocatalytic activity, we detail the synthesis of semiconductor metal-oxides-based nanospheres as well as core@shell nanostructures as photocatalysts under visible light irradiation. The significant advance in this study is the synthesis of ferrite-based superparamagnetic heterostructures for the removal of anionic and cationic dyes. The effect of porosity, superparamagnetic nature, electronic band structure, narrower bandgap and the morphology of the catalysts have paved the way for the higher adsorption and photocatalytic efficiency of the nanoparticles.

Keywords Superparamagnetism · Heterostructures · Ferrites · Adsorption · Organic dye degradation · Band structure · Pore hierarchy

---

B. Thomas · L. K. Alexander (✉)
Department of Physics, University of Calicut, Kozhikode 673635, Kerala, India
e-mail: LKA@uoc.ac.in

S. S. Muthu and A. Khadir (eds.), *Advanced Oxidation Processes in Dye-Containing Wastewater*, Sustainable Textiles: Production, Processing, Manufacturing & Chemistry, https://doi.org/10.1007/978-981-19-0882-8_7

1 Introduction-Importance of Removal of Dyes From Water Bodies

Dyes are chemical compounds giving colour to substances. They can absorb light of a wavelength in the visible spectrum (400–700 nm), have a colour-giving group called chromophores. Dyes can be classified as natural and synthetic dyes. Natural dyes are taken from natural sources like plants, animals, flowers, fruits, insects, minerals, soil, etc. They are not risky to living organisms. Examples of natural sources are saffron, Lac insects (red), Murex snail (purple), etc. Synthetic dyes are artificially made dyes. Synthetic Dyes are generally used widely in modern-day industrial processes—major users belong to cloth manufacturing units, plastic and printing industries, leather tanning, etc. [3]. Synthetic dyes can be classified as cationic (basic dyes) and anionic (acidic) dyes [9]. The chromophore in a dye molecule contains positive and negative charges in cationic and anionic dyes, respectively. Examples of anionic dyes are Reactive Orange 6, Methyl Orange, Congo Red, Brilliant Red HE-3B, etc. Methylene Blue, Crystal violet, Rhodamine B, etc., are examples of cationic dyes. Dyes are highly resistant to degradation because of their complex structure. The wastewater along with washed-out dyes from the above-mentioned industries, when released into the neighbouring water bodies without remediation processes, is reported to severely damage the quality of water bodies. It affects aquatic ecosystems, aquatic life and, in turn, other living beings. The presence of dyes in water is very toxic because it contains acetic acid, nitrates, sulphur, naphthol, heavy metals, etc., as the components. The increase of turbidity, foul smell and bad appearance are the main consequences of the presence of dyes in water bodies. Dyes reduce the entry of light into water bodies, thereby affecting the photosynthesis process of aquatic plants. Some chemicals evaporated from dye solutions blow out in the atmosphere and cause serious allergy problems to living organisms. Most of the synthetic dyes are classified as genotoxic, mutagenic and carcinogenic—leading to detrimental effects in the reproductive and immune system of human beings [15, 32, 39].

2 Photocatalysis and Role of Photocatalysts in Dye Removal

Photocatalysis is an advanced oxidation process that takes place in the presence of UV–Visible radiation. In advanced oxidation processes, reactive oxygen species like superoxide radicals (O^{2-}), hydroxyl radicals ($OH^{\cdot}$), electrons (e-) and holes (h^+) are the reactive species responsible for the complete mineralization of organic compounds (Fig. 1). Photocatalysis is an attractive route for the remediation of water [8, 42]. The technique utilizes semiconductors as photocatalysts. Depending on the bandgap of the photocatalyst, the incident photons generate electron–hole pairs which drive the catalytic process. The solar spectrum comprises only 5% of UV light and about 45% of visible light; see Fig. 2. Therefore, the fabrication of visible light active narrow bandgap semiconductors has got great importance in photocatalysis. The

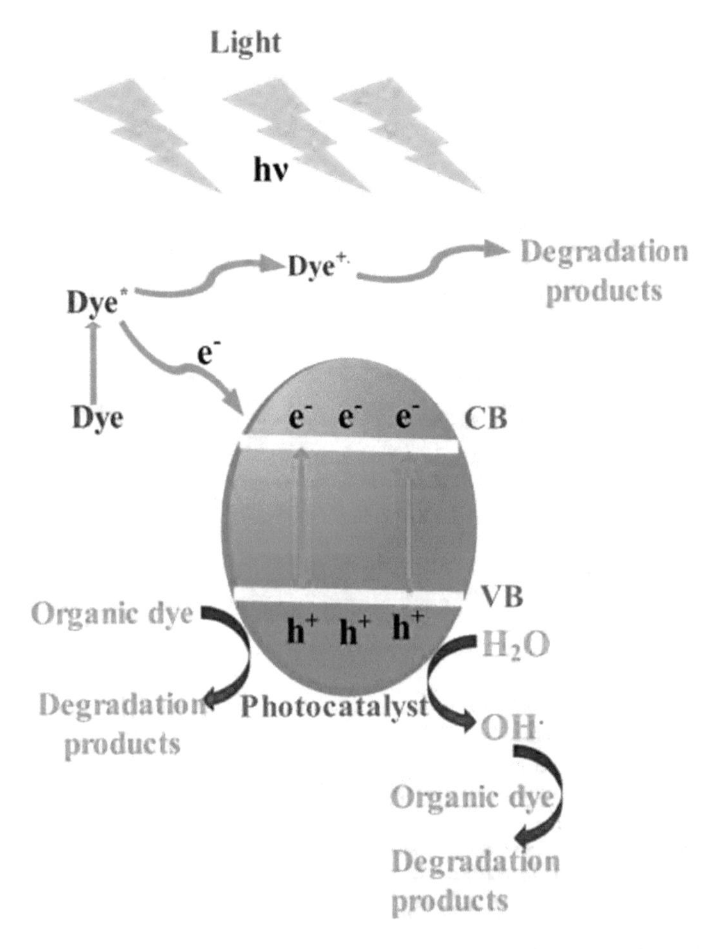

Fig. 1 General mechanism of photocatalysis for water remediation (CB and VB denote conduction band and valance band respectively)

remediation of water includes the removal of impurities like organic dyes, inorganic heavy metal ions, and various organic compounds from

water bodies. Specific to dye molecules, photocatalysis has numerous advantages when compared with other water purification techniques because it ensures the destruction of organic hazardous dye molecules from water. When the photocatalyst is exposed to light, electron–hole pairs are created which initiates the oxidation process [60].

For the commercial adoption of photocatalysts for the removal of dyes from the water system, the catalyst needs to be assessed based on the following considerations:

i. Stability
ii. Remediation Efficiency
iii. Catalyst Reusability

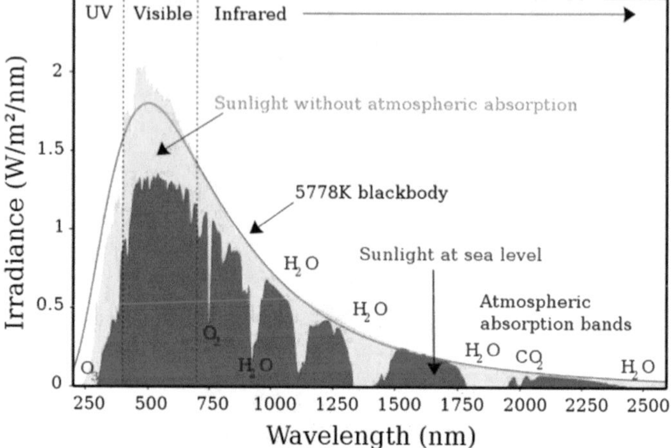

Fig. 2 Solar radiation spectrum. It shows the importance of photocatalysts having activity in visible light frequencies. (The image is adapted from https://commons.wikimedia.org/wiki/File:Solar_spectrum_en.svg)

iv. Catalyst Recyclability
v. Cost of manufacture
vi. Activity on a mixture of dyes
vii. Toxicity of the catalyst and by-products.

The photocatalysts should be less toxic, thermally and chemically stable, and highly resilient to photo corrosion. In the commercial application of photocatalysts for water purification, the process should be facile and economical [2]. The reuse of the photocatalysts should make the process more cost-effective. Conventional methods, especially filtration and centrifugation have the drawbacks in water purification process because of the difficulty in the separation of the photocatalysts after the water purification process [38, 54]. Conventional approaches also have a tremendously slow degradation rate and it could distress the quality of water. Moreover, it yields hazardous and toxic constituents as by-products. In addition to the above-mentioned problems, the disposal of such types of materials into the surroundings could be very dangerous. Therefore it is commercially important and environmentally sensitive to develop photocatalysts that can be recovered, rejuvenated and reused. In search of suitable photocatalysts which can satisfy the above criteria, the ferrite-based superparamagnetic nanoparticles (MNP) gains enormous interest.

2.1 Ferrite-Based Superparamagnetic Nanoparticles (MNPs)

Magnetic nanoparticles (MNPs) are formed from pure magnetic materials like Fe, Co, Ni, and their alloys [61]. Mainly, Fe-based oxides, such as magnetite (Fe_3O_4), maghemite (γFe_2O_3), are used for the synthesis of magnetic nanoparticles. Some magnetic compounds, when confined to the nano-size regime, show some superior magnetic characteristics. When the dimension of the magnetic material is reduced to the nano level, magneto-crystalline anisotropy energy (E_A) of the magnetic particles falls to the order of thermal energy (K_BT). If the size of the particle is extremely small, then K_BT will overcome the energy barrier E_A and a rapid rotation of magnetic moments occur that leads to the desertion of hysteresis curve in the magnetization (M) vs. applied field (H) loop and this phenomenon is termed as superparamagnetism [16, 36]. The distinguishing and attractive magnetic properties of superparamagnetic materials are shown in Fig. 3. The features such as characterization by higher magnetization, negligible remanence, and coercivity values lead to a quick response to an applied external magnetic field and they can act like a normal nanoparticle in the absence of a magnetic field. They are single-domain in nature, therefore, the total

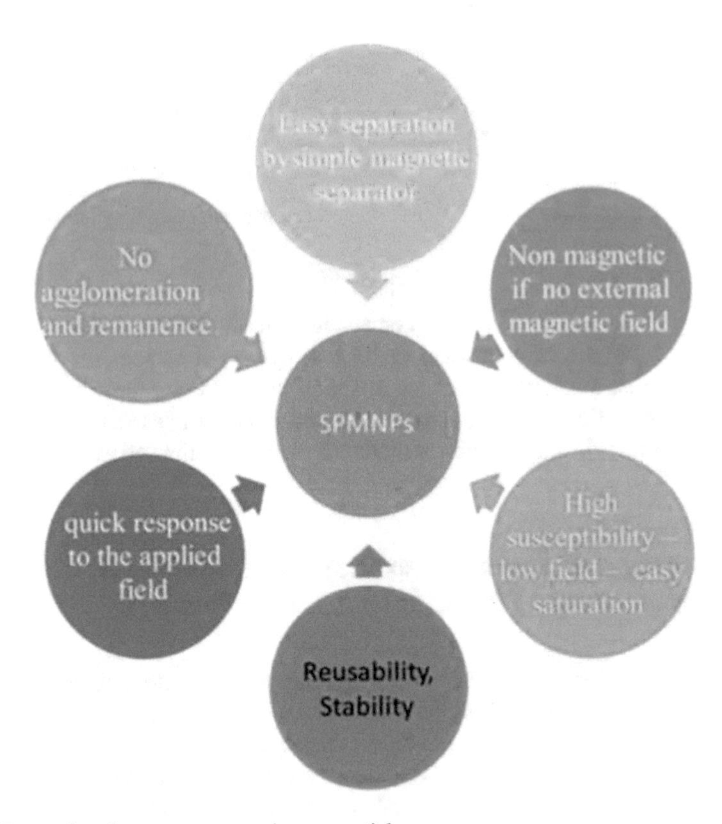

Fig. 3 Properties of superparamagnetic nanoparticles

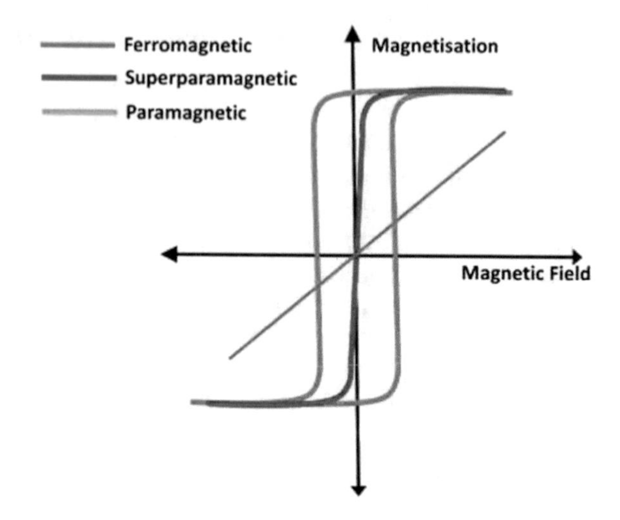

Fig. 4 Comparision of magnetic materials based on hysteresis plots. The slope of the *M* versus *H* plot gives susceptibility. A hysteresis plot discusses a class of magnetic material based on (i) Spontaneous magnetization (M_s): the net magnetization that exists inside a uniformly magnetized microscopic volume in the absence of a field, (ii). Saturation magnetization (M_{sat}): the maximum induced magnetic moment that can be obtained in a magnetic field (H_{sat})—a field beyond which further increase in magnetization does not occur, and (iii). Coercivity of remanence (H_r): The reverse field, when applied and then removed, reduces the saturation remanence (M_r) to zero. In normal paramagnets, in the presence of a field, there is a partial alignment of the atomic magnetic moments in the direction of the field, resulting in a net positive magnetization and positive susceptibility. The magnetization is zero when the field is removed. The ferromagnets will have a significant area under the hysteresis plot in contrast to the negligible area in superparamagnets as explained elsewhere in this manuscript

magnetic moment can be deliberated as a sum of individual magnetic moments of every atom. MNP are of great interest for researchers of various disciplines such as photocatalysis, magnetic resonance imaging, biomedicine, ferrofluids and environmental applications. The superparamagnetic nanoparticles offer some unique attractions for photocatalysis and resultant water remediation scheme (Fig. 4).

3 Advantages of Ferrite-Based Superparamagnetic Nanoparticles as Photocatalysts in Dye Removal

The utilization of superparamagnetic magnetic nanoparticle photocatalysts has numerous advantages in the field of the water purification process. Magnetic separation offers a very appropriate method for separating the catalysts by applying external magnetic fields and reusing the recycled catalyst. In addition, the ferrites offer the benefits of visible light-based photocatalysts due to their narrower bandgap. Moreover, it can be used in combination with wide bandgap materials to extend the activity of the photocatalyst from UV to visible region [50, 67]. All the unique features make MNP attractive as commercially viable photocatalysts. MNPs are demonstrated for the elimination of various types of organic and inorganic pollutants. It holds a higher adsorption capacity for the removal of organic dyes [4, 87]. The deployment of superparamagnetic nanoparticles as photocatalysts would resolve the agglomeration of magnetic nanoparticles because a superparamagnetic particle has the magnetic property only in the presence of an external magnetic field. After the removal of the field, it acts like a nonmagnetic material.

4 Single Constituent Ferrite Photocatalysts

A variety of single constituent magnetic nanomaterials such as $Fe_3O_4, \alpha - Fe_2O_3, \beta Fe_2O_3$ etc. concerning its photocatalytic activity are reported. Pal and Sharon [58] reported Fe_2O_3 for the removal of phenol under visible light. Later, reported the hollow Fe_2O_3 nanoparticle as a photocatalyst but its lower saturation magnetization value prevents its separation using an external magnetic field. In addition, the use of single-phase magnetic nanoparticles as photocatalysts causes the photo dissolution of the magnetic oxides. This reasons for the reduction of the stability and photocatalytic activity of the photocatalysts [43]. Gupta et al. [25, 26] have presented $MFe_2O_4(M = Co, Ni, Cu, Zn)$ nanoparticles for the dye removal at the natural pH. Oliveira et al. [57] have studied the degradation of dyes using $ZnFe_2O_4$ synthesized at different temperatures. A list of various ferrite-based photocatalysts is shown in Table 1. The removal of Rhodamine B dye using $Zn_xCo_{1-x}Fe_2O_4$(x = 0.0, 0.2, 0.4, 0.6, 0.8and1.0) nanospheres of diameter 5 nm were deliberated by Singh et al. [72].

Photocatalytic removal of textile dyes using ferrite nanoparticles such as $CoFe_2O_4, ZnFe_2O_4, MnFe_2O_4, NiFe_2O_4, MgFe_2O_4$, etc., has been reported. In

Table 1 List of various ferrite-based photocatalysts

Ferrite	References
$M_xCo_{1-x}Fe_2O_4$	[51]
$CoFe_2O_4$	[83]
$ZnFe_2O_4$	[19]
$Ni_{0.4}Cu_{0.6}Fe_2O_4$	[30]
$CuFe_2O_4$	[78]
$Cu_{1.25}Fe_{1.75}O_4$	[96]
MFe_2O_4(M = Ni, Cu, Zn)	[23]
$ZnFe_2O_4$	[19]
$CoZnFe_2O_4$	[10]
$TiCuFe_2O_4$	[7]

2020, [13] synthesized pure cobalt ferrite using the solution combustion method for the removal of Methylene Blue dye. A series of $CoFe_2O_4$ photocatalysts were fabricated by a microwave/wet chemistry route and systematically studied for the removal of a group of dyes viz. Methylene Blue, Rhodamine B (cationic dyes), Congo Red and Methyl Orange (anionic dyes), under sunlight irradiation [75]. Photocatalytic degradation of a group of dyes like Methyl Red, Bromo Green, Methylene Blue and Methyl Orange using MFe_2O_4 (M = Co, Ni, Cu, Zn) were studied [25, 26]. Shetty et al. [71] have reported a comparative study of $ZnFe_2O_4$, $NiFe_2O_4$ and $CuFe_2O_4$ nanoparticles for the removal of Malachite Green dye under visible light radiation. Rashad et al. [63] have studied cubic $CuFe_2O_4$ for the removal of Methylene Blue dye synthesized at different temperatures. The photodegradation of the RR198 and RR120 dyes was carried out using $ZnFe_2O_4$ nanoparticles. The dependence of catalyst dose and dye concentration on photocatalytic efficiency was evaluated [49]. The study of photocatalytic removal of Methylene Blue dye using $MFe_2O_4(M = Mn, Co)$ was done by Liu et al. [44]. Bhukal et al. [11] have reported the photocatalytic removal of Methyl Orange dye using $Co_{0.6}Zn_{0.4}Mn_xFe_2 - xO_4$ and they studied the variation of efficiency with Mn^{3+} doping. Priya et al. have investigated the photocatalytic removal of Methylene Blue dye using $CuFe_2O_4$. Also, they analyzed the factors affecting photocatalytic efficiency [62]. The magnetic $CuFe_2O_4$ catalyst was used for the degradation of Acid Red B azo dye from water. Fe_3O_4 nanoparticle was used for the degradation of azo dyes Congo Red and Methyl Red. The photocatalytic efficiency was studied in the presence and absence of H_2O_2 [73].

Ferrites can be fabricated in diverse morphologies. Some of them are nanospheres, nanosheets, nanotubes, nanowires, core@shell nanoparticles, etc. The morphology of the catalysts is an important factor that decides the photocatalytic activity. Thomas and Alexander [79] have reported the synthesis of $Co_xZn_{1-x}Fe_xO_4(x = 0, 0.5, 1)$ magnetic nanospheres by the solvothermal method. The synthesized nanospheres were used for the removal of binary impurities like Methyl Orange, which is an anionic dye and hexavalent Cr (VI) heavy metal ion. Reddy et al. [65] have fabricated $CoFe_2O_4$ nanospheres of average grain size of 180 nm by hydrothermal method. $Co_{1-x}Zn_xFe_2O_4$ nanospheres were synthesized by the solvothermal method and its magnetic properties (M_s = 65 emu/g) were studied [31]. $NiFe_2O_4$ nanospheres of diameter 7–200 nm were fabricated by varying their experimental parameters like reaction time and precursors [20]. Rod-shaped $NiFe_2O_4$ nanoparticles were fabricated by Cao et al. [14] with a larger coercivity and smaller saturation magnetization.

5 Challenges to Be Solved Before MNP Can Be Adopted as Efficient Photocatalysts

The single constituent MNP has the advantages offered by the superparamagnetic nanoparticles. But when it comes to the credible and repeatable applicability of MNPs for photocatalytic removal of dye, there are certain challenges to be solved.

Photodissolution

Photodissolution is a problem associated with magnetic photocatalysts, leading to leaching out of the magnetic nanoparticles in water [43]. This results in an undesirable decline in the stability and photocatalytic activity of the magnetic nanoparticles.

Agglomeration

Agglomeration is the adhesion of nanoparticles due to the presence of a weak force between them leading to the formation of particles in the micrometre range. This originates from the higher surface area to volume ratio in nanoparticles [59]. Agglomeration reduces the surface area and hence the decline in the efficiency and the stability of the catalyst [31].

The bandgap of the catalyst

The bandgap of the photocatalyst is a noteworthy factor because its value determines whether the catalyst is visible or UV light active. From a technological perspective, it is better to use visible light active photocatalyst since a major proportion of sunlight is in the frequencies of visible light. Therefore, it is desirable to extend the bandgap of the photocatalyst to enable it to capture the maximum range of solar radiation.

Control of the porosity of the catalyst

The porosity of the catalysts has a serious role in their photocatalytic ability. In general, a photocatalyst should also be a good adsorbent. As a result, the details about the porosity of the photocatalyst have significance for its design, both from technological and theoretical perspectives. The nature of porosity in catalysts is a crucial parameter as it determines the various factors including surface area. Therefore. it is necessary to study the factors that affect the porosity comprising temperature, pH, surfactants, precursors, and solvents used for the synthesis of nanoparticles [17, 27]. The choice of applicable fabrication method is also essential. The complete explanation behind the creation of pores in the nanocatalysts and the factors influencing porosity is one of the main challenges to be resolved. Some studies indicated that the presence of mesopore–macropore combination nanostructure in photocatalyst would improve catalytic activity [40, 79].

Removal efficiency on a mixture of dyes solution

The waterbody polluting dyes from industries could have a mixture of dyes along with other inorganic and organic pollutants. The presence of one of the pollutants could improve or reduce photocatalytic efficiency [79]. So the investigations on the activity of catalyst on a mixture of dyes in the water system are quite important. A recent report by Gogoi et al. on a mixture of dyes is quite noticeable in this regard [22].

In addition to the challenges discussed above, a better understanding of the percentage contributions of different mechanisms governing the photocatalytic process would be essential for the production of the nanocatalysts from an industrial viewpoint. MNP based-nanocomposites are suggested as better photocatalysts to solve many of the challenges mentioned above. Among the nanocomposites, surface modified MNP offer some promising characteristics.

6 Surface Modified Ferrite Photocatalysts

As mentioned earlier, there are several challenges to be overcome for the utilization of magnetic nanoparticles as photocatalysts. This difficulty has been overcome by the surface modification of the magnetic component by semiconductors, metals, organic materials, etc., by forming the heterojunctions of binary or ternary composites. Surface modification is a process in which the surface of a nanomaterial is altered by carrying a change in its physicochemical properties which is different from its original properties utilizing other materials like metals, semiconductors, polymers, etc. [6, 85, 99]. It can be done with organic and inorganic materials and improve surface activities and biocompatibility [85]. Surface modification [12, 48] of magnetic nanoparticles is one of the solutions to avoid photo dissolution and agglomeration. As stated before, the catalyst must have an adequate bandgap for absorbing visible light. Therefore, the tuning of the bandgap towards the visible region has an impact on real-world commercial applications. It can be achieved by functionalization with noble metals like palladium [81] silver, etc., and the formation of core@shell nanostructures. Fan et al. [20] have reported visible light active ($Zn_{1-x}Co_xFe_2O_4$) nanoparticles for the degradation of Methylene Blue dye. Thomas and Alexander [79, 80], have investigated the visible light active $CoZnFe_2O_4$ nanospheres and $CoZnFe_2O_4@SrTiO_3$ core–shell nanocomposites for the removal of Methyl Orange and binary Congo Red-Methylene Blue dye impurities, respectively. Wu et al. [90] have developed a ternary photocatalytic system $ZnO/Fe_3O_4/g-C_3N_4$ and explored it for the photocatalytic removal of mono-azo dyes. Therein, the role of heterojunction between ZnO and $g-C_3N_4$ nanocomposites were well studied under visible light. Recently, degradation studies on a large variety of industrial dyes, are reported by Gogoi et al. [22]—utilizing the bandgap of CoFe2O4-gC3N4 nanocomposites.

Various researchers tried to modify the surface of ferrites through various fabricating methods and applied them for the removal of anionic and cationic dyes. In 2019, [18] prepared nanostructures of superparamagnetic cobalt ferrite by sol–gel method, and its surface was modified with Strontium titanate ($SrTiO_3$). It was used for the elimination of acid black and acid brown azo dyes under visible and UV light irradiation. The photocatalytic removal of Congo red anionic dye using novel multifunctional $NiFe_2O_4/ZnO$ hybrids under solar light irradiation was reported by Zhu et al. [102]. Similarly, [101]. have reported the removal of dyes from wastewater using $NiFe_2O_4$ functionalized with multi-walled carbon nanotubes. The surface modification of Fe_3O_4 nanoparticles was done by SiO_2 nanoparticles by the formation of core@shell $Fe_3O_4@SiO_2$ nanocomposites and applied for the removal of Methylene Blue dye in neutral pH [93]. Similarly, the removal of Methylene Blue dye using $Fe_3O_4@TiO_2$ by Abbas et al. [1] and $Fe_3O_4/chitosan/TiO_2$ nanocomposites [91] were reported. In the same way, binary composites Fe_3O_4/ZnO [70] and ternary composites $Fe_3O_4/ZnO/CuWO_4$ were reported for the degradation of Rhodamine B in visible light radiation. The photocatalytic activity and the stability of Fe_3O_4 nanoparticles were modified by the deposition of nanoparticles on $g - C_3N_4$ sheets and checked its activity against the removal of Rhodamine B dye under visible light [95]. All these studies pointed out the advantages of surface modification of ferrite photocatalysts. Besides, the authors testified to better recyclability of photocatalysts without the loss of their activity [37]. Ternary core@shell type nanoparticles have been also reported for the removal of azo dyes. For example, $Co_{0.53}Mn_{0.31}Fe_{2.16}O_4@TiO_2$ [53] and $MnFe_2O_4@PANI@Ag$ were reported with higher degradation efficiency. The doping of various metal ions into ferrites was also reported [5]. For instance, [34] have reported the synthesis of Ni^{2+} doped/$MnFe_2O_4$ and applied it for the removal of indigo carmine dye. In another study, $CoFe_2O_4$ photocatalyst was modified by sulphur (S) and used for the degradation of Orange II dye. The $CoFe_2O_4$ nanoparticle was modified using $g - C_3N_4$ and used for the removal of multiple dye impurities from water [94]. Similarly, a core with multi shells like $CoFe_2O_4@TiO_2@SiO_2$ nanocomposites was used for the degradation of Methylene Blue dye under UV light [28]. The surface modification of ferrites using rare earth elements was also reported. The Sm^{+3} doped $ZnFe_2O_4$ was used for the removal of Methyl Orange dye [64]. The surface modification can improve the parameters like bandgap, surface area, etc., thereby enhancing its photocatalytic performance. Surendra [76] have reported the removal of malachite green using noble metal Ag modified $CuFe_2O_4$ nanoparticles.

Core–shell nanoparticles

Core–shell nanocomposites comprise an inner core and an outer shell components of different materials. So it exhibits the unique properties of both the core and shell materials and has found applications in the fields of adsorption, photocatalysis, hydrogen production, sensing, etc. Core@shell materials can be categorized based on their composition into organic@organic, organic@inorganic, inorganic@inorganic, etc. The outer layer shell material can be used to modify the properties of the inner material. [80] have reported the photocatalytic removal of Congo Red and Methylene Blue dyes using $Co_{0.7}Zn_{0.3}Fe_2O_4@SrTiO_3$ core@shell nanoparticles of thickness of about 1 nm. A mechanism of the adsorption process was also proposed (Fig. 5). There are reports of the synthesis of magnetic nanoparticles-based core@shell nanostructures but those are fabricated using expensive solvents [33, 88]. Zhang et al. [98] reported $Fe_3O_4@TiO2$ core@shell nanoparticles via solvothermal technique. It was used for the removal of Rhodamine B dye using LED

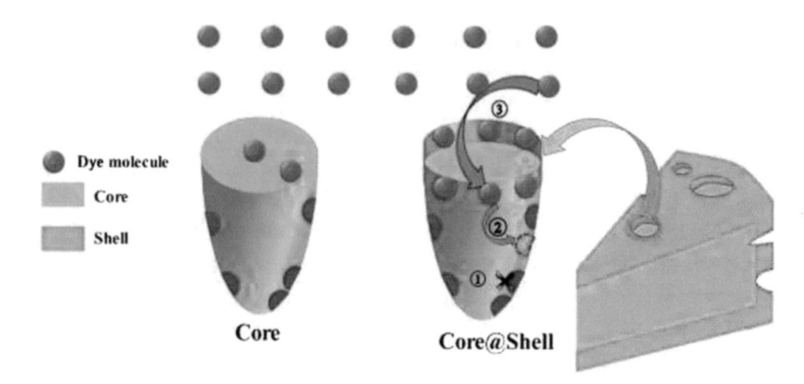

Fig. 5 Mechanism of adsorption in core@shell nanoparticles [80]

light with a saturation magnetization of 2 emu/g. Fe_3O_4@ZnO core–shell nanocomposite was synthesized by thermal decomposition method using triethylene glycol as the solvent [84]. Wang et al. [86] synthesized core@shell Fe_3O_4@SiO_2@ZnO-Ag nanocomposites and carried out the removal of Rhodamine B dye in the presence of UV light. The Fe_3O_4has a microsphere morphology with the diameter of 325 nm achieved by a solvothermal method and coated with SiO_2 shell by Stober process and outer layer of ZnO coated by co-precipitation method. Over the past years, various types of magnetic-based photocatalysts are fabricated comprising single-phase magnetic components, surface modified magnetic nanoparticles including MNP–semiconductor composites, MNP–metal modified nanoparticles, MNP–organic framework composites, core@shell nanomaterials, etc., as photocatalysts. Gholizadeh and Abharya [21] have reported the removal of Methyl Orange and Methylene Blue dye using Fe_3O_4@GO core–shell nanocomposite. The surface of the magnetic nanoparticle was functionalized using 3-Aminopropyl triethoxysilane and coated with graphene oxide nanosheets. Luo et al. [47] have studied the visible light active $Fe_3O_4 - GCN$ (graphite carbide nitrogen) nanocomposite for the removal of organic impurities. Persulfate (PS) is a strong oxidizing agent and it was used to promote the degradation of the organic impurities under visible light irradiation. Intending to make the synthesis process cost-effective, a new route of production of core@shell nanocomposites has to be introduced, wherein they used industrially produced magnetic nanoparticles[55]. Kim et al. [35] have presented the synthesis of $NiCe@SiO_2$ multi–yolk-shell nanotube as a catalyst. Suresh et al. [77] have studied about MFe_2O_4 and their binary composites for wastewater treatment. $MnFe_2O_4$/coal fly ash nanocomposites. Mushtaq et al. are used as a photocatalyst for dye degradation in the presence of sunlight [52] Suharyadi et al [74]. have reported the photocatalytic activity of $CoFe_2O_4@ZnO$ core@shell nanoparticles synthesized at different molar ratios of both core and shell nanoparticles [74].

A list of various core@shell photocatalysts and ferrite-based core@shell nanocomposites for the removal of organic dyes are presented in Table 2.

Table 2 List of various core@shell based photocatalysts for dye removal

Core@shell nanocomposites	Dyes	References
CdS/ZnS	Methyl Orange	
TiO_2/polyaniline	Methyl Orange	[56]
$ZnO@ZnS$	Rose Bengal	[66]
$SiO_2/g-C_3N_4$	Rhodamine B	[41]
$CoFe_2O_4@ZnO$	Methylene Blue	[74]
$Co_3O_4@ZrO_2$	Rhodamine B	[69]
$Fe_3O_4@TiO_2$	Rhodamine B dye	[92]
$Fe_3O_4@TiO_2$	Methylene Blue	[100]
$SiO_2/g-C_3N_4$	Rhodamine B dye	[41]
$Fe_3O_4@SiO_2@TiO_2@CoPcS$	Methylene Blue	[89]
$\gamma-Fe_2O_3@SiO_2@Ce$-doped-TiO_2	Rhodamine B dye	[29]

7 Conclusion

Briefly, ferrite-based magnetic nanoparticles offer themselves as credible candidates in the field of photocatalysis, especially for the removal of dyes from the water system. This review article delivers insight into the utilization of ferrite-based heterostructures for the removal of various anionic and cationic dyes. Ferrite-based photocatalysts exhibit excellent catalytic performance due to their higher surface area, narrower bandgap appropriate for visible light activity, superparamagnetic nature suitable for the separation and reuse of the catalyst. In this review, we tried to develop a review on ferrite-based magnetic nanoparticle heterostructures for the removal of dyes. They include a basic understanding of the formation of magnetic-based heterostructures and their application, especially for the removal of organic dyes from water. All the major developments made to date for photocatalytic dye removal are merely remarkable but their commercial and real-world application is a little far away from reality. Assuredly, there is bounteous room for the research to evade the above-mentioned challenges. In addition, ferrite-based heterostructures hold boundless applications in various fields, and investigation in this route has opened up countless possibilities.

References

1. Abbas M, Rao BP, Reddy V, Kim C (2014) Fe_3O_4/TiO_2 core/shell nanocubes: Single-batch surfactantless synthesis, characterization and efficient catalysts for methylene blue degradation. Ceram Int 40:11177–11186. https://doi.org/10.1016/j.ceramint.2014.03.148
2. Adesina AA (2004) Industrial exploitation of photocatalysis: progress, perspectives and prospects. Catal Surv Asia 8:s265-273. https://doi.org/10.1007/s10563-004-9117-0

3. Aljeboree AM, Alshirifi AN, Alkaim AM (2017) Kinetics and equilibrium study for the adsorption of textile dyes on coconut shell activated carbon. Arab J Chem 10:S3381–S3393. https://doi.org/10.1016/j.arabjc.2014.01.020
4. Ambashta RD, Sillanpaa M (2010) Water purification using magnetic assistance: a review. J Hazard Mater 180:38–49. https://doi.org/10.1016/j.jhazmat.2010.04.105
5. Amir M, Kurtan U, Baykal A, Sözeri H (2016) $MnFe_2O_4$@PANI@Ag heterogeneous nanocatalyst for degradation of industrial aqueous organic pollutants. J Mater Sci Technol 32:134–141. https://doi.org/10.1016/j.jmst.2015.12.011
6. Andreeva D, Tabakova T, Idakiev V, Christov P, Giovanoli R (1998) Au/α-Fe_2O_3 catalyst for water–gas shift reaction prepared by deposition–precipitation. Appl Catal Gen 169:9–14. https://doi.org/10.1016/S0926-860X(97)00302-5
7. Arifin N, Karim KR, Abdullah H (2019) Synthesis of titania doped copper ferrite photocatalyst and its photoactivity towards methylene blue degradation under visible light irradiation. Bull Chem React Eng Catal 14:219–227. https://doi.org/10.9767/bcrec.14.1.3616.219-227
8. Belver C, Bedia J, Avilés AG, Garzón MP, Rodriguez JJ (2019) Semiconductor photocatalysis for water purification. Micro Nano Technol 581–651. https://doi.org/10.1016/B978-0-12-813926-4.00028-8
9. Benkhaya S, Mrabbet S, Harfi AE (2020) A review on classifications, recent synthesis and applications of textile dyes. Inorg Chem Commun 115:107891. https://doi.org/10.1016/j.inoche.2020.107891
10. Bhukal S, Shivali SS (2014a) magnetically separable copper substituted cobalt-zincNanoferrite photocatalyst with enhanced photocatalytic activity. Mater Sci Semicond Process 26:467–476. https://doi.org/10.1016/j.mssp.2014.05.023
11. Bhukal S, Bansal S, Singhal S (2014b) Magnetic Mn substituted cobalt zinc ferrite systems: Structural, electrical and magnetic properties and their role in photo-catalytic degradation of methyl orange azo dye. Phys B Condensed Matter 445:48–55.https://doi.org/10.1016/j.physb.2014.03.088
12. Bucak S, Yavuzturk B, Demir A (2012) Magnetic nanoparticles: synthesis, surface modifications and application in drug delivery (chapter-7). IntechOpen, pp 165–200
13. Cao Z, Zuo C (2020) Direct synthesis of magnetic $CoFe_2O_4$ nanoparticles as recyclable photofenton catalysts for removing organic dyes. ACS Omega 5(35):22614–22620. https://doi.org/10.1021/acsomega.0c03404
14. Cao X, Guo X, Dong H, Meng J, Sun J (2014) Study on the synthesis and magnetic properties of rod-shaped $NiFe_2O_4$ ferrites via precipitation–top tactic reaction employing $Na_2C_2O_4$ and NaOH as precipitants. Mater Res Bull 49:229–236. https://doi.org/10.1016/j.materresbull.2013.08.017
15. Carneiro PA, Umbuzeiro GA, Oliveira DP, Zanoni MVB (2010) Assessment of water contamination caused by a mutagenic textile effluent/dyehouse effluent bearing disperse dyes. J Hazard Mater 694–699
16. Cullity BD, Graham CD (2009) Ferromagnetism (chapter-4), In: Introduction to magnetic materials. Wiley, pp 115–149
17. Dapsens PY, Hakim SH, Su B, Shanks BH (2010) Direct observation of macropore self-formation in hierarchically structured metal oxides. Chem Commun 46:8980–8982. https://doi.org/10.1039/C0CC02684K
18. Eskandari N, Nabiyouni G, Masoumi S, Ghanbari D (2019) Preparation of a new magnetic and photo-catalyst $CoFe_2O_4$–$SrTiO_3$ perovskite nanocomposite for photo-degradation of toxic dyes under short time visible irradiation. Compos B Eng 176:107343. https://doi.org/10.1016/j.compositesb.2019.107343
19. Fan G, Gu Z, Yang L, Li F (2009) Nanocrystalline zinc ferrite photocatalysts formed using the colloid mill and hydrothermal technique. Chem Eng J 155:534–541. https://doi.org/10.1016/j.cej.2009.08.008
20. Fan G, Tong J, Li F (2012) Visible-light-induced photocatalyst based on cobalt-doped zinc ferrite nanocrystals. Ind Eng Chem Res 51:13639–13647. https://doi.org/10.1021/ie201933g

21. Gholizadeh A, Abharya A (2020) Structural, optical and magnetic feature of core-shell nanostructured Fe_3O_4@GO in photocatalytic activity. Iran J Chem Chem Eng 39(2):49–58. https://doi.org/10.30492/ijcce.2020.34296
22. Gogoi D, Makkar P, Ghosh NN (2021) solar light-irradiated photocatalytic degradation of model dyes and industrial dyes by a magnetic $CoFe_2O_4$–gC_3N_4 S-scheme heterojunction photocatalyst. ACS Omega 6(7):4831–4841
23. Goyal A, Bansal S, Singhal S (2014) Facile reduction of nitrophenols: comparative catalytic efficiency of MFe2O4 (M [Ni, Cu, Zn) nano ferrites. Int J Hydrogen Energy 39:4895–4908. https://doi.org/10.1016/j.ijhydene.2014.01.050
24. Gubin SP, Koksharov YA, Khomutov GB, Yurkov GY (2005) Magnetic nanoparticles: preparation, structure and properties. Russ Chem Rev 74:489–520. https://doi.org/10.1070/RC2005v074n06ABEH000897
25. Gupta NK, Ghaffari Y, Kim S (2020a) Photocatalytic degradation of organic pollutants over MFe_2O_4 (M = Co, Ni, Cu, Zn) nanoparticles at neutral pH. Sci Rep 10:4942. https://doi.org/10.1038/s41598-020-61930-2
26. Gupta NK, Ghafari Y, Kim S, BaeJ KKS, Saifuddin M (2020b) Photocatalytic degradation of organic pollutants over MFe_2O_4 (M= Co, Ni, Cu, Zn) Nanoparticles at Neutral pH. Sci Rep 10:4942. https://doi.org/10.1038/s41598-020-61930-2
27. Hakim H, Shanks BH (2009) A comparative study of macroporous metal oxides synthesized via a unified approach. Chem Mater 21:2027–2038. https://doi.org/10.1021/cm801691g
28. Harraz FA, Mohamed RM, Rashad MM, Wang YC, Sigmund W (2014) Magnetic nanocomposite based on titania-silica/cobalt ferrite for photocatalytic degradation of methylene blue dye Ceram. Int 40:375–384. https://doi.org/10.1016/j.ceramint.2013.06.012
29. He M, Li D, Jiang D, Chen M (2012) Magnetically separable γ-Fe_2O_3@SiO_2@Ce-doped TiO_2 core-shell nanocomposites: fabrication and visible-light-driven photocatalytic activity. J Solid State Chem 192: 139–143.https://doi.org/10.1016/j.jssc.2012.04.004
30. Hong Y, Ren A, Jiang Y, He J, Xiao L, Shi W (2015) Sol-gel synthesis of visible-light-driven $Ni_{(1-x)}Cu_{(x)}Fe_2O_4$ photocatalysts for degradation of tetracycline. Ceram Int 41:1477–1486. https://doi.org/10.1016/j.ceramint.2014.09.082
31. Hou C, Yu H, Zhang Q, Li Y, Wang H (2010) Preparation and magnetic property analysis of monodisperse Co–Zn ferrite nanospheres. J alloy Compd 491:431–435. https://doi.org/10.1016/j.jallcom.2009.10.217
32. Ismail M, Akhtar K, Khan MI, Kamal T, Khan MA, Asiri AM, Seo J, Khan SB (2019) Pollution, toxicity and carcinogenicity of organic dyes and their catalytic bio-remediation. Curr Pharm Des 25:3653–3671. https://doi.org/10.2174/1381612825666191021142026
33. Jacinto MJ, Wizbiki M, Justino LC, Silva VC (2016) Platinumsupported mesoporous silica of facile recovery as a catalyst for hydrogenation of polyaromatic hydrocarbons under ultramild conditions. J Sol-Gel Sci Technol 77(2):298–305. https://doi.org/10.1007/s10971-015-3854-6
34. Jesudoss SK, Vijaya JJ, Kennedy LJ, Rajan PI, Al lohedan HA, Ramalingam RJ, Kaviyarasu K, Bououdina M (2016) Studies on the efficient dual performance of Mn1-x Ni $_x$ Fe_2O_4 spinel nanoparticles in photodegradation and antibacterial activity. J Photochem Photobiol B Biol 165:121–132. https://doi.org/10.1016/j.jphotobiol.2016.10.004
35. Kim S, Crandall BS, Lance MJ, Cordonnier N, Lauterbach J, Sasmaz E (2019) Activity and stability of NiCe@SiO_2 multi–yolk-shell nanotube catalyst for tri-reforming of methane. Appl Catal B: Environ 259:118037. https://doi.org/10.1016/j.apcatb.2019.118037
36. Knobel M, Nunes WC, Socolovsky LM, Biasi ED, Vargas JM, Denardin JC (2008) Superparamagnetism and other magnetic features in granular materials: a review on ideal and real systems. J Nanosci Nanotechnol 8:2836–2857. https://doi.org/10.1166/jnn.2008.15348
37. Kumar S, Surendar T, Kumar B, Baruah A, Shanker V (2013) Synthesis of magnetically separable and recyclable g-C_3N_4-Fe_3O_4 hybrid nanocomposites with enhanced photocatalytic performance under visible-light irradiation. J Phys Chem C 117:26135–26143
38. Lazar M, Varghese S, Nair S (2012) Photocatalytic water treatment by titanium dioxide: recent updates. Catalysts 2:572–601. https://doi.org/10.3390/catal2040572

39. Lellis B, Favaro-Polonio CZ, Pamphile JA, Polonio JC (2019) Effects of textile dyes on health and the environment and bioremediation potential of living organisms. Biotechnol Res Innov 3:275–290. https://doi.org/10.1016/j.biori.2019.09.001
40. Li M, Li X, Jiang G, He G (2015) Hierarchically macro–mesoporous ZrO_2–TiO_2 composites with enhanced photocatalytic activity. Ceram Int 41:5749–5757. https://doi.org/10.1016/j.ceramint.2014.12.161
41. Lin B, Xue C, Yan X, Yang G, Yang G, Yang B (2015) Facile fabrication of novel SiO_2/g-C_3N_4 core-shell nanosphere photocatalysts with enhanced visible-light activity. Appl Surf Sci 357:346–355. https://doi.org/10.1016/j.apsusc.2015.09.041
42. Lin L, Jiang W, Chen L, Xu P, Wang H (2020) Treatment of produced water with photocatalysis: recent advances, affecting factors and future research prospects. Catalysts 10:924. https://doi.org/10.3390/catal10080924
43. Liu S (2012) Magnetic semiconductor nano-photocatalysts for the degradation of organic pollutants. Environ Chem Lett 10:209–216. https://doi.org/10.1007/s10311-011-0348-9
44. Liu ST, Zhang AB, Yan KK, Ye Y, Chen XG (2014) Microwave-enhanced catalytic degradation of methylene blue by porous MFe_2O_4 (M=Mn, Co) nanocomposites: pathways and mechanisms. Sep Purif Technol 135:35–41. https://doi.org/10.1016/j.seppur.2014.07.049
45. Liu H, Liu T, Zhang Z, Dong X, Liu Y, Zhu Z (2015) Simultaneous conversion of organic dye and Cr(VI) by SnO_2/rGO micro composites. J Mol Catal A Chem 410:41–48. https://doi.org/10.1016/j.molcata.2015.08.025
46. Liu Y, Lan K, Bagabas AA, Zhang P, Gao W, Wang J, Sun Z, Fan J, Elzatahry AA, Zhao D (2016a) Ordered Macro/Mesoporous TiO_2 hollow microspheres with highly crystalline thin shells for high-efficiency photoconversion. Small 12:860–867. https://doi.org/10.1002/smll.201503420
47. Luo W, Hu F, Hu Y et al (2019) Persulfate enhanced visible-light photocatalytic degradation of organic pollutants by constructing magnetic hybrid heterostructure. J Alloy Compd 806:1207–1219. https://doi.org/10.1016/j.jallcom.2019.07.329
48. Ma D, Guan J, Dénommée S, Enright G, Veres T, Simard B (2006) Multifunctional nanoarchitecture for biomedical applications. Chem Mater 18:1920–1927. https://doi.org/10.1021/cm052067x
49. Mahmoodi NM (2013) Zinc ferrite nanoparticle as a magnetic catalyst: synthesis and dye degradation. Mater Res Bull 48:4255–4260
50. Makovec D, Sajko M, Selisnik A, Drofenik M (2012) Low-temperature synthesis of magnetically recoverable, superparamagnetic, photocatalytic, nanocomposite particles. Mater Chem Phys 136:230–240. https://doi.org/10.1016/j.matchemphys.2012.06.058
51. Maksoud MIA, El-Sayyad GS, Ashour AHH, El-Batal AI, Elsayed MA, Gobara M, El-Khawaga AM, Abdel-Khalek EKK, El-Okr MMM (2019) Antibacterial, antibiofilm, and photocatalytic activities of metals-substituted spinel cobalt ferrite nanoparticles. Microb Pathog 127:144–158. https://doi.org/10.1016/j.micpath.2018.11.045
52. Mushtaq F, Zahid M, Mansha A (2020) $MnFe_2O_4$/coal fly ash nanocomposite: a novel sunlight-active magnetic photocatalyst for dye degradation. Int J Environ Sci Technol 17:4233–4248. https://doi.org/10.1007/s13762-020-02777-y
53. Neris AM, Schreiner WH, Salvador C, Silva UC, Chesman C, Longo E, Santos IMG (2018) Photocatalytic evaluation of the magnetic core@shell system (Co, Mn)Fe_2O_4 @TiO_2 obtained by the modified Pechini method. Mater Sci Eng B 229:218–226. https://doi.org/10.1016/j.mseb.2017.12.029
54. Ngo HH, Vigneswaran S, Hu JY, Thirunavukkarasu O, Viraraghavan T (2002) A comparison of conventional and non-conventional treatment technologies on arsenic removal from water. Water Sci Tech Water Supply. 2(5):119–125. https://doi.org/10.2166/ws.2002.0159
55. Noval VE, Carriazo JG (2019) Fe_3O_4-TiO_2 and Fe_3O_4-SiO_2 Core-shell powders synthesized from industrially processed magnetite (Fe_3O_4) microparticles. Mater Res 22(3). https://doi.org/10.1590/1980-5373-mr-2018-s0660
56. Olad A, Behboudi S, Entezami AA (2012) Preparation, characterization and photocatalytic activity of TiO_2/polyaniline core-shell nanocomposites. Bull Mater Sci 5(35):801–809

57. Oliveira TP, Marques GN, Castro MAM, Costa RCV, Rangel JHG, Rodrigues SF, Santos CC, Oliveira MM (2020) Synthesis and photocatalytic investigation of $ZnFe_2O_4$ in the degradation of organic dyes under visible light. J Mater Res Technol 9(6):15001–15015. https://doi.org/10.1016/j.jmrt.2020.10.080
58. Pal B, Sharon M (2000) Preparation of iron oxide thin film by metal-organic deposition from Fe (III)-acetylacetonate: a study of photocatalytic properties. Thin Solid Films 379(1–2):83–88. https://doi.org/10.1016/S0040-6090(00)01547-9
59. Pellegrino F, Pellutiè L, Sordello F, Minero C, Ortel E, Hodoroaba VD, Maurino V (2017) Influence of agglomeration and aggregation on the photocatalytic activity of TiO_2 nanoparticles. Appl Catal B 216:80–87. https://doi.org/10.1016/j.apcatb.2017.05.046
60. Phaniendra A, Jestadi DB, Periyasamy L (2015) Free radicals: properties, sources, targets, and their implication in various diseases. Indian J Clin Biochem 30(1):11–26. https://doi.org/10.1007/s12291-014-0446-0
61. Poole CP, Owens FJ (2003) Properties of individual nanoparticles (chapter-4). In: Introduction to nanotechnology. John Wiley & Sons, pp 72–102
62. Priya R, Stanly S, Anuradha R, Sagadevan S (2019) Evaluation of photocatalytic activity of copper ferrite nanoparticles. Mater Res Express 6:095014. https://doi.org/10.1088/2053-1591/ab2d15
63. Rashad MM, Mohamed RM, Ibrahim MA, Ismail LFM, Abdel-aal EA (2012) Magnetic and catalytic properties of cubic copper ferrite nano-powders synthesized from secondary resources. Adv Powder Technol 23:315–323. https://doi.org/10.1016/j.apt.2011.04.005
64. Rashmi SK, Bhojya Naik HS, Jayadevappa H, Viswanath R, Patil SB, Madhukara M (2017) Solar light responsive Sm-Zn ferrite nanoparticle as efficient photocatalyst. Mater Sci Eng B 225:86–97. https://doi.org/10.1016/j.mseb.2017.08.012
65. Reddy MP, Mohamed AMA, Zhou XB, Du S, Huang Q (2015) A facile hydrothermal synthesis, characterization and magnetic properties of mesoporous $CoFe_2O_4$ nanospheres. J Magn Magn Mater 388:40–44. https://doi.org/10.1016/j.jmmm.2015.04.009
66. Sadollahkhani A, Kazeminezhad I, Lu J, Nur O, Hultman L, Willander M (2014) Synthesis, structural characterization and photocatalytic application of ZnO@ZnS core-shell nanoparticles. RSC Adv 4:36940–36950
67. Seo B, Lee C, Yoo D, Kofinas P, Piao Y (2017) A magnetically recoverable photocatalyst prepared by supporting TiO_2 nanoparticles on a superparamagnetic iron oxide nanocluster core@fibrous silica shell nanocomposite. RSC Adv 7:9587–9595. https://doi.org/10.1039/C6RA27907D
68. Shan A, Wu X, Lu J, Chen C, Wang R (2015) Phase formations and magnetic properties of single crystal nickel ferrite ($NiFe_2O_4$) with different morphologies. CrystEngComm 17:1603–1608. https://doi.org/10.1039/C4CE02139H
69. Shanmuganathan R, Felix FL, Alharbi NS, Brindhadevi K, Pugazhendhi A (2020) Core/shell nanoparticles: synthesis, investigation of antimicrobial potential and photocatalytic degradation of Rhodamine B. J Photochem Photobiol B 202:111729. https://doi.org/10.1016/j.jphotobiol.2019.111729
70. Shekofteh-Gohari M, Habibi-Yangjeh A (2016) Fabrication of novel magnetically separable visible-light-driven photocatalysts through photosensitization of Fe_3O_4/ZnO with CuWO4. J Ind Eng Chem 44:174–184. https://doi.org/10.1016/j.jiec.2016.08.028
71. Shetty K, Renuka L, Nagaswarupa HP, Nagabhushana H, Anantharaju KS, Rangappa D, Prashantha SC, Ashwini K (2017) A comparative study on $CuFe_2O_4$, $ZnFe_2O_4$ and $NiFe_2O_4$: morphology, impedance and photocatalytic studies. Mater Today Proc 4:11806–11815. https://doi.org/10.1016/j.matpr.2017.09.098
72. Singh C, Jauhar S, Kumar V, Singh J, Singhal S (2015) Synthesis of zinc substituted cobalt ferrites via reverse micelle technique involving in situ template formation: a study on their structural, magnetic, optical and catalytic properties. Mater Chem Phys 156:188–197. https://doi.org/10.1016/j.matchemphys.2015.02.046
73. Solomon V, Lydia IS, I Merlin JP, Venuvanalingam P (2012) Enhanced photocatalytic degradation of azo dyes using nano Fe_3O_4. J Iran Chem Soc 9:101–109. https://doi.org/10.1007/s13738-011-0033-8

74. Suharyadi E, Muzakki A, Nofrianti A, Istiqomah NI, Kato T, Iwata S (2020) Photocatalytic activity of magnetic core-shell $CoFe_2O_4$@ZnO nanoparticles for purification of methylene blue. Mater Res Express 7:085013 https://doi.org/10.1088/2053-1591/abafd1
75. Sun M, Han X, Chen S (2019) Synthesis and photocatalytic activity of nano-cobalt ferrite catalyst for the photo-degradation various dyes under simulated sunlight irradiation. Mater Sci Semicond Process 91:367–376. https://doi.org/10.1016/j.mssp.2018.12.005
76. Surendra BS (2017) Green engineered synthesis of Ag-doped $CuFe_2O_4$: characterization, cyclic voltammetry and photocatalytic studies. Adv Mater Devices. https://doi.org/10.1016/j.jsamd.2018.01.005
77. Suresh R, Rajendran S, Senthil Kumar P, Dai-Viet NV, Ponce LC (2021) Recent advancements of spinel ferrite based binary nanocomposite photocatalysts in wastewater treatment. Chemosphere 274:129734. https://doi.org/10.1016/j.chemosphere.2021.129734
78. Tehrani-Bagha AR, Gharagozlou EF (2016) Catalytic wet peroxide oxidation of a reactive dye by magnetic copper ferrite nanoparticles. J Environ Chem Eng 4:1530–1536. https://doi.org/10.1016/j.jece.2016.02.014
79. Thomas B, Alexander LK (2018) Enhanced synergetic effect of Cr(VI) ion removal and anionic dye degradation with superparamagnetic cobalt ferrite meso–macroporous nanospheres. Appl Nanosci 8:125–135. https://doi.org/10.1007/s13204-018-0655-6
80. Thomas B, Alexander LK (2019) Nanoreactor based enhancement of photocatalysis with $Co_{0.7}Zn_{0.3}Fe_2O_4$@$SrTiO_3$ core-shell nanocomposites. J Alloy Compd 788:257–266. https://doi.org/10.1016/j.jallcom.2019.02.190
81. Thomas B, Alexander LK (2020) Removal of Pb^{2+} and Cd^{2+} toxic heavy metal ions driven by Fermi level modification in $NiFe_2O_4$–Pd nano hybrids. J Solid State Chem 288:121417. https://doi.org/10.1016/j.jssc.2020.121417
82. Tiwari JN, Tiwari RN, Kim KS (2012) Zero-dimensional, one-dimensional, two-dimensional and three-dimensional nanostructured materials for advanced electrochemical energy devices. Prog Mater Sci 724–803. https://doi.org/10.1016/j.pmatsci.2011.08.003
83. Vijayalakshmi S, Elaiyappillai E, Johnson PM (2020) Multifunctional magnetic $CoFe_2O_4$ nanoparticles for the photocatalytic discoloration of aqueous methyl violet dye and energy storage applications. J Mater Sci Mater Electron 31:10738–10749. https://doi.org/10.1007/s10854-020-03624-z
84. Wan J, Li H, Chen K (2009) Synthesis and characterization of Fe_3O_4@ZnO core-shell structured nanoparticles. Mater Chem Phys 114(1):30–32. https://doi.org/10.1016/j.matchemphys.2008.10.039
85. Wang Z, Shen B, Aihua Z, He N (2005) Synthesis of Pd/Fe_3O_4 nanoparticle-based catalyst for the cross-coupling of acrylic acid with iodobenzene. Chem Eng J 113:27–34
86. Wang J, Yang J, Li X (2015) Synthesis of Fe_3O_4@SiO_2@ZnO–Ag core-shell microspheres for the repeated photocatalytic degradation of rhodamine B under UV irradiation. J Mol Catal A: Chem 406:97–105. https://doi.org/10.1016/j.molcata.2015.05.023
87. Warner MG, Warner CL, Addleman RS, Yantasee W (2011) Magnetic nanomaterials for environmental applications (chapter 9). Wiley, pp 311–344
88. Wu W, He Q, Jiang C (2008) Magnetic iron oxide nanoparticles: synthesis and surface functionalization strategies. Nanoscale Res Lett 3:397–415. https://doi.org/10.1007/s11671-008-9174-9
89. Wu S, Wu J, Jia S, Chang Q, Ren H, Liu Y (2013) Cobalt(II) phthalocyanine-sensitized hollow Fe_3O_4@SiO_2@TiO_2 hierarchical nanostructures: fabrication and enhanced photocatalytic properties. Appl Surf Sci 287:389–396
90. Wu Z, Chen X, Liu X, Yang X, Yang Y (2019) A ternary magnetic recyclable ZnO/Fe_3O_4/gC_3N_4 composite photocatalyst for efficient photodegradation of monoazo dye. Nanoscale Res Lett 14(1):147. https://doi.org/10.1186/s11671-019-2974-2
91. Xiang Y, Wang H, He Y, Song G (2015) Efficient degradation of methylene blue by magnetically separable Fe_3O_4/chitosan/TiO_2 nanocomposites. Desalin Water Treat 55:1018–1025
92. Xin T, Ma M, Zhang H, Gu J, Wang S, Liu M, Zhang Q (2014) A facile approach for the synthesis of magnetic separable Fe_3O_4@TiO_2, core-shell nanocomposites as highly

recyclable photocatalysts. Appl Surf Sci 288:51–59. https://doi.org/10.1016/j.apsusc.2013.09.108

93. Yang ST, Zhang W, Xie J, Liao R, Zhang X, Yu B, Wu R, Liu X, Li H, Guo Z (2015) Fe_3O_4@SiO_2 nanoparticles as a high-performance Fenton-like catalyst in a neutral environment. RSC Adv 5:5458–5463. https://doi.org/10.1039/c4ra10207j
94. Yao Y, Wu G, Lu F, Wang S, Hu Y, Zhang J, Huang W, Wei F (2016) Enhanced photo-Fenton-like process over Z-scheme $CoFe_2O_4$/g-C_3N_4 Heterostructures under natural indoor light. Environ Sci Pollut Res 23:21833–21845. https://doi.org/10.1007/s11356-016-7329-2
95. Ye S, Qiu LG, Yuan YP, Zhu YJ, Xia J, Zhu JF (2013) Facile fabrication of magnetically separable graphitic carbon nitride photocatalysts with enhanced photocatalytic activity under visible light. J Mater Chem A 1:3008. https://doi.org/10.1039/c2ta01069k
96. Zaharieva K, Rives V, Tsvetkov M, Cherkezova-Zheleva Z, Kunev B, Trujillano R, Mitov I, Milanova M (2015) Preparation, characterization and application of nanosized copper ferrite photocatalysts for dye degradation under UV irradiation. Mater Chem Phys 160:271–278. https://doi.org/10.1016/j.matchemphys.2015.04.036
97. Zhang L, Wang H, Zhang Z, Qin F, Liu W, Song Z (2011) Preparation of monodisperse polystyrene/silica core-shell nano-composite abrasive with controllable size and its chemical mechanical polishing performance on copper. Appl Surf Sci 258:1217–1224. https://doi.org/10.1016/j.apsusc.2011.09.074
98. Zhang J, Li L, Shi R, Mei J, Xiao Z, Ma W (2019) An efficient approach for the synthesis of magnetic separable Fe_3O_4@TiO_2 core-shell nanocomposites and its magnetic and photocatalytic performances. Mater Res Express 6(10):105014.https://doi.org/10.1088/2053-1591/ab3531
99. Zhang DH, Li GD, Li JX, Chen JS (2008) One-pot synthesis of Ag–Fe_3O_4 nanocomposite: a magnetically recyclable and efficient catalyst for epoxidation of styrene. Chem Commun 3414. https://doi.org/10.1039/B805737K
100. Zheng J, Wu Y, Zhang Q, Li Y, Wang C, Zhou Y (2016) Direct liquid phase deposition fabrication of waxberry-like magnetic Fe_3O_4@TiO_2 core-shell microspheres. Mater Chem Phys 181:391–396. https://doi.org/10.1016/j.matchemphys.2016.06.074
101. Zhu H, Jiang R, Huang S, Yao J, Fu F, Li J (2015) Novel magnetic $NiFe_2O_4$/multi-walled carbon nanotubes hybrids: facile synthesis, characterization, and application to the treatment of dyeing wastewater. Ceram Int 41:11625–11631
102. Zhu H, Jiang R, Fu Y, Li R, Yao J (2016) Novel multifunctional $NiFe_2O_4$/ZnO hybrids for dye removal by adsorption, photocatalysis and magnetic separation. Appl Surf Sci 369:1–10. https://doi.org/10.1016/j.apsusc.2016.02.025

Metal Oxide-Based Nanomaterials for the Treatment of Industrial Dyes and Colorants

Kalya Tulasidas Vadiraj and Harikaranahalli Puttaiah Shivaraju

Abstract Industries are always the backbone of a country and among them, the textile industry has excelled and far stretched throughout their world. The materials used in the textile industry are hazardous and toxic dyes and various other corrosive chemicals. The handling of these wastes and effluents has been a problem for industries as well as others related to them. In recent days, dye degradation using transition metal oxide nanoparticles has been carried out. Their favorable bandgap and better efficiency to utilize solar energy conversion have allowed it to use it as a photocatalyst. These nanostructured materials play a crucial role in the performance of photocatalytic activity because of that large surface area, effective light absorption and electron transportation facility. The various synthetic methods of these metal oxides have been understood, and functionalization strategies and characteristics are well studied and applied in the field. In this chapter, focus is given to the challenges and future metal oxide nanoparticles in the field of photocatalysis.

1 Introduction

Metal oxides play an important role in the treatment of waste and pollutants along with strategic activities energy generation, conversion and storage, which display unique functional properties based on crystal structure, morphology, intrinsic defects, doping components and compositions determining properties like optical, electrical, catalytic activities. The synthesis and handling procedures determine the structure and morphological characteristics displayed by metal oxide nanoparticles. Metal oxides have tunable bandgap and electronic structure can provide a multi-functional

K. T. Vadiraj · H. P. Shivaraju (✉)
Department of Environmental Sciences, JSS Academy of Higher Education and Research, Mysuru 570 015, India
e-mail: shivarajuenvi@gmail.com

H. P. Shivaraju
Center for Water, Food and Energy, Dombaranahalli Post, GREENS Trust, Turuvekere Taluka, Tumkur District, Harikaranahalli 572215, India

S. S. Muthu and A. Khadir (eds.), *Advanced Oxidation Processes in Dye-Containing Wastewater*, Sustainable Textiles: Production, Processing, Manufacturing & Chemistry, https://doi.org/10.1007/978-981-19-0882-8_8

possibility in various fields of environment [1, 2]. By controlling, the bandgap and surface properties help in application in heterogeneous photocatalysis. Metal oxides display a variety of structural diversity by tuning the synthesis end handling procedures. Good number of work and research has already been carried out in synthesizing the metal oxide nanoparticles with controlled structures like nano-rods, nano-spheres, core–shell nanostructures etc. along with elevated catalytic efficiency removal of pollutants [3].

Effluents released by industries producing like pharmaceutical, textiles and paint-based products are highly toxic and not easily degradable, for which metals and metal oxide nanoparticles have displayed remarkable results through photocatalytic properties. These metal oxides provide environmental remediation by facilitating oxidation–reduction process through capturing the photons from light leading to the degradation of hazardous pollutants. These metal oxide nanoparticles are capable of degrading organic pollutants like dyes, poly-aromatic hydrocarbons and inorganic metal species heavy metals, radioactive compounds and other inorganic contaminants, which are needed to be removed from the water bodies. Metal oxide nanoparticles are highly effective in monitoring the levels of pollutants in different mediums like soil, air and water [4, 5].

Dye molecules are removed by various procedures like coagulation, ion exchange, membrane separation and biomolecule absorption techniques where the dyes are disintegrated partially or unchanged. These methods generation of secondary pollutants along with causing high equipment and maintenance cost for the industry. Thus, photocatalytic dye degradation by nanoparticles is highly effective, workable and can be scaled up to industrial size. Transition metal oxide nanoparticles are chemically superior with better physical and optical properties and these materials are widely used for nano-catalytic activity that providing multiple oxidation state, high efficiency, recyclability and low cost makes them better alternative to traditional processes. These metal oxide nanoparticles are highly effective in utilizing solar energy in ultraviolet and visible region to inject electron into the dye molecule during degradation. This chapter focuses on remediation methods contaminated by textile dyes, photocatalytic activity of metal oxide nanoparticles and their use in wastewater treatment. This chapter also deals with mode of synthesis and current application status of transition metal oxide nanoparticles in water treatment [6].

2 Types of Dyes and Sources

Textile industry is one of the oldest industries in the world, and coloring of clothes has been practiced for many centuries. Various coloring compositions are used in textile industries and these coloring agents are called dyes, which are commonly organic molecules with a chromophore in their molecular structure. These dyes are naturally available as well as synthesized in the industries in large scale. In many cases, these dyes are complex in nature, stable in any conditions, with carcinogenic and toxic properties, non-biodegradable and sometimes bio-accumulative in nature.

Table 1 Important textile dyes in various effluents and their health effects

Dyes	Sources	Effects	References
Congo red	Paper, textile, leather	Lymphoma	[8]
Methyl orange	Paper, leather, textile, printing	Tumors and carcinogenic	[9]
Methylene blue	Textile, paper, cosmetics, plastics, Antidote for methemoglobinemia	Eye irritation, nausea, vomiting, skin disease and anemia	[10, 11]
Disperse red 1	Textile (fabric, polyester, cotton)	Rashes, skin irritation and carcinogenic	[12]
Reactive black	Textile industry, nylon printing, printing paper	Wheezing, Asthma and blindness	[13]

Dye molecules absorb certain wavelengths of radiation and reflecting the remaining radiation of other wavelengths making that cloth or textile look colored. Dyes are also used in industries like biomedicine, food product, paint, cosmetics, paper, printing, image and photography. According to a report, global dye and pigment industry has a market size of USD 32.9 billion in 2020 with compound annual growth rate (CAGR) of 5.1% [7].

More than 100,000 types of dyes are available in the market and one third of the total production is consumed by textile industries. The process in industries are conventional that leads to produce waste up to 20% and these waste are contaminate the groundwater causing several health issues.

Different classes of dyes are available in the market depending upon the requirement of the industry like natural dyes, synthetic dyes, acid dyes, vat dyes, printing inks, azo dyes etc. Dyes are hazardous in nature that can cause skin irritation, burning and other gastronomical problems [4, 6]. Dyes have various classifications and classes along with the different sources and effect on human beings, and some of them are shown in Table 1.

2.1 Environmental Hazards

Chemicals used in the textile industry are hazardous, carcinogenic and toxic in nature are the main sources of pollution. Industries use over 3000 chemicals have to convert raw yarns into finished goods. The wastes generated from these industries are the reasons for water, air and soil pollution and can also end up as a solid waste in a landfill. Most of the dyes are reasons for water pollution especially groundwater where a small pinch dye of less than 1 ppm can change the color of the water and show drastic effect on the complete ecosystem. These dyes can change their effect on water quality depending upon the environmental conditions like temperature, pressure and other chemical properties of water like dissolved gases, pH, etc. These colors reduce the transparency of water disturbing the photosynthetic activity in the water, which directly reduces the dissolved oxygen. When effluent from the textile industry enters into soil ecosystem, which reduces the fertility by hindering the growth of microorganism insects and plants [14].

3 Conventional Dye Degradation Process

Conventional methods of dye degradation include filtration, coagulation, flocculation, oxidation, microbial process, biosorption and adsorption processes are used traditionally by industries (Fig. 1) [15]. These processes might convert dye completely different or simpler compounds but, in certain cases, the byproducts of the dyes can cause secondary pollution in water sources.

3.1 Coagulation Process

Coagulation is an important process collide of a smaller size picture distributed in a liquid medium is a change to large particles. This process helps by capturing the toxic molecules and purifying the water. Here, dyes are in the form of collides getting certain charges and coagulants chemicals are introduced to the water, having opposite charge that can neutralize the suspended solids and form larger particles that can

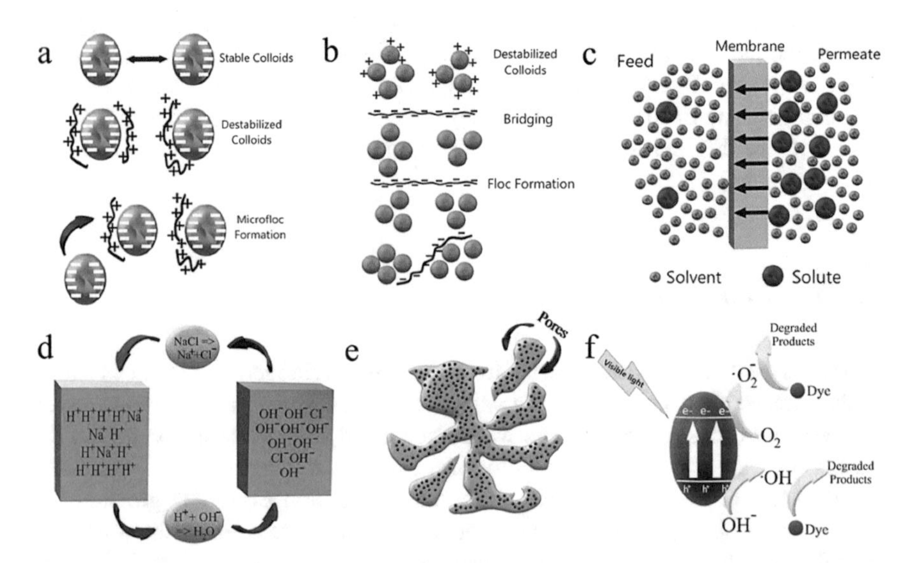

Fig. 1 Dye degradation techniques: **a** coagulation, **b** flocculation, **c** membrane filtration, **d** ion exchange, **e** adsorption, **f** advanced oxidation

be trapped in the settling tanks [13]. The main principle behind coagulation is destabilizing the electrostatic force of charged particle and neutralizes them. Unsettled negatively charged dye molecules are usually suspended in the water that are neutralized by positively charged coagulant molecules can form larger particles known as micro floc. Aluminum-based coagulants are used for this process. These flocs are agglomerated particles, which are neutral in nature, insoluble polymers, that can be easily settled and separated [16].

3.2 Membrane Separation

It is a filtration technique where selective separation of contaminants is carried out through defects, pores and minute gaps in the membrane. This process usually is carried out by the principles of reverse osmosis (pore size <1 nm), nanofiltration (pore size 1–2 nm), ultrafiltration (pore size 2–100 nm) and microfiltration (pore size >0.1 μm) depending upon the class of membrane and the pollutant. The membrane class can be determined according to the pore size and the ability to filter the materials like pathogens, microorganisms, macromolecules, dissolved solids and salt in the inflow water. This method is effective to 99% as the pollutants are separated from the water but also generates second pollutants or sludge, which require high capital cost to handle [8, 17–19].

3.3 Ion Exchange Process

Ion exchange is a chromatographic technique where exchange of ions occurs between electrolyte and the medium. Same principle is applied in the wastewater treatment where the charged dyes are replaced with the ions in the resin and cleans the waters. Primarily resins are treated with sodium chloride or potassium chloride to create cationic or anionic raisins. Surface of resins acts as a site in ion exchange, react with the effluent water to soften the water. This is an eco-friendly method cost-effective and applicable to any scale, but only good with selected dyes makes it less applicable in dyes and textile industry [20].

3.4 Adsorption on Activated Carbon

Adoption is a surface activity where the concentrated dyes adhere on the surface of the adsorbent in a very short duration. These adsorbents have activated sites with interstitial spaces and pores, which trap the dye molecules. This method is effective only at lower concentrations and cannot remove the color completely from the water at higher concentrations, which, in turn, transfers the load to aquatic ecosystems.

The second disadvantage of this process is sledge generation that can reduce the adsorption along with high capital and maintenance cost [21].

3.5 Advanced Oxidation Process (AOP)

Advance oxidation is a catalytic process where organic molecules are removed using hydroxyl radical. This process includes the excitation of electrons upon exposure to ultraviolet or visible radiation causing an electron–hole pair, reacts with dye making the molecule unstable and degrades them; this is also called as photocatalytic degradation technique. This method generates non-hazardous molecules by producing hydroxyl radical and other superoxide anions during the process. This technique depends upon the intensity of radiation, bright light is better than dim light and the metal oxide semiconductors are recyclable them to reuse any number of times [22].

4 Photocatalytic Degradation of Dyes

Photochemical degradation of dyes is a heterogeneous oxidation process where dye molecule is degraded into non-hazardous compounds like salts, carbon dioxide and water. This is catalyzed by semiconductor by generating oxidizing radical upon irradiation of ultraviolet or visible light. During photochemical degradation, hole-electron pair is generated; it plays an important role in the production of hydroxyl radical. Photochemical degradation starts with the dye molecules adsorbed onto their metal oxide nanoparticle surface. Upon illumination, the surface of the semiconductor picks up the photons and excites over the bandgap and results in the transition of electron (e^-) from valence band to conduction bands creating a hole (h^+) in valance band. This excited electron binds with the oxygen that creates a superoxide ion (O_2^-) while the hole generated at valence band reacts with water molecule to form hydroxyl radicals (OH^*). The generated superoxide ions and free hydroxyl radical attack dye molecule and break them into nonhazardous ecofriendly material [23]. The mechanism of reaction is shown below

$$\text{Photocatalyst} + h\upsilon \rightarrow e^-{}_{CB} + h^+{}_{VB}$$

$$O_2 + 2e^-{}_{CB} \rightarrow O_2{}^{*-}$$

$$h^+ + H_2O \rightarrow H^+ + OH^-$$

$$O_2{}^{*-} + H^+ \rightarrow HO_2{}^{*-}$$

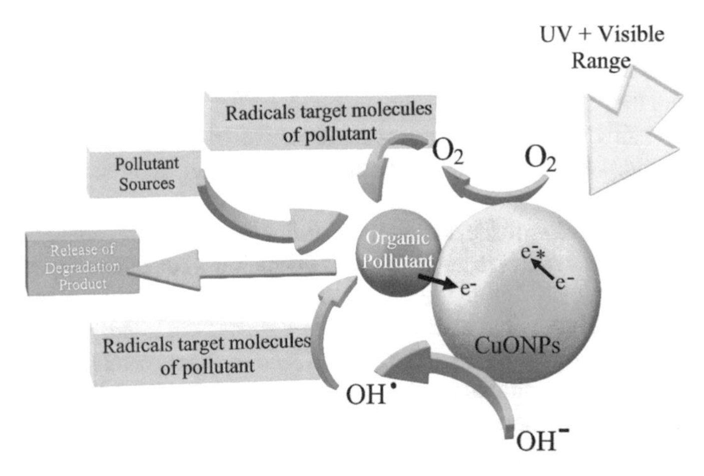

Fig. 2 Schematic representation of organic dye degradation by metal oxide (CuO) nanoparticles [26]

$$\text{Dye} + \text{HO}_2{}^{*-} \rightarrow \text{CO}_2 + \text{H}_2\text{O}$$

During the photocatalytic degradation, crystal defects on transition metal oxide semiconductor play an important role as trapping sites for oxidation and reduction reactions [24]. These semiconductors in quantum size have a higher number of defects because of increased surface area and enough active sites [24, 25] (Fig. 2).

4.1 Factors Influencing Dye Degradation

Catalysis is a process of increasing rate of chemical reaction by the additional certain substances called catalyst, which are not consumed during the reaction but can be used repeatedly. This is the important aspect in photocatalytic degradation of dyes where these metal oxide nanoparticles are excited by inducing light can increase the rate of degradation of dyes. Important properties of photocatalysts are electron mobility, bandgap energy, biological inertness, transparency, non-toxicity, eco-friendliness and high surface area [27, 28]. Transition metal oxide nanoparticles have always shown higher efficiency in photocatalytic activities by their following properties:

1. Bandgap energy: Nanocatalysts have a wide optical bandgap, which gives advantage in terms of absorbing energy from both ultraviolet and visible lights. Along with energy absorption, these wide bandgaps also help them to maintain electron–hole pair without recombination during photoexcitation process.
2. Electron dynamics: Nanocatalysts display better electron transitions with less resistivity provided during the excitation of electron from valance band to conduction band resulting in effective photocatalytic activities.

3. Transparency: Metal oxide nanoparticles show better reflectivity along with high optical transparency at visible and infrared region, which provides an advantage in irradiation.
4. Photosensitivity: Semiconductors are sensitive to the specific wavelength of the radiation during the photoexcitation during the generation of electrons and holes.
5. Stability: Catalysts are the material, which does not involve in chemical reaction but increases the rate of reaction. Metal oxides show a better stability chemically, thermodynamically during the treatment process.
6. Chemical and biological inertness: Catalyst should not interact with degrading chemicals or biological matter during the reaction. These catalysts only increase the rate of reaction without involving are generating any chemical species during their chemical reaction.
7. Toxicity: Catalyst should be non-toxic during the handling or at the disposal in any manner. Toxicity adversely impacts on ecosystems, cannot be used or considered for the treatment of water
8. High surface area: Metal oxide nanoparticles provide a large surface area compared to the bulk materials. This acts as an advantage for photocatalytic activity with better availability active sites for dye molecules [29–35].

These properties of metal oxide nanocomposites have greatly improved with the decrease in size from bulk to micro and two nano dimensions. Photocatalytic activity of metal oxide are improved at the higher rates due to increase in surface area with availability of traps and defects for more radical formation, helping in mass transportation of electrons to reactant molecule facilitates diffusion and increases the formation of super oxide radical during dye degradation [29, 36].

5 Metal Oxide Nanoparticle-Based Dye Degradation

Photocatalytic degradation is mediated by metal oxide-based materials, which are characterized with different nanostructures. Metal oxides nanoparticles with definite structure and functionalized sites are important to tune the bandgap and surface adsorption properties of these heterogeneous photocatalysts [37–39]. In nanoparticles maintaining the structure, bandgap, optoelectronic properties are the fundamental for employing in the water treatment. These metal oxide nanoparticles are synthesized by various methods like co-precipitation, deposition, hydrothermal method, sol–gel method, soak-deoxidized air oxidation, and impregnation method allows nanoparticles with diverse structural and functional properties [40–44].

5.1 Hydrothermal Technique

Hydrothermal synthesis has shown high degree of efficiency in synthesizing transition metal oxide photocatalysts at higher temperature, in a confined volume to generate high pressure. This is a simple ecofriendly technique, which does not need calcination or aging of nanoparticles. The characteristics of nanoparticles can be modified by varying the reactant parameters like temperature, solvent, surfactant, metal precursors, reducing agents and reaction time. Metal precursors used in the hydrothermal process are commonly metal salts containing chlorides, nitrates or sulfate, dissolved in aqueous or organic solvents and mixed with a reducing agent like base with constant steering can start seeding. The mixture is then transferred to PTFE-lined stainless steel autoclave and maintain temperature at 200 °C or above with the reaction timing from 2 to 48 h. The nanoparticles synthesized are collected by filtering, dried overnight or centrifugation for applications. Metal oxides like titanium dioxide synthesized by a hydrothermal process at 200 °C can degrade rhodamine B dye [45]. Manganese oxide (MnO_2) nanoparticle synthesized by hydrothermal process degraded methylene blue dye with 90% efficiency [46].

5.2 Co-precipitation Process

Co-precipitation method is a simple, cost-effective and fast technique for the synthesis of metal oxide nanoparticles at room temperature. Properties of nanoparticles synthesized by this technique depend on the concentration of precursor, precipitating agent, temperature, type of solvent, pH and reaction time. This process has four major steps: nucleation, growth, coarsening and agglomeration. All these steps occur simultaneously during the reaction resulting in the formation of nanoparticles. Nanoparticles synthesized are small in size, uniform shaped with a narrow size distribution. Generally, nitrates, acetates, chlorides are used as precursors dissolved in aqueous or organic solvents to prepare metal oxide nanoparticles. Bases like sodium hydroxide, potassium hydroxide or ammonium hydroxide are used as precipitating agent, when introduced into the solution mixture nanoparticles precipitated and can be separated by filtration, gravity separation or centrifugation.

Typically, zinc oxide was synthesized using zinc acetate as a precursor at room temperature, sodium hydroxide was added dropwise to the solution resulting in ZnO nanoparticle formation as precipitate [47]. Zinc oxide synthesized by this method showed effective degradation of congo red dye in aqueous medium under solar radiation, and degradation time was 120 min [48]. Transition metal oxides like tin oxide, nickel oxide, copper oxide can be synthesized by this method [49].

5.3 Thermal Decomposition Method

High temperature is employed to decompose the metal precursor with solvent having a higher boiling point. In this method, solvents used are mostly organic and have high boiling point more than 150 °C and refluxed in various time periods for better results. The nanoparticle properties majorly depend on reflux time and temperature. Nanoparticles synthesized by this procedure need not undergo calcination or annealing. This method has good control over size and particle shape with high crystallinity in monodispersed nanoparticles. During the metal oxide synthesis, organometallic precursors are dispersed in an organic solvent and boiled to decomposition temperature while maintaining inert atmosphere. Along with the reaction mixture, surfactants are added to regulate the shape and size of the nanoparticles. Due to the presence of various ions and organic constituents, the formation of nanoparticle is a complex process in thermal decomposition method.

Many other methods for synthesizing nanoparticles have already been reported like sol–gel method. Sol–gel method has the steps like hydration and condensation of molecular precursor in a solution resulting in the formation of metal oxide nanoparticles. The nanoparticles obtained are called sol and evaporating solvent resulting in the formation of gel. Most metal precursors in this method are salts like chlorides and alkoxides dissolved in water, and acid is introduced to hydrolyze the precursor salts [50]. Different metal oxides' mode of synthesis application is listed in Table 2.

Table 2 Various synthesis methods of metal oxide nanoparticles

Metal oxide and morphology	Synthetic method	Application	Reference
ZnO 3D hierarchical flower nanosheets	Solvothermal	Degradation of dyes like triphenylmethane	[5]
Fe_2O_3 UiO66 cubically arranged nanoparticle composits	Sonication	Adsorption	[51]
Cu/ZnO impregnated Al_2O_3	Impregnation	Carbon monoxide removal	[52]
Ce/Fe bimetallic oxides with flowerlike 3D hierarchy	Hydrothermal	Heavy metal remediation like As (V) and Cr (VI)	[53]
$MTiO_3$ (Perovskite titanates) with 3D leaf architecture	Heating grinding and photodeposition	Photoreduction of CO_2	[54]
Fe_2O_3/WO_3 core shell architecture	Solvothermal and oxidation route	Photocatalysis of dye materials	[55]

6 Metal Oxide Nanoparticles in Dye Degradation

Photocatalytic degradation is mediated by the metal oxide semiconductors like titanium dioxide (TiO_2), zinc oxide (ZnO), etc. Most of these semiconductors are highly active in ultraviolet radiation since their bandgap matches with the wavelength [56]. For example, titanium dioxide is active in ultraviolet radiation since the bandgap is 3.2 eV. Titanium oxide nanoparticles are highly used in the water treatment due to its high photon absorption capacity. Apart from this zinc oxide, other metal oxides like zirconium oxide (ZrO_2) iron oxide (Fe_2O_3), tin oxide (SnO_2), copper oxide (CuO), silver oxide (AgO) and bismuth oxide (Bi_2O_3) are also employed for the dye degradation to overcome the drawbacks of titanium dioxide [9, 11, 57–62]. Typical metal oxides used in the dye degradation and their bandgaps are shown in Table 3.

Transition metal oxide nanoparticles have various factors to improve the photocatalytic activities such as shape, size, surface structures and phase of the catalysts. These properties can be easily changed by varying the process of synthesis and modifying the parameters like temperature, pH, types of precursors, various solvents, surfactants, precipitating agents and quantity. Table 4 has the list of important metal oxides used in photocatalytic degradation of dyes.

In recent days, transition metal oxides are being coupled up with other transition metal oxides creating metal oxide composites. These metal oxide composites have improved properties when compared to their initial metal oxides and desired bandgaps with better lattice points in the composites make them better photocatalysts. Metal oxide composite nanoparticles have shown good electronic excitation, number of super oxide molecule formation and better degradation of hazardous dyes. Examples of some metal oxide composites, dye degraded and efficiency are shown in Table 5.

7 Conclusions

Researchers have provided promising results on structure and functional variations of metal oxide nanoparticles which play a vital role in environmental remediation and solving the problems of effluent treatment. They have shown exceptionally good efficiency in degrading industrial pollutants along with their ability to scale up to industrial level aspects.

The present chapter focused on the synthetic methods, photocatalytic applications of various metal oxides like titanium oxide, tin oxide, zinc oxide, zirconium oxide etc. Various aspects including morphology, nanostructures composite, functionalized materials have been understood and discussed the scalability for real-time applications. These metal oxides are vital in degrading hazardous dyes and can be better alternatives for industries along with the invention of new catalytic methods for environmental remediation. The scalability from laboratory to industry is one of the important features that needs to be understood to make considering these

Table 3 Various metal oxides used in photocatalytic activities with bandgap and corresponding activation wavelength

Metal oxide in photocatalysis of dyes	Bandgap and corresponding wavelengths (nm)
TiO_2 (Anatase)	3.2 (387)
TiO_2 (Rutile)	3.0 (415)
ZnO	3.36 (370)
WO_2	2.76 (450)
CuO	1.2 (1035)
Cu_2O	2.2 (565)
MgO	5.90
Mn_2O_4	3.28 (380)
CeO_2	3.19 (390)
Fe_2O_3	2.3 (540)
Fe_3O_4	2.25 (550)
ZrO_2	3.87 (320)
Ag_2O	1.4 (885)
$SrTiO_3$	3.25 (380)
Bi_2WO_6	3.13 (395)
CdO	2.20 (560)
CoO	2.01 (620)
Cr_2O_3	3.50 (355)
HgO	1.90 (650)
MnO	3.60 (345)
PbO	2.80 (440)
PdO	1.00 (1240)
Sb_2O_3	3.00 (415)
SnO	4.20 (295)
SnO_2	3.50 (355)
V_2O_5	2.80 (440)

metal oxide-based photocatalysts. Few important points are to be considered before commercializing metal oxide-based photocatalysts and are summarized below:

1. Cost and availability of the precursor metals and accessibility of auxiliary material for their synthesis
2. Innate toxicity of metal or metal oxide and the auxiliary materials are used in the synthesis, which can cause pollution while production.
3. Synthesis of metal oxide nanoparticles with high energy efficiency and management.
4. Producing metal oxides with well-tuned bandgap for better light absorption and high specific surface area for better catalytic activity in a large scale.
5. Better performance under natural light source like sun.

Table 4 List of important metal oxides and dyes degraded by photocatalysis

Metal oxide nanoparticles	Dyes degraded	Methods	References
TiO_2	Methyl orange Methylene blue Congo red Indigo Indigo carmine Acetophenone Nitrobenzene Malachite green	Ultraviolet radiation was effective in degrading most of the dyes. Titanium oxide in anatase phase was employed	[63–66]
ZnO flowers	Methyl orange Congo red Eosin B Chicago sky blue Rhodamine B	Asymmetric Zn(ii) complex showed better results with methyl orange and treatment under sunlight has shown a better results	[57, 66, 67]
Fe_2O_3	Methylene blue Methyl orange Methyl red Congo red Remazol yellow RR Phenol Crystal violet	Iron oxide showed a more than 90% dye degradation under ultraviolet irradiation (<200 nm), iron oxide showed better results in the presence hydrogen peroxide	[11, 68–71]
AgO	Methylene blue Tartrazine Carmoisine Brilliant blue	Silver and silver oxide nanoparticles show better photocatalytic activity on hazardous chemical dyes	[61, 72–74]

6. Should be a broad spectrum photocatalyst without changing performance.
7. Recyclable photocatalysts are environmentally friendly and easy to recover [86–89].

Even with good research support, these metal oxide systems have always been costly in manufacturing and applications. Zinc oxide and titanium oxides are the only metal oxide nanoparticles that were found in commercial applications while iron oxide-based materials are being explored for their magnetic properties.

Table 5 Metal oxide composites used in dye degradation with efficiency

Catalyst	Synthetic method	Target dye	Efficiency	Reference
Cu-VO	Microwave treatment	methylene blue	93%	[75]
Cu_2O/TiO_2, and Cu_2O/ZnO	Solvothermal method	Congo red	100% within 10 min	[76]
ZnO/RGO	(ZnO) Coprecipitation, (RGO) Modified Hummer's method	Basic blue	98%	[77]
$CaMgO_2$	Ultrasound assisted co-precipitation method	Methylene blue	~95% within 60 min	[78]
$CuFeO_2$ and Mg-doped $CuFeO_2$	Chemical solution route	Methylene blue	92.7% after 90 min	[79]
Ag-$BiVO_4$	Solution combustion synthesis	Methylene blue	~100% in 120 min	[80]
$BaTiO_3$	Electrospinning	Rhodamine B	97.6% within 110 min	[81]
ZnO-Cu_xO	Precipitation method	Rhodamine B	73.5% within 120 min	[82]
$ZnSnO_3$	Sol–gel synthesis	Rhodamine B	98.1%	[83]
NiO-CdO-ZnO	Homogeneous co-precipitation method	rhodamine B and methylene blue	99% in 60 min for RhB and 98% in 60 min for MB	[84]
Pd-γ-Al_2O_3	Co-precipitation method	Bromocresol Green (BCG) Bromothymol Blue Methylene blue Methylene orange	Up to 100% efficient in aqueous medium under UV radiation	[85]

References

1. Danish MSS, Bhattacharya A, Stepanova D, Mikhaylov A, Grilli ML, Khosravy M et al (2020) A systematic review of metal oxide applications for energy and environmental sustainability. Metals (Basel) 10(12):1604. https://www.mdpi.com/2075-4701/10/12/1604
2. Grilli ML (2020) Metal oxides. Metals (Basel) 10(6):820. https://www.mdpi.com/2075-4701/10/6/820
3. Gautam S, Agrawal H, Thakur M, Akbari A, Sharda H, Kaur R et al (2020) Metal oxides and metal organic frameworks for the photocatalytic degradation: a review. J Environ Chem Eng 8(3):103726
4. Fan J, Chen D, Li N, Xu Q, Li H, He J et al (2018) Adsorption and biodegradation of dye in wastewater with Fe_3O_4@MIL-100 (Fe) core–shell bio-nanocomposites. Chemosphere 1(191):315–323
5. Pei C, Han G, Zhao Y, Zhao H, Liu B, Cheng L et al (2016) Superior adsorption performance for

triphenylmethane dyes on 3D architectures assembled by ZnO nanosheets as thin as ~1.5 nm. J Hazard Mater 318:732–741
6. Rani B, Thamizharasan G, Nayak AK, Sahu NK (2020) Degradation mechanism of organic dyes by effective transition metal oxide. In: Photocatalysts in advanced oxidation processes for wastewater treatment. Wiley, pp 197–228. https://onlinelibrary.wiley.com/doi/abs/10.1002/9781119631422.ch7
7. Global Dyes and Pigments Market Size Report, 2021–2028. https://www.grandviewresearch.com/industry-analysis/dyes-and-pigments-market
8. Purkait MK, Maiti A, DasGupta S, De S (2007) Removal of congo red using activated carbon and its regeneration. J Hazard Mater 145(1–2):287–295
9. Channei D, Inceesungvorn B, Wetchakun N, Ukritnukun S, Nattestad A, Chen J et al (2014) Photocatalytic degradation of methyl orange by CeO_2 and Fe-doped CeO_2 films under visible light irradiation. Sci Rep 4(1):1–7. www.nature.com/scientificreports
10. Pathania D, Sharma S, Singh P (2017) Removal of methylene blue by adsorption onto activated carbon developed from Ficus carica bast. Arab J Chem 1(10):S1445–S1451
11. Lassoued A, Lassoued MS, Dkhil B, Ammar S, Gadri A (2018) Photocatalytic degradation of methylene blue dye by iron oxide (α-Fe_2O_3) nanoparticles under visible irradiation. J Mater Sci Mater Electron (Springer New York LLC) 29:8142–8152
12. da Silva LL, de Souza MB, de Aragão UG, Pupo Nogueira RF (2016) Monitoring ecotoxicity of disperse red 1 dye during photo-Fenton degradation. Chemosphere 1(148):511–517
13. Aguedach A, Brosillon S, Morvan J, Lhadi EK (2005) Photocatalytic degradation of azo-dyes reactive black 5 and reactive yellow 145 in water over a newly deposited titanium dioxide. Appl Catal B Environ 57(1):55–62
14. Yaseen DA, Scholz M (2019) Textile dye wastewater characteristics and constituents of synthetic effluents: a critical review. Int J Environ Sci Technol (Center for Environmental and Energy Research and Studies) 16:1193–1226. https://doi.org/10.1007/s13762-018-2130-z
15. Sonune A, Ghate R (2004) Developments in wastewater treatment methods. Desalination 167(1–3):55–63
16. Muruganandam L, Kumar MPS, Jena A, Gulla S, Godhwani B (2017) Treatment of waste water by coagulation and flocculation using biomaterials. IOP Conf Ser Mater Sci Eng (Institute of Physics Publishing) 032006. https://iopscience.iop.org/article/10.1088/1757-899X/263/3/032006
17. Molinari R, Pirillo F, Falco M, Loddo V, Palmisano L (2004) Photocatalytic degradation of dyes by using a membrane reactor. Chem Eng Process Process Intensif 43(9):1103–1114
18. Ezugbe EO, Rathilal S (2020) Membrane technologies in wastewater treatment: a review. J Membr Sci 10(5):89. https://www.mdpi.com/2077-0375/10/5/89/htm
19. Yang G, Zhang J, Peng M, Du E, Wang Y, Shan G et al (2021) A mini review on antiwetting studies in membrane distillation for textile wastewater treatment. Process 9(2):243. https://www.mdpi.com/2227-9717/9/2/243/htm
20. Raghu S, Ahmed BC (2007) Chemical or electrochemical techniques, followed by ion exchange, for recycle of textile dye wastewater. J Hazard Mater 149(2):324–330
21. Tan KB, Vakili M, Horri BA, Poh PE, Abdullah AZ, Salamatinia B (2015) Adsorption of dyes by nanomaterials: recent developments and adsorption mechanisms. Sep Purif Technol (Elsevier B.V.) 150:229–242
22. Ghime D, Ghosh P (2020) Advanced oxidation processes: a powerful treatment option for the removal of recalcitrant organic compounds. In: Advanced oxidation processes - applications, trends, and prospects. IntechOpen
23. Konstantinou IK, Albanis TA (2004) TiO_2-assisted photocatalytic degradation of azo dyes in aqueous solution: kinetic and mechanistic investigations: a review. Appl Catal B Environ (Elsevier) 49:1–14
24. Chen D, Wang Z, Ren T, Ding H, Yao W, Zong R et al (2014) Influence of defects on the photocatalytic activity of ZnO. J Phys Chem C 118(28):15300–15307. https://pubs.acs.org/doi/abs/10.1021/jp5033349

25. Zhang X, Qin J, Xue Y, Yu P, Zhang B, Wang L et al (2014) Effect of aspect ratio and surface defects on the photocatalytic activity of ZnO nanorods. Sci Rep 4(1):1–8. www.nature.com/scientificreports
26. Akintelu SA, Folorunso AS, Folorunso FA, Oyebamiji AK (2020) Green synthesis of copper oxide nanoparticles for biomedical application and environmental remediation. Heliyon (Elsevier Ltd.) 6:e04508. https://doi.org/10.1016/j.heliyon.2020.e04508
27. Gnanaprakasam A, Sivakumar VM, Thirumarimurugan M (2015) Influencing parameters in the photocatalytic degradation of organic effluent via nanometal oxide catalyst: a review. Indian J Mater Sci 17(2015):1–16
28. Ajmal A, Majeed I, Malik RN, Idriss H, Nadeem MA (2014) Principles and mechanisms of photocatalytic dye degradation on TiO_2 based photocatalysts: a comparative overview. RSC Adv R Soc Chem 4:37003–37026. https://pubs.rsc.org/en/content/articlehtml/2014/ra/c4ra06658h
29. Gupta R, Eswar NKR, Modak JM, Madras G (2017) Effect of morphology of zinc oxide in ZnO-CdS-Ag ternary nanocomposite towards photocatalytic inactivation of E. coli under UV and visible light. Chem Eng J 307:966–980
30. Tomkiewicz M (2000) Scaling properties in photocatalysis. Catal Today 58(2–3):115–123
31. Zhou C, Lai C, Zhang C, Zeng G, Huang D, Cheng M et al (2018) Semiconductor/boron nitride composites: synthesis, properties, and photocatalysis applications. Appl Catal B Environ 15(238):6–18
32. Gladis F, Schumann R (2011) Influence of material properties and photocatalysis on phototrophic growth in multi-year roof weathering. Int Biodeterior Biodegrad 65(1):36–44
33. Liu G, Zhen C, Kang Y, Wang L, Cheng H-M (2018) Unique physicochemical properties of two-dimensional light absorbers facilitating photocatalysis. Chem Soc Rev 47(16):6410–6444. https://pubs.rsc.org/en/content/articlehtml/2018/cs/c8cs00396c
34. Wang Y, Wang Q, Zhan X, Wang F, Safdar M, He J (2013) Visible light driven type II heterostructures and their enhanced photocatalysis properties: a review. Nanoscale 5(18):8326–8339. https://pubs.rsc.org/en/content/articlehtml/2013/nr/c3nr01577g
35. Gao X, Yao Y, Meng X (2020) Recent development on BN-based photocatalysis: a review. Mater Sci Semicond Process 120:105256
36. Becker J, Raghupathi KR, St. Pierre J, Zhao D, Koodali RT (2011) Tuning of the crystallite and particle sizes of ZnO nanocrystalline materials in solvothermal synthesis and their photocatalytic activity for dye degradation. J Phys Chem C 115(28):13844–13850. https://pubs.acs.org/doi/abs/10.1021/jp2038653
37. Yumashev A, Ślusarczyk B, Kondrashev S, Mikhaylov A (2020) Global indicators of sustainable development: evaluation of the influence of the human development index on consumption and quality of energy. Energies 13(11):2768. https://www.mdpi.com/1996-1073/13/11/2768
38. Yumashev A, Mikhaylov A (2020) Development of polymer film coatings with high adhesion to steel alloys and high wear resistance. Polym Compos 41(7):2875–2880. https://onlinelibrary.wiley.com/doi/abs/10.1002/pc.25583
39. Danish MSS, Estrella LL, Alemaida IMA, Lisin A, Moiseev N, Ahmadi M et al (2021) Photocatalytic applications of metal oxides for sustainable environmental remediation. Metals (Basel) 11(1):1–25. https://doi.org/10.3390/met11010080
40. Fujishima A, Honda K (1972) Electrochemical photolysis of water at a semiconductor electrode. Nature 238(5358):37–38. https://www.nature.com/articles/238037a0
41. Aydoğan Ş, Grilli ML, Yilmaz M, Çaldiran Z, Kaçuş H (2017) A facile growth of spray based ZnO films and device performance investigation for Schottky diodes: determination of interface state density distribution. J Alloys Compd 1(708):55–66
42. Yilmaz M, Grilli ML, Turgut G (2020) A bibliometric analysis of the publications on in doped ZnO to be a guide for future studies. Metals (Basel) 10(5):598. https://www.mdpi.com/2075-4701/10/5/598
43. Masetti E, Grilli ML, Dautzenberg G, Macrelli G, Adamik M (1999) Analysis of the influence of the gas pressure during the deposition of electrochromic WO_3 films by reactive r.f. sputtering of W and WO_3 target. Sol Energy Mater Sol Cells 56(3–4):259–269

44. Grilli ML, Kaabbuathong N, Dutta A, Bartolomeo E Di, Traversa E (2002) Electrochemical NO_2 sensors with WO_3 electrodes for high temperature applications. J Ceram Soc Jpn 110(1279):159–162. http://joi.jlc.jst.go.jp/JST.Journalarchive/jcersj1988/110.159?from=CrossRef
45. Asiltürk M, Sayilkan F, Erdemoğlu S, Akarsu M, Sayilkan H, Erdemoğlu M et al (2006) Characterization of the hydrothermally synthesized nano-TiO_2 crystallite and the photocatalytic degradation of Rhodamine B. J Hazard Mater 129(1–3):164–170
46. Cheng G, Yu L, Lin T, Yang R, Sun M, Lan B et al (2014) A facile one-pot hydrothermal synthesis of β-MnO_2 nanopincers and their catalytic degradation of methylene blue. J Solid State Chem 1(217):57–63
47. Mounika T, Belagali SL, Vadiraj KT (2020) Synthesis and characterization of zinc oxide quantum dots (QD) using acidic precursor. Emerg Mater Res 9(2):1–6. https://www.icevirtuallibrary.com/doi/10.1680/jemmr.19.00152
48. Adam RE, Pozina G, Willander M, Nur O (2018) Synthesis of ZnO nanoparticles by co-precipitation method for solar driven photodegradation of Congo red dye at different pH. Photon Nanostruct Fundam Appl 1(32):11–18
49. Gnanasekaran L, Hemamalini R, Saravanan R, Ravichandran K, Gracia F, Agarwal S et al (2017) Synthesis and characterization of metal oxides (CeO_2, CuO, NiO, Mn_3O_4, SnO_2 and ZnO) nanoparticles as photo catalysts for degradation of textile dyes. J Photochem Photobiol B Biol 1(173):43–49
50. Nikam AV, Prasad BLV, Kulkarni AA (2018) Wet chemical synthesis of metal oxide nanoparticles: a review. CrystEngComm 20(35):5091–5107. https://pubs.rsc.org/en/content/articlehtml/2018/ce/c8ce00487k
51. Zhan X-Q, Yu X-Y, Tsai F-C, Ma N, Liu H-L, Han Y et al (2018) Magnetic MOF for AO7 removal and targeted delivery. Crystals 8(6):250. http://www.mdpi.com/2073-4352/8/6/250
52. Tanaka Y, Utaka T, Kikuchi R, Sasaki K, Eguchi K (2003) CO removal from reformed fuel over $Cu/ZnO/Al_2O_3$ catalysts prepared by impregnation and coprecipitation methods. Appl Catal A Gen 238(1):11–18
53. Wen Z, Ke J, Xu J, Guo S, Zhang Y, Chen R (2018) One-step facile hydrothermal synthesis of flowerlike Ce/Fe bimetallic oxides for efficient As(V) and Cr(VI) remediation: performance and mechanism. Chem Eng J 1(343):416–426
54. Zhou H, Guo J, Li P, Fan T, Zhang D, Ye J (2013) Leaf-architectured 3D hierarchical artificial photosynthetic system of perovskite titanates towards CO_2 photoreduction into hydrocarbon fuels. Sci Rep 3(1):1–9. www.nature.com/scientificreports
55. Xi G, Yue B, Cao J, Ye J (2011) Fe_3O_4/WO_3 hierarchical core-shell structure: high-performance and recyclable visible-light photocatalysis. Chem A Eur J 17(18):5145–5154. http://doi.wiley.com/10.1002/chem.201002229
56. Viswanathan B (2018) Photocatalytic degradation of dyes: an overview. Curr Catal 7(2):99–121. http://www.eurekaselect.com/158447/article
57. Shinde DR, Tambade PS, Chaskar MG, Gadave KM (2017) Photocatalytic degradation of dyes in water by analytical reagent grades ZnO, TiO_2 and SnO_2: a comparative study. Drink Water Eng Sci 10(2):109–117. https://dwes.copernicus.org/articles/10/109/2017/
58. Mohammad A, Kapoor K, Mobin SM (2016) Improved photocatalytic degradation of organic dyes by ZnO-nanoflowers. ChemistrySelect 1(13):3483–3490. https://doi.org/10.1002/slct.201600476
59. Koelsch M, Cassaignon S, Guillemoles JF, Jolivet JP (2002) Comparison of optical and electrochemical properties of anatase and brookite TiO_2 synthesized by the sol-gel method. Thin Solid Films (Elsevier) 312–319
60. Shaabani B, Alizadeh-Gheshlaghi E, Azizian-Kalandaragh Y, Khodayari A (2014) Preparation of CuO nanopowders and their catalytic activity in photodegradation of Rhodamine-B. Adv Powder Technol 25(3):1043–1052
61. Vanaja M, Paulkumar K, Baburaja M, Rajeshkumar S, Gnanajobitha G, Malarkodi C et al (2014) Degradation of methylene blue using biologically synthesized silver nanoparticles. Bioinorg Chem Appl 2014

62. Muruganandham M, Amutha R, Lee GJ, Hsieh SH, Wu JJ, Sillanpää M (2012) Facile fabrication of tunable Bi_2O_3 self-assembly and its visible light photocatalytic activity. J Phys Chem C 116(23):12906–12915. https://pubs.acs.org/doi/abs/10.1021/jp302343f
63. Iskandar F, Nandiyanto ABD, Yun KM, Hogan CJ, Okuyama K, Biswas P (2007) Enhanced photocatalytic performance of brookite TiO_2 macroporous particles prepared by spray drying with colloidal templating. Adv Mater 19(10):1408–1412. http://doi.wiley.com/10.1002/adma.200601822
64. Bakardjieva S, Šubrt J, Štengl V, Dianez MJ, Sayagues MJ (2005) Photoactivity of anatase-rutile TiO_2 nanocrystalline mixtures obtained by heat treatment of homogeneously precipitated anatase. Appl Catal B Environ 58(3–4):193–202
65. Colón G, Hidalgo MC, Munuera G, Ferino I, Cutrufello MG, Navío JA (2006) Structural and surface approach to the enhanced photocatalytic activity of sulfated TiO_2 photocatalyst. Appl Catal B Environ 63(1–2):45–59
66. Kansal SK, Kaur N, Singh S (2009) Photocatalytic degradation of two commercial reactive dyes in aqueous phase using nanophotocatalysts. Nanoscale Res Lett 4(7):709–716. http://www.nanoscalereslett.com/content/4/7/709
67. Tripathy N, Ahmad R, Song JE, Park H, Khang G (2017) ZnO nanonails for photocatalytic degradation of crystal violet dye under UV irradiation. AIMS Mater Sci 4(1):267–276. http://www.aimspress.com/article/doi/10.3934/matersci.2017.1.267
68. Khurram R, Wang Z, Ehsan MF, Peng S, Shafiq M, Khan B (2020) Synthesis and characterization of an α-Fe_2O_3/ZnTe heterostructure for photocatalytic degradation of Congo red, methyl orange and methylene blue. RSC Adv 10(73):44997–45007. https://pubs.rsc.org/en/content/articlehtml/2020/ra/d0ra06866g
69. Balu P, Asharani IV, Thirumalai D (2020) Catalytic degradation of hazardous textile dyes by iron oxide nanoparticles prepared from Raphanus sativus leaves' extract: a greener approach. J Mater Sci Mater Electron 31(13):10669–10676. https://link.springer.com/article/10.1007/s10854-020-03616-z
70. Bhuiyan MSH, Miah MY, Paul SC, Aka TD, Saha O, Rahaman MM et al (2020) Green synthesis of iron oxide nanoparticle using Carica papaya leaf extract: application for photocatalytic degradation of remazol yellow RR dye and antibacterial activity. Heliyon 6(8):e04603
71. Nathan VK, Ammini P, Vijayan J (2019) Photocatalytic degradation of synthetic dyes using iron (III) oxide nanoparticles (Fe_2O_3-Nps) synthesised using Rhizophora mucronata Lam. IET Nanobiotechnol 13(2):120–123. https://pubmed.ncbi.nlm.nih.gov/31051441/
72. David L, Moldovan B (2020) Green synthesis of biogenic silver nanoparticles for efficient catalytic removal of harmful organic dyes. Nanomaterials 10(2):202. https://www.mdpi.com/2079-4991/10/2/202
73. Marimuthu S, Antonisamy AJ, Malayandi S, Rajendran K, Tsai PC, Pugazhendhi A et al (2020) Silver nanoparticles in dye effluent treatment: a review on synthesis, treatment methods, mechanisms, photocatalytic degradation, toxic effects and mitigation of toxicity. J Photochem Photobiol B Biol 205:111823
74. Laouini SE, Bouafia A, Tedjani ML (2021) Catalytic activity for dye degradation and characterization of silver/silver oxide nanoparticles green synthesized by aqueous leaves extract of phoenix Dactylifera L. https://www.researchsquare.com
75. Nandanwar S, Borkar S, Cho JH, Kim HJ (2020) Microwave-assisted synthesis and characterization of solar-light-active copper–vanadium oxide: evaluation of antialgal and dye degradation activity. Catalysts 11(1):36. https://www.mdpi.com/2073-4344/11/1/36
76. Mohammed AM, Mohtar SS, Aziz F, Aziz M, Ul-Hamid A, Wan Salleh WN et al (2021) Ultrafast degradation of congo red dye using a facile one-pot solvothermal synthesis of cuprous oxide/titanium dioxide and cuprous oxide/zinc oxide p-n heterojunction photocatalyst. Mater Sci Semicond Process 122:105481
77. Al Aqad KM, Basheer C (2021) Photocatalytic degradation of basic blue dye using zinc nanoparticles decorated graphene oxide nanosheet. J Phys Org Chem 34(1)
78. Karuppusamy I, Samuel MS, Selvarajan E, Shanmugam S, Sahaya Murphin Kumar P, Brindhadevi K et al (2021) Ultrasound-assisted synthesis of mixed calcium magnesium oxide

($CaMgO_2$) nanoflakes for photocatalytic degradation of methylene blue. J Colloid Interface Sci 584:770–778
79. Chang YH, Wang H, Siao TF, Lee YH, Bai SY, Liao CW et al (2021) A new solution route for the synthesis of $CuFeO_2$ and Mg-doped $CuFeO_2$ as catalysts for dye degradation and CO_2 conversion. J Alloys Compd 854:157235
80. Basavalingaiah KR, Udayabhanu, Harishkumar S, Nagaraju G, Chikkahanumantharayappa (2020) Uniform deposition of silver dots on sheet like $BiVO_4$ nanomaterials for efficient visible light active photocatalyst towards methylene blue degradation. FlatChem 19:100142
81. Liu D, Jin C, Shan F, He J, Wang F (2020) Synthesizing $BaTiO_3$ nanostructures to explore morphological influence, kinetics, and mechanism of piezocatalytic dye degradation. ACS Appl Mater Interfaces 12(15):17443–17451. https://pubs.acs.org/doi/abs/10.1021/acsami.9b23351
82. Nandi P, Das D (2020) ZnO-CuxO heterostructure photocatalyst for efficient dye degradation. J Phys Chem Solids 143:109463
83. Chen J, Luo W, Yu S, Yang X, Wu Z, Zhang H et al (2020) Synergistic effect of photocatalysis and pyrocatalysis of pyroelectric $ZnSnO_3$ nanoparticles for dye degradation. Ceram Int 46(7):9786–9793
84. Munawar T, Iqbal F, Yasmeen S, Mahmood K, Hussain A (2020) Multi metal oxide NiO-CdO-ZnO nanocomposite–synthesis, structural, optical, electrical properties and enhanced sunlight driven photocatalytic activity. Ceram Int 46(2):2421–2437
85. Kumar AP, Bilehal D, Tadesse A, Kumar D (2021) Photocatalytic degradation of organic dyes: Pd-γ-Al_2O_3 and PdO-γ-Al_2O_3 as potential photocatalysts. RSC Adv 11(11):6396–6406. https://pubs.rsc.org/en/content/articlehtml/2021/ra/d0ra10290c
86. Ammar SH, Kareem YS, Ali AD (2018) Photocatalytic oxidative desulfurization of liquid petroleum fuels using magnetic CuO-Fe_3O_4 nanocomposites. J Environ Chem Eng 6(6):6780–6786
87. Thiruppathi M, Senthil Kumar P, Devendran P, Ramalingan C, Swaminathan M, Nagarajan ER (2018) Ce@TiO_2 nanocomposites: an efficient, stable and affordable photocatalyst for the photodegradation of diclofenac sodium. J Alloys Compd 25(735):728–734
88. Nasr O, Mohamed O, Al-Shirbini AS, Abdel-Wahab AM (2019) Photocatalytic degradation of acetaminophen over Ag, Au and Pt loaded TiO_2 using solar light. J Photochem Photobiol A Chem 1(374):185–193
89. Cerrón-Calle GA, Aranda-Aguirre AJ, Luyo C, Garcia-Segura S, Alarcón H (2019) Photoelectrocatalytic decolorization of azo dyes with nano-composite oxide layers of ZnO nanorods decorated with Ag nanoparticles. Chemosphere 1(219):296–304

Operational Parameters in Dye Decolorization via Sonochemical and Sonoenzymatic Treatment Processes

Maneesh Kumar Poddar, Priyanka Prabhakar, and Hari Mahalingam

Abstract In the last decade, the use of ultrasound in the development of novel technologies for mainly the food processing and industrial biotechnology has increased exponentially. At the same time, its use in the material sciences and engineering field has also gained considerable interest with a natural outcome of these efforts being directed towards the development of new and more efficient technologies for wastewater treatment. This chapter gives a brief overview of the fundamentals in an ultrasound-assisted process including mechanistic details of the process with respect to dye wastewater treatment and systematically reviews the influence of the key operational parameters in the sonochemical and sonoenzymatic process. Several observations can be made from this study however the key findings are as follows: higher degradation is generally favored by low frequencies but this depends on the amount of free radicals generated; sono-enzymatic process can give more complete degradation than sonochemical process however the timescales are longer in the former. Finally, this chapter ends with some insights into the challenges and future perspectives.

Keywords Dyes · Decolorization · Degradation · Sonochemical · Sonoenzymatic · Advanced oxidation process · Operational parameters · Power · Frequency · Cavitation

1 Introduction

Discharge of untreated or partially treated dye wastewaters from various chemical industries such as textiles, leather, food, pulp and papers lead to serious health and environmental problems due to the dye's synthetic and complex structures as well as

M. K. Poddar · P. Prabhakar · H. Mahalingam (✉)
Department of Chemical Engineering, National Institute of Technology Karnataka (NITK), Surathkal, Mangalore, D.K. 575025, Karnataka, India
e-mail: mhari@nitk.edu.in

M. K. Poddar
e-mail: maneesh.poddar@nitk.edu.in

S. S. Muthu and A. Khadir (eds.), *Advanced Oxidation Processes in Dye-Containing Wastewater*, Sustainable Textiles: Production, Processing, Manufacturing & Chemistry, https://doi.org/10.1007/978-981-19-0882-8_9

its toxic and hazardous nature. Among the various dyes, azo dyes contribute almost 70% of the total industrially dyes used and show high resistance against the decolorization and degradation during their treatment due to their fused aromatic structures and synthetic origin [17, 53]. Numerous conventional methods such as coagulation/flocculation, adsorption, membrane separation, ozonation, ion-exchange, katox treatment, etc. used for removal and/or degradation of dyes are inefficient, expensive and unable to treat the waste waters effectively over a wide range of concentrations or handle multiple dye contamination.

In the past decades, the sonochemical route for treatment of dyes present in wastewater was extensively used as an advanced oxidation (AOPs) process with formation of powerful oxidization agents. Sonication or the use of ultrasound refers to the formation of highly reactive hydroxyl radicals ($OH^{\bullet}$) along with other radical species of oxygen ($O^{\bullet}$), hydrogen ($H^{\bullet}$) and hydroperoxyl ($HO^{\bullet}{}_2$) through the phenomenon of cavitation [11, 26]. Under the irradiation of the ultrasound wave into aqueous solutions consisting of dye molecules, these radicals can, through the series of chemical reactions, decolorize and degrade the dye molecules [20, 61]. Often sonication alone does not help in complete decolorization of the dye molecules and requires a conjugation with other AOP technologies e.g., sono-photolysis, sonocatalysis, sonoenzymatic, etc. [1].

Among the various AOPs coupled with sonication, sonoenzymatic method is one of the highly effective and the most popular due to the conversion of dyes molecules into harmless end-products using very low amounts of enzyme. The interaction between the ultrasound waves and the enzymes is physical in nature and enhanced by the generation of intense microconvections in the form of microturbulence and shockwaves in the bulk liquid medium [43, 54]. Sonication also increases the enzymatic reaction rate by acting on different targets via altering the substrate and enzyme macromolecules that ultimately accelerate the enzyme activity and product yield.

Enzymes such as laccase, horseradish peroxidase, lignin peroxidase are the commonly used in sonoenzymatic dye decolorization. However, in presence of sonication, the enzyme activity and its stability reduce due to the very high local temperatures (~5000 K) and pressures (~500 bar) generated during bubble collapse. To overcome these issues, use of low power sonication, addition of stabilizing agents and enzyme immobilization techniques have been adapted and have shown a promising and fruitful results in maintaining enzyme activity with high decolorization [2, 35].

This review assesses the effect of sonochemical and sonoenzymatic methods for dye decolorization and degradation in wastewater systems. A literature research on the decolorization mechanism and factors controlling the dyes decolorization and degradation efficiency used in both methods is discussed. Also, a summary of the literature, including the effect of various operating parameters on dyes decolorization and degradation using both techniques is presented in Table 1.

Table 1 Summary of the effect of operational parameters used in dye decolorization/degradation

Name of the dye	Sonochemical operating conditions	Other experimental conditions	Major findings and conclusion	Reference
(A) Sonochemical route				
Direct scarlet 4BS (a type of azo dye)	Frequency: 28, 40 kHz Power: 500 W	Temperature: 30–50 °C pH: 2–7 Reaction time: 0–120 min Initial dye concentration: 100 mg/L Exfoliated graphite: 0–1 g/L	• Combined effect of sonication and exfoliated graphite improved decolorization efficiency • High decolorization efficiency at 50 °C and 28 kHz than 30 °C and 28 kHz • Decolorization efficiency decreased with increase in pH • Optimum conditions: Temp 50 °C; frequency 28 kHz; pH 2; exfoliated graphite 0.6 g/l	[33]
C.I. reactive black 5 (non-biodegradable azo dye)	Frequency: 20–817 kHz Power: 50–150 W	Temperature: 25 °C pH: 2–10 Reaction time: 0–60 min Initial dye concentration: 5–300 mg/L Gas type: Argon (Ar)	• High decolorization at 817 kHz • Acidic pH accelerated the decolorization rate • Decolorization efficiency strongly depends on operating parameters	[58]
Orange II (a type of azo dye)	Frequency: 20–1176 kHz Power: 2.04–22.07 W	Temperature: 20 ± 3 °C Initial pH 6.3 Reaction time: 120 min Initial dye concentration: 0.05 g/dm^3	• Maximum decolorization (49%) at 850 kHz. and power of 22.07 W • In-situ generation of H_2O_2 increased the decolorization efficiency	[17]

(continued)

Table 1 (continued)

Name of the dye	Sonochemical operating conditions	Other experimental conditions	Major findings and conclusion	Reference
Naphthol blue black (an acidic diazo dye)	Frequency: 585,860 and 1140 kHz Intensity: 3.58 W/cm^2	Temperature: 25 °C pH: 2–10 Reaction time: 0–90 min dye concentration: 3–120 mg/L Gas types: Ar, air and N_2	• Low degradation at high frequency • Rise in intensity increased the degradation • High degradation in acidic pH • High degradation in presence of Ar gas	[20]
Coomassie brilliant blue (CBB)	Frequency: 200–1000 kHz Power density: 3.5–19.6 W mL^{-1}	Temperature: 25 ± 1 °C pH: 3–8 Reaction time: 0–30 min Initial dye concentration: 1×10^5 M Effect of various additives like Fe^{2+}, H_2O_2, $S_2O_8{}^{2-}$, SDS and humic acid	• High degradation in acidic pH of 3 • US/H_2O_2 system enhanced the CBB degradation up to 98% • Reduction in degradation in SDS, humic acid systems due to $OH^{\bullet}$ scavenging effect	[47]
C.I. acid orange 8 (an aryl-azo-naphthol dye)	Frequency: 300 kHz; Power: 25 W	Temperature: −4 °C pH: 5 Sonication time: 30 min Initial dye concentration: 20, 30 and 40 μM System: US, O_3 and US/O_3	• Bleaching rate decreased with increase in initial dye concentration • US/O_3 system showed highest bleaching • Decolorization followed pseudo-first order reaction kinetics	[24]
Basic blue 41	Frequency: 35 kHz Power: 160 W	Temperature: 25 °C pH: 4.5–8 Reaction time: 180 min Initial dye concentration: 15–240 mgl^{-1} Catalyst used: nano titanium oxide	• 89.5% decolorization at initial dye concentration of 15 mg/L • Decolorization declined to 33% when initial dye concentration raised to 240 mg/L • US process combined with TiO_2 and H_2O_2 fond to be effective in decolorization	[1]

(continued)

Table 1 (continued)

Name of the dye	Sonochemical operating conditions	Other experimental conditions	Major findings and conclusion	Reference
C.I. reactive blue 19	Frequency: 850 kHz Power: 190–760 W/dm^3	Temperature: 20 °C pH: 11.4 and 12 Reaction time: 180 min Initial dye concentration: 132 and 917 mg/dm^3 Systems: ultrasound, activated carbon and US/activated carbon	• Sonication Improved the decolorization by adsorption mechanism • 36% decolorization with only sonolysis at 132 mg/dm^3 of initial dye concentration • 99% decolorization with combine use of ultrasound and activated carbon	[50]
Acid red 88 (mono-azo textile dye)	Frequency: 213 kHz Power: 16–64 mWm/L	Temperature: 25 ± 2 °C pH: 2.7 Initial dye concentration: 0.025–0.09 mM Systems: US + Fe^{3+}, UV + Fe^{3+} and US + Fe^{3+} + UV	• Degradation rate increased with an increase in the ultrasound power • Sonophotocatalysis (US + Fe^{3+} + UV) revealed high degradation rate	[34]
Malachite green	Frequency: 20 kHz Power: 60 W	Temperature: 20–40 °C pH: 5–9 Reaction time: 10–30 min Initial dye concentration 1000 mg/L Ozone concentration of 0.17 g/L	• 100% decolorization using sonochemical assisted ozone oxidation process • Optimum conditions: Temp 39.81 °C; power 60 W; pH 5.29; ozone conc; 0.17 g/L	[61]
Acid green 20	Frequency: 40 kHz Power density: 0.75–1.25 W/mL	Temperature: 20 °C pH: 2–6 H_2O_2: 0.8–2.4 mM	• 96. 8% decolorization using sonication • Optimum conditions using response su rface methodology: power density 1.08 W/mL; pH 4.85; H_2O_2 1.94 mM	[60]
(B) Sono-enzymatic route				

(continued)

Table 1 (continued)

Name of the dye	Sonochemical operating conditions	Other experimental conditions	Major findings and conclusion	Reference
Acid orange 5 Acid orange 52 Direct blue 71; etc.	Frequency: 850 kHz Power: 60–120 W	Enzyme: Laccase Temperature: 30 °C Initial dye concentration: 100 μM	• Combined use of ultrasound and enzyme increased the dyes degradation • Power >90 W, enzyme showed deactivation	[48]
Acid orange 52 Direct blue 71 Reactive black 5; etc.	Frequency: 850 kHz Power: 90–120 W	Enzyme: Laccase Temperature: 40 °C pH: 4.5	• Combined use of ultrasound and enzyme increased decolorization • Power >120 W caused deactivation of Enzyme	[54]
Indigo carmine	Frequency: 20 kHz Power: 7–50 W	Enzyme: Laccase Temperature: 50 °C pH: 5 Initial dye concentration: 10 mg/L Sonication time: 0–40 min	• Combined use of ultrasound (7 W) and enzyme showed high bleaching efficiency in 30 min • Polyvinyl alcohol used as stabilizing agent for the enzyme	[7]
Acid red	Frequency: 35 kHz Power: 35 W	Enzyme: Horseradish peroxidase Temperature: 25 °C pH: 6.5, 7 Sonication time: 0–60 min Enzyme amount 2U/ml Poly ethylene glycol: 4–626.8 g/mol	• Optimum pH for free and immobilized HRP were found 7 and 6.5 respectively • Combined use of ultrasound and enzyme with poly ethylene glycol showed 61.2% decolorization	[35]
Acid red and malachite green	Frequency: 20 kHz Power: 35 W	Enzyme: Horseradish peroxidase Temperature: 20 °C Pressure: 1.3 bar pH: 6 Sonication time: 0–6 min Enzyme amount 2U/ml Poly ethylene Glycol: 4–626.8 g/mol	• Combined use of ultrasound and enzyme at high pressure (1.3 bar) improved the degradation of dyes • Shock wave produced during sonication had adverse effect on enzyme activity	[43]

2 Fundamentals of the Sonochemical Treatment Process

The sonochemical treatment process is based on the use of Ultrasound which refers to a sound wave with a frequency (>20 kHz) higher than the upper audible limit of human hearing. When an ultrasound wave is irradiated into the liquid medium, it travels through a series of alternating compression and rarefaction cycles (sinusoidal variation of pressure waves) and gives rise to the phenomenon of *acoustic cavitation* i.e., the nucleation, growth, oscillation and the implosive collapse of the cavitation bubble. The moment at which the acoustic pressure during the rarefaction cycle of the ultrasound wave is sufficiently below the system static pressure, the liquid molecules evaporate to form the cavities which expand rapidly to their actual sizes. Also, a large pressure gradient exists at the bubble–liquid interface, allowing the diffusion of vapor molecules towards the core of the bubble. During the first half of the rarefaction cycle (maximum negative pressure), the bubble rapidly expands to its maximum size and implosively collapses in the consecutive compression cycle at maximum pressure (Fig. 1). The collapse of the cavitation bubble is extremely violent and energetic resulting in a very high local temperature (~5000 K) and pressure (~500 bar) inside the bubble core, at the interfaces and the so-called "*local hotspots*" [27, 52]. At these extreme conditions, the vapor/gases (water vapor, dissolved gases like O_2, N_2, etc.) trapped in the bubbles, fragment into highly reactive $OH^{\bullet}$ species which degrade the pollutants through various chemical reactions and this is manifested as the chemical effect of ultrasound or sonochemical treatment process. The chemical effect of the ultrasound induces highly reactive hydroxyl radicals ($OH^{\bullet}$) into the system which are capable of oxidizing and degrading the toxic and/or hazardous organic pollutants such as complex dyes structures into biodegradable and less harmful compounds. In virtue of the high oxidation potential of $OH^{\bullet}$ (~2.8 V) and its effective role in the degradation of emerging contaminants (ECs) present in wastewater, the sonochemical treatment process is also categorized in the list of advanced oxidation processes (AOPs) [4]. On the other hand, the physical effect of ultrasound and consequently, the collapse of short-lived cavitation bubbles is the generation of strong microturbulence in the system which eliminates the mass transfer resistance and accelerates the reaction kinetics by several folds [3]. Detailed mechanistic features of ultrasound involving

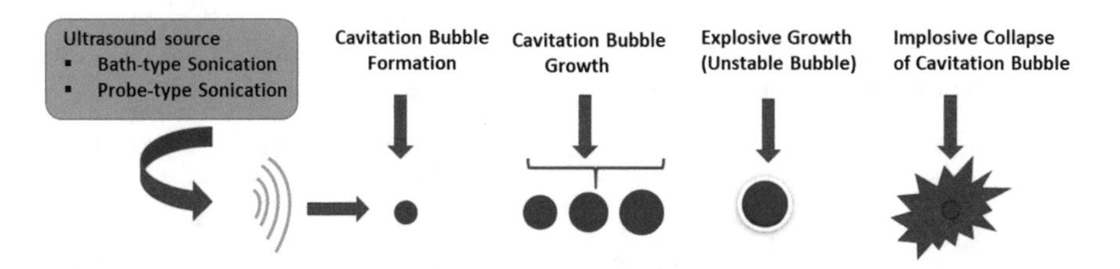

Fig. 1 A schematic of bubble formation, growth and its implosion collapse

its chemical and physical effects for the degradation of dyes and other emerging contaminants (ECs) can be found in the various textbooks and popular reviews [13, 38].

3 Dye Decolorization Using Sonochemical or Ultrasound Process

Over the past few decades, the sonochemical route for degradation of synthetic dyes, classified as soluble (acid, base, mordant and reactive) and insoluble types (azo, metal complex, azo metal complex, phythalocyanine, etc.) has gained immense popularity due to its operational simplicity, faster reaction kinetics and high degradation efficiency. Both the pyrolysis reaction inside the cavitation bubble and the hydroxyl radicals ($OH^{\bullet}$) formation during the collapse of cavitation bubble are the important contributing factors in enhancing dye decolorization and degradation. A detailed reaction mechanism, involving these factors under the effect of ultrasound is provided in the following section.

3.1 Mechanistic Features of Sonochemical Treatment in Dye Decolorization

The reaction mechanism of dyes decolorization under the influence of ultrasound can be described in a way similar to the degradation of other organic pollutants in wastewater discharged from various industries such as textiles industries, leather industries, paper industries, food industries, etc. As stated earlier, the intense conditions (temperature ~5000 K, pressure ~500 bar) produced during the acoustic cavitation results in the fragmentation of the vapor/gases (water vapor, dissolved gases like O_2, N_2, etc.) trapped in the bubble into highly reactive $OH^{\bullet}$. When the dye molecules in wastewater interact with the $OH^{\bullet}$ species, it gets degraded or decolorized through chemical reactions. This effect of ultrasound in the formation of free radicals is well illustrated by Eqs. (1)–(4) [10]. Formation of other radicals ($O^{\bullet}$, $N^{\bullet}$, $HO_2^{\bullet}$, etc.) also contribute to the dye degradation but at much lower rates than that of the $OH^{\bullet}$, due to the high oxidizing power of $OH^{\bullet}$ over other radicals.

$$H_2O \xrightarrow{\text{ultrasound}} H^{\bullet} + OH^{\bullet} \tag{1}$$

$$O_2 \xrightarrow{\text{ultrasound}} 2O^{\bullet} \tag{2}$$

$$N_2 \xrightarrow{\text{ultrasound}} 2N^{\bullet} \tag{3}$$

$$H^{\bullet} + O_2 \overset{\text{ultrasound}}{\rightarrow} HO_2 \quad (4)$$

Generally, the decolorization/degradation of dyes or any organic pollutants under the effect of sonochemical treatment occurs in three reaction zones (i) interior of the bubble or cavity, (ii) bubble–liquid interface, and (iii) bulk liquid. In the first zone i.e., the interior of the cavitation bubble, degradation of the volatile components and hydrophobic molecules occurs in this zone either by pyrolysis or in reaction with $OH^{\bullet}$ formed inside the bubble. The temperature and pressure in this zone are the highest lying in the range of 5000–10,000 K and 1000 atm respectively [13]. The reactions in the second zone are at the bubble–liquid interface and take place between the target compounds (the dye molecules) present in the bulk liquid and $OH^{\bullet}$ produced after the implosive collapse of cavitation bubble. The collapse of cavitation bubble renders various physical effects on the degradation process via generating strong micro-convections and shock waves into the liquid medium. This increases the reaction rate between the target pollutants and $OH^{\bullet}$ via increasing the mass transfer rate at the bubble–liquid interfaces via energetic collision into the medium. The temperature and pressure reached in this zone are ~2000 K and 1.0 atm respectively and is less than the temperature inside the bubble due to the minimum or the absence of pyrolysis effect at interfaces [13]. The third and final zone of the degradation of target compounds occur in the bulk liquid. In this zone, the reaction occurs in the presence of mostly $OH^{\bullet}$ that diffuses out from the bubble¯liquid interfaces after the bubble collapse. This available $OH^{\bullet}$ reacts with the pollutants present in the bulk liquid and degrades the pollutants along with formation of various intermediates. A schematic of all three reaction zones with corresponding chemical reactions are shown in Fig. 2. Also, the

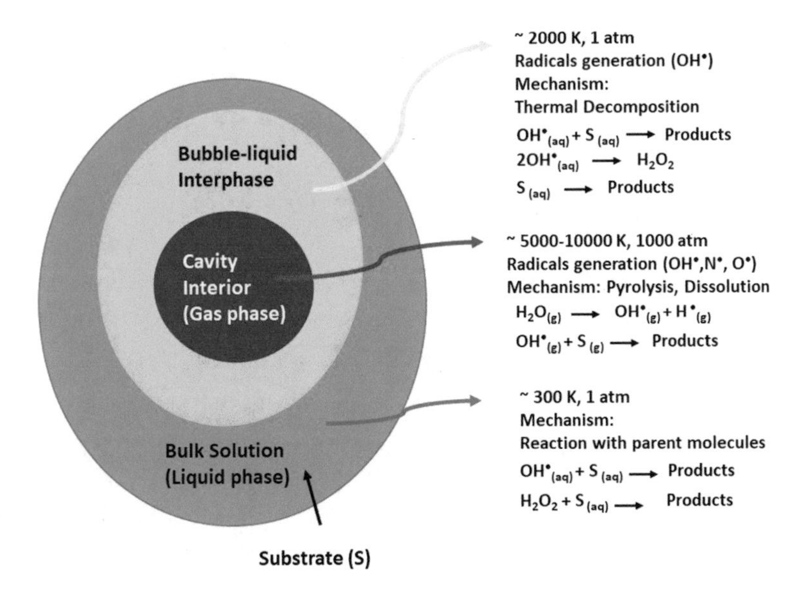

Fig. 2 Reaction zones involved in sonochemical degradation of pollutants present in wastewater

detailed reaction mechanism involved for degradation of dyes and/or other emerging pollutants present into the wastewater are expressed in Eqs. (5)–(13) [13, 25]. The superoxide radicals ($O_2^{\bullet-}$) formed in Eq. (10) also contribute to the degradation but a much smaller proportion than the $OH^{\bullet}$. However, the $O_3^{\bullet-}$ formed from Eq. (11) is highly unstable in acidic pH and quickly converts into $OH^{\bullet}$ therefore speeding up the degradation mechanism. The large values of the reaction rate constant (k) in almost all reactions mentioned in Eqs. (5)–(13) confirms the formation of active radicals ($OH^{\bullet}$, $O_2^{\bullet}$, $HO_2^{\bullet}$, etc.) under the influence of ultrasound [10, 13]. However, in a few cases, the concentration of in-situ generated hydroxyl radicals ($OH^{\bullet}$) are high and it recombines or is quenched with interaction of other $OH^{\bullet}$ present in the medium, and this leads to reduction in the degradation efficiency. To avoid this undesirable side effect and to gain high selectivity of $OH^{\bullet}$, the degradation process is always recommended to be performed under optimum sonication process parameters for maximum degradation efficiency of target pollutants like dyes or any other emerging contaminants present in wastewater.

$$HO^{\bullet} + HO^{\bullet} \rightarrow H_2O_2 \quad k = 5\text{ - }8 \times 10^9\ M^{-1}s^{-1} \tag{5}$$

$$H_2O_2 + HO^{\bullet} \rightarrow HO_2^{\bullet} + H_2O_2 \quad k = 2.7 \times 10^7\ M^{-1}s^{-1} \tag{6}$$

$$HO_2^{\bullet} + HO_2^{\bullet} \rightarrow H_2O_2 + O_2 \quad k = 8.3 \times 10^5\ \text{-}\ 2.6 \times 10^7\ M^{-1}s^{-1} \tag{7}$$

$$HO_2^{-} + HO^{\bullet} \rightarrow HO_2^{\bullet} + OH^{-} \quad k = 7.5 \times 10^9\ M^{-1}s^{-1} \tag{8}$$

$$H_2O_2 + HO^{\bullet} \rightarrow HO_2^{\bullet} + H_2O \quad k = 2.7 \times 10^7\ M^{-1}s^{-1} \tag{9}$$

$$HO_2^{\bullet} \rightleftarrows H^{+} + O_2^{\bullet-} \quad pK_a = 4.8 \tag{10}$$

$$O_3 + O_2^{\bullet-} \rightarrow O_3^{\bullet-} + O_2 \quad k = 1.6 \times 10^6\ M^{-1}s^{-1} \tag{11}$$

$$O_3^{\bullet-} + H^{+} \rightarrow HO^{\bullet} + O_2 \quad k = 5.2 \times 10^2\ M^{-1}s^{-1} \tag{12}$$

$$O_2^{\bullet-} + H_2O \rightleftarrows HO^{\bullet} + OH^{-} \quad k = 1.0 \times 10^8\ s^{-1} \tag{13}$$

3.2 *Effect of Process Parameters on Dye Decolorization Under Sonochemical Treatment System*

The dye decolorization efficiency obtained in an ultrasound-assisted system depends closely on the various operating process parameters such as applied frequency, ultrasound intensity, process temperature, initial dyes concentration, solution pH, and dissolved gas ratio. To realize a low-cost and energy-efficient dye decolorization with maximum decolorization efficiency, the detailed information and effects of individual operating parameters on the final outputs are of utmost importance. Hence a brief discussion of all these process parameters is discussed in the subsequent sections.

3.2.1 Effect of Ultrasound Frequency

Ultrasound frequency refers to the repetition of acoustic cycles (in the form of compression and rarefaction) per unit time. Generally, ultrasound frequencies used in the degradation of the dye vary from 20 to 10,000 kHz and can be classified as a low (20–100 kHz) and high frequency (5,000–10,000 kHz) ultrasound wave. Variation in ultrasound frequency changes the compression and rarefaction time cycle and corresponding its cavitation bubble growth. For example, as the ultrasound frequency increases, the time duration of rarefaction cycles decreases; it inhibits the bubble growth rate as a result, formation of smaller bubble sizes occurs during the maximum half of the rarefaction cycle. Further at the peak of the compression cycle, not all of these bubbles collapse due to limited available time at high frequencies and stay for many acoustic cycles. At maximum compression, these bubbles collapse with less intensity which results in reduction of $OH^{\bullet}$ and micro convection intensity [37, 41, 45]. The effect of variation in ultrasound frequency (585, 860, and 1140 kHz) for degradation of naphthol blue-black (NBB), an acidic diazo dye, has been reported by [20]. The study showed that the degradation of NBB is minimum at a higher frequency (1140 kHz) as compared to the maximum degradation achieved at low ultrasound frequency (585 kHz) (Fig. 3). The numerical investigation of single bubble dynamics with variation in ultrasound frequency confirms that at low ultrasound frequencies, the bubble shows a very high expansion and compression ratio (R_{max}/R_o) during their consecutive rarefaction and compression cycle and stay for a longer duration. The collapse of these large size bubbles generates intense energy with a large amount of $OH^{\bullet}$ and accelerates the degradation of targeting compounds [39, 40]. Also at low ultrasound frequencies, the amount of in-situ generated H_2O_2 along with $OH^{\bullet}$ is higher than that generated at high frequencies and hence, the decolorization efficiency is improved several-fold [9, 20]. The increase in dye decolorization efficiency with a decrease in ultrasound frequency is not always true and depends closely on the availability of the actual amount of active $OH^{\bullet}$ without the involvement of any quenching effect [8].

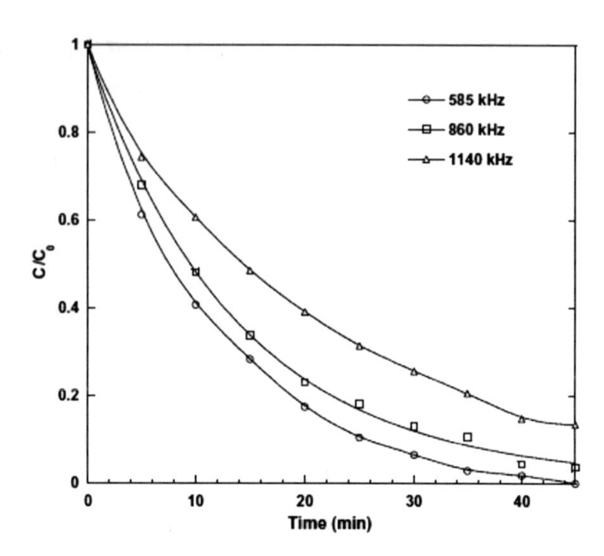

Fig. 3 Effect of ultrasound frequency on sonochemical degradation of naphthol blue black (NBB) an acidic diazo dye in water (Reprinted from Ultrasonics sonochemistry, 26, Ferkous H; Oualid H and Slimane M, Sonochemical degradation of naphthol blue black in water: Effect of operating parameters, 40–47, Copyright (2015), with permission from Elsevier)

3.2.2 Effect of Ultrasound Intensity

The intensity (I) of the ultrasound wave is directly proportional to the square of the acoustic pressure ($I = P^2{}_A/2\rho c$), where P_A is the acoustic pressure amplitude, ρ is the liquid density and c is the sound velocity into the liquid medium. With the increase in ultrasound intensity, the magnitude of P_A decreases during the rarefaction cycle of ultrasound wave and is minimum during the first half of the rarefaction cycle ($-P_{a\ min}$). At constant frequency, with further increase in intensity, bubble grows faster and attain maximum size at the peak of the rarefaction cycle ($-P_{a\ min}$). Further, the collapse of this large bubble occurs in the consequent compression cycle with the generation of intense energy with a greater amount of $OH^{\bullet}$ which accelerates the degradation rate of targeting compounds [5, 46]. An increase in acoustic energy also improves the transmittance of sonication energy into the reacting system followed by enhancement in the concentration of $OH^{\bullet}$ and intensification in dye degradation [31]. However, the intensity (P_a) irradiated above an optimum value cannot bring any significant changes in the degradation efficiency and therefore, it is always emphasized to find an optimum ultrasound intensity (I) at which maximum decolorization of dye molecules or any targeting pollutants could be obtained [28].

3.2.3 Effect of Dissolved Gas Concentration

Dissolved gases are present in the liquid medium in the form of either atmospheric air (O_2, N_2) or externally supplied gases (Ar, He, Ne, CO_2, etc.) and provide additional nucleation cites (weak spots) required for the cavitation bubble formation. Presence of dissolved gases reduces the magnitude of cavitation threshold i.e., the minimum energy needed to form the cavity in the liquid medium. At a low cavitation threshold, the sum of the liquid–vapor pressure (P_v) and dissolved gas pressure (P_g) overcome the tensile forces of liquid molecules and leads to the cavitation phenomenon. In presence of dissolved gases, the bubbles are formed even at low ultrasound intensity and also provide the cushioning effect to the bubble. This can further reduce the bubble collapse frequency and reduces the energy transfer into the liquid medium [38]. The intensity of the cavitation threshold varies with the types of gases as dissolved into the liquid medium during sonication [32, 39]. Gases with high polytropic index ($\gamma \sim 1.66$ for Ar, He, Ne) have more cavitation effect than the gases with low γ $\gamma \sim 1.4$ for N_2 and O_2 [49].

3.2.4 Effect of Solution pH

Solution pH plays a vital role in the degradation of dye or any pollutants due to changes in the solution ionization behavior. pH value governs the hydrophilic and hydrophobic characteristics of the targeting pollutants. In acidic pH and ultrasound-assisted systems, the pollutants with hydrophilic characters undergo protonation and improve their hydrophobicity. Further, these hydrophobic pollutants transport towards the bubble–liquid interface and/or into the bubble core (showing high hydrophobic characteristics) and degrade at a much faster rate under the influence of pyrolysis and $OH^•$, formed immediately after the collapse of cavitation bubbles. However, the hydrophilic molecules present far away from the bubble⁻liquid interface react with possible available $OH^•$, available in bulk liquid, and show a slower degradation rate. On the other hand, in alkaline pH, the solubility of the pollutants increases and exhibit a slower degradation rate as compared to acidic pH [30, 51]. Vajnhandl and Marechal [58] used the ultrasound method in the degradation of azo dye reactive black (RB) 5 in a pH range of 2–10. Figure 4 shows that the highest decolorization of RB 5 dye is at pH 2. This is attributed to the increase in protonation of negatively charged SO_3^- and strong hydrophobic characteristics of RB 5 dye in an acidic environment.

3.2.5 Effect of Temperature

Under an ultrasound-assisted system, the degradation efficiency of dyes or any organic pollutants shows significant variation with the reaction temperature. According to Arrhenius theory, the degradation rates of an irreversible chemical reaction always increase with the increase in reaction temperature but this theory is not completely true in case of ultrasound-assisted systems. In case of ultrasound, the rate of reaction initially increases with an increase in temperature and thereafter decreases with further rise in temperature. This phenomenon is attributed to the decrease in the cavitation intensity and bubble temperature (T_{max}) at high temperatures. In addition, various side reactions and intermediates formed at high temperatures also cause a

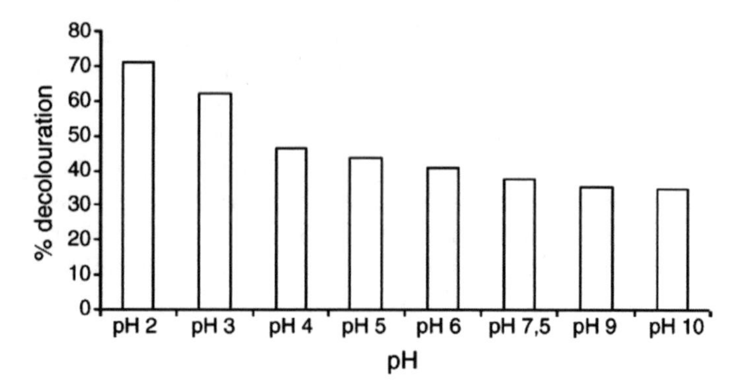

Fig. 4 Effect of pH on decolorization efficiency of azo dye reactive black (RB) 5 after 1 h of sonication (Reprinted from J. Hazardous Materials, 141, Vajnhandl S; Marechal AM, Case study of the sonochemical decolouration of textile azo dye Reactive Black 5, 329–335, Copyright (2007), with permission from Elsevier)

reduction in decolorization efficiency [19, 44]. Therefore, in case of sonication, an optimum temperature needs to be found for achieving the highest degradation rate.

3.2.6 Effect of Substrates or Initial Dye Concentration

It is well known that degradation of various organic pollutants treated under sonochemical effect mostly follows pseudo-first order or zero order reaction kinetics [26]. At constant ultrasound frequency and input power supply, the degradation efficiency decreases with an increase in initial dye concentration [20]. At high initial concentration, the competition between the intermediates and dyes molecules for the maximum utilization of $OH^{\bullet}$ produced from cavitation of the bubble is very high, as a result of which there is a sharp reduction in the degradation rate. Other than the concentration of the initial dye, the degradation kinetics also depends on the amount of $OH^{\bullet}$ radicals generated in the system and its reaction with the dye molecules [42].

3.3 Synergetic Effect of Sonochemical Method Coupled with Other Technologies in Dye Degradation

Although the sonochemical approach used in dye decolorization or degradation is quite fast and efficient but complete decolorization of dye molecules is not achieved in most cases. The possible reasons could be due to the limited availability of $OH^{\bullet}$ and its reaction with dye molecules and/or inefficient supply of ultrasound power into the system. To overcome these issues, the sonochemical technique is often coupled with other treatment processes such as (i) ultrasound and ozone (US/O_3), (ii) sonophotolysis, (iii) sono-photo-ferrioxalate hybrid AOP system, (iv) sono-enzymatic process, and so on [24, 54]. Among the coupled processes listed above, the sono-enzymatic is more popular and efficient due to the enzyme's eco-friendly nature, high decolorization/degradability even at low enzyme amount, faster degradation of intermediates, and increase in mass transfer between enzymes and substrates under the effect of sonication. The discussion of the sono-enzymatic process for the dye decolorization with emphasis on the influence of the important operational parameters is discussed in the next section.

4 Dye Decolorization Using Sono-Enzymatic Treatment Process

The sono-enzymatic process refers to the use of both ultrasound and enzymes for treatment of the wastewater. This approach involves using ultrasound throughout the enzymatic reaction. The use of ultrasound is very popular in enzyme-catalyzed biological manufacturing processes. The combination of sonication and enzymatic treatment is reported to give enhanced effects as compared to individual techniques with the added advantage of being an eco-friendly approach [16, 18, 54]. Ultrasound has been known to enhance the enzymatic reactions by acting on different targets. It can alter both substrate and enzyme macromolecules, which can be helpful for accelerating enzyme activity and product yield [59]. The sono-enzymatic method has been used for treatment of dyes in wastewater from textile industries and pharmaceutical recalcitrant compounds. Enzymes that utilize H_2O_2 as an electron acceptor are generally used in the sono-enzymatic process. Some of the enzymes commonly used for decolorization through the sono-enzymatic route are peroxidase enzymes such as horseradish peroxidase and laccase. To overcome the limitations such as stability, enzyme recovery and difficulty in product separation; enzyme immobilization approaches have been tried. An example is Horseradish peroxidase immobilized in Polyethylene glycol [35]. On the other hand, enzyme inactivation can occur by ultrasound due to bubble collapse, increase in local pressure and temperature. Under these extreme conditions, rupture of hydrogen bonds and van der Waal's interaction can change the active conformation of the enzyme. However, for low ultrasound frequencies and atmospheric static pressure, sono-enzymatic treatment appears to be more effective than the individual techniques.

4.1 Sono-Enzymatic Mechanism in Dye Decolorization

The interaction between the generated waves and the enzyme is purely physical in nature. The enhanced degradation is due to the intense convection created in the bulk liquid medium which in turn improves the interactions between the dye and enzyme molecules. The enzyme's oxidation potential is utilized for the degradation of the dyes and this is influenced by the convection levels present. The cavitation phenomena is reported to contribute by itself to the degradation of dyes but is often antagonistic to the enzyme action. The shock waves produced by the cavitation bubbles can alter and denature the molecular arrangement in the enzyme which leads to enzyme deactivation. Therefore, immobilization of enzymes can help to prevent deactivation of enzymes by phenoxy radicals formed from organic pollutants and also prevents attachment of intermediate products during the course of decolorization [43]. Sometimes, immobilized enzymes show less activity than the free enzyme because of localization of higher concentration of enzyme in immobilized form, while the free enzyme will be homogeneously mixed in bulk medium. But the net decolorization of the dye at the end of the treatment is higher which is attributed to the favorable environment provided by the immobilization support to the enzyme [35]. Polyethylene glycol and polyvinyl alcohol can be used to immobilize horseradish peroxidase and laccase enzyme respectively. Azo dye such as acid red is readily degraded by both enzyme and ultrasound because of the nature of N=N group. The enzyme converts the dye into cation radical that is susceptible to nucleophilic attack by H_2O_2 or water molecules [23]. High pressure can help in diminishing the cavitation effect and use of high pressure in the sono-enzymatic system results in higher degradation rates. Intense convection and fast movement of the dye and enzyme molecule facilitate the increase in probability of interaction [43]. The enthalpy change is negative for both sonochemical and sono-enzymatic (at atmospheric static pressure) processes due to the high energy and spontaneous reaction of the radicals. On the other hand, enzymatic and sono-enzymatic (at elevated static pressure) have positive enthalpy values [36].

4.2 Factors Affecting Dye Decolorization by Sono-Enzymatic Process

The activity of enzyme and the efficiency of the sono-enzymatic process is majorly governed by certain parameters such as enzyme concentration, substrate concentration, pH, temperature, ultrasound frequency and ultrasound intensity. To get energy efficient and maximum decolorization rates through this route; optimization of these parameters is necessary. Hence a brief discussion of all these process parameters is discussed in the subsequent section.

4.2.1 Enzyme Types Used in Dye Decolorization

Peroxidase and laccase are the enzymes majorly used for sono-enzymatic decolorization of dyes [35, 43]. Peroxidases are made up of hemoproteins which act as a catalyst for the reactions in the presence of H_2O_2. Peroxidases have three substrate binding sites which are denoted by heme d and c edges and an exposed tryptophan residue [22]. Laccases belong to the multicopper oxidase protein family or copper containing polyphenol oxidases family. Laccases have been widely used in dye treatment processes because of their non-specific oxidation capacity, discretional requirement of co-factors. Also, they do not need the available oxygen as an electron acceptor [55].

4.2.2 Effect of Enzyme and Substrate Concentration

Enzyme concentration plays a major role in determining the effects on its activity by sonication. At low enzyme concentrations, random distribution of enzyme molecules takes place which results in ineffective interaction with micro-streams produced due to continuous acoustic cavitation. Energy transfer process can be hindered if excess enzyme beyond optimum concentration is available. Beyond optimum concentration of enzyme, aggregates will be formed due to acoustic cavitation [29]. Substrate concentration is also an important parameter that has to be governed upon. In the case of horseradish peroxidase, hydrogen peroxide acts as a substrate. At low concentrations the turnover rate is low whereas in high concentration, the enzyme is deactivated due to highly reactive hydrogen peroxide substrate. Therefore, optimal substrate concentration is necessary for balancing the enzyme stability and its performance with substrate loading [15].

4.2.3 Effect of Temperature and pH

The immediate or micro environment around the enzyme molecules majorly impacts the enzyme activity. Therefore, pH and temperature play an important role in stabilizing the enzyme activity. Optimal temperature is required to break the strong interactions such as van der Waals force, dipole interactions and hydrogen bonding. Effect of temperature study on decolorization of acid red dye by horseradish peroxidase showed that extent of decolorization increases marginally with increase in temperature [35]. Higher temperatures may denature the enzyme. Regarding the effect of pH, ultrasound disrupts the weak interactions and causes conformational changes in the protein structures of the enzyme. These conformational changes results in an altered activity of the enzyme. The effect of pH on decolorization of acid red dye study showed that optimum pH is required to obtain maximum decolorization efficiency [35].

4.2.4 Effect of Ultrasonic Intensity

Ultrasonic intensity is an important parameter to get maximum decolorization rates. As noted earlier, to reduce intense cavitation and for stable cavitation, low frequency and mild intensity are required. High intensity ultrasound shows detrimental effects on the enzyme activity as significant shear is caused in the medium [56]. Secondary active conformation of the enzyme can be modified if there is sudden increase in the localized pressure and temperature at high intensity [6]. Studies on effect of ultrasound intensity on decolorization of azo dyes show increase in power from 60 to 120 W results in enzyme inactivation and decolorization occurs solely due to sonolysis. However higher energy levels result in higher degradation but enzyme inactivation can be compensated by higher enzyme addition [48].

4.2.5 Effect of Ultrasound Frequency

The ultrasound energy absorbed by the enzyme molecules is a function of ultrasound wavelength. Also, the stability of the enzyme is affected resulting in change of the catalytic activity [57]. High ultrasound frequencies generate excessive amount of heat caused by the large number of bubble collapse in the reaction medium. Decrease in half life time of enzyme is reported when frequency of 80–500 kHz is used in decolorization of indigo carmine dye through sono-enzymatic route [7].

5 Comparative Evaluation of Sonochemical Versus Sono-Enzymatic Treatment Process for Decolorization of Dyes

Sonolysis by itself has the better effect on the reaction, whereas cavitation does not favour the enzyme action. The shock waves produced by the cavitation bubble can denature the enzyme molecules and reduce their activity [43]. Sono-enzymatic route requires exact optimum conditions to be maintained throughout the process. If the operating parameters is not maintained at optimal conditions, there is a higher chance of enzyme inactivation. However, many sonochemical processes still show degradation even if the operating parameters are not maintained around the optimal conditions. Complete decolorization of dye molecules is not achieved in most of the cases through the sonochemical approach; the sono-enzymatic route can overcome this limitation because of its specificity, the enzymes being able to facilitate the oxidation of an array of substrates with hydrogen peroxide as an electron acceptor and also, being an environmentally friendly process [43]. The activation energy required for the sono-enzymatic process is lesser than that for the enzymatic part alone due to the spontaneous reactions prompted by the radicals produced due to the transient cavitation phenomena. The frequency factor in the kinetics of sono-enzymatic treatment is greater than that for sonolysis due to the volumetrically uniform enzymatic reactions [12]. Coming to the cost analysis, the cost of sono-enzymatic process is more than that of the sonochemical process as there is a high chance of enzyme inactivation throughout the process. But several additives such as polyethylene glycol and polyvinyl alcohol can be used for immobilization of low purity enzymes thus increasing their stability and preventing inactivation of enzymes. In a preliminary cost analysis, it has been shown that use of polyethylene glycol contributes to a significant reduction in the process cost due to reduced enzyme requirements. However, it needs to be noted that the lowest cost estimate for the enzyme-catalyzed process is much higher than the Fenton process, for example [18].

6 Current Challenges and Future Perspectives

The key challenge in the application of sonochemical and sonoenzymatic processes for the treatment of dye wastewaters is the scaleup of these technologies to industrial scale. So far, the studies have been confined to the laboratory at very small reactor volumes generally on the order of 250 ml to 3 L. Effective solutions are sorely required at industrial scale because of rapid industrialization and a burgeoning population putting tremendous strain on rapidly declining fresh water reserves. More studies are required to establish the effectiveness of such technologies at slightly larger volumes (on the order of ~10 L), then pilot plant and finally industrial scale. A constraint in such studies may be posed by the currently available configurations of the ultrasonics technology. Advances in electronics miniaturization may help to reduce the size of the components in this technology thus enabling the design of more localized technology. Another factor could be the use of 3D printing to redesign the ultrasonics equipment. It needs to be noted that the factors listed above could also help in reducing the power required per unit volume of solution to be treated. This (power requirement) is still a key constraint and ultimately impacts the cost of scaleup.

Since ultrasound does not interact with the material at a molecular level, the use of different spatial configurations or the use of multiple probes in the ultrasonics equipment can be explored to realize more efficient treatment. Also, the use of loop reactors can be considered to improve the overall degradation performance in the system. Additionally, the degradation obtained in sonochemical or sono-enyzmatic treatment processes should be critically appraised by evaluating the chemical oxygen demand (COD) or total organic carbon content (TOC) reduction and delineation of the degradation end-products using appropriate analytical techniques such as liquid chromatography coupled with mass spectrometry etc.

Finally, since it is noted that the sono-enzymatic process is an example of the so-called *hybrid* process, other such novel combinations should be investigated for evaluating their potential in dye wastewater treatment. For example, the use of ultrasound and microwave can be considered [14]. Another example is the use of ultrasound along with other advanced oxidation technologies such as Fenton [21], photocatalysis etc.

7 Conclusions

In the preceding sections, mechanistic investigation, as well as the influence of operating parameters of the sonochemical and sonoenzymatic techniques for decolorization and/or degradation of dye wastewater have been discussed. The study revealed that both the sonochemical and sonoenzymatic processes require optimum ultrasonic (operating frequency and power) and experimental (temperature, pH, dissolved gas amount, initial substrate amount, enzyme amount) conditions for maximum dye degradation. The study revealed that the sonochemical process is mainly governed by pyrolysis and free radical reactions. The optimization studies in the sonochemical process attributed that low ultrasound frequency, low substrate amount and acidic pH are favorable conditions for maximum degradation. It is also confirmed that the initial rate for dye degradation using the sonochemical route is high but requires high energy for the complete degradation of dye molecules. On the other side, the sonoenzymatic process is fast, eco-friendly and capable to degrade the dye compounds completely. The simultaneous catalytic action of enzyme and hydrogen peroxide produced from sonochemical reaction degrade and/or decolorize the dyes compounds either by precipitation or conversion into other harmless products. The study also revealed that the shock waves generated by cavitation during the sonoenzymatic process caused a reduction in enzyme activity followed by a lowering in degradation rate. An optimum process condition with additives can control the enzyme deactivation and increased dye degradation. We believe that the above discussion in dye decolorization and degradation of wastewater using sonochemical and sonoenzymatic approach will provide an insight into the intricacies of both the process and will give important inputs for further research in this area.

Acknowledgements One of the authors (PP) thanks the ministry of Education, Government of India, for the scholarship.

References

1. Abbasi M, Asl NR (2008) Sonochemical degradation of basic blue 41 dye assisted by nano TiO_2 and H_2O_2. J Haz Mater 153:942–947. https://doi.org/10.1016/j.jhazmat.2007.09.045
2. Abou-Okeil A, El-Shafie A, El Zawahry MM (2010) Ecofriendly laccase–hydrogen peroxide/ultrasound-assisted bleaching of linen fabrics and its influence on dyeing efficiency. Ultrason Sonochem 17(2):383–390. https://doi.org/10.1016/j.ultsonch.2009.08.007
3. Adewuyi YG (2001) Sonochemistry: environmental science and engineering applications. Ind Eng Chem Res 40:4681–4715. https://doi.org/10.1021/ie010096l
4. Andreozzi R, Caprio V, Insola A, Marotta R (1999) Advanced oxidation processes (AOP) for water purification and recovery. Catal Today 53(1):51–59. https://doi.org/10.1016/S0920-5861(99)00102-9
5. Asaithambi P, Sajjadi B, Aziz ARA, Daud WMABW (2017) Ozone (O_3) and sono (US) based advanced oxidation processes for the removal of color, COD and determination of electrical energy from landfill leachate. Separ Purif Technol 172:442–449. https://doi.org/10.1016/j.seppur.2016.08.041

6. Basto C, Silva CJ, Gübitz G, Cavaco-Paulo A (2007) Stability and decolourization ability of trametes villosa laccase in liquid ultrasonic fields. Ultrason Sonochem 14(3):355–362. https://doi.org/10.1016/j.ultsonch.2006.07.005
7. Basto C, Tzanov T, Cavaco-paulo A (2007) Combined ultrasound-laccase assisted bleaching of cotton. Ultrason Sonochem 14:350–354. https://doi.org/10.1016/j.ultsonch.2006.07.006
8. Basturk E, Karatas M (2014) Advanced oxidation of reactive blue 181 solution: a comparison between fenton and sono-fenton process. Ultrason Sonochem 21:1881–1885. https://doi.org/10.1016/j.ultsonch.2014.03.026
9. Beckett MA, Hua I (2001) Impact of ultrasonic frequency on aqueous sonoluminescence and sonochemistry. J Phys Chem A 105:3796–3802. https://doi.org/10.1021/jp003226x
10. Camargo-Perea AL, Rubio-Clemente A, Peñuela GA (2020) Use of ultrasound as an advanced oxidation process for the degradation of emerging pollutants in water. Water 12(4):1–23. https://doi.org/10.3390/w12041068
11. Chakma S, Moholkar VS (2015) Sonochemical synthesis of mesoporous $ZrFe_2O_5$ and its application for degradation of recalcitrant pollutants. RSC Adv 5:53529–55354. https://doi.org/10.1039/C5RA06148B
12. Chakma S, Moholkar VS (2016) Investigations in sono-enzymatic degradation of ibuprofen. Ultrason Sonochem 29:485–494. https://doi.org/10.1016/j.ultsonch.2015.11.002
13. Chauhan R, Dinesh GK, Alawa B, Chakma S (2021) A critical analysis of sono-hybrid advanced oxidation process of ferrioxalate system for degradation of recalcitrant pollutants. Chemosphere 277:130324. https://doi.org/10.1016/j.chemosphere.2021.130324
14. Cintas P, Tagliapietra S, Caporaso M, Tabasso S, Cravotto G (2015) Enabling technologies built on a sonochemical platform: challenges and opportunities. Ultrason Sonochem 25:8–16. https://doi.org/10.1016/j.ultsonch.2014.12.004
15. Danielson AP, Van-Kuren DB, Bornstein JP, Kozuszek CT, Berberich JA, Page RC et al (2018) Investigating the mechanism of horseradish peroxidase as a RAFT-initiase. Polymers 10(7):741–754. https://doi.org/10.3390/polym10070741
16. Delgado-Povedano MM, De Castro ML (2015) A review on enzyme and ultrasound: a controversial but fruitful relationship. Anal Chim Acta 889:1–21. https://doi.org/10.1016/j.aca.2015.05.004
17. Dükkancı M, Vinatoru M, Mason TJ (2012) Sonochemical treatment of orange II using ultrasound at a range of frequencies and powers. J Adv Oxid Technol 15:277–283. https://doi.org/10.1515/jaots-2012-0205
18. Entezari MH, Pétrier C (2003) A combination of ultrasound and oxidative enzyme: sonobiodegradation of substituted phenols. Ultrason Sonochem 10:241–246. https://doi.org/10.1016/S1350-4177(03)00087-7
19. Farias J, Albizzati ED, Alfano OM (2009) Kinetic study of the photo-Fenton degradation of formic acid: combined effects of temperature and iron concentration. Catal Today 144(1–2):117–123. https://doi.org/10.1016/j.cattod.2008.12.027
20. Ferkous H, Hamdaoui O, Merouani S (2015) Sonochemical degradation of naphthol blue black in water: effect of operating parameters. Ultrason Sonochem 26:40–47. https://doi.org/10.1016/j.ultsonch.2015.03.013
21. Gujar SK, Gogate PR (2021) Application of hybrid oxidative processes based on cavitation for the treatment of commercial dye industry effluents. Ultrason Sonochem 75:105886. https://doi.org/10.1016/j.ultsonch.2021.105886
22. Gumiero A, Murphy EJ, Metcalfe CL, Moody PCE, Raven EL (2010) An analysis of substrate binding interactions in the heme peroxidase enzymes: a structural perspective. Arch Biochem Biophy 500(1):13–20. https://doi.org/10.1016/j.abb.2010.02.015
23. Goszczynski S, Paszczynski A, Pasti-Grigsby MB, Crawford RL, Crawford DL (1994) New pathway for degradation of sulfonated azo dyes by microbial peroxidases of phanerochaete chrysosporium and streptomyces chromofuscus. J Bacteriol 176(5):1339–1347. https://doi.org/10.1128/jb.176.5.1339-1347.1994
24. Gultekin I, Ince NH (2006) Degradation of aryl-azo-naphthol dyes by ultrasound, ozone and their combination: effect of α-substituents. Ultrason Sonochem 13:208–214. https://doi.org/10.1016/j.ultsonch.2005.03.002

25. Gurol MD, Akata A (1996) Kinetics of ozone photolysis in aqueous solution. AIChE J 42:3283–3292. https://doi.org/10.1002/aic.690421128
26. Gogate PR, Pandit AB (2004) A review of imperative technologies for wastewater treatment II: hybrid methods. Adv Environ Res 8:553–597. https://doi.org/10.1016/S1093-0191(03)00031-5
27. Hart EJ, Henglein A (1985) Free radical and free atom reactions in the sonolysis of aqueous iodide and formate solutions. J Phys Chem 89(20):4342–4347. https://doi.org/10.1021/j100266a038
28. Henglein A, Gutierrez M (1990) Chemical effects of continuous and pulsed ultrasound: a comparative study of polymer degradation and iodide oxidation. J Phys Chem 94:5169–5172. https://doi.org/10.1021/j100375a073
29. Jadhav SH, Gogate PR (2014) Ultrasound assisted enzymatic conversion of non edible oil to methyl esters. Ultrason Sonochem 21(4):1374–1381. https://doi.org/10.1016/j.ultsonch.2013.12.018
30. Jiang Y, Petrier C, Waite TD (2002) Effect of pH on the ultrasonic degradation of ionic aromatic compounds in aqueous solution. Ultrason Sonochem 9(3):163–168. https://doi.org/10.1016/S1350-4177(01)00114-6
31. Kanthale P, Ashokkumar M, Grieser F (2008) Sonoluminescence, sonochemistry (H_2O_2 yield) and bubble dynamics: frequency and power effects. Ultrason Sonochem 15:143–150. https://doi.org/10.1016/j.ultsonch.2007.03.003
32. Lee JH, Poddar MK, Yerriboina NP, Ryu HY, Han KM, Kim TG, Hamada S, Wada Y, Hiyama H, Park JG (2019) Ultrasound-induced break-in method for an incoming polyvinyl acetal (PVA) brush used during post-CMP cleaning process. Polym Testing 78:105962. https://doi.org/10.1016/j.polymertesting.2019.105962
33. Li J, Li M (2007) Decolorization of azo dye direct scarlet 4BS solution using exfoliated graphite under ultrasonic irradiation. Ultrason Sonochem 14:241–245. https://doi.org/10.1016/j.ultsonch.2006.04.005
34. Madhavan J, Selvam P, Kumar S, Anandan S, Grieser F, Ashokkumar M (2010) Degradation of acid red 88 by the combination of sonolysis and photocatalysis. Sep Purif Technol 74(3):336–341. https://doi.org/10.1016/j.seppur.2010.07.001
35. Malani RS, Khanna S, Moholkar VS (2013) Sonoenzymatic decolorization of an azo dye employing immobilized horse radish peroxidase (HRP): a mechanistic study. J Hazard Mat 256:90–97. https://doi.org/10.1016/j.jhazmat.2013.04.023
36. Malani RS, Khanna S, Chakma S, Moholkar VS (2014) Mechanistic insight into sono-enzymatic degradation of organic pollutants with kinetic and thermodynamic analysis. Ultrason Sonochem 21(4):1400–1406. https://doi.org/10.1016/j.ultsonch.2014.01.028
37. Mason TJ (1990) Sonochemistry: the uses of ultrasound in chemistry. Royal Soc Chem Cambridge, United Kingdom. https://doi.org/10.1002/ange.19911030749
38. Mason TJ, Lorimer JP (2002) Applied sonochemistry: uses of power ultrasound in chemistry and processing. Wiley VCH Verlag GmbH & Co. https://doi.org/10.1002/352760054X
39. Merouani S, Hamdaoui O, Rezgui Y, Guemini M (2015) Sensitivity of free radicals production in acoustically driven bubble to the ultrasonic frequency and nature of dissolved gases. Ultrason Sonochem 22:41–50. https://doi.org/10.1016/j.ultsonch.2014.07.011
40. Merouani S, Hamdaoui O, Rezgui Y, Guemini M (2015) A method for predicting the number of active bubbles in sonochemical reactors. Ultrason Sonochem 22:51–58. https://doi.org/10.1016/j.ultsonch.2014.07.015
41. Naddeo V, Belgiorno V, Ricco D, Kassinos D (2009) Degradation of diclofenac during sonolysis, ozonation and their simultaneous application. Ultrason Sonochem 16:790–794. https://doi.org/10.1016/j.ultsonch.2009.03.003
42. Okitsu K, Iwasaki K, Yobiko Y, Bandow H, Nishimura R, Maeda Y (2005) Sonochemical degradation of azo dyes in aqueous solution: a new heterogeneous kinetics model taking into account the local concentration OH radicals and azo dyes. Ultrason Sonochem 12:255–262. https://doi.org/10.1016/j.ultsonch.2004.01.038

43. Patidar R, Khanna S, Moholkar VS (2012) Physical features of ultrasound assisted enzymatic degradation of recalcitrant organic pollutants. Ultrason Sonochem 19(1):104–118. https://doi.org/10.1016/j.ultsonch.2011.06.005
44. Perez JS, Soriano-Molina P, Rivas G, Sanchez JG, Lopez JC, Sevilla JF (2017) Effect of temperature and photon absorption on the kinetics of micropollutant removal by solar photo-Fenton in raceway pond reactors. Chem Eng J 310:464–472. https://doi.org/10.1016/j.cej.2016.06.055
45. Pétrier C, Francony A (1997) Ultrasonic waste-water treatment: incidence of ultrasonic frequency on the rate of phenol and carbon tetrachloride degradation. Ultrason Sonochem 4:295–300. https://doi.org/10.1016/s1350-4177(97)00036-9
46. Rao Y, Yang H, Xue D, Guo Y, Qi F, Ma J (2016) Sonolytic and sonophotolytic degradation of carbamazepine: kinetic and mechanisms. Ultrason Sonochem 32:371–379. https://doi.org/10.1016/j.ultsonch.2016.04.005
47. Rayaroth MP, Aravind UK, Aravinda Kumar CT (2015) Sonochemical degradation of coomassie brilliant blue: effect of frequency, power density, pH and various additives. Chemosphere 119:848–855. https://doi.org/10.1016/j.chemosphere.2014.08.037
48. Rehorek A, Tauber M, Gubitz G (2004) Application of power ultrasound for azo dye degradation. Ultrason Sonochem 11:177–182. https://doi.org/10.1016/j.ultsonch.2004.01.030
49. Rooze J, Rebrov EV, Schouten JC, Keurentjes JTF (2013) Dissolved gas and ultrasonic cavitation-a review. Ultrason Sonochem 20(1):1–11. https://doi.org/10.1016/j.ultsonch.2012.04.013
50. Sayan E, Edecan ME (2008) An optimization study using response surface methods on the decolorization of reactive blue 19 from aqueous solution by ultrasound. Ultrason Sonochem 15:530–538. https://doi.org/10.1016/j.ultsonch.2007.07.009
51. Stock NL, Peller J, Vinodgopal K, Kamat PV (2000) Combinative sonolysis and photocatalysis for textile dye degradation. Environ Sci Technol 34(9):1747–1750. https://doi.org/10.1021/es991231c
52. Suslick KS (1989) The chemical effects of ultrasound. Sci Am 260(2):80–87
53. Srinivasan A, Viraraghavan T (2010) Decolorization of dyewastewaters by biosorbents: a review. J Environ Manage 10:1915–1929. https://doi.org/10.1016/j.jenvman.2010.05.003
54. Tauber MM, Guebitz GM, Rehorek A (2004) Degradation of azo dyes by laccase and ultrasound treatment. Appl Environ Microbiol 71(5):2600–2607. https://doi.org/10.1128/AEM.71.5.2600-2607.2005
55. Telke AA, Ghodake GS, Kalyani DC, Dhanve RS, Govindwar SP (2011) Biochemical characteristics of a textile dye degrading extracellular laccase from a Bacillus sp. ADR. Bioresour Technol 102(2):1752–1756. https://doi.org/10.1016/j.biortech.2010.08.086
56. Tian ZM, Wan MX, Wang SP, Kang JQ (2004) Effects of ultrasound and additives on the function and structure of trypsin. Ultrason Sonochem 11(6):399–404. https://doi.org/10.1016/j.ultsonch.2003.09.004
57. Tsikrika K, Chu BS, Bremner DH, Lemos MA (2018) The effect of different frequencies of ultrasound on the activity of horseradish peroxidase. LWT 89:591–595. https://doi.org/10.1016/j.lwt.2017.11.021
58. Vajnhandl S, Marechal AM (2007) Case study of the sonochemical decolouration of textile azo dye reactive black 5. J Haz Mater 141:329–335. https://doi.org/10.1016/j.jhazmat.2006.07.005
59. Wang D, Yan L, Ma X, Wang W, Zou M, Zhong J, Liu D (2018) Ultrasound promotes enzymatic reactions by acting on different targets: enzymes, substrates and enzymatic reaction systems. Int J Biol Macromol 119:453–461. https://doi.org/10.1016/j.ijbiomac.2018.07.133
60. Zhang Z, Zheng H (2009) Optimization for decolorization of azo dye acid green 20 by ultrasound and H_2O_2 using response surface methodology. J Haz Mater 172(2–3):1388–1393. https://doi.org/10.1016/j.jhazmat.2009.07.146
61. Zhou X, Guo W, Yang S, Ren N (2012) A rapid and low energy consumption method to decolorize the high concentration triphenylmethane dye wastewater: operational parameters optimization for the ultrasonic-assisted ozone oxidation process. Bioresource Technol 105:40–47. https://doi.org/10.1016/j.biortech.2011.11.089

Nanoceramic Based Composites for Removal of Dyes from Aqueous Stream

Saptarshi Roy and Md. Ahmaruzzaman

Abstract Water is an essential necessity for all living beings, but the world is suffering from a major crisis of clean and safe drinking water. Water pollution has become one of the most aggravated problems throughout the world threatening the sustainability of human race and other life forms due to the rapid pace of civilization and industrialization. A long history exists of discharge of hazardous dyes into the water bodies by selfish human activities since the Industrial Revolution, but no effort has been completely successful in curbing the activities that result in the degradation of our environment. These pollutants are harmful, carcinogenic and have adverse health effects to all forms of life. Thus, remarkable efforts have been geared up to obtain clean water by exploiting science and technology. Among the potential methods for the treatment of wastewater, adsorption by solid materials is worth mentionable. Its simplicity in operation and satisfactory efficiency in the elimination of contaminants make it a good candidate for adsorption. Ceramic based nanomaterials have attracted scientists and have been widely utilized in remediating dye laden waste water because of their sustainable nature, magnetically recyclability, stability and maximum pollutant binding capacities. This chapter provides an understanding of the different treatment methods and outlines the possible utility of nanoceramic based composite materials in the treatment of waste water contaminated with dyes. Summary of the global scenario and the recent developments of nanoceramics as adsorbents have been reviewed. It is expected that this book chapter provides a scope to the future researchers and opens new scientific avenues in the areas of dye removal.

Keywords Dyes · Hydroxyapatite · Wastewater · Nanoceramics · Adsorption

---

S. Roy · Md. Ahmaruzzaman (✉)
Department of Chemistry, National Institute of Technology, Silchar, Assam 788010, India
e-mail: mda2002@gmail.com

S. S. Muthu and A. Khadir (eds.), *Advanced Oxidation Processes in Dye-Containing Wastewater*, Sustainable Textiles: Production, Processing, Manufacturing & Chemistry, https://doi.org/10.1007/978-981-19-0882-8_10

1 Introduction

Rapid modernization and industrialization have resulted in drastic changes in the environment that is detrimental to both human and animal life forms. With the growth of industries at an alarming rate, a large number of wastes are being generated and are eventually released into the water bodies as industrial effluents. This has become a problem of global concern in the developing countries affecting the standard and the aesthetic of the water resources [1, 23, 50]. Presently, one of the urgent issues faced by the global community is the shortage of clean and pure drinking water. Among the various wastes emitted by the industries, dyes provide a major contribution among the discharged pollutants. They are widely utilized in the manufacture of cosmetics, pharmaceutical, paint, paper, textile, carpet, agro-based products and in the tannery industries [18, 36]. A total of 80% of the manufactured dyes are used by the fabric and textile-based industries as colorant compelling them to be the highest consumers of dyes. On an average, about 7.0×10^5 metric tons of dyes is produced annually worldwide with an estimated loss of 10–15% of the dye in the effluent during processing operations [7, 14, 15, 28, 42, 62, 65]. As per the Ministry of Textiles, India alone produces about 28.89 lakh kg of dye and is the second-largest exporter of dye after China [32, 33]. The inevitable use of chemicals in various industrial processes is essential for the growth of the economy and cannot be eliminated making them an utmost necessity for its safe use and disposal. Dyes are hazardous not only to human health but also cause irreversible damage to the aquatic and floral environments. They are mutagenic, allergic and are also identified as a carcinogen to all life forms [29, 40]. They find their way through the pores of the skin, by inhalation or by ingestion causing diseases like kidney problems, skin irritation, mutation or even cancer if exposed for a longer duration [2]. These dyes are highly toxic to microbial and mammalian populations. Thus, it is very much essential to remove these effluents before their discharge into the aqueous systems. Dyes are stable to light and completely resistant to biodegradable processes due to their complex chemical chromogen-chromosphere structure [48]. These colored effluents are water soluble and mixes with water inhibiting the growth of microorganisms and light transmittance [11, 51]. Their presence in the water bodies destroys the food source of aquatic organisms by blocking the photosynthetic activities of aquatic flora [31]. They are insusceptible to attack by moderate oxidative agents that are usually used in conventional treatment processes. These dyes and pigments possess acidic properties due to which remediation of waste water from colorant based industries is a major challenge for the research community. They are projected to be one of the most difficult classes of contaminants to be eliminated from the industrial wastewater.

Consequently, researchers have reported many treatment alternatives. Presently, many adsorbent materials have been exploited for eliminating dyes from pollutant laden textile wastewater without any great degree of success due to low selectivity and ineffective separation after the experiment. Further, in-depth advanced research has to be invested for easy recovery of the adsorbents so as to recycle and reuse them and also enhance their removal efficiency. Intensive search has been carried out for adsorbents that are stable, magnetically recyclable, possess high selectivity for pollutant binding and are also cost and eco-friendly. Ceramic based materials have emerged as prospective candidates in this regard. Their various attractive characteristics have drawn the fancy of the research community for their utilization in elimination of dyes.

2 Definition of Dyes

Dyes are colored chemical compounds comprising of chromophores and auxochromes. Chromophores are functional groups that are accountable for imparting color and auxochromes magnify the color intensity of the dyes [38]. Mauveine, an organic aniline dye, was the first dye discovered in 1856 by William Henry Perkin.

Dyes are the main raw materials in textiles, tannery, printing, food, cosmetics, and plastics and are considered the main culprits of colored wastewater. They are generally derived from natural sources or are synthesized. Plants such as indigo and saffron are main sources of dye. Moreover, insects such as beetles and lac scale insects, some species of shellfish and minerals like ferrous sulfate and ochre are other some sources. Synthetic dyes are mainly accountable for causing hazardous environmental problems because of their non-biodegradable complex chemical structure [49].

Dyes can be broadly classified into ionic and non-ionic dyes. Ionic dyes are further divided into cationic and anionic dyes. Dyes that give positively charged ions in the aqueous solution are termed as cationic dyes. Anionic dyes form negatively charged ions. Some examples of anionic dyes are acid dyes, reactive dyes, and direct dyes. Malachite green, Rhodamine B, and methylene blue belong to cationic dyes. Disperse dyes and vat dyes are grouped under non-ionic dyes.

3 Techniques of Dye Removal

The release of dyes into the water bodies is a serious emerging issue which may pose a significant hazard to our environment. Several conventional methodologies have been proposed for the remediation of dye effluents prior to their release into the aqueous systems. These techniques have been categorically examined under chemical, physical, and biological treatment as listed below.

Chemical treatment includes ozonation, Fenton's reagent, and photochemical processes whereas electro kinetic coagulation, ion exchange, adsorption, and membrane filtration are some of the physical methods that have been adopted. Anaerobic and aerobic degradation are the primary biological methods utilized in dye removal strategy (Fig. 1). The scientific community is therefore continuously

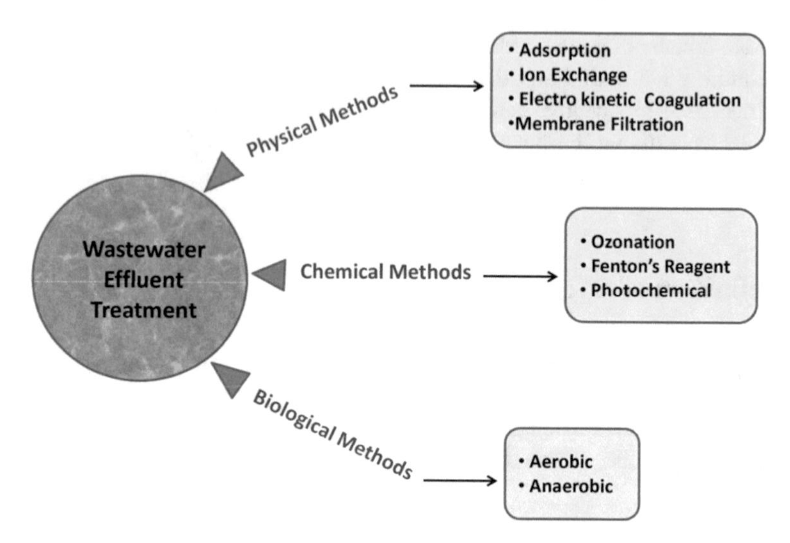

Fig. 1 Different methods of dye removal

trying to provide tangible technological solutions for the potential treatment of dye discharge for ensuing quality life of mankind.

3.1 Physical Methods

3.1.1 Adsorption

The strategy of adsorption has acquired enormous favor because of their high capability and feasibility in the removal of pollutants. It is governed by various factors namely interaction of the dye and sorbent, surface area of the adsorbent, size of the particle, pH, temperature, and contact time. Commonly used adsorbents are activated carbon, peat, fly ash, silica gel, ceramic materials, etc.

Activated carbon is the most frequently used sorbent for the elimination of dyes from effluents [24, 35]. Removal performance is affected by the nature of the carbon and the effluent wastewater. The rates of removal can be magnified by increasing the doses. Repeated regeneration and reuse brings down the performance and efficacy of the sorbent material. Activated carbon is compatible for its application in the removal of a particular pollutant and hence not usually reliable. Moreover, it is expensive and gives rise to problems of waste disposal.

Peat, another adsorbent material, is well suited for the removal of transition metals and polar compounds from wastewater contaminated with dye stuffs. It does not demand for activation, unlike activated carbon, and is also budget-friendly [43]. Moreover, it presents better adsorption capability due to its large surface area. Woodchips require longer contact time and are only effective for acid dyes [37]. Silica gel is used for removing basic dyes but is not commercially preferred because of problems of binding and fouling. Other materials namely natural clays, corn cobs, and rice husks are superior because of their extensive availability, cheapness, and do not require regeneration.

3.1.2 Ion Exchange

Ion exchange is an effective tool for the remediation of aqueous wastewater which aids to lower the magnitude of toxic pollutants for the purpose of its reuse. Another notable criterion of this technique is that it separates as well as concentrates the unwanted materials [52]. Ion exchange resins are usually present as beads and are highly ionic, insoluble, and covalently cross-linked. Macroporous resins provide large effective surface area and good exchange sites which facilitates the treatment process. While on the other hand, microporous resins allow the diffusion of solute ions through the particles for its interaction with the exchange sites. These resins are less fragile, requires less care while handling and possess high capacity and selectivity which makes them proper materials for dye sorption from textile effluents. Moreover, this methodology embraces no adsorbent damage and regain of the solvent after use.

Despite its enormous advantages, ion exchange technique has not been extensively employed for the remediation of wastewater considering that ion exchangers are unable to assist a broad variety of dyes. Besides, organic solvents are toxic and expensive thereby heightening the cost of the treatment process.

3.1.3 Electro Kinetic Coagulation

Electro kinetic coagulation is an economically feasible method for wastewater treatment and requires a direct current source between the metal electrodes submerged in water [4]. It administers a convenient, reliable, and budget friendly method compared to the conventional chemical coagulation techniques employed for the elimination of dyes from wastewater. Furthermore, no secondary pollution is generated. The major disadvantages associated are the requirement of secondary treatment by flocculation and filtration. Besides, the removal of high amount of sludge produced during the treatment process ultimately results in the increase of the cost [13].

3.1.4 Membrane Filtration

Membrane filtration technique has the capability to concentrate and eliminate dyes from the wastewater laden with contaminants [34, 64]. Characteristic features such as high temperature resistance, tolerance toward chemical environment and microbial attack are mentionable. But the problem of disposal of the concentrated residues leads to the possibility of membrane clogging and increase in the treatment cost. This method can be limited to recycling water in presence of minimum concentration of dyes. High pressure during the operation and accumulation of dyes and organic matters on the surface of the membrane result in membrane fouling.

3.2 Chemical Methods

3.2.1 Ozonation

The technique of ozonation originated in the early 1970s and involves the oxidation of chlorinated hydrocarbons, phenols, and other aromatic compounds with the help of a powerful oxidizing agent, ozone. The reaction mechanism primarily involves two steps.

Step 1: Ozone undergoes reaction selectively with the unsaturated bonds of dye molecules at pH 5–6.

Step 2: Ozone undergoes decomposition with the generation of hydroxyl radicals. The radicals react non-selectively with the organic compounds at pH above 8.

Ozonation has been successfully implemented in the elimination of dyes from textile effluents in industries. Chromophores are usually conjugated double bond groups present in the dyes that are broken down into smaller molecules reducing the coloration. The reactive dyes are found to be degraded to a high extent with ozone while the outcome obtained are moderate for the basic dyes and negligible in case of disperse dyes. The crucial advantage of this technique is the application of gaseous ozone with no increase in the water volume and no generation of sludge or hazardous metabolites making it a potential effective tool of dye removal. Nonetheless, its operational cost and continuous ozonation because of its short half-life are some of the hurdles associated with this process. An estimated removal efficiency of 98% has been found for Rhodamine-B dye with ozonization method [55].

3.2.2 H_2O_2-Fe (II) Salts (Fenton's Reagent)

Fenton's reagent is an effective Advanced Oxidation Process (AOP) of treating wastewaters for the removal of several dyes. It is a solution of hydrogen peroxide and ferrous ions as a catalyst which helps to oxidize the organic contaminants present in the effluent waters [45]. The decomposition of Fenton's reagent takes place in three consecutive steps which can be represented by the following reactions.

The Fe (III) ion undergoes complexation with hydrogen peroxide to form iron (III) peroxo complex. It then decomposes producing Fe (II) and finally oxidizes to yield the oxidant and hydroxyl radicals. These hydroxyl radicals are mainly responsible for the oxidative degradation of the compounds.

$$Fe^{3+} + H_2O_2 \rightleftharpoons Fe^{III}(HO_2)^{2+} + H^+ \tag{1}$$

$$Fe^{III}(HO_2)^{2+} \rightarrow Fe^{2+} + HO_2 \quad (2)$$

$$Fe^{2+} + H_2O_2\cdot \rightarrow Fe^{3+} + {}^{\bullet}OH + \overline{O}H \quad (3)$$

The dye eliminating capability of this process is influenced by the concentration of H_2O_2 and Fe^{2+} ions which helps to generate the oxidant and hydroxyl radicals for dye degradation. This method of oxidation completely degrades the pollutants into innocuous and non-toxic compounds such as CO_2, H_2O and salts of inorganic compounds. Additionally, this method is convenient to implement, completely reacts with the pollutants and inexpensive reagent makes it a cost-effective treatment. Optimum pH should also be maintained in between 3 and 5. Major drawbacks associated with this method are toxic reagents, generation of a large excess of ferric hydroxide sludge and the difficulties associated with its disposal restrict the practical utilization of this process in large industrial scale. This type of oxidative method was reported to be potentially effective for Malachite Green dye presenting a removal efficiency of 99% [17]. The removal efficiency was found to be 98.2% for Crystal violet dye [54].

3.2.3 Photochemical

The process of photochemical degradation breaks down the dye molecules by UV radiation in the presence of H_2O_2 [41]. UV light initiates the destruction of peroxide molecules resulting in the production of large concentrations of hydroxyl radicals which are ultimately responsible for the oxidation of the organic compounds.

$$H_2O_2 + h\upsilon \rightarrow 2OH^{\bullet}$$

The dye elimination rate is affected by the intensity of the UV radiation, pH, and the chemical design of the dye. This method greatly reduces the foul odors of the effluents and can be performed at ambient temperature and pressure, thus reducing the operational cost of the treatment. Moreover, it presents superiority as no sludge is generated. Photochemical degradation is employed in the effective removal of Methyl Orange dye, Reactive red, and direct green in wastewater treatment [5, 10, 58]. Besides, degradation of some photocatalyst generates toxic by-product which is a major drawback of this method.

Elimination of brightly colored and water-soluble dyes from the contaminated water bodies are troublesome as they are likely to bypass the traditional treatment techniques unaffected [60]. The above described methods have their own inherent limitations based on the cost, disposal, sludge production, and dye separation efficiency [3, 8, 56]. The biological treatment was not convenient for large fluxes of industrial treatments as the dye removal efficiency was found to be relatively low and depended on climate change and the reactor's size. On the other hand, the chemical treatment targeted coagulation of dyes with chemical reagents ending up generating

Table 1 Merits and demerits of the various separation techniques of dyes

Separation techniques	Advantages	Disadvantages
Physicochemical methods		
Adsorption	High adsorption capacity for dyes	Low surface area and high cost of adsorbents
Ion exchange	No loss of sorbents	Ineffective for disperse dyes
Electro kinetic coagulation	Economically feasible	High sludge generation and requires secondary treatments
Membrane filtration	Effective for all dyes with high quality effluents	Sludge production, membrane fouling and suitable for low volume treatment
Chemical methods		
Ozonation	No sludge generation	High operational cost and short half-life (~20 min)
Fenton's reagent	Low cost reagent and efficient method	Disposal problems and sludge generation
Photochemical	Low operational cost and economically feasible	Toxic by products
Biological methods		
Aerobic degradation	Use of by products as sources of energy	More treatment and formation of methane and hydrogen sulphide
Anaerobic degradation	Low operational cost	Provide suitable environment for growth of microorganisms and very slow process

a huge amount of sludge requiring a further remediation. Physical treatments particularly adsorption technique exhibited superior dye removal efficiency at a cheap cost and ease of operation. Moreover, it presented low sensitivity to toxic pollutants avoiding the complications associated with removal of hazardous substances from various medium. For this excellence, research is flourishing to develop new novel adsorbents of low cost and high effectiveness for its employment in dye removal in wastewater treatment. A brief overview of the merits and demerits of these methods is enlisted in Table 1 [8, 25].

4 Advantages of Adsorption

The conventional methods used in the remediation of wastewater have been encumbered with numerous drawbacks of high cost, consumption of hazardous chemicals, and production of sludge in bulk and are not environmentally sound. Recently, attempts have been made to use eco-friendly methods with relatively good efficacy for wastewater treatment. Among them, adsorption technique has been identified as

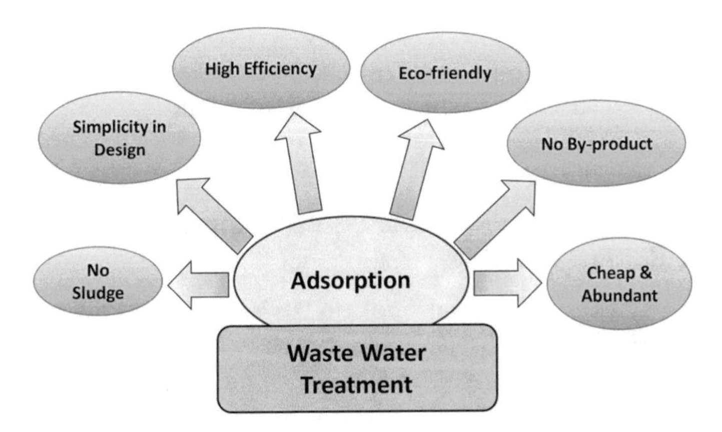

Fig. 2 Features of adsorption technique

an improved and sophisticated candidate attracting the fancy of the research community due to its high efficiency, ease of application, minimum use of chemicals, and its wide applicability for different kinds of dyes. Moreover, this method has been reputed to be eco-friendly, environmentally benign with no energy consumption and sludge generation (Fig. 2).

Adsorption is a convenient separation technique that binds the adsorbate molecules on the adsorbent surface by physical or chemical forces. This separation methodology realizes considerable applicability in the elimination of dyes from wastewater and is argued to be the finest method compared to others. Adsorption is demonstrated as an effective and reliable technique for the elimination of contaminants, particularly when the adsorbent is abundant and inexpensive. It is a mass removal methodology by which the adsorbent can judiciously eliminate dissolved contaminants from a solvent medium by the accumulation of the adsorbate particles on the surface or inter phase medium. Adsorption can be categorized as chemical or physical based on the kind of forces involved in the mechanism. In the physisorption, the pollutants are attached on the surface of the adsorbent by physical forces such as hydrogen bonding, dipole–dipole interactions, and Vander Waals forces. Although factors such as the nature of the adsorbent, i.e., molecular size and structure, weight, concentration of the solution, polarity, surface area, and particle size determine the magnitude of adsorption [47]. Chemisorption, on the contrary, includes the formation of chemical bonds to the surface of the adsorbent. The effectiveness of this method relies on the cost of the adsorbent, ease of operation, their abundance, and surface area. Gisi et al. grouped the low-cost sorbent materials into five broad categories: (i) agricultural and household wastes, (ii) industrial by-products, (iii) sludge, (iv) sea materials, (v) soil and ore, and (vi) novel low-cost materials and investigated their removal efficiency for different contaminants [9].

The search for novel low-cost materials having good efficiency has led to many alternative sources. So far, numerous nano materials have been employed as adsorbents among which ceramic nanocomposites are promising contestants in regard to their enhanced surface area, free active valencies, and porosity in their structure. The materials have displayed their superiority to eliminate dyes from contaminated water emerging as a potential adsorbent material in the last decade.

5 Application of Nanoceramic Based Composites in the Removal of Dyes

5.1 Methylene Blue

Chitosan is a notable adsorbent and is potent in the adsorption of anionic or reactive dyes because of the presence of amino groups as coordination sites [61]. However, it displays poor adsorption behavior for basic dyes. Taking this into consideration, Samandari et al. designed an efficient and a low-cost adsorbent by embedding Fe doped hydroxyapatite into chitosan beads suited to interact with both metal and cationic dye [46]. The synthesized chitosan/Fe-substituted hydroxyapatite composite was utilized for the elimination of methylene blue dye from wastewaters. The dye adsorption behavior was studied in various conditions to achieve maximum adsorption capacity. It was found that the removal efficiency was directly proportional to the hydroxyapatite content of the sorbent beads and the highest adsorption of MB was found to be 1324 mg g^{-1} in optimal conditions. Moreover, it was revealed that the adsorption of the dye to the composite coincided well with the Langmuir adsorption model suggesting the monolayer adsorption via chemisorption. The data indicated that the prepared sorbent material is feasible, cost-effective, and highly efficient for the previously mentioned dye with the potentiality of regeneration and stability for its utilization in water treatment plants. Usually adsorbents are specific to the treatment of only a definite type of pollutant but cannot be utilized in the treatment of mixed pollutants. Consequently, the preparation of such dual purpose adsorbent for the elimination of heavy metals and toxic dyes captivates the fancy of the researchers which is of appreciable demand in the water treatment technology in days to come. Addressing this issue, Wang et al. synthesized a unique and novel adsorbent by combining the benefits of magnetic hydroxyapatite and oxidized multi-walled carbon Nanotubes [57]. The fabricated (mHAP-oMWCNTs) nanocomposite displayed better adsorption capability for the potential elimination of Pb (II) and methylene blue (MB) than the constituent parent materials. This could undoubtedly be attributed to the ion-exchange characteristic of HAP and the numerous oxidized functional groups on the surface of oMWCNTs. The maximum removal capability was recorded to be 328.4 mg g^{-1} for MB from the Langmuir isotherm. Moreover, good magnetic separability and reusability of the synthesized material will permit it to serve as an encouraging adsorbent for the remediation of complex wastewater resulting in enhanced purification efficiency for the benefit of the community. Another unique carbon-based hydroxyapatite nanocomposite was proposed by Hassan et al. as a competent adsorbent material for the elimination of methylene blue (MB) in wastewater treatment [19]. The influence of several factors such as the initial concentration of MB, HAp/G's dosage, and pH of the solution on the removal ability of the nanocomposite was explored and considered to be following pseudo-second-order kinetics. Interestingly, the fabricated sorbent displayed superior removal efficiency of 333 mg g^{-1} (99.9%) for the dye which excelled several previously reported achievements of other sorbents. One major drawback associated with HAP nanoparticles is that they are very unstable and have a tendency to aggregate quickly in aqueous medium. This leads to the creation of bigger particles or flocs, thereby compromising with the aqueous dispersibility and adsorption ability of contaminants [20, 30]. Besides, implementation of HAP in bulk amount is restricted due to secondary contamination and cost of the material. To overcome this problem, initiatives have been taken to upgrade the colloidal stability by modifying HAP with stabilization methods. One such promising novel inexpensive support material is biochar (BC). A reed derived biochar (BC) supported nanoscale HAP was designed by Li et al. to improve the elimination of Methylene blue (MB) dye from aqueous systems [27]. The synthesized BC-HAP composite manifested much greater surface area, pore volume, and excellent adsorption capacity than BC or HAP, which can be clearly chalked down to its modification of colloidal stability and surface area, along with reactivity of BC-HAP. Results displayed good dispersibility of HAP on BC with a dramatic reduction in the aggregation. Increase in pH of the solution and temperature led to increase in dye adsorption while the adsorption decreased with increase in the ionic strength. The uptake of MB on BC-HAP can be completely attributed to the Lewis acid–base interaction, hydrogen-bonding, electrostatic attraction, and π–π dispersion interaction. Consequently, the BC-HAP nanocomposite could serve as an eco-friendly and encouraging adsorbent for MB removal due to its superior efficiency and excellent reusability. Humic acid is another sustainable and suitable alternative for surface alteration of nanomaterials. It is familiar for its chelating behavior toward several pollutants; however, it cannot be used alone as an adsorbent because it dissolves in water. Wei et al. prepared a novel HA-nHAP sorbent by surface immobilizing of nano-hydroxyapatite (nHAP) by HA and investigated its efficiency for the practical adsorption of Cu (II) and Methylene blue (MB) [59]. At optimal conditions, the removal rates of MB were almost 100%, displaying good electrostatic interaction and surface complexation. It was unveiled from desorption studies that the HA-nHAP adsorbent could be successfully regenerated and reutilized establishing its practical applicability in dye removal.

5.2 *Congo Red*

The application of chitosan for dye removal has gained attention due to its great affinity for most dyes. Nonetheless, widespread application has been limited because of its low specific gravity and mechanical strength, tendency to agglomerate, and inadequate solubility. To pave the way measures have been endeavored for boosting the mechanical strength and the efficiency of adsorption by grafting CS on the surface of rigid inorganic materials. Herein, a hydroxyapatite-CS (HAp-CS) nanocomposite was proposed by Hou and his team by embedding HAp into CS and employed it for the potential elimination of Congo red (CR) dye from aqueous systems [22]. Detailed studies of the effects of several parameters such as dosage of the adsorbent, contact time, initial pH, and dye concentration on CR adsorption were executed making use of batch experiments at ambient temperature. The fabricated composite material with a composition of 50 wt% of CS manifested an adsorption ability of 769 mg g^{-1} for the dye which is much superior to any other previously reported adsorbents. This can undoubtedly be attributed to surface complexation, ion exchange, and hydrogen bonding. The utilization of modified hydroxyapatite as a versatile sorbent for the elimination of Congo red dye from aqueous medium at room temperature and with no alteration in pH was first developed by Srilakshmi and his team [53]. The Ag doped calcium hydroxyapatite (CaHAp) composite unveiled high adsorption capacity. CaHAp and Ag-doped CaHAp composites were prepared by the method of aqueous precipitation. Interestingly, it was unveiled that with the elevation of the Ag content, the removal capacity of the adsorbent greatly enhanced with a 49.89–267.81 mg g^{-1} for 50–300 ppm in aqueous medium for Ag (10): CaHAp adsorbent. It is also envisioned that the fabricated unique material can be applied as an inexpensive adsorbent for successful dye elimination from factory effluents. Zein derived from corn is abundantly present as a secondary product of bioethanol. It is a heterogeneous combination of several amino acids and is reported to be biodegradable and biocompatible with high removal efficiency for the elimination of toxic dyes as an efficient sorbent material [63]. To magnify the removal efficiency and rate of elimination of Congo red from aqueous streams, Nasab et al. proposed an efficient adsorbent by attaching nHAp with Zein to form Zein/nanohydroxyapatite (Zein/nHAp) [16]. The magnification of the surface area and the amount of reactive sites significantly resulted in enhancing the removal percentage. The impact of various factors such as pH, temperature, contact time, and adsorbent dosage on the uptake of Congo red was carefully explored. The synthesized nanocomposite afforded a highest elimination percentage of 99.48% at optimum conditions and the kinetics of the adsorption fitted well with the pseudo-second-order model. Regeneration studies revealed that the Zein/nHAp composite can be remarkably reutilized thus widening its applicability as an encouraging adsorbent for the elimination of Congo red in real samples.

5.3 Rhodamine B

Rhodamine B (Rb) is a major contaminant found in the industrial effluents. It is toxic and may cause irritation, swelling, and redness in the eyes and skin. Decolorization of this toxic dye from the effluent waters is essential to satisfy the standard permissible limits set by the EPA. On this account, the call for novel and nature-friendly methods to eliminate contaminants from industrial discharge is critical and has captivated the interest of the scientific community. The maximum surface area, interchangeable hydroxyl groups, and enhanced reactivity of hydroxyapatite (HAp) have been continuously explored for its industrial benefits. Nonetheless, the separation of the HAp powder has always been burdensome. To tackle this problem, Oladipo et al. developed a nanocomposite material by encapsulating Hap into the polymeric network of alginate to enhance the mechanical strength of alginate thereby, boosting its industrial applications [39]. The prepared effective adsorbent with advanced structural integrity and magnified stability has been utilized for the elimination of Rhodamine B in batches. The maximum uptake ability was noted to be 480 mg g^{-1}, and it was observed that in the existence of humic acid, the rate of adsorption greatly intensified. Outcomes from the sorption studies suggested that the removal efficiency of n-HApAg beads reduces in the third cycles of regeneration.

Furthermore, the benefits of silver embedded biopolymer-based composite materials have been investigated. Apart from delivering antimicrobial properties, embedding restricts the leaching of silver ions into the water making the treatment reliable and inexpensive. One such adsorbent, chitosan/Ag-substituted hydroxyapatite nanocomposite beads, synthesized by Li et al. were employed to remove Rhodamine B dye from aqueous solutions [26]. The metal-substituted nano hydroxyapatite powder can successfully interact with chitosan which leads to the magnification of the characteristics of the nanocomposites for its utilization in various environmental applications. Batch adsorption studies were conducted by altering the ratio of Ag-HA to Cs, pH of the solution, contact time, and the initial concentration of the pollutants. The maximum Langmuir adsorption capability for the dye was found to be 127.61 mg/g and the percentage removal efficiency of untreated tap and river water varied between 86.7 and 94.4%. The main objective of this study was to come up with an environmentally sound and inexpensive multifunctional nanocomposite adsorbent for its practical application in wastewater treatment plants.

5.4 *Other Dyes*

To develop the efficiency and to facilitate the ease of separation of nHAP after the adsorption process is over, it was important to modify them into composites. Graphene oxide (GO) owns a hydrophilic negative charge density that supports the effective removal of specific pollutants and dyes by electrostatic interaction. Taking this into consideration, Prabhu et al. synthesized a nano composite by doping graphene into hydroxyapatite (nHAp@GO) by a green in-situ one-pot approach and employed it for the potential removal of Congo Red (CR) and Trypan Blue (TB) from aqueous medium [44]. Doping of GO on the surface of nHAP greatly magnified the surface area by 2.5 times. The composite displayed exceptional adsorption capacities of 41 mg g^{-1} and 48.5 mg g^{-1} for TB and CR respectively, which was several times higher than those of bare parent nHAP. Hydrogen bonding, π–π stacking, electrostatic, and hydrophobic interactions were mainly responsible for the adsorption of the dyes on the nHAP@GO material. Furthermore, the developed material sustained efficiency even after three cycles of recycling, thus demonstrating a multi-functional material for practical applications in the depollution of hazardous diazo dyes. Another unique and cost-effective biopolymer—inorganic porous composite (Hs–Cs–HAp) has been reported utilizing waste material by a convenient, economic and scalable approach [6]. Optimum dye removal capacities of 168 mg g^{-1} of Sunset Yellow and 3.8 mg g^{-1} of MB have been observed suggesting higher binding efficiency toward the acidic dye due to its polycationic characteristics. Moreover, it presented exceptionally high capacity to remove four commercially used dyes from their solutions and was successfully utilized to decolorize effluents from leather processing. Therefore, this study proved that industrial waste materials could undoubtedly be used for the large-scale remediation of colored wastewater generated by the industrial sector without any loss of efficiency even after five times of recycling. The loading of Hap nanoparticles into the surface of GO possibly incorporates the benefits of high removal capability of GO to serve as an excellent material for water purification. So, a unique polysaccharide-based hydrogel adsorbent (NHA) was fabricated to investigate the uptake capacity of Malachite Green (MG) from aqueous medium [21]. The porosity, biocompatibility, and biodegradability of the prepared material were analyzed in detail. It was revealed that the dye removal process was spontaneous, endothermic, and feasible from the temperature dependence data with a highest removal efficiency of 297 mg g^{-1}. Moreover, the adsorption isotherm data matched with the Langmuir isotherm and maintained its efficiency even after repeated cycles of adsorption–desorption. In accordance with the data, the proposed nano adsorbent is eco-friendly and can be employed as a potential candidate for the uptake of cationic dyes from dye pollutant water.

MgO is a destructive absorbent that breaks down contaminant molecules prior to adsorption thus lowering the toxicity of the treated solution. Exploiting this, Foroutan et al. modified bio-hydroxyapatite (Bio-HAp) from waste poultry bone with magnesium oxide (MgO) nanoparticles (Bio-HAp/MgO) and utilized it for the removal of methyl violet (MV) from aqueous environment [12]. It displayed higher specific surface area of 14.7 m^2 g^{-1}, and the influence of several operational factors on the removal capacity was investigated. The adsorption of the dye on the mesoporous composite was exothermic and spontaneous and had good reusability up to five cycles ideal for real time applications for purifying textile wastewater.

6 Future Prospective

Nanoceramics as an adsorbent has attained outstanding interest from the researchers across the world because of its attractive features of bioactivity, eco-friendly and low-cost nature, excellent performance of adsorption, minimum solubility in water and enhanced stability under harsh circumstances. To enhance the uptake and catalytic efficiencies, many functional groups have been grafted for modification by co-doping and impregnation on the surface of the ceramic materials. These modified functionalized materials have the novelty of easy regeneration, fast recovery rate and strong binding ability for the competitive adsorption of contaminants. Hence, its potentiality must be employed in the existence of organic, inorganic, and other hazardous pollutants for broad-scale commercial implementation. Also, further investigations should be conducted concerning the utilization of ceramic materials in catalytic degradation of contaminants. Moreover, the unique approach of one-pot synthesis of the materials should be encouraged and a comprehensive cost-friendly treatment of industrial effluents should be cultivated. To encourage the applicability of ceramic nanomaterials based water filtration in the forthcoming days, certain factors should be contemplated: (i) the development of advanced nano composites for superior adsorption, (ii) design of inexpensive ceramic based filtration units for the easy affordability to the poor people, (iii) effects of pH, temperature need to be considered for optimization, and (iv) complete analysis of the nanomaterials to avert any secondary pollution and significant health issues. The conclusive objective of the entire research directed toward water remediation is to serve the people with their basic right of clean consumable water. Reformation in the present available techniques can symbolize a new perspective in the arena of water remediation in the coming decades.

7 Conclusion

This chapter primarily outlines the prospects of nano ceramic-based composites as vigorous materials to resolve the continuing issue of water scarcity that the global community is facing presently. The exercising of ceramic based nanomaterials for the remediation of water laden with contaminants can be a promising and encouraging substitute to the traditional techniques, which tolerates the drawback of high sludge generation as by-products. Ceramics are potent materials of interest due to their novel electronic structure, capability of light absorption and porous structure. Consequently, ceramic nano materials have been perpetually been exploited by scientists as superior adsorbents to alleviate the hazardous contaminants from wastewater. Minimum cost of fabrication and enhanced surface area grants them better utility for their utilization as adsorbents.

Furthermore, these nano materials are designed so as to regenerate and reuse them thus, eliminating any minute possibility of negative impacts to the environment. To further enhance the activity of the ceramic nano materials, several modifications have been executed such as synthesis of composites and doping. Vital factors such as size of the particles, crystallinity, pH, initial pollutant concentration, and ionic strength of the competitive ions illustrate substantial effects of the adsorption process. Altogether, ceramic based nanomaterials can be considered as highly efficient and advantageous in the arena of environmental remediation. Advanced studies and in-depth research concerning the construction of recyclable, magnetic, and high adsorptive ceramic materials will not only permit the effective treatment of pollutant–laden wastewater treatment in commercial scale but also aid the people with clean and safe drinking water in the days to come.

References

1. Ahmad R, Kumar R (2010) Conducting polyaniline/iron oxide composite: a novel adsorbent for the removal of amido black 10B. J Chem Eng Data 55:3489–3493. https://doi.org/10.1021/je1001686
2. Akarslan F, Demiralay H (2015) Effects of textile materials harmful to human health. Acta Phys Pol A 128:407–408. https://doi.org/10.12693/APhysPolA.128.B-407
3. Anirudhan TS, Ramachandran M (2015) Adsorptive removal of basic dyes from aqueous solutions by surfactant modified bentonite clay (organoclay): kinetic and competitive adsorption isotherm. Process Saf Environ Prot 95:215–225. https://doi.org/10.1016/j.psep.2015.03.003
4. Aoudj S, Khelifa A, Drouiche N, Hecini M, Hamitouche H (2010) Electrocoagulation process applied to wastewater containing dyes from textile industry. Chem Eng Process 49:1176–1182. https://doi.org/10.1016/j.cep.2010.08.019
5. Cao M, Lin J, Lu J, You Y, Liu T, Cao R (2011) Development of a polyoxometallate-based photocatalyst assembled with cucurbit[6] uril via hydrogen bonds for azo dyes degradation. J Hazard Mater 186:948–951. https://doi.org/10.1016/j.jhazmat.2010.10.119
6. Chatterjee S, Gupta A, Mohanta T, Mitra R, Samanta D, Mandal AB, Majumder M, Rawat R, Singha NR (2018) Scalable synthesis of hide substance–chitosan–hydroxyapatite: novel biocomposite from industrial wastes and its efficiency in dye removal. ACS Omega 3:11486–11496. https://doi.org/10.1021/acsomega.8b00650
7. Christie R (2007) Environmental aspects of textile dyeing. Wood Head Publishing, Cambridge
8. Dawood S, Sen TK (2013) Review on dye removal from its aqueous solution into alternative cost effective and non-conventional adsorbents. J Chem Process Eng 1:1–7. https://doi.org/10.17303/JCE.2014.105
9. De Gisi S, Lofrano G, Grassi M, Notarnicola M (2016) Characteristics and adsorption capacities of low-cost sorbents for wastewater treatment: a review. Sustain Mater Technol 9:10–40. https://doi.org/10.1016/j.susmat.2016.06.002
10. Fan J, Hu X, Xie Z, Zhang K, Wang J (2012) Photocatalytic degradation of azo dye by novel Bi-based photocatalyst Bi4TaO8I under visible-light irradiation. Chem Eng J 179:44–51. https://doi.org/10.1016/j.cej.2011.10.029
11. Forgacs E, Cserhati T, Oros G (2004) Removal of synthetic dyes from wastewaters: a review. Environ Int 30:953–971. https://doi.org/10.1016/j.envint.2004.02.001
12. Foroutan R, Peighambardoust SJ, Aghdasinia H, Mohammadi R, Ramavandi B (2020) Modification of bio-hydroxyapatite generated from waste poultry bone with MgO for purifying methyl violet-laden liquids. Environ Sci Pollut Res 27:44218–44229. https://doi.org/10.1007/s11356-020-10330-0
13. Gahr F, Hermanutz F, Opperman W (1994) Ozonation – an important technique to comply with new German laws for textile wastewater treatment. Water Sci Technol 30:255–263. https://doi.org/10.2166/wst.1994.0115
14. Garg VK, Kumar R, Gupta R (2004) Removal of malachite green dye from aqueous solution by adsorption using agro-industry waste: a case study of *Prosopis cineraria*. Dyes Pigm 62:1–10. https://doi.org/10.1016/j.dyepig.2003.10.016
15. Geetha P, Latha MS, Mathew K (2015) Biosorption of malachite green dye from aqueous solution by calcium alginate nanoparticles: equilibrium study. J Mol Liq 212:723–730. https://doi.org/10.1016/j.molliq.2015.10.035
16. Ghanavati Nasab S, Semnani A, Teimouri A, Kahkesh H, Momeni Isfahani T, Habibollahi S (2018) Removal of congo red from aqueous solution by hydroxyapatite nanoparticles loaded on zein as an efficient and green adsorbent: response surface methodology and artificial neural network-genetic algorithm. J Polym Environ 26:3677–3697. https://doi.org/10.1007/s10924-018-1246-z
17. Hameed BH, Lee TW (2009) Degradation of malachite green in aqueous solution by Fenton process. J Hazard Mater 164:468–472. https://doi.org/10.1016/j.jhazmat.2008.08.018
18. Hashem A, Akasha RA, Ghith A, Hussein DA (2007) Adsorbent based on agricultural wastes for heavy Mmetal and dye removal: a review. Energy Educ Sci Technol 19:69–86

19. Hassan MA, Mohammad A, Salaheldin TA, El Anadouli BE (2018) A promising hydroxyapatite/graphene hybrid nanocomposite for methylene blue dye's removal in wastewater treatment. Int J Electrochem Sci 13:8222–8240
20. He F, Zhao D, Liu J, Roberts CB (2007) Stabilization of Fe-Pd nanoparticles with sodium carboxymethyl cellulose for enhanced transport and dechlorination of trichloroethylene in soil and groundwater. Ind Eng Chem Res 46:29–34. https://doi.org/10.1021/ie0610896
21. Hosseinzadeh H, Ramin S (2018) Fabrication of starch-graft-poly(acrylamide)/graphene oxide/hydroxyapatite nanocomposite hydrogel adsorbent for removal of malachite green dye from aqueous solution. Int J Biol Macromol 106:101–115. https://doi.org/10.1016/j.ijbiomac.2017.07.182
22. Hou H, Zhou R, Wu P, Wu L (2012) Removal of congo red dye from aqueous solution with hydroxyapatite/chitosan composite. Chem Eng J 211–212:336–342. https://doi.org/10.1016/j.cej.2012.09.100
23. Janaki V, Vijayaraghavan K, Oh BT, Lee KJ, Muthuchelian K, Ramasamy AK, Kamala-Kannan S (2012) Starch/polyaniline nanocomposite for enhanced removal of reactive dyes from synthetic effluent. Carbohydr Polym 90:1437–1444. https://doi.org/10.1016/j.carbpol.2012.07.012
24. Jedynak K, Wideł D, Rędzia N (2019) Removal of Rhodamine B (a basic dye) and acid yellow 17 (an acidic dye) from aqueous solutions by ordered mesoporous carbon and commercial activated carbon. Colloids Interfaces 3:30. https://doi.org/10.3390/colloids3010030
25. Kausar A, Iqbal M, Javed A, Aftab K, Nazli ZH, Bhatti HN, Nouren S (2018) Dyes adsorption using clay and modified clay: a review. J Mol Liq 256:395–407. https://doi.org/10.1016/j.molliq.2018.02.034
26. Li L, Iqbal J, Zhu Y, Zhang P, Chen W, Bhatnagar A, Du Y (2018) Chitosan/Ag-hydroxyapatite nanocomposite beads as a potential adsorbent for the efficient removal of toxic aquatic pollutants. Int J Biol Macromol 120:1752–1759. https://doi.org/10.1016/j.ijbiomac.2018.09.190
27. Li Y, Zhang Y, Zhang Y, Wang G, Li S, Han R, Wei W (2018) Reed biochar supported hydroxyapatite nanocomposite: characterization and reactivity for methylene blue removal from aqueous media. J Mol Liq 263:53–63. https://doi.org/10.1016/j.molliq.2018.04.132
28. Lian L, Guo L, Guo C (2009) Adsorption of congo red from aqueous solutions onto Ca-bentonite. J Hazard Mater 161:126–131. https://doi.org/10.1016/j.jhazmat.2008.03.063
29. Lin JX, Zhan SL, Fang MH, Qian XQ, Yang H (2008) Adsorption of basic dye from aqueous solution onto fly ash. J Environ Manage 87:193–200. https://doi.org/10.1016/j.jenvman.2007.01.001
30. Liu W, Tian S, Zhao X, Xie W, Gong Y, Zhao D (2015) Application of stabilized nanoparticles for in situ remediation of metal-contaminated soil and groundwater: a critical review. Curr Pollut Rep 1:280–291. https://doi.org/10.1007/s40726-015-0017-x
31. Manatunga DC, de Silva RM, de Silva KMN, de Silva N, Premalal EVA (2018) Metal and polymer-mediated synthesis of porous crystalline hydroxyapatite nanocomposites for environmental remediation. R Soc Open Sci 171557:1–15. https://doi.org/10.1098/rsos.171557
32. Mathur N, Bakre P, Bhatnagar P (2006) Assessing mutagenicity of textile dyes from Pali (Rajasthan) using AMES bioassay. Appl Ecol Environ Res 4:111–118
33. Ministry of Textile Govt of India (2018) Annual Report 2017–18
34. Mishra G, Tripathy M (1993) A critical review of the treatments for decolourization of textile effluent. Colourage 40:35–38
35. Nasser MM, El-Geundi MS (1991) Comparative cost of colour removal from textile effluents using natural adsorbents. J Chem Technol Biotechnol 50:257–264. https://doi.org/10.1002/jctb.280500210
36. Nguyen VC, Pho QH (2014) Preparation of Chitosan coated magnetic hydroxyapatite nanoparticles and application for adsorption of reactive blue 19 and Ni^{2+} ions. Sci World J 2014:1–9. https://doi.org/10.1155/2014/273082
37. Nigam P, Armour G, Banat IM, Singh D, Marchant R (2000) Physical removal of textile dyes from effluents and solid-state fermentation of dye-adsorbed agricultural residues. Bioresour Technol 72:219–226. https://doi.org/10.1016/S0960-8524(99)00123-6

38. Nusrat T, Siddiqui SI, Rathi G, Chaudhry SA, Inamuddin AAM (2020) Nano-engineered adsorbent for the removal of dyes from water: a review. Curr Anal Chem 16:14–40. https://doi.org/10.2174/1573411015666190117124344
39. Oladipo AA, Gazi M (2015) Uptake of Ni^{2+} and rhodamine B by nano-hydroxyapatite/alginate composite beads: batch and continuous-flow systems. Toxicol Environ Chem 98:189–203. https://doi.org/10.1080/02772248.2015.1115506
40. Ozdemir O, Turan M, Turan AZ, Faki A, Engin AB (2009) Feasibility analysis of color removal from textile dyeing wastewater in a fixed-bed column system by surfactant-modified zeolite (SMZ). J Hazard Mater 166:647–654. https://doi.org/10.1016/j.jhazmat.2008.11.123
41. Peralta-Zamora P, Kunz A, Games de Moraes S, Pelegrini R, de Campos MP, Reyes J, Duran N (1999) Degradation of reactive dyes I. A comparative study of ozonation, enzymatic and photochemical processes. Chemosphere 38:835–852. https://doi.org/10.1016/S0045-6535(98)00227-6
42. Pi Y, Li X, Xia Q, Wu J, Li Y, Xiao J, Li Z (2018) Adsorptive and photocatalytic removal of persistent organic pollutants (POPs) in water by metal-organic frameworks (MOFs). Chem Eng J 337:351–371. https://doi.org/10.1016/j.cej.2017.12.092
43. Poots VJP, McKay G, Healy JJ (1976) The removal of acid dye from effluent using natural adsorbents. I. Peat. Water Res 10:1061–1066
44. Prabhu SM, Khan A, Hasmath Farzana M, Hwang GC, Lee W, Lee G (2018) Synthesis and characterization of graphene oxide-doped nano-hydroxyapatite and its adsorption performance of toxic diazo dyes from aqueous solution. J Mol Liq 269:746–754. https://doi.org/10.1016/j.molliq.2018.08.044
45. Robinson T, McMullan G, Marchant R, Nigam P (2001) Remediation of dyes in textile effluent: a critical review on current treatment technologies with a proposed alternative. Bioresour Technol 77:247–255. https://doi.org/10.1016/s0960-8524(00)00080-8
46. Saber-Samandari S, Saber-Samandari S, Nezafati N, Yahya K (2014) Efficient removal of lead (II) ions and methylene blue from aqueous solution using chitosan/Fe-hydroxyapatite nanocomposite beads. J Environ Manage 146:481–490. https://doi.org/10.1016/j.jenvman.2014.08.010
47. Salleh MAM, Mahmoud DK, Karim WAWA, Idris A (2011) Cationic and anionic dye adsorption by agricultural solid wastes: a comprehensive review. Desalination 280:1–13. https://doi.org/10.1016/j.desal.2011.07.019
48. Shertate RS, Thorat P (2015) Biotransformation of textile dyes: a bioremedial aspect of marine environment. Am J Environ Sci 10:489–499. https://doi.org/10.3844/ajessp.2014.489.499
49. Shindy HA (2017) Problems and solutions in colors, dyes and pigments chemistry: a review. Chem Int 3:97–100
50. Siddiqui SI, Rathi G, Chaudhry SA (2018) Acid washed black cumin seed powder preparation for adsorption of methylene blue dye from aqueous solution: thermodynamic, kinetic and isotherm studies. J Mol Liq 264:275–284. https://doi.org/10.1016/j.molliq.2018.05.065
51. Singh A, Dutta DP, Ramkumar J, Bhattacharya K, Tyagi AK, Fulekar MH (2013) Serendipitous discovery of super adsorbent properties of sonochemically synthesized nano $BaWO_4$. RSC Adv 3:22580–22590. https://doi.org/10.1039/C3RA44350G
52. Slokar YM, Le Marechal AM (1998) Methods of decoloration of textile wastewaters. Dyes Pigm 37:335–356. https://doi.org/10.1016/S0143-7208(97)00075-2
53. Srilakshmi C, Saraf R (2016) Ag-doped hydroxyapatite as efficient adsorbent for removal of congo red dye from aqueous solution: synthesis, kinetic and equilibrium adsorption isotherm analysis. Micropor Mesopor Mat 219:134–144. https://doi.org/10.1016/j.micromeso.2015.08.003
54. Su C, Wang Y (2011) Influence factors and kinetics on crystal violet degradation by Fenton and optimization parameters using response surface methodology. In: International conference on environmental and agricultural engineering, vol 15, pp 76–80
55. Thao NT, Nga HTP, Vo NQ, Nguyen HDK (2017) Advanced oxidation of Rhodamine-B with hydrogen peroxide over Zn-Cr layered double hydroxide catalysts. J Sci-Adv Mater Dev 2:317–325. https://doi.org/10.1016/j.jsamd.2017.07.005

56. Toor M, Jin B, Dai S, Vimonses V (2015) Activating natural bentonite as a cost-effective adsorbent for removal of congo-red in wastewater. J Ind Eng Chem 21:653–661. https://doi.org/10.1016/j.jiec.2014.03.033
57. Wang Y, Hu L, Zhang G, Yan T, Yan L, Wei Q, Du B (2017) Removal of Pb(II) and methylene blue from aqueous solution by magnetic hydroxyapatite-immobilized oxidized multi-walled carbon nanotubes. J Colloid Interface Sci 494:380–388. https://doi.org/10.1016/j.jcis.2017.01.105
58. Wawrzyniak B, Morawski AW (2006) Solar-light-induced photocatalytic decomposition of two azo dyes on new TiO_2 photocatalyst containing nitrogen. Appl Catal B 62:150–158. https://doi.org/10.1016/j.apcatb.2005.07.008
59. Wei W, Han X, Zhang M, Zhang Y, Zhang Y, Zheng C (2020) Macromolecular humic acid modified nano-hydroxyapatite for simultaneous removal of Cu (II) and methylene blue from aqueous solution: experimental design and adsorption study. Int J Biol Macromol 150:849–860. https://doi.org/10.1016/j.ijbiomac.2020.02.137
60. Willmott N, Guthrie J, Nelson G (2008) The biotechnology approach to colour removal from textile effluent. J Soc Dye Colour 114:38–41. https://doi.org/10.1111/j.1478-4408.1998.tb01943.x
61. Wu FC, Tseng RL, Juang RS (2000) Comparative adsorption of metal and dye on flake and bead-types of chitosans prepared from fishery wastes. J Hazard Mater 73:63–75. https://doi.org/10.1016/s0304-3894(99)00168-5
62. Wu Y, Chen L, Long X, Zhang X, Pan B, Qian J (2018) Multi-functional magnetic water purifier for disinfection and removal of dyes and metal ions with superior reusability. J Hazard Mater 347:160–167. https://doi.org/10.1016/j.jhazmat.2017.12.037
63. Xu H, Zhang Y, Jiang Q, Reddy N, Yang Y (2013) Biodegradable hollow zein nanoparticles for removal of reactive dyes from wastewater. J Environ Manag 125:33–40. https://doi.org/10.1016/j.jenvman.2013.03.050
64. Xu Y, Lebrun RE (1999) Treatment of textile dye plant effluent by nanofiltration membrane. Separ Sci Technol 34:2501–2519. https://doi.org/10.1081/SS-100100787
65. Young L, Yu J (1997) Ligninase-catalysed decolorization of synthetic dyes. Water Res 31:1187–1193. https://doi.org/10.1016/S0043-1354(96)00380-6

Electroflocculation for Wastewater Treatment of Textile Industry: Overview and Process Variables Effects

Sofia Caroline Moraes Signorelli, Josiel Martins Costa, and Ambrósio Florêncio de Almeida Neto

Abstract Wastewater from the textile industry presents several organic and inorganic pollutants, contributing to 54% of the dyes released into the environment in the world. Textile dyes are an environmental and health problem resulting from the advance of industrialization. The consequence is a reduction in the dissolved oxygen level, harming aquatic life. In addition, chemical and biological oxygen demand levels are altered. Regarding the effects on the human organism, azo dyes indicated mutagenic and carcinogenic potential. In this context, chemical, biological, physical, and electrochemical techniques have been studied to treat wastewater from the textile industry. Biological processes generally have long operating times and are ineffective in removing toxic compounds. The chemical coagulation technique produces large amounts of sludge and has slow kinetics. Electroflocculation has been an alternative due to its versatility and high dye removal efficiency. Therefore, this review discussed the effect of variables such as bubble formation, electrode arrangement, distance between electrodes, solution pH, and temperature on the electroflocculation process.

Keywords Decolorization · Degradation · Dye removal · Electrochemical technique · Electrocoagulation · Electroflocculation parameters · Electroflotation · Textile dye · Textile treatment · Wastewater treatment

1 Introduction

Every human action generates environmental impacts. However, in recent decades, they have caused intense degradation of ecosystems and the accumulation of synergistic effects that intensify the adverse effects on the environment on a global scale. Among the ecosystems, the water ecosystem has been highly impacted.

S. C. M. Signorelli (✉) · J. M. Costa · A. F. de Almeida Neto
Laboratory of Electrochemical Processes and Anticorrosion, Department of Product and Process Design, School of Chemical Engineering, University of Campinas, Avenida Albert Einstein, 500, Campinas, SP 13083-852, Brazil
e-mail: sofia.cms@gmail.com

S. S. Muthu and A. Khadir (eds.), *Advanced Oxidation Processes in Dye-Containing Wastewater*, Sustainable Textiles: Production, Processing, Manufacturing & Chemistry, https://doi.org/10.1007/978-981-19-0882-8_11

For many years, water was considered inexhaustible; however, factors such as industrial and population growth [13] generated an increase in contamination and demand for water, in addition to the destruction of native forests, generating an increase in local temperatures altering rain cycles [18]. Deforestation also affects river springs, causing a decrease in their volume or their depletion.

The contamination of water ecosystems is generated through various human activities, such as agricultural production [26], mining activities [71], discharge of domestic effluents [43], and industrial activity [25], among others. Besides, these occupations involve the contamination of effluents by biological and chemical contaminants such as pesticides and fertilizers [26], antibiotics [43], toxic metals [25, 71], and dyes [13].

In this review, the highlights are the effluents generated by the textile industry, an industry of great economic importance worldwide. In Brazil, according to the Brazilian Association of Textile Industries (Abit), the textile and clothing chain was responsible for revenues of US$ 51.58 billion, BRL 3.1 million in investments, 1.5 million direct jobs, and 8 million indirect jobs in 2017. As for the market on a global scale, according to Global Industry Analysts, Inc. (GIA), the global market for textile and apparel is estimated to reach US$ 1600 billion in 2025.

Nonetheless, the fabric processing chain involves a significant water expenditure. A medium-sized textile industry needs an average of 1.6 million liters of water per day, of which 16% is destined for dyeing and 8% for stamping [31]. The water enters the process from the supply station. It leaves contaminated by dyes, with high levels of chemical oxygen demand (COD), biochemical oxygen demand (BOD), color, turbidity, total suspended solids, and toxicity, among others, needing proper treatment, following current environmental legislation [28].

Therefore, this review aims to address the main trends for treating and removing textile dyes in the last decade, highlighting the potential of electroflocculation to the detriment of other processes. The electroflocculation presents simple instrumentation and operation [16]. It removes suspended particulate material [19] and can treat complex matrices [38].

2 Textile Production and Tailings

The fabric processing chain, presented in Fig. 1, involves a significant water expenditure. A medium-sized textile industry, endowed with a production of about 8000 kg of fabric per day, uses approximately 1.6 million liters of water daily, of which 16% are destined for the dyeing stage and 8% for stamping [31].

Water enters the process, being mixed and reacted with strong bases and acids such as sodium hydroxide, sulfuric acid, and chlorine-based products since the stage of pre-treatment [58]. Additionally, the production of fabrics has as one of its main steps the dyeing of the fabrics produced. This step characterizes the textile industry as one of the biggest pollution generators [28]. It involves the use of dyes, in addition

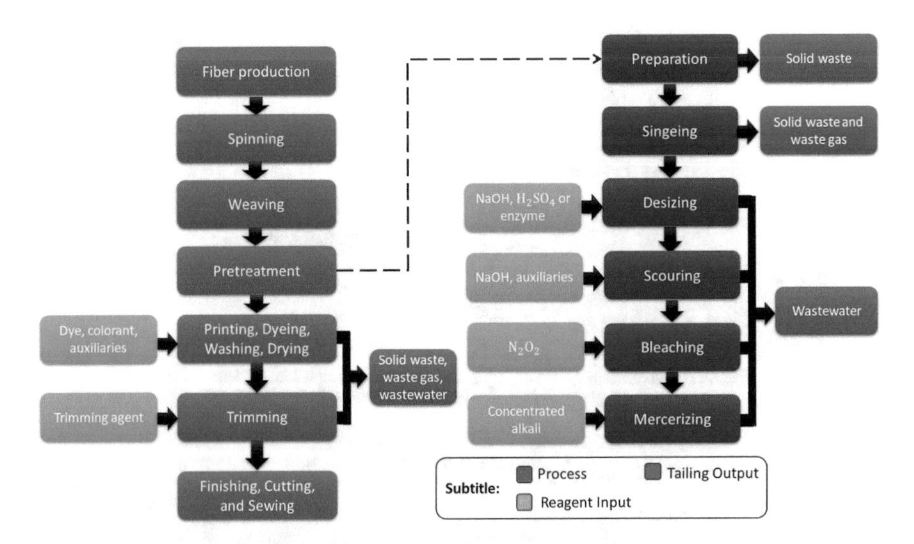

Fig. 1 Fabric processing chain

to solvents and auxiliary reagents such as surfactants, acids, bases, electrolytes, and catalysts, among others [58].

Thus, wastewater from the textile process exits with high levels of contaminants needing treatment [28].

During the fabrics manufacturing process, approximately 20–40% of dyes are lost to industrial effluents [20]. These compounds in wastewater cause severe environmental impacts due to their toxicity to aquatic life and mutagenicity to humans [6]. Furthermore, the dyes increase the BOD of the receiving waters and reduce the reoxygenation process, preventing photoautotrophic organisms' growth [22].

3 Types of Textile Dyes in the Industry

There are several types of dyes applied in the textile industry. Their solubility in water can classify them: (1) soluble anionic dyes (acid, reactive, direct, and mordant) or (2) cationic (basic) and insoluble dyes (azo, disperse, sulfur, vat, and solvent) as reported by Samsami et al. [66], or by its chromophore group, as shown in Fig. 2.

Textile dyes provide color, toxicity, suspended solids, and increased water turbidity, requiring treatment. However, it can be disposed of in the environment following the legislation of the country. Additionally, in general, textile dyes cause alterations in the cellular functioning of humans and animals, compromising primary functions such as reproduction, breathing, and osmoregulation, in addition to causing allergic reactions [34].

Among the existing dyes, azo dyes stand out, such as Orange II, CI Disperse Blue 373, Direct Blue 6, others. According to Fig. 2, they are characterized by the group

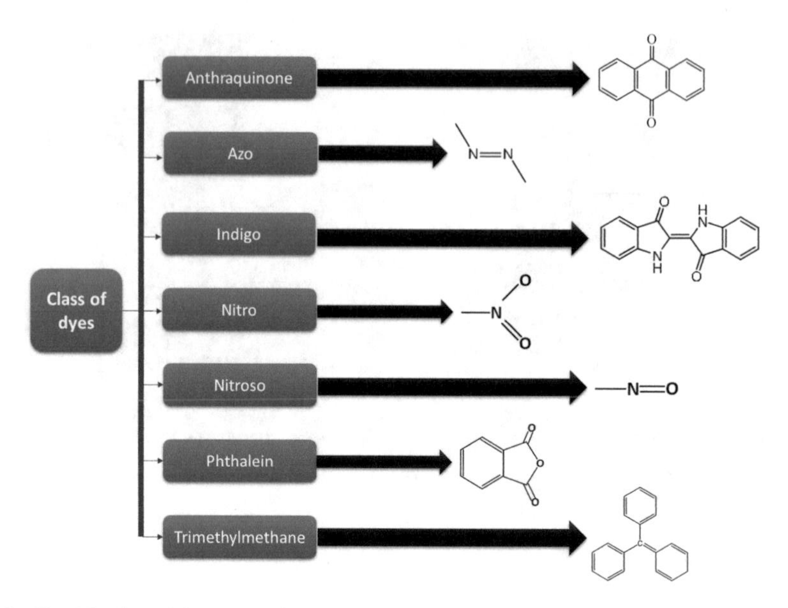

Fig. 2 Classification of dyes according to their chromophore group

R–N=N–R′ and represent 70% of all textile dyes produced by the industry. Azo dyes are highly demanded because they are low cost, ensure high stability, and yield various colors [70]. However, the electronegativity of the azo group protects the dye from the action of oxygenases, decreasing the efficiency of conventional wastewater treatment. On the other hand, anaerobic conditions demonstrate the relative ease of reductive cleavage of the azo bond, producing aromatic amines [6].

4 Dye Removal Treatment Technologies

Several studies on physical, chemical, and biological procedures for wastewater treatment from the textile industry aim to promote the separation of contaminants from the effluent in a short period without causing secondary contamination [66]. The studies and their respective treatment techniques are listed in Table 1.

The most applied treatment methods are the physical as they are simple, efficient processes that require less addition of chemical reagents compared to others [35]. Additionally, physical treatments are more accessible to control than biological ones as they do not depend on the growth of living organisms [32].

Adsorption method needs to control variables such as pH and temperature. However, adsorbent materials are dye-specific or expensive, such as activated carbon [75]. While microfiltration, nanofiltration, ultrafiltration, and reverse osmosis depend on the periodic maintenance of membranes and generate concentrated sludge [75].

Table 1 Treatment methods for dye-contaminated effluents

	Methodology	Principle	Reference
Biological processes	Bacteria-assisted biodegradation	Action of enzymes of microbial system or dye biosorption	[36]
	Algae-assisted biodegradation	Dye metabolization and biosorption on the surface of algal biomass	[64]
	Fungi-assisted biodegradation	Dyes biodegradation through the action of enzymes derived from the intracellular and extracellular metabolism of fungi	[12]
	Enzyme-assisted biodegradation	Enzyme extracts are applied to degrade dyes	[50]
	Aerobic-anaerobic combination	Activated sludge acts by breaking down complex dye molecules	[46]
Chemical processes	Advanced Oxidation Process (AOPs)	The hydroxyl radical acts as the main oxidant involved	[62]
	Coagulation—Flocculation	Reaction with a coagulating agent, floccule formation	[15]
	Electrocoagulation	Dye-coagulant interaction. The coagulant is formed from the dissolution of ions from a metallic electrode on an external electrical current action	[54]
	Electro-Fenton reaction	Electrolysis produces hydrogen peroxide and Fe^{2+}. Hydrogen peroxide reacts to produce hydroxyl radicals that oxidize the dyes. The catalytic action is given by the reduction of Fe^{3+} radicals	[37]
	Catalytic ozonation	The ozonization is catalyzed by a metallic catalyst or metallic oxide immobilized on supports such as activated carbon, alumina, silica, among others. It mineralizes dyes through a direct mechanism with the ozone molecule or indirectly, producing hydroxyl radicals	[23]

(continued)

Table 1 (continued)

	Methodology	Principle	Reference
	Photocatalytic reaction	Dye molecules are adsorbed on the photocatalyst membrane. The photocatalysts are excited by high-energy photons to form electron–hole pairs. The pairs electron–hole react with the oxygen and hydroxyl groups in water, generating hydroxyl radicals, oxygen radicals, and hydrogen peroxide that reacts oxidizing the dyes	[42]
Physical processes	Adsorption	Physical/chemical interaction between the contaminant and the adsorbent material	[39]
	Microfiltration/nanofiltration	Membrane separation	[73]
	Reverse osmosis	Under pressure, water passes through a thin membrane, leaving contaminants on one side and water on the other	[21]
	Micellar-enhanced ultrafiltration	A surfactant with a concentration higher than the critical micellar is added to the dye solution, forming micelles that solubilize the dyes. The ultrafiltration membranes reject these micelles, and a stream of treated water leaves the membrane	[78]

Biological methods are also commonly used in the textile industry due to their ease of application and low cost [63], while chemical processes such as the advanced oxidation process involve high costs. However, the most common biological treatments require other combined methods, such as physical and chemical, as they are efficient for COD removal but still return effluents with toxicity and color [57]. Furthermore, these methods are limited by the growth rate of microorganisms, which causes instability in the system [32]. Therefore, they are usually applied by combining aerobic and anaerobic processes [3].

According to Table 1, there is a considerable range of wastewater treatment methodologies in the textile industry, with disadvantages such as generating harmful gases to the environment and human health, high cost, and difficulty maintaining. Thus, new treatment techniques still need to be studied to obtain more efficient and cleaner processes. Therefore, the electroflocculation process is an alternative treatment.

5 Electroflocculation for Wastewater Treatment

Electroflocculation is the combination of electrocoagulation and electroflotation processes. Electrocoagulation involves the generation of metal ions due to the metal electrode oxidation, from which the coagulating agent is generated. Flotation involves generating hydrogen and oxygen gases from the water electrolysis, transporting the coagulated material to the reactor surface [65].

The technique involves simple instrumentation and operation. The electrochemical reactor is a tank containing the effluent, with metal sacrificial electrodes inside. The application of electric current to the electrodes generates electrochemical reactions, resulting in coagulant and gas release [55]. The treatment involves four events listed in Eqs. (1)–(6):

(1) Electrode oxidation forming metal ions [29]:

$$\text{Anode: } M^0_{(s)} \rightarrow M^{n+} + ne^- \tag{1}$$

(2) Hydrolysis of water with the formation of gas bubbles [29]:

$$2H_2O \rightarrow 4H^+ + O_{2(g)} + 4e^- \tag{2}$$

$$2H_2O + 2e^- \rightarrow 2OH^- + H_{2(g)} \tag{3}$$

$$H_2O \leftrightarrow H_{2(g)} + \frac{1}{2}O_{2(g)} \tag{4}$$

(3) Destabilization of colloidal particles [24]:

$$\text{Cathode: } M^{n+} + ne^- \rightarrow M^0_{(s)} \tag{5}$$

$$M^{n+} + nOH^- \rightarrow M(OH)_{n(s)} \tag{6}$$

(4) (4) Coagulant-dye aggregation, as illustrated in Fig. 3.

However, depending on the metal used as electrode and the pH of the medium, other complexes can be formed. For example, aluminum electrodes result in several complexes, as per Eqs. (7)–(12) [9].

$$Al_{(s)} \leftrightarrow Al^{3+}_{(aq)} + 3e^- \tag{7}$$

$$Al^{3+} + 30H^- \leftrightarrow Al(OH)_{3_{(s)}} \tag{8}$$

$$Al(OH)_3 \leftrightarrow Al(OH)_4^- + H^+ \tag{9}$$

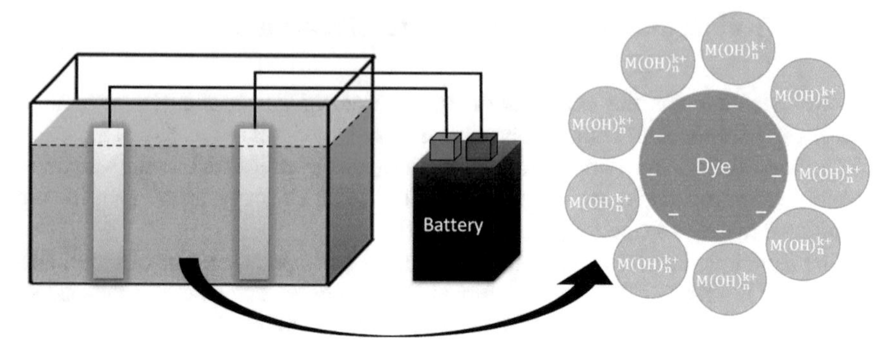

Fig. 3 Illustrative scheme regarding colloidal particles and floccules formation with dye molecules

$$Al(OH)_3 + H^+ \leftrightarrow Al(OH)_2^+ \quad (10)$$

$$Al(OH)_2^+ + H^+ \leftrightarrow Al(OH)^{2+} \quad (11)$$

$$Al(OH)^{2+} + H^+ \leftrightarrow Al^{3+} \quad (12)$$

The process must be carried out under conditions that minimize coalescence between the bubbles. Coalescence prevents aggregation of the bubbles with the coagulant dye, reducing the removal efficiency [47].

Electroflocculation is a quick and safe process. Besides, it has the advantage of consuming nutrients vital to the bacteria present in the effluent, causing the effluent to disinfect [41] and the generation of a sludge that can be used as fuel [40] and in the manufacturing of bricks [5, 8].

On the other hand, metal ions from electrodes in treated effluent are commonly found, which is a disadvantage of the technique to be overcome. Furthermore, the consumption of electricity can make the process economically unfeasible [69]. Thus, numerous studies have been carried out to overcome such difficulties and generate high dye removal efficiencies in short periods, requiring low energy consumption, as shown in Table 2.

Abdulhadi et al. [1] studied the influence of current density and electrode spacing on electrocoagulation-flotation for red dye removal. The increased current density from 2 to 6 mA/cm^2 potentiated dye removal from 87 to 98%. On the other hand, increasing the electrode distance from 5 to 15 mm decreased removal efficiency from 96 to 80%. According to most studies in Table 2, NaCl has been used as an electrolyte.

Table 2 Electroflocculation parameters in the dye treatment

Effluent type	Initial effluent concentration (mg/L)	Effluent volume (L)	Electrolyte	Effective area of anodes (cm^2)	Electric current/area (A/m^2)	Metal of the anode; cathode	Treatment time (min)	Color removal (%)	Reference
Synthetic with the dye RB-19	400	1	NaCl 1.5 g/L	27.7	216.6	TiO_2-RuO_2-IrO_2; stainless steel	80	100	[60]
Synthetic with the dye green 50	100	3	NaCl 1 g/L	1256.6	16.7	Al; Al	21	100	[17]
Real	–	3.1	Nd	–	53	Al; Al	110	65	[45]
Synthetic with the dye optilan MF e novacron MF	–	0.2	Nd	46.83	80	Fe; Al	10	80	[54]
Synthetic with the dye reactive orange 84	300	1.5	NaCl	47.7	130	Fe; Fe	40	89	[77]

(continued)

Table 2 (continued)

Effluent type	Initial effluent concentration (mg/L)	Effluent volume (L)	Electrolyte	Effective area of anodes (cm^2)	Electric current/area (A/m^2)	Metal of the anode; cathode	Treatment time (min)	Color removal (%)	Reference
Synthetic with the dye: reactive orange 84	300	1.5	NaCl	47.7	110	Stainless steel; Stainless steel	40	99.8	[77]
Synthetic with the dye reactive blue 4	100	0.5	NaCl 1 g/L	28	107.1	Fe; Stainless steel	12	84	[74]
Synthetic with the dye reactive blue 4	100	0.5	NaCl 1 g/L	28	107.1	Fe; Steel wool	12	98	[74]
Synthetic with the dye reactive black 5	50	0.5	NH_4Cl	480	8.33	Al; Al	12.5	99.9	[33]
Synthetic with the dye reactive black 5	50	0.5	KCl	480	8.33	Fe; Fe	12.5	99.9	[33]

(continued)

Table 2 (continued)

Effluent type	Initial effluent concentration (mg/L)	Effluent volume (L)	Electrolyte	Effective area of anodes (cm^2)	Electric current/area (A/m^2)	Metal of the anode; cathode	Treatment time (min)	Color removal (%)	Reference
Synthetic with the dye brilliant green dye	250	1	KCl	20	107.57	Fe; Fe	27.9	96.1	[49]
Synthetic with the dye eiochrome black	200	1	NaCl 2 g/L	60	–	Al; Al	60	98.5	[10]
Synthetic with the dye reactive red 120	20	1	NaCl	280	60	Al; Al	12	96	[1]
Synthetic with the dye methylene blue	30	0.25	NaCl 11.7 g/L	25.13	50	Al; Ti	30	99.37	[44]
Synthetic with the dye brilliant breen	100	1	NaCl 0.2 g/L	72	41	Fe; Fe	30	99.59	[53]

6 Effects of Process Variables on Electroflocculation

According to studies [4, 30, 72], factors such as electrode arrangement, pH, temperature, electrode material, electric current density, electrolyte concentration, contaminant concentration, dimensions, and electrode roughness directly influence the efficiency and costs of electroflocculation, whether in the electrocoagulation or flotation stage.

6.1 Bubble Formation

The formation of bubbles plays an essential role in the electroflocculation process. However, the quantity, size, homogeneity, and detachment of bubbles depend on factors such as the dimension and roughness of the electrodes, pH, applied electric current density, electrolyte concentration [4, 30, 72], and electrode material [67]. Besides, smaller bubble sizes increase recovery due to greater surface area [61] and stability of the coagulant-dye-bubble aggregate [76].

Higher electrode roughness indices form smaller bubbles [30]. However, increasing the mesh opening for stainless steel mesh electrodes increased the diameter of the bubbles [72].

There is no consensus on the influence of electric current density on the diameter of bubbles. Although studies reported that increasing current increased the diameter of the bubbles, causing coalescence [48, 67], other studies stated the opposite [4, 59, 72].

The bubble size also depends on the electrode material due to the difference in the superpotential of oxygen and hydrogen generation between the materials, associated with the activation energy for charge transfer [67].

6.2 Arrangement of Electrodes

The arrangement and spacing between the electrodes are related to the difference in potential required for the passage of electrical current, according to Eqs. (13) and (14).

$$\delta = \frac{i}{A_s} \tag{13}$$

$$U = \frac{\delta \cdot d}{k} \tag{14}$$

$$R = \frac{d}{A_s k} \tag{15}$$

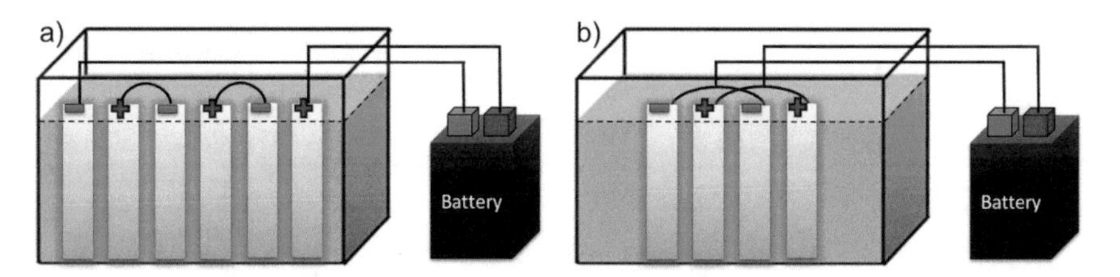

Fig. 4 Illustrative scheme of the monopolar array: **a** series; **b** parallel

where, δ—Electric current density (A/m^2); i—Electrical current; A_s—Surface area of the electrodes (m^2), which act as an anode, submerged in the electrolyte solution; U—Voltage required to carry out the electroflocculation (V), d—Spacing between the electrodes (m), k—Solution conductivity (S/m), and R—Solution resistance.

The serial arrangement is exhibited in Fig. 4a. It provides the system with greater electrical resistance than the parallel arrangement displayed in Fig. 4b, requiring a greater potential difference [51].

6.3 Distance Between Electrodes

There is a direct relation between the distance of the electrodes, dye removal efficiency, and chemical oxygen demand [2, 27, 52]. The relation can be observed from Eqs. (14) and (15). Increasing the distance between the electrodes causes intensification of the resistance and potential difference, influencing even the energy costs of the process.

Two points can explain the relation, according to Akhtar et al. [2]. Decreasing the distance between the electrodes causes an increase in pores in order to intensify the formation of oxides and the chemical dissolution process, increasing the efficiency of the process. However, a small distance results in the absence of agitation between the electrodes, decreasing the aggregation and production of flakes. Akhtar et al. [2] emphasize that the ideal distance, aiming at greater dye removal efficiencies, is around 3 cm. The distance is not a consensus according to the studies presented in Table 3.

6.4 Solution pH

Due to the behavior of the chemical speciation of each metal, the prevalence of metal complexes derived from the oxidation–reduction of the electrodes involved in electroflocculation varies according to the pH. Thus, depending on the metal used as an electrode, the ideal pH of the process will vary. It ensures that the complexes form in greater quantity and quality coagulants and floccules with the dyes [7].

Table 3 Distance between electrodes in electroflocculation

Distance between electrodes (cm)	Reference
1.0	[44]
1.0	[53]
0.5	[1]
2.5	[49]
3.0	[2]
1.0	[77]
1.0	[33]

Table 4 Ideal initial pH values for efficient color removal

Metal (anode/cathode)	Ideal pH_0	Reference
Fe/Al	9.0	[54]
Fe/Fe	7.0	[77]
Fe/Fe	8.5	[49]
Al/Al	8.0	[1]
Al/Ti	7.0	[44]
Al/Al	3.0	[56]
Fe/Fe	4.0	[53]
Fe/Fe	10.0	[11]

Regardless of the initial pH, it tends to increase along the electroflocculation reaching alkaline values [56], resulting from the cathodic reactions presented generically in Eq. (3). According to Chenik et al. [11], the pH tends to a buffer value that balances the production and consumption of OH^- ions. Its value depends on the dye in the composition.

Table 4 presents the initial pH values (pH_0) adopted as ideal, aiming at the highest percentage of color removal and their respective metals applied as electrodes in several studies.

6.5 *Temperature*

Temperature variation does not significantly affect the performance of electroflocculation, as verified by Donneys-Victoria et al. [14]. Although the increase in temperature causes an increase in the dissolution of the anode, it leads to the intensification of effective shocks between the colloids, favoring the formation of large aggregates, which could be deposited on the surface of the electrodes, reducing the dosage of metallic cations [68].

6.6 Cost Analysis

As presented in Sect. 5, electroflocculation costs can make the process economically unfeasible, especially considering the electricity costs involved. However, Yuksel et al. [77], through optimized process conditions, achieved low treatment costs. They obtained a cost of US$ 0.58 per m^3 of treated effluent using stainless steel as electrodes and US$ 0.77 per m^3 using iron plates.

The operating costs of electroflocculation can be estimated in a simplified way using Eqs. (16)–(18).

$$C_{energy} = \frac{U \times i \times t}{V} \tag{16}$$

$$C_{electrode} = \frac{\text{Electrode mass loss}}{V} \tag{17}$$

$$\text{Cost of operation} = C_{energy} \times Co_{energy} + C_{electrode} \times Co_{electrode} \tag{18}$$

where C_{energy}—total electrical energy consumption of the process, $C_{electrode}$—consumed mass of electrode per volume of treated effluent, t—electroflocculation time, Co_{energy}—electricity cost, $Co_{electrode}$—cost of metal electrodes.

7 Conclusion

Chemical, physical, and biological processes have been discussed in this review for dye removal from textile industry wastewater in recent years. All techniques have advantages, disadvantages, and limitations. Removal efficiency can be maximized by combining techniques in hybrid or integrated processes. Electroflocculation is an example that associates coagulation and flocculation during the process. Its versatility, safety, cost, and high dye removal efficiency make it an attractive method. However, studies focus on the use of synthetic solutions and laboratory-scale. Therefore, pilot-scale studies with real effluent should be carried out to explore the technique for effective wastewater treatment.

Acknowledgements Sofia Caroline Moraes Signorelli and Josiel Martins Costa thank the financial support of the Conselho Nacional de Desenvolvimento Científico e Tecnológico, process number 138839/2019-0 and 150172-2020-6, respectively.

References

1. Abdulhadi BA, Kot P, Hashim KS, Shaw A, Khaddar RA (2019) Influence of current density and electrodes spacing on reactive red 120 dye removal from dyed water using electrocoagulation/electroflotation (EC/EF) process. IOP Conf Ser Mater Sci Eng 584:012035. https://doi.org/10.1088/1757-899x/584/1/012035
2. Akhtar A, Aslam Z, Asghar A, Bello MM, Raman AAA (2020) Electrocoagulation of congo red dye-containing wastewater: optimization of operational parameters and process mechanism. J Environ Chem Eng 8(5):104055. https://doi.org/10.1016/j.jece.2020.104055
3. Al-Alwani MAM, Ludin NA, Mohamad AB, Kadhum Abd AH, Mukhlus A (2018) Application of dyes extracted from Alternanthera dentata leaves and Musa acuminata bracts as natural sensitizers for dye-sensitized solar cells. Spectrochim Acta Part A Mol Biomol Spectrosc 192:487–498. https://doi.org/10.1016/j.saa.2017.11.018
4. Alam R (2015) Fundamentals of electro-flotation and electrophoresis and applications in oil sand tailings management. Civil and Environmental Engineering, The University of Western Ontario
5. Aoudj S, Khelifa A, Drouiche N, Hecini M, Hamitouche H (2010) Electrocoagulation process applied to wastewater containing dyes from textile industry. Chem Eng Process 49(11):1176–1182. https://doi.org/10.1016/j.cep.2010.08.019
6. Arora S (2014) Textile dyes: it's impact on environment and its treatment. J Bioremed Biodegrad 05(03). https://doi.org/10.4172/2155-6199.1000e146
7. Barrera-Díaz C, Bernal-Martínez LA, Natividad R, Peralta-Hernández JM (2012) Synergy of electrochemical/O_3 process with aluminum electrodes in industrial wastewater treatment. Ind Eng Chem Res 51(27):9335–9342. https://doi.org/10.1021/ie3004144
8. Barrera-Díaz C, Bilyeu B, Roa G, Bernal-Martinez L (2011) Physicochemical aspects of electrocoagulation. Sep Purif Rev 40(1):1–24. https://doi.org/10.1080/15422119.2011.542737
9. Cañizares P, Martínez F, Jiménez C, Lobato J, Rodrigo MA (2006) Comparison of the aluminum speciation in chemical and electrochemical dosing processes. Ind Eng Chem Res 45(26):8749–8756. https://doi.org/10.1021/ie060824a
10. Cestarolli DT, das Graças de Oliveira A, Guerra EM (2019) Removal of eriochrome black textile dye from aqueous solution by combined electrocoagulation–electroflotation methodology. Appl Water Sci 9(4). https://doi.org/10.1007/s13201-019-0985-x
11. Chenik H, Elhafdi M, Dassaa A, Essadki AH, Azzi M (2013) Removal of real textile dyes by electrocoagulation/electroflotation in a pilot external-loop airlift reactor. J Water Resour Prot 05(10):1000–1006. https://doi.org/10.4236/jwarp.2013.510104
12. Dayi B, Duishemambet Kyzy A, Akdogan HA (2019) Characterization of recuperating talent of white-rot fungi cells to dye-contaminated soil/water. Chin J Chem Eng 27(3):634–638. https://doi.org/10.1016/j.cjche.2018.05.004
13. do Vale-Júnior E, da Silva DR, Fajardo AS, Martínez-Huitle CA (2018) Treatment of an azo dye effluent by peroxi-coagulation and its comparison to traditional electrochemical advanced processes. Chemosphere 204:548–555. https://doi.org/10.1016/j.chemosphere.2018.04.007
14. Donneys-Victoria D, Bermúdez-Rubio D, Torralba-Ramírez B, Marriaga-Cabrales N, Machuca-Martínez F (2019) Removal of indigo carmine dye by electrocoagulation using magnesium anodes with polarity change. Environ Sci Pollut Res 26(7):7164–7176. https://doi.org/10.1007/s11356-019-04160-y
15. Dotto J, Fagundes-Klen MR, Veit MT, Palácio SM, Bergamasco R (2019) Performance of different coagulants in the coagulation/flocculation process of textile wastewater. J Clean Prod 208:656–665. https://doi.org/10.1016/j.jclepro.2018.10.112
16. Dura A, Breslin CB (2019) Electrocoagulation using stainless steel anodes: simultaneous removal of phosphates, orange II and zinc ions. J Hazard Mater 374:152–158. https://doi.org/10.1016/j.jhazmat.2019.04.032
17. El-Ashtoukhy E-SZ, Amin NK (2010) Removal of acid green dye 50 from wastewater by anodic oxidation and electrocoagulation—a comparative study. J Hazard Mater 179(1–3):113–119. https://doi.org/10.1016/j.jhazmat.2010.02.066

18. Ellison D, Morris CE, Locatelli B, Sheil D, Cohen J, Murdiyarso D, Springgay E et al. (2017) Trees, forests and water: cool insights for a hot world. Glob Environ Chang 43:51–61. https://doi.org/10.1016/j.gloenvcha.2017.01.002
19. Emamjomeh MM, Sivakumar M (2009) Review of pollutants removed by electrocoagulation and electrocoagulation/flotation processes. J Environ Manag 90(5):1663–1679. https://doi.org/10.1016/j.jenvman.2008.12.011
20. Essawy AA, Ali AE-H, Abdel-Mottaleb MSA (2008) Application of novel copolymer-TiO_2 membranes for some textile dyes adsorptive removal from aqueous solution and photocatalytic decolorization. J Hazard Mater 157(2–3):547–552. https://doi.org/10.1016/j.jhazmat.2008.01.072
21. Garud RM, Kore SV, Kore VS, Kulkarni GS (2011) A short review on process and applications of reverse osmosis, vol 1, pp 233–238. http://www.environmentaljournal.org/1-3/ujert-1-3-2.pdf
22. Ghaly A, Ananthashankar R, Alhattab M, Ramakrishnan V (2014) Production, characterization and treatment of textile effluents: a critical review. Chem Eng Process Technol 5:182. https://doi.org/: https://doi.org/10.4172/2157-7048.1000182
23. Ghuge SP, Saroha AK (2018) Catalytic ozonation of dye industry effluent using mesoporous bimetallic Ru-Cu/SBA-15 catalyst. Process Saf Environ Prot 118:125–132. https://doi.org/10.1016/j.psep.2018.06.033
24. GilPavas E, Dobrosz-Gómez I, Gómez-García M-Á (2019) Optimization and toxicity assessment of a combined electrocoagulation, H_2O_2/Fe^{2+}/UV and activated carbon adsorption for textile wastewater treatment. Sci Total Environ 651:551–560. https://doi.org/10.1016/j.scitotenv.2018.09.125
25. Hoang H-G, Lin C, Tran H-T, Chiang C-F, Bui X-T, Cheruiyot NK, Lee C-W et al (2020) Heavy metal contamination trends in surface water and sediments of a river in a highly-industrialized region. Environ Technol Innov 20:101043. https://doi.org/10.1016/j.eti.2020.101043
26. Hu C, Deng Z, Xie Y, Chen X, Li F (2015) The risk assessment of sediment heavy metal pollution in the east Dongting lake wetland. J Chem 2015:1–8. https://doi.org/10.1155/2015/835487
27. Huda N, Raman AAA, Bello MM, Ramesh S (2017) Electrocoagulation treatment of raw landfill leachate using iron-based electrodes: effects of process parameters and optimization. J Environ Manage 204:75–81. https://doi.org/10.1016/j.jenvman.2017.08.028
28. Islam MM, Mahmud K, Faruk O, Billah S (2011) Assessment of environmental impacts for textile dyeing industries in Bangladesh. In: International conference on green technology and environmental conservation (GTEC-2011). https://www.academia.edu/21276822/Assessment_of_environmental_impacts_for_textile_dyeing_industries_in_Bangladesh. Accessed 2 June 2021
29. Jiménez C, Sáez C, Cañizares P, Rodrigo MA (2016) Optimization of a combined electrocoagulation-electroflotation reactor. Environ Sci Pollut Res 23(10):9700–9711. https://doi.org/10.1007/s11356-016-6199-y
30. Jiménez C, Talavera B, Sáez C, Cañizares P, Rodrigo MA (2010) Study of the production of hydrogen bubbles at low current densities for electroflotation processes. J Chem Technol Biotechnol 85(10):1368–1373. https://doi.org/10.1002/jctb.2442
31. Kant R (2012) Textile dyeing industry an environmental hazard. Nat Sci 04(01):22–26. https://doi.org/10.4236/ns.2012.41004
32. Katheresan V, Kansedo J, Lau SY (2018) Efficiency of various recent wastewater dye removal methods: a review. J Environ Chem Eng 6(4):4676–4697. https://doi.org/10.1016/j.jece.2018.06.060
33. Keyikoglu R, Can OT, Aygun A, Tek A (2019) Comparison of the effects of various supporting electrolytes on the treatment of a dye solution by electrocoagulation process. Colloid Interface Sci Commun 33:100210. https://doi.org/10.1016/j.colcom.2019.100210
34. Khan S, Malik A (2014) Environmental and health effects of textile industry wastewater. In: Malik A, Grohmann E, Akhtar R (eds) Environmental deteri-oration and human health

natural and anthropogenic determinants. Springer, Netherlands, Dordrecht, pp 55–71. Scientific Research Publishing (2014). https://www.scirp.org/(S(lz5mqp453edsnp55rrgjct55))/reference/ReferencesPapers.aspx?ReferenceID=2113098. Accessed 3 June 2021

35. Khan NA, Bhadra BN, Jhung SH (2018) Heteropoly acid-loaded ionic liquid@metal-organic frameworks: effective and reusable adsorbents for the desulfurization of a liquid model fuel. Chem Eng J 334:2215–2221. https://doi.org/10.1016/j.cej.2017.11.159
36. Khan R, Bhawana P, Fulekar MH (2012) Microbial decolorization and degradation of synthetic dyes: a review. Rev Environ Sci Bio/Technol 12(1):75–97. https://doi.org/10.1007/s11157-012-9287-6
37. Khataee AR, Vatanpour V, Amani Ghadim AR (2009) Decolorization of C.I. acid blue 9 solution by UV/nano-TiO_2, Fenton, Fenton-like, electro-Fenton and electrocoagulation processes: a comparative study. J Hazard Mater 161(2–3):1225–1233. https://doi.org/10.1016/j.jhazmat.2008.04.075
38. Krithika R, Verma R, Shrivastav P (2012) Antioxidative and cytoprotective effects of andrographolide against CCl4-induced hepatotoxicity in HepG2 cells. Hum Exp Toxicol 32(5):530–543. https://doi.org/10.1177/0960327112459530
39. Kumar R, Rashid J, Barakat MA (2014) Synthesis and characterization of a starch–AlOOH–FeS_2 nanocomposite for the adsorption of congo red dye from aqueous solution. RSC Adv 4(72):38334–38340. https://doi.org/10.1039/c4ra05183a
40. Kushwaha JP, Srivastava VC, Mall ID (2010) Organics removal from dairy wastewater by electrochemical treatment and residue disposal. Sep Purif Technol 76(2):198–205. https://doi.org/10.1016/j.seppur.2010.10.008
41. Kyzas GZ, Matis KA (2016) Electroflotation process: a review. J Mol Liq 220:657–664. https://doi.org/10.1016/j.molliq.2016.04.128
42. Lee S-L, Ho L-N, Ong S-A, Wong Y-S, Voon C-H, Khalik WF, Nordin N et al (2018) Exploring the relationship between molecular structure of dyes and light sources for photodegradation and electricity generation in photocatalytic fuel cell. Chemosphere 209:935–943. https://doi.org/10.1016/j.chemosphere.2018.06.157
43. Leung HW, Minh TB, Murphy MB, Lam JCW, So MK, Martin M, Richardson BJ et al (2012) Distribution, fate and risk assessment of antibiotics in sewage treatment plants in Hong Kong, South China. Environ Int 42:1–9. https://doi.org/10.1016/j.envint.2011.03.004
44. Liu N, Wu Y (2019) Removal of methylene blue by electrocoagulation: a study of the effect of operational parameters and mechanism. Ionics 25(8):3953–3960. https://doi.org/10.1007/s11581-019-02915-8
45. Sen SM, Pal D, Kumar Prajapati A (2019) Electrocoagulation treatment of textile dyeing effluent using aluminium electrodes. SSRN Electron J. https://doi.org/10.2139/ssrn.3366890
46. Manavi N, Kazemi AS, Bonakdarpour B (2017) The development of aerobic granules from conventional activated sludge under anaerobic-aerobic cycles and their adaptation for treatment of dyeing wastewater. Chem Eng J 312:375–384. https://doi.org/10.1016/j.cej.2016.11.155
47. Mansour LB, Chalbi S (2006) Removal of oil from oil/water emulsions using electroflotation process. J Appl Electrochem 36(5):577–581. https://doi.org/10.1007/s10800-005-9109-4
48. Mansour L, Kesentini I (2008) Treatment of effluents from cardboard industry by coagulation–electroflotation. J Hazard Mater 153(3):1067–1070. https://doi.org/10.1016/j.jhazmat.2007.09.061
49. Mariah GK, Pak KS (2020) Removal of brilliant green dye from aqueous solution by electrocoagulation using response surface methodology. Mater Today Proc 20:488–492. https://doi.org/10.1016/j.matpr.2019.09.175
50. Mojsov K, Andronikov D, Janevski A, Kuzelov A, Gaber S (2016) The application of enzymes for the removal of dyes from textile effluents. Adv Technol 5(1):81–86. https://doi.org/10.5937/savteh1601081m
51. Mollah M, Morkovsky P, Gomes J, Kesmez M, Parga J, Cocke D (2004) Fundamentals, present and future perspectives of electrocoagulation. J Hazard Mater 114(1–3):199–210. https://doi.org/10.1016/j.jhazmat.2004.08.009

52. Naje A, Chelliapan S, Zakaria Z, Abbas S, Lumpur M (2015) Electrochemical science enhancement of an electrocoagulation process for the treatment of textile wastewater under combined electrical connections using titanium plates. Int J Electrochem Sci 10:4495–4512. http://electrochemsci.org/papers/vol10/100604495.pdf
53. Nandi BK, Patel S (2017) Effects of operational parameters on the removal of brilliant green dye from aqueous solutions by electrocoagulation. Arab J Chem 10:S2961–S2968. https://doi.org/10.1016/j.arabjc.2013.11.032
54. Núñez J, Yeber M, Cisternas N, Thibaut R, Medina P, Carrasco C (2019) Application of electrocoagulation for the efficient pollutants removal to reuse the treated wastewater in the dyeing process of the textile industry. J Hazard Mater 371:705–711. https://doi.org/10.1016/j.jhazmat.2019.03.030
55. Pajootan E, Arami M, Mahmoodi NM (2012) Binary system dye removal by electrocoagulation from synthetic and real colored wastewaters. J Taiwan Inst Chem Eng 43(2):282–290. https://doi.org/10.1016/j.jtice.2011.10.014
56. Palanisamy S, Nachimuthu P, Awasthi MK, Ravindran B, Chang SW, Palanichamy M, Nguyen DD (2020) Application of electrochemical treatment for the removal of triazine dye using aluminium electrodes. J Water Supply Res Technol AQUA 69(4):345–354. https://doi.org/10.2166/aqua.2020.109
57. Pan Y, Wang Y, Zhou A, Wang A, Wu Z, Lv L, Zhu T et al (2017) Removal of azo dye in an up-flow membrane-less bioelectrochemical system integrated with bio-contact oxidation reactor. Chem Eng J 326:454–461. https://doi.org/10.1016/j.cej.2017.05.146
58. Parisi ML, Fatarella E, Spinelli D, Pogni R, Basosi R (2015) Environmental impact assessment of an eco-efficient production for coloured textiles. J Clean Prod 108:514–524. https://doi.org/10.1016/j.jclepro.2015.06.032
59. Rahmani F, Khalfan M, Maqsood T, Noor MA, Alshanbri NM (2013) How can trust facilitate the implementation of early contractor involvement (ECI)? In: Proceedings procurement systems: selected papers presented at the CIB world building congress construction and society. International Council for Building, Queensland, Australia, pp 74–85
60. Rajkumar D, Song BJ, Kim JG (2007) Electrochemical degradation of reactive blue 19 in chloride medium for the treatment of textile dyeing wastewater with identification of intermediate compounds. Dyes Pigm 72(1):1–7. https://doi.org/10.1016/j.dyepig.2005.07.015
61. Ralston J, Dukhin SS (1999) The interaction between particles and bubbles. Colloids Surf A 151(1–2):3–14. https://doi.org/10.1016/s0927-7757(98)00642-6
62. Rehman F, Sayed M, Khan JA, Shah NS, Khan HM, Dionysiou DD (2018) Oxidative removal of brilliant green by $UV/S_2O_8^{2-}$, UV/HSO_5^- and UV/H_2O_2 processes in aqueous media: a comparative study. J Hazard Mater 357:506–514. https://doi.org/10.1016/j.jhazmat.2018.06.012
63. Robinson T, McMullan G, Marchant R, Nigam P (2001) Remediation of dyes in textile effluent: a critical review on current treatment technologies with a proposed alternative. Biores Technol 77(3):247–255. https://doi.org/10.1016/s0960-8524(00)00080-8
64. Robledo-Padilla A, Silva-Núñez A-N, Castillo-Zacarías R-M, Parra-Saldívar R (2020) Evaluation and predictive modeling of removal condition for bioadsorption of indigo blue dye by *Spirulina platensis*. Microorganisms 8(1):82. https://doi.org/10.3390/microorganisms8010082
65. Rubach S, Saur IF (1997) Onshore testing of produced water by electroflocculation. Filtr Sep 34(8):877–882. https://doi.org/10.1016/s0015-1882(97)81411-5
66. Samsami S, Mohamadizaniani M, Sarrafzadeh M-H, Rene ER, Firoozbahr M (2020) Recent advances in the treatment of dye-containing wastewater from textile industries: overview and perspectives. Process Saf Environ Prot 143:138–163. https://doi.org/10.1016/j.psep.2020.05.034
67. Sarkar MdSKA, Evans GM, Donne SW (2010) Bubble size measurement in electroflotation. Miner Eng 23(11–13):1058–1065. https://doi.org/10.1016/j.mineng.2010.08.015
68. Shammas NK, Pouet M-F, Grasmick A (2010) Wastewater treatment by electrocoagulation-flotation. Flotat Technol 199–220. https://doi.org/10.1007/978-1-60327-133-2_6

69. Singh G, Reddy AS (2012) Electroflocculation on textile dye wastewater. Thapar.edu. http://hdl.handle.net/10266/1943
70. Solís M, Solís A, Pérez HI, Manjarrez N, Flores M (2012) Microbial decolouration of azo dyes: a review. Process Biochem 47(12):1723–1748. https://doi.org/10.1016/j.procbio.2012.08.014
71. Sun W, Ji B, Khoso SA, Tang H, Liu R, Wang L, Hu Y (2018) An extensive review on restoration technologies for mining tailings. Environ Sci Pollut Res 25(34):33911–33925. https://doi.org/10.1007/s11356-018-3423-y
72. Sun W, Ma L, Hu Y, Dong Y, Zhang G (2011) Hydrogen bubble flotation of fine minerals containing calcium. Min Sci Technol (China) 21(4):591–597. https://doi.org/10.1016/j.mstc.2011.01.002
73. Tahri N, Masmoudi G, Ellouze E, Jrad A, Drogui P, Ben Amar R (2012) Coupling microfiltration and nanofiltration processes for the treatment at source of dyeing-containing effluent. J Clean Prod 33:226–235. https://doi.org/10.1016/j.jclepro.2012.03.025
74. Wei M-C, Wang K-S, Huang C-L, Chiang C-W, Chang T-J, Lee S-S, Chang S-H (2012) Improvement of textile dye removal by electrocoagulation with low-cost steel wool cathode reactor. Chem Eng J 192:37–44. https://doi.org/10.1016/j.cej.2012.03.086
75. Yagub MT, Sen TK, Afroze S, Ang HM (2014) Dye and its removal from aqueous solution by adsorption: a review. Adv Coll Interface Sci 209:172–184. https://doi.org/10.1016/j.cis.2014.04.002
76. Yoon R-H (2000) The role of hydrodynamic and surface forces in bubble–particle interaction. Int J Miner Process 58(1–4):129–143. https://doi.org/10.1016/s0301-7516(99)00071-x
77. Yuksel E, Eyvaz M, Gurbulak E (2011) Electrochemical treatment of colour index reactive orange 84 and textile wastewater by using stainless steel and iron electrodes. Environ Prog Sustain Energy 32(1):60–68. https://doi.org/10.1002/ep.10601
78. Zaghbani N, Hafiane A, Dhahbi M (2007) Separation of methylene blue from aqueous solution by micellar enhanced ultrafiltration. Sep Purif Technol 55(1):117–124. https://doi.org/10.1016/j.seppur.2006.11.008

ZnO Nanocomposites in Dye Degradation

M. V. Manohar, Amogha G. Paladhi, Siji Jacob, and Sugumari Vallinayagam

Abstract Substances like dyes having substantial colouring capacity are used in textiles which releases the effluents into natural streams by evading waste water treatment. Pollution caused by such non-biodegradable dyes like making the water unfit for human activities, harming aquatic life, causing diseases in humans etc. has become the major concern. Advanced Oxidative Processes (AOPs) by photocatalysis are being employed to remove these dyes and bring a considerable reduction in the contamination. Various semiconductor nanoparticles are widely used for photocatalysed degradation of dyes, out of which ZnO nanoparticle is one of the effective catalysts for this purpose. ZnO is considered above all other metal oxides due to its stability, low cost, high photosensitivity and optical properties. ZnO is combined with metal, metal oxides etc. in order to overcome the recombination of generated charge carriers and increase its photocatalytic and sonocatalytic efficiency. ZnO is produced by several methods like hydrothermal synthesis, solvothermal synthesis, one step flaming process etc. Characterization and confirmation of the synthesized nanoparticles are carried out by techniques such as X-ray diffraction (XRD), UV–Visible analysis, Scanning Electron Microscopy (SEM), Transmission Electron Microscopy (TEM), Raman Spectroscopy, Brunauer-Emmer-Teller (BET) technique, Field Emission Scanning Electron Microscopy (FE-SEM), Energy Dispersal X-ray analysis (EDX), Fourier-Transform Infrared (FTIR) spectroscopy Analysis, Energy Dispersal X-Ray spectroscopy (EDS) etc. Photocatalytic and sonocatalytic dye degradation depends on pH, size of the ZnO nanocomposite and calcination process. In this review different methods of ZnO synthesis, nanocomposite synthesis of ZnO with metals, characterization of the ZnO nanoparticles and dye degradation processes have been discussed.

M. V. Manohar
JSS Medical College (Deemed to be University), Mysuru, Karnataka, India

A. G. Paladhi · S. Jacob
CHIRST (Deemed to be University), Bengaluru, Karnataka, India

S. Vallinayagam (✉)
Department of Biotechnology, Vel Tech Rangarajan Dr.Sagunthala R&D Institute of Science and Technology, Chennai 600062, India
e-mail: sugumariv@veltech.edu.in

S. S. Muthu and A. Khadir (eds.), *Advanced Oxidation Processes in Dye-Containing Wastewater*, Sustainable Textiles: Production, Processing, Manufacturing & Chemistry, https://doi.org/10.1007/978-981-19-0882-8_12

Keywords ZnO nanocomposites · Dye degradation · Photocatalysis · Sonocatalysts · Nanoparticles · Dye decolourization · ZnO co-dope · Doping · Nanospheres · Dye · Calcination · Methylene blue · Methyl orange · Congo red · Azo Rhodamine B · Acid red B · Coomassie brilliant blue

1 Introduction

The water in natural bodies is contaminated due to the release of residual dyes from textile industries, dye producing and other industries. These residual dyes produced from textile industries are not easily biodegradable. Thus, contaminated water should be treated for the removal of these non-biodegradable dyes that are harmful to the biome [47, 50]. The dye affects the properties of water like colour, odour and physiochemical properties. There are different techniques involved in the treatment of dye-contaminated water. The principles for dye degradation are basically adsorption of dye by physical or chemical coagulations that can help the removal of textile dyes [47]. These are the advanced techniques used for the degradation of dyes which involves H_2O_2, O_3 in the influence of irradiation that causes oxidative degradation of contaminants. The above process is termed as Advanced Oxidative Processes (AOPs). The technique of photocatalysis using semiconductor metal oxides is recently introduced AOPs that help in the easy removal of the dye by the principle of surface adhesion due to charge. ZnO (Zinc oxide), TiO_2 (Titanium dioxide), Fe_2O_3 (Hematite), WO_3 (Tungsten oxide) and ZnS (Zinc sulphate) are few light-sensitive semiconductors that use the principle of redox reactions due to their surface morphology. The morphology is due to the chemical combination possessing reliable characters of empty conduction bands and of filled valence bands [7]. As they are light-sensitive semiconductors, striked with irradiations of UV or natural light or other compatible source of light that can excite the electrons from valence band to conduction band forming an electron hole in the valence band. The electron and hole helps in the adhering of desired molecules (dye) to its surface due to the formed valency. The electron sharing assures the binding of dye molecules. Simultaneously in the process due to UV irradiation, ultrasound treatment splits the aqueous solution into hydroxyl ions [15] and superoxide radicals due to its lesser Electromotive Force [48]. ZnO having emV compared to other semiconductor photocatalysts is a very efficient nanoparticle that has significantly high molecular size compared to TiO_2 and has high catalytic capacity [7]. ZnO is found to be the most eligible nanoparticle among other semiconductor photocatalysts because of its property of harvesting the majority of the solar spectrum for its photocatalytic activity, also it is known to be very stable in the presence of sunlight [7].

These metal oxide semiconductor photocatalysts use UV light that is present in the spectrum of visible light where the percentage of UV light in visible light spectrum is very minimum, i.e. 5–7% [10]. To reduce or minimize the water cleavage due to the photocatalytic activity of ZnO under UV irradiation different transition metals are used as dopants. ZnO is usually doped with transition metals like Cu, W, Ni and N to reduce the photoreactivity by producing more number of lone pairs of electrons. ZnO is known to be efficient for the degradation of dye in natural biomes because of its property to produce H_2O_2 more efficiently [6]. ZnO has more number of active sites that helps in the adhesion of the majority of the dye molecules due to its property of surface reactivity [33]. The aim is to study the preparation of ZnO, doping or co-doping, degradation of textile dyes, influence of different parameters that affects dye degradation and to review different studies and experiments based on the use of ZnO to degrade textile dyes.

2 Experimental Approach

2.1 Properties of ZnO Nanoparticles and Nanosphere

Among metal oxides, ZnO is an ideal photocatalyst. ZnO is stable, affordable and environment-friendly [22]. ZnO despite being photocatalyst also possesses the properties of sonocatalyst [58].

The assemblage and irregularly formed spatial arrangement in ZnO molecules are comparatively lower than that of adhered metal oxides. The above property of irregularly formed spatial arrangement and assemblage is a beneficial activity of ZnO as it provides increased active sites to adhere dye molecules more efficiently. The impurities like 2% Ni and 2% Al are doped with ZnO which causes the increase in the surface area of the catalyst which in turn facilitates the adsorption of dye adhering to the surface of doped ZnO catalyst. Hence, the doping of ZnO improvises the property of dye degradation [40].

2.1.1 Photocatalytic Property and Activity

The photocatalytic activity of ZnO can be elevated by doping ZnO with specific transition metals [21].

The Methyl Orange (MO) and Congo Red (CR) dyes are degraded using the photocatalytic activities of ZnO in visible-light spectra, where MO and CR are ideally used textile dyes which are also organic pollutants. The photocatalytic activity of ZnO is efficient in the presence of visible light for dye degradation. Ni is doped in a preferred quantity to elevate the efficiency of photocatalysis, if the doping of Ni exceeds the threshold quantity, it in turn causes the declination in the performance of ZnO resulting in the formation of oxygen vacancies on the grain and surface boundaries where the development of the nanoparticles is inhibited causing a stress field that further acts as the scattering centers to holes and electrons that avoids the recombination rate of hole–electron pairs. It was observed that these scattering centers enhance the photocatalytic activity [40, 50]. If Ni is doped in high quantities than threshold, it decreases the surface area for photocatalytic activity, because Ni occupies majority of voids in the composition of ZnO nanoparticle, due to which the photon adsorption of ZnO declines [40]. The reactive species like Oxygen O_2 and hydroxide (OH^-) are significant in the process of photodegradation of dye [27].

2.1.2 Sonocatalytic Properties of ZnO

During sonolysis, water under high temperature and pressure, that occurs in short intervals generates free radicles, H^+ and OH^- having high oxidative potential. The free radicles thus produced react with each other to form Hydrogen Peroxide (H_2O_2) and Hydrogen (H_2) [17]. Hydrogen and Hydrogen Peroxide are produced during sonochemical reactions due to the ultra-sound that is used to process the reactions [2, 3]. These species are also called Sono-generated charge carriers that hinder the sonocatalytic efficiency [50]. The ZnO (doped or co-doped) must be aggregated with some other sonocatalytic semiconductors that can help the completion of effective degradation of dye which are organic pollutants [24]. Thus, semiconductors like TiO_2, WO_3, Fe_2O_3 or other efficient metal oxides can be co-doped for the preparation of nanocomposites [16].

2.2 *Preparation*

2.2.1 Synthesis of ZnO Nanospheres

The ZnO catalyst which is 99% pure is available in the market and can be used for photocatalysis or dye degradation without any further treatment/as such [7].

$ZnOCl_2 \cdot 8H_2O$, $ZnCl_2$ or $AlCl_3 \cdot 6H_2O$, $ZnCl_2$ or $AlCl_3 \cdot 6H_2O$ is used for the synthesis of ZnO nanoparticle using ultracentrifugation and hydrothermal synthesis method, where water or 1,2 ethanediol and aqueous NaOH of 5M is used to dissolve the above-mentioned complex molecule and calcinated at high temperature of 150 if dissolved in 1,2 ethanediol or calcinated at 90 °C if the complex is dissolved in water. The dissolved contents are centrifuged at 6000 rpm. The washing is done using H_2O and 2-propanol to eliminate the excess chloride ions, the process is otherwise known as peptization. Peptization helps in the prevention of unnecessary aggregation of nanounits which may result in declined results. Calcination using various temperatures is done for the nanounits that contain microaggregations that are separated and the disintegration of micro-aggregates is carried out in water or 2-propanol using ultrasound or sonocatalysis for multiple number of times [31].

The ZnO nanoparticles were prepared using hydrothermal synthesis procedure. ZnO was added to Teflon liner with a different mol% of WO_3, a base (NaOH) and N-butylamine were blended gently and then kept sealed in a common autoclave for 12 h at high calcinating temperatures of 120 °C, then the compound was extracted and it was washed multiple times using distilled water and dried at room temperature [10].

2.2.2 Synthesis of Co-doped Nanospheres

The currently used dopes are the materials that possess multiple properties. Due to these multiple properties, dopes are used as composite materials that help to elevate the properties of individual compound containing nanospheres [5]. To exemplify, WO_3–ZnO is more efficient than ZnO alone. The process of doping different species or compounds with individual nanospheres to enhance the efficiency without manipulating the original properties is called Co-doping.

CuO/ZnO nanocomposites are used where CuO is doped with ZnO nanocomposites for thermal decomposition under visible-light-driven photocatalytic dye degradation of textile dyes like Methylene Blue (MB) and Methyl Orange (MO) synthesized by Li et al. [27]. To assess the photocatalytic activity of ZnO/CuO nanocomposites, azo dye Rhodamine B (RhB), which is an organic pollutant discarded from textile industries, was experimentally used under water using the photodegradation process where Xe-lamp radiation is used [27]. To investigate photocatalytic dye degradation of Methyl Orange (MO) and Methylene Blue (MB) dyes by sunlight irradiations using carbothermal evaporation, CuO/ZnO nanocomposites were used by Kuriakose et al. [25].

Yu et al. used WO_3/ZnO nanoparticles as photocatalyst by varying the concentration of WO_3 by precipitation grinding method where the above composite is desiccated at varied temperatures [17, 56]. The composite of WO_3 and ZnO was prepared by Adhikari et al. using known quantities of WO_3 and ZnO nanoparticles with varied ratios and determined that the ratio of ideal quantity of nanocuboid WO_3 and ZnO is 1:9 for the mixture to be efficient to photocatalytically decompose Methylene Blue (MB) and Methyl Orange (MO) accordingly [1]. Lam et al. prepared a very effective WO_3–ZnO nanopowders using hydrothermal decomposition method and used the above nanoparticles for the degradation of 2,4-D (2,4-dichlorophenoxyacetic acid) using sunlight and revealed the use of WO_3 and calcined temperature resulted in effective photocatalytic activity [26]. WO_3 and ZnO nanopowders with varying WO_3 concentrations using a simple aqueous medium at lesser temperatures to determine the photocatalytic activities of WO_3 and ZnO nanoparticulations were used for the dye degradation. The dye degraded using the above method was Methyl Orange by Xie et al. [53].

Similarly, Ma et al. have formulated N/ZnO photocatalyst using thermal treatment protocols to elucidate the photocatalytic activity [20, 30]. The photocatalytic degradation of Methylene Blue was comparatively observed under visible range and UV light using N/ZnO nanoparticles by hydrothermal method. N/ZnO nanocomposites were composed and synthesized by Prabakaran et al. [37]. Sudrajat et al. used combustion reaction to prepare N/ZnO nanocomposites to determine the photocatalytic activity in Methylene Blue dye degradation [46].

Ni/ZnO nanoparticles were used as precursors to elevate the activity using co-gel or co-precipitation techniques and are compared to individual components to study dye-degradation [11, 27]. Chakrabarti et al. have compared the dye degradation capacity of ZnO nanoparticles to degrade organic dyes like Methylene Blue and Eosin Y, where 16 W lamp was used as the source of light and use of ZnO was accompanied by UV irradiation [7].

3 Characterization Techniques

3.1 X-Ray Diffraction (XRD)

It is one of the easy approaches for the determination of crystallite size of the powder samples. As the peak broadens, we can obtain a precise quantification of the particular sample [51]. Crystal structure of deposited samples was determined by using the X-ray diffraction patterns obtained with CuKα radiation (λ = 1.5406 Å) from a Bruker D 2 Phaser in the range of 20–50 °C by Y. M. Hunge et al. The XRD pattern of the WO_3–ZnO nanocomposite could be indexed to the monoclinic and hexagonal crystal structures of WO_3 and ZnO, respectively. The crystallite size (D) was calculated using Debye–Scherrer's formula $D = 0.9\ \lambda/\beta \cos\theta$ (where λ = 1.5406 Å wavelength of the CuKα line, β is the Full Width at Half-Maximum (FWHM) for corresponding peak in radians and θ is Bragg's angle). Mean crystalline size of ultrasound-assisted WO_3 and WO_3–ZnO nanocomposite was found to be 69 and 26 nm, respectively. Decreasing the crystallite size means increasing the specific surface area and hence increasing the sonocatalytic degradation rate [17].

Jun Wang et al. found the average size of the ZnO nanoparticles from the XRD patterns to be 33 nm according to the scherrer's equation. The ZnO nanoparticles were used without any special treatments for the degradation process [52].

Karanpal Singh et al. characterized green synthesized ZnO naoparticles using *Punica granatum* using PANalytical X-ray diffractometer at the range of angle from 0° to 100°. The XRD pattern revealed that the nanoparticles synthesized had a crystalline and wurtzite hexagonal structure which was confirmed comparing with JCPDS data sheet/ICDD no. 36–1451. The diameter was calculated according to the Debye–Scherrer's Equation and the average diameter was estimated as 20 nm [44].

3.2 UV–Visible Analysis

Synthesis of nanoparticles can also be confirmed by UV–Visible analysis. This technique works based on the principle of Beer-Lambert's law. The analyte concentration can be determined from the absorbance at a particular wavelength. The UV–visible absorption spectrum of green synthesized ZnO-Nanoparticles using *P. granatum* in the range of 200–800 nm in a quartz cuvette using Shimadzu UV Spectrophotometer. The absorbance peak was centered near 382 nm, indicating the reduction of zinc nitrate hexahydrate into ZnO nanoparticles [44].

Optical absorbance spectra of ZnO, Green AgNPs, ZnO–AgNPs–5 (5% green synthesized AgNPs) and ZnO–AgNPs–10 (10% green synthesized AgNPs) were analysed by using UV–vis (DRS)-NIR V-77O (JASCO, Japan spectrophotometer) in the wavelength range from 200 to 800 nm. At the end of the experimentation process, a black coloured powder which was Ag and the white coloured powder ZnO was obtained. As both the nanopowders were insoluble in MQ (Milli-Q)water, they were subjected to sonication for 30 min to obtain a well-dispersed solution. Then the completely dispersed supernatant of the solution was taken into a quartz cuvette and exposed to UV–visible radiation. Here, maximum absorption of ZnONPs and AgNPs was observed at 410 and 390 nm which is in agreement with the reported values. With increasing the particle size, the absorption band shifts towards red. The inter-particle distance and the surrounding media also affect the absorption bands [19].

3.3 Scanning Electron Microscopy (SEM)

The surface morphology of green synthesized silver nanoparticles by Gangura leaves (AgNPs), ZnO powder, ZnO–AgNPs (5% green synthesized AgNPs) and ZnO–AgNPs (10% green synthesized AgNPs) were studied by Pranav Jadhav et al. using a Scanning Electron Microscope (Model JEOL-JSM-6360, Japan) which was operated at an acceleration voltage of 20 kV. The energy dispersive X-ray spectroscopy (EDAX) was carried out to check elements present in composites by using Oxford instruments INCA with SEM (S4800), Hitachi Japan. The nanoparticles were found to be spherical in shape with the diameter in the range of 1.1–1.5 μm. The shape, size distribution and morphology of the prepared ZnO–AgNPs–5% and ZnO–AgNPs–10% composites revealed that the major percentage of composites is in nanorods shape with the size lying in the range between 0.3 and 0.4 mm. The SEM images showed pebble-shaped ZnO nanopowder and rod-shaped AgNPs. The ZnO–AgNPs composites with different concentrations show an intermediate structure. The rod-shaped structures were found broken in the image which may have happened during grinding process [19].

Synthesized ZnO (N), Ag/ZnO (A) and Pt/ZnO (P) samples for 3 h (N3, A3 and P3) and 6 h (N6, A6 and P6) of reaction time, respectively, was subjected to analysis by L. Muñoz-Fernandez et al. using Scanning Electron Micrographs and EDS spectra. The micrograph of pure ZnO sample (N3) revealed the formation of quasi-aligned ZnO nanowires with a flat hexagonal top. Still an agglomerated structure was observed in hexagonal cross-sectional view of the top of the nanowires. The presence of Zinc and Oxygen peaks in the selected area of semi-quantitative analysis confirms the presence of pure zinc oxide according to XRD results. The micrograph of Ag/ZnO sample (A3) suggested that particles of noble metal mainly have a spherical morphology. Estimation of the real morphology was quite difficult due to the dense agglomeration and the heterogeneous sizes of ZnO.

Sample N3 and A3 presented quasi-aligned ZnO nanowires morphology. Micrograph of a Pt/ZnO sample (P3) also implied the nanowire morphology, which corresponds to ZnO.

EDS spectra displayed that chemical composition was based on pure Zinc, Oxygen and Platinum, being in good agreement with the XRD result. The micrographs of samples synthesized for 6 h were named N6, A6 and P6, respectively. These samples exhibit similar morphology of nanowires that was obtained by the samples synthesized for 3 h, which again corresponded to ZnO particles. But in comparison, the growth of wires was rather uncontrolled, non-aligned and irregular when a longer reaction time was used. The differences in the morphology may be due to the kinetical growth control of certain crystal surfaces under different surfactants and electrostatic interaction between the positively charged facet of the ZnO crystal and charged chemical species. The nucleation of ZnO crystal and the preferred direction of crystal growth can be influenced this way by the surfactants. Cationic behaviors of surfactant like CTAB with similar molecular charge to the plane cause repulsion on the related plane boundary which can be the cause for change in the morphology [32].

3.4 Transmission Electron Microscopy (TEM)

This microscopic technique is used to describe the size and shape of nanoparticles. It uses a beam of electrons to focus the material and produce a highly magnified and clear image of it. TEM enables easy characterization of the image in its morphological features, compositions and crystallization information is also detailed. Jun Wang et al. found that the treated nanosized ZnO powders had a nanometer scale with the size around 30–45 nm, which was in accordance with the size calculated using Scherrer's equation (33 nm) [52].

Karanpal Singh et al. conducted the TEM analysis of green synthesized ZnO nanoparticles using *P. granatum*. The green synthesized ZnO-NPs were found to be polycrystalline with spherical structure and the size of the nanoparticles was in the range of 10–30 nm which correlated to the value obtained from the XRD data [44].

3.5 Raman Spectroscopy

The Fourier Transform Raman (FT-Raman) spectra of the prepared samples were collected in the spectral range of 100–1000 cm^{-1} using an FT-Raman spectrometer that applies Nd:YAG laser source with an excitation wavelength of 1064 nm. Raman spectroscopy is an inelastic light scattering method, where the energy transfers to and from the system under investigation are characteristic of electronic, optical, vibrational or even magnetic properties [9]. The structure and the symmetry of samples are studied using this method. The chief characteristic peaks for WO_3 were observed at 712.14 and 804.10 cm^{-1}. The peak centered at 436.38 cm^{-1} is the main characteristic peak of ZnO. The appearance of strong peak around 712.14, 804.10 and 436.38 cm^{-1} in a way suggested the formation of WO_3–ZnO nanocomposites [17]. The Raman spectrum of layered WO_3/ZnO thin films were analysed by Y. M. Hunge et al. at room temperature in the range of 100 cm^{-1}. The Vibrational modes of the WO_3 sample, located at 272.60, 715.88 and 806.95 cm^{-1}, are due to the stretching and bending vibrations between O–W–O bonds of layered WO_3/ZnO thin films. The peak at 132.13 cm^{-1} corresponds to the lattice vibrations in the WO_3 thin films. The two peaks were observed at 102.01 and 439.07 cm^{-1} in the layered WO_3/ZnO thin film, which are related to the Zn sub-lattice and the oxygen sub-lattice, respectively. The peak centered at 439.07 cm^{-1} confirmed that the synthesized samples have a hexagonal wurtzite phase [18].

3.6 Brunauer-Emmer-Teller (BET) Techniques

The surface area was analysed by nitrogen adsorption/desorption at 77 K using Gas Sorption System (Micro-metrics, Instruments, ASAP 2420). Inert gas adsorption technique is used to measure the specific surface area, pore size distribution and heats of adsorption. Surface area affects the dissolution rates, adsorption capacity and electron/ion current density, representing the free energy that is available for the bonding. Pore size also determines the performance of the material by affecting its diffusion rates, molecular sieving properties and surface area per unit volume [43]. So it is necessary to analyse, understand and confirm a particular material for better efficiency. Gas Sorption Analyzer records various pressures of gas in the sample cell due to adsorption and desorption. The instrument then calculates the amount (as STP volume) of gas adsorbed/desorbed. Surface area, pore size is calculated using computer software. Y. M. Hunge et al. measured nitrogen adsorption–desorption isotherms to examine the porous structure of WO_3 and WO_3–ZnO nanocomposites by nitrogen adsorption/desorption at 77 K using a Gas Sorption System. BET surface area for WO_3–ZnO nanocomposite was 57.1 m^2 g^{-1} which comparatively was higher than the pure WO_3 (20.6 m^2 g^{-1}). When observed, the results showed mesoporous structures of ZnO doped with WO_3 (ZnO/WO_3 nanospheres). With the increased surface area, WO_3/ZnO nanospheres are assured sonocatalytic nanoparticles that help in the degradation by providing more active sites on the surface due to the increased adsorption of dye molecules. This helps to elevate the effect of sonocatalysis [17].

3.7 Field Emission Scanning Electron Microscopy (FE-SEM)

Electron microscopy has been popular since 1944, when scientists Keith R. Porter and Albert Claude first used an electron microscope to observe the morphology and composition of cells. Imaging of the cells extends our scientific knowledge with respect to the morphology and the organization of organelles in cells. This technique also contributed to the modern improvements in molecular biology and medicine. Emergence of nanotechnology around 1959 has remodeled many of the tools and technologies required to view and interpret nanoenvironments [12].

As instrumentation and technology have evolved, scanning electron microscope is turning out to be a potent aid to analyse heterostructures and doping distributions on the nanometre scale. The introduction of Field-Emission Electron Guns coupled with the design of detector systems has made it possible to obtain high-resolution images over a wide range of acceleration voltages in the SEM. Scanning Electron imaging of any specimen may depend on microscopic parameters such as accelerating voltage, specimen tilt angle, beam current etc. It may also depend on the surface and bulk properties of the material [49]. FE-SEM is used to visualize and analyse every small topographic detail and thus used to determine the particle dimensions and morphology [44]. The morphological characterization of WO_3 and WO_3–ZnO nanocomposite was studied by Y. M. Hunge et al. using a MIRA3 XMU TESCAN Field Emission Scanning Electron Microscope (FE-SEM). The samples showed uniform and compact surface morphology. The morphology of WO_3 structure exhibited an irregular distribution of nanoparticles, whereas the surface of WO_3–ZnO nanocomposite showed round-shaped nanoparticles which efficiently increase the surface area for redox reaction usable for degradation of organic molecules. The effect of ultrasound causes empty spaces termed as cavities or voids that appear on the surface that effect the surface morphology and uniformity of the samples [17]. All the FE-SEM images of Green synthesized ZnO nanoparticles using *P. granatum* was found to be spherical shaped and the average particle diameter was 20 nm which corresponds to the XRD result [44].

3.8 Energy Dispersal X-Ray Analysis (EDX)

EDX is a technique that uses X-ray radiations to examine the composition of elements present in materials. It is a technique used to analyse the surface morphology of the sample where a beam of electron hits another electron present in the inner shell due to which it causes excitation. Due to the excitation of the electron present in the inner shell is ejected resulting in the formation of electron–hole or void in the composition of electrons in the outer shell of element. EDX systems are mainly used in electron microscopy instruments viz., SEM (Scanning electron Microscopy) or TEM (Transmission Electron Microscopy) instruments. The property of this electron excitation helps these instruments to produce the images of specimen that are microscopic. The data collected by EDX analysis contains a graphical representation where the peaks represent the respective spectra related to the corresponding elements. This helps us to know the actual composition of sample that is being analyse. The mapping of sub atomic particles like electrons in the sample helps in analyzing images that are obtained by SEM and TEM. The element analysis or chemical characterization of green synthesized ZnO nanoparticles was done using EDX by Karanpal Singh et al. and the EDX spectrum showed the presence of ZnO and O^- ions in ZnO nanoparticles synthesized using *P. granatum*. The elemental analysis of the ZnO powder indicated that 76% of Zinc and 15% of Oxygen were present which inferred good purity and very little impurities [44].

3.9 Fourier-Transform Infrared (FTIR) Analysis

To obtain the analysis of interfaces to analyse the adsorption of functional groups on the surface of nanoparticles, FTIR analysis is used. The molecular data analysed with the help of FTIR helps the investigator to interpret the conformational and structural changes of the self-aggregated functional groups that are adhered on the surface of nanoparticles. By analyzing the vibrations and rotations of molecules that are influenced by IR (Infra-Red radiation), a particular wavelength to which FTIR analyses spatial arrangement. There are few advantages that attribute for a better assessment in this technique viz., high accuracy, high stability, high signal–to-noise ratio and high energy. Karanpal Singh et al. carried out FTIR spectroscopy to confirm the formation of Zn–O bond and to identify the phytoconstituents of *P. granatum* that are capped on ZnO–NPs surface. Using a Bruker Alpha FTIR spectrometer, spectrum of green synthesized ZnO–NPs was analysed. The spectral peaks 3610 and 3822 cm^{-1} are due to O–H stretching. The peak around 2354 cm^{-1} is due to C–H stretch. The peak around 1512 cm^{-1} is due to the C=O stretching. The peak at 1683 cm^{-1} corresponds to ZnO bending deformation vibrations. The strong vibrational bands at 610 cm^{-1} are assigned to the stretching modes for the formation of ZnO nanoparticles. Thereby in this study it is clear that the ZnO–NPs were devoid of being aggregated by the phytoconstituents and also confirms the stability of the surface of nanoparticles during the synthesis [44].

3.9.1 Energy Dispersal X-Ray Spectroscopy (EDS)

EDS (Energy Dispersal X-Ray Spectroscopy) is a prescribed technique used for the identification of materials. A primary beam is used to obtain characteristic X-rays by placing the EDS systems on SEM (Scanning Electron Microscope). The analysis of the sample is done using the X-rays obtained, which are very particular to the samples. The intensity of the primary beam and the material of the samples are two factors on which the spatial arrangement or resolution of EDS depends [13]. Deducing the structure of atoms, identification and the site that are void can be occupied to know the magnetic properties [38]. The chemical composition of WO_3–ZnO nanocomposite was examined using Energy Dispersive X-ray Spectroscopy (EDS) system attached to an FE-SEM (M MIRA3 XMU Tescon Orsay Holding) and the EDS pattern studies of WO_3–ZnO nanocomposite prepared by sonochemical route ruminate the presence of W, Zn and O elements. The atomic percentage of the W and the Zn elements in the WO_3–ZnO nanocomposite were found to be 32.56% and 22.18%, respectively. Thus, EDS results in addition to Raman and XRD results confirm the formation of WO_3–ZnO nanocomposites [17].

4 Methods of Dye Degradation

4.1 Photocatalytic Degradation

Karanpal Singh et al. carried out photocatalytic degradation of Coomassie Brilliant Blue R-250 dye using Green synthesis of Zinc Oxide nanoparticles (ZnO–NPs) using *P. granatum* leaf extract. The process was carried out in an aqueous solution under sunlight. The powdered dye was dissolved in deionized water to which ZnO nanoparticles were added and sonicated. The mixture was stirred continuously for 30 min in the presence of sunlight. Four millilitres of the solution was centrifuged to pellet down at regular time intervals to check the absorbance at 600 nm which was the optimum absorbance peak of the dye. It was clearly observed that with time the intensity of blue colour of the dye solution decreased gradually and became light at last. ZnO–NPs could degrade the pollutant drastically in 3 h [44].

Pranav Jadhav et al. investigated the degradation efficiency of ZnO nanomaterials doped with Green synthesized silver (Ag) nanoparticles using Gongura leaves on Methylene Blue by photocatalytic degradation process. To 500 mL of distilled water, 5 mg of MB dye was added slowly for the preparation of dye. The dye preparation was completely carried out in dark. The prepared dye solution and the synthesized nanomaterials such as ZnO, ZnO–AgNPs–5 (5% green synthesized AgNPs using Gongura leaves) and ZnO–AgNPs–10 (10% green synthesized AgNPs using Gongura leaves) were added in separate beakers as a catalyst along with pre-prepared dye sample in the proportion of 1:1 (i.e. 50 mL of dye and 50 mL of catalyst) ratio. One beaker containing only dye solution was maintained as control. The beakers containing catalysts were stirred on magnetic stirrer for 30 min so that the dispersion is uniform. The photocatalytic treatment was carried out in the presence of sunlight at the ambient temperature in between 25 and 30 °C. The 1 mg/mL of concentration of ZnO nanoparticle was used for dye solution. The photocatalytic degradation progress was monitored at different intervals. The percentage of degradation was calculated by Degradation percentage (%) = $(C_0 - C_e/C_0) \times 100$, where C_0 and C_e are the dye initial concentration and dye concentration after the photocatalytic treatment, respectively. The dye solution before the addition of the catalyst was exposed to sunlight and the spectrophometric analysis of the sample after 15 min interval revealed that 37.76% dye was degraded under the sunlight which took about 120 min. The absorption spectra were recorded in the wavelength range between 400 and 800 nm. The absorption peak of Methylene Blue dye was found to be 664 and 614 nm in the visible region. Peaks are found to be decreased with an increase in time intervals due to redox reactions. Once ZnO was added to the Methylene Blue as catalyst, 91.67% degradation was observed after 120 min. Composites of ZnO–AgNPs–5 and ZnO–AgNPs–10 were added to the Methylene Blue, showing 99.21% and 97.71% dye degradation within 75 min, respectively. Composites of ZnO–AgNPs–5% proportion showed the better results of all as mentioned above [19].

4.2 Sonocatalytic Degradation

The sonocatalytic degradation is a novel technology for treating wastewater. It is an advantageous method over photocatalytic degradation as it uses low-power ultrasonic irradiation as the excitation energy for the degradation to take place. Ultraviolet light costs a lot of energy to induce semiconductor catalyst and also the photocatalytic degradation is not feasible for low transparent organic wastewater [52].

Y. M. Hunge et al. in his work used sonocatalytic procedure for Brilliant Blue dye degradation. The sonocatalytic degradation was executed using an ultrasonic bath under 200 W output power and 40 kHz frequency in the presence of prepared WO_3–ZnO nanocomposite. As less number of OH^- radicals were formed in the sonolysis degradation alone, WO_3–ZnO nanocomposite prepared with ultrasonic bath was an effective support for enhancing the sonocatalytic degradation efficiency. All reactions were performed under normal temperature and pressure. In this method, the prepared samples (0.2 g) and oxidants were added into 100 ml of 0.5 mM Brilliant Blue solution and then it is filled into the reaction container (150 ml). The reaction mixture was stirred for 15 min to attain adsorption–desorption equilibrium. A 4-ml sample was pipetted from the reaction mixture and centrifuged for 10 min. Once the catalyst was settled down the absorbance of Brilliant Blue solution was measured using an ultraviolet–visible (UV–vis) spectrophotometer [17].

The phenomenon of luminescence caused due to sonocatalysis is because of acoustic cavitation that results in the formation of beam of wavelength 420 nm. These beams of short wavelength generated has sonoluminiscence induce WO_3/ZnO nanoparticles to produce lone pair of electrons that causes electron holes. Thus, generated electrons are excited from the conduction band of ZnO to the conduction band of WO_3 and the holes are shifted from valence bands of WO_3 to valence band of ZnO. This exchange of holes and electrons must be avoided to improve the dye degradation efficiency of nanoparticles (WO_3/ZnO nanocatalyst). The formation of holes has its negative impact on the adsorption of dye on the surface of photocatalytic nanospheres by degrading the nanospheres in the aqueous solution by producing OH from H_2O molecules. The presence of degraded nanoparticles in the solution causes the activation of sites that are responsible for producing acoustic cavitation. The bubbles produced due to cavitation settle on the surface, which causes the formation of more cavitation bubbles [17].

During the course of sonocatalytic degradation, UV–Visible absorption spectra of Brilliant Blue stain were recorded at different time intervals between 450 and 850 nm wavelengths. There was a frequent decrease in the absorbance peak which clearly indicates the Brilliant Blue degradation due to the decomposition process. Degradation was calculated using the formula Degradation formula (%) = $(Ext_0 - Ext/Ext_0) \times 100$, where Ext_0 is initial and Ext final concentration of Brilliant Blue dye. WO_3 nanoparticles alone degraded around 60% of brilliant blue in 40 min, whereas 90% of degradation was observed in 40 min when WO_3–ZnO nanocomposite was administered. The slope of the plot in (Ex_t/Ext_0) versus reaction time was apparently found to be first-order reaction and its kinetics in both the cases. The slope of this plot gives a value of the rate constant (k) and is found to be $k = 7.01 \times 10^{-3}\ s^{-1}$ for WO_3–ZnO nanocomposite and $0.346\ cm^{-3}\ s^{-1}$ for WO_3 nanoparticles [17].

Jun Wang et al. in the sonacatalytic degradation process of acid red B and Rhodamine B dye, ZnO powder was used as such by heating it at 300 °C for 2 h. The activated ZnO powder was mixed with acid red B and Rhodamine B solutions, respectively. After a proper stirring to make a good dispersion, was placed in an ultrasound apparatus and irradiated. The UV–vis spectra of acid red B and Rhodamine B solutions during degradation were determined at regular intervals. The experimental conditions were initial concentration of 10 mg/L ZnO nano powder, TiO_2 addition amount of 1.0 mg/L, initial acidity of pH 7.0, systemic temperature of 25.0 ± 0.2 °C, ultrasonic irradiation time of 60 min and total volume of 50 mL were collected to analyse the sonocatalytic activity of nanosized ZnO powder. It was found that the characteristic absorption peaks of both Acid Red B and Rhodamine B solutions at 510 nm and 560 nm, respectively, declined under ultrasonic irradiation in the presence of nanosized ZnO powder. The peaks declined as a result of decomposition and showed great results when compared to other samples with onefold ultrasonic degradation and onefold adsorption of the nanosized ZnO powder. In comparison with Acid Red B and Rhodamine B, the enhancement effect of nanosized ZnO powder on the sonocatalytic degradation of Acid Red B was much better than that of Rhodamine B. Degradation ratio of Acid Red B in the presence of nanosized ZnO powder was 71.2% in 60 min ultrasonic irradiation, while it was only 39.1% for Rhodamine B at the same time. In the absence of ultrasound irradiation, the degradation percentage of Acid Red B and Rhodamine B was 22.3% and 8.2%, respectively. Here also the reaction kinetics was inferred as first-order reaction. The rate constant for Acid Red B is much greater than that for Rhodamine B, which is $0.016\ min^{-1}$ and $0.0066\ min^{-1}$, respectively, and both confirmed pseudo first-order kinetic reactions. As initial concentration of dye waste water increases, the degradation ratio gradually decreases because of the increase in dye molecules on the surface of ZnO particles. This will hinder OH^- ions from being absorbed by ZnO nanoparticles; thus lowering the formation rate of hydroxyl radicals (OH^-), and consequently, it affects negatively the degradation efficiency. Degradation ratios of both Acid Red B and Rhodamine B increased with the increase of nanosized ZnO powder up to 1.0 g/L after which Rhodamine showed a slight decrease in the degradation ratio which may be due to the nearness and aggregation of nanosized ZnO particles [52] (Fig. 1).

5 Effects of Parameters on ZnO Dye Degradation

5.1 *Effect of pH on the Photocatalytic Activity*

The effect of pH on the adsorption efficiency of the materials used for adsorption and the removal capacity of the adsorbate or adsorbent was evaluated [39]. The

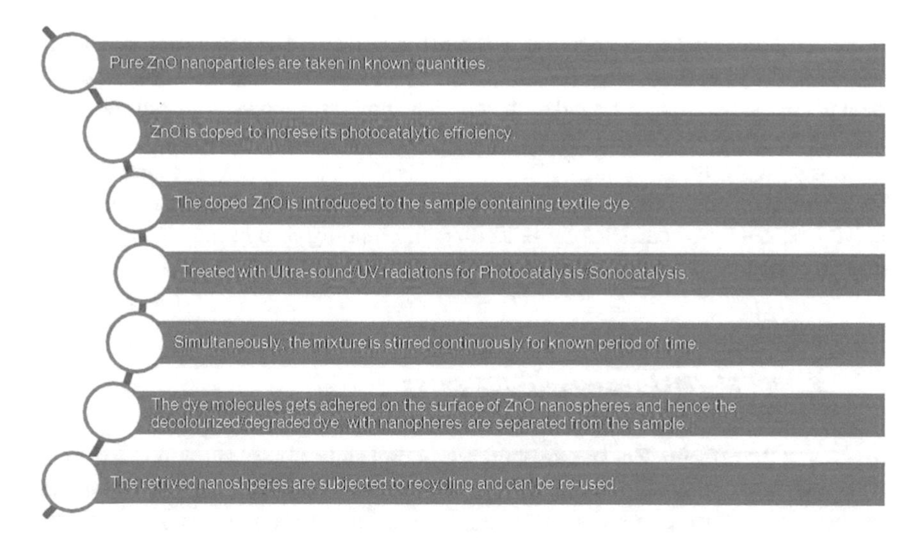

Fig. 1 Schematic representation of process of degradation

studies where the influence of pH was considered as criteria that can alter the photocatalytic process were conducted in the pH range of 4–10 using ZnO nanosphere catalysts dissolved in dye solution in an ideal reaction time. The adjustments of pH in the solution were maintained using acidic and basic solutions like HCl and NaOH, respectively. The rate of degradation and bleaching of dye are influenced or affected by the change in pH. The rate of degradation and bleaching of dye were observed to be high or efficient at pH 6 which is a bit acidic. The graphical elucidation resulted due to the variation of pH has shown that the efficiency of photodegradation was maintained high and raised till pH 6. After a slight constant curve, the graph declined when there is an increase in the pH number (that is, >pH 6 viz. 7–14). In the above study, the effect of pH in dye degradation is dependent on semiconductor oxide (CuO, WO_3, NiO etc.) that causes acid–base equilibrium which regulates the surface dynamics of nanocatalyst activities [47, 54]. Hence, the experiments use pH 6 as an optimum pH to carry out photocatalytic dye degradation. The effect of pH in the removal or adsorption efficiency (surface chemistry of nanocatalysts) is due to the variation of electrostatic force of attraction on ZnO nanospheres where the dye molecules adhere. Due to the variation of pH in the medium, the increase in the pH to optimum levels (i.e., pH 6) attracts the dye molecules and decrease in the optimum pH or any pH which is more than pH 6 causes the removal of dye and in turn decreases the activity of nanoparticles as the electrostatic force of attraction is directly proportional to the adsorption and removal of dye molecules on the surface of nanoparticles [57]. At the basic pH, that is, pH 9, the surface of ZnO nanoparticle possesses positive charges [8]. Hence, due to the property of electrostatic force of attraction, any pollutant that carries a negative charge (anion pollutant) moves to the adsorbent surface of nanoparticles [45]. The pKa value of dye with elevated pH of the solution, the dye exists with negative ion for and gets adhered to the adsorbent surface of nanoparticles. Thus, the increase in the pH to 6, the rate of adsorption of anionic pollutant or dye and the degradation efficiency of ZnO nanoparticles increases. At ideal pH 6, the rate of dye degradation and bleaching rate increases. The above property can be accredited to the decreased oxidation potential of OH^- radical that has an impact on pH as hydroxyl radical contributes to increasing the pH of the solution [29]. Simultaneously, the increased hydroxyl ions also affect the surface adhering activity of ZnO nanoparticles by inhibiting the binding of dye molecules that are anionic in nature to the surface of photocatalyst [41].

5.2 Effect of ZnO Nanospheres Dose

To observe the effect of ZnO nanospheres or photocatalysts concentration on the rate of degradation and decolouration of dye molecules, a variable range of photocatalyst nanospheres were introduced to the solution for a certain period of time at the optimum pH 6 with certain initial dye concentration where the range of concentration of nanosphere lies in between 50 and 300 mg/L. The findings and reports proved that the rate of degradation and rate of decolourization were elevated due to the increase in the concentration of photocatalytic nanospheres to a higher range compared to initial concentration. In further increase in the concentration of catalyst, there is no raise in the degradation and decolourization of dye with the increase in concentration of catalyst. When the concentration was further more enhanced, the decrease in the photocatalyst activity causes less decolourization and degradation of dye. Henceforth, this shows there exists an optimum concentration of photocatalyst for its enhanced or ideal activity. Until a certain concentration, the decolourization and degradation is directly proportional to the activity of photocatalytic nanospheres. The increased release of hydroxyl radicals are due to the catalytic activity of nanospheres cribbles the rate of degradation and decolourization. When the concentration of nanoparticles is increased, the solution turns turbid, a milky white turbidity can be observed which prevents the penetration of UV light that causes decreased effect of UV irradiation. This phenomenon causes decreased efficiency in dye degradation. Thus, the concentration of nanosphere where it shows the ideal activity in decolourizing and degrading the dye becomes a critical criterion to be considered for further investigation. The ideal concentration of ZnO nanospheres is determined to be 100 mg/L with the corresponding and correlating factors that are known [14, 55].

5.3 Effect of Calcination (Elevated Temperature) on the Optical Activity of ZnO Nanospheres

Range of temperatures °C was used to determine the effect of calcination on the photocatalytic activity of ZnO nanospheres. The effects of calcinated photocatalyst are increased with the elevation of temperatures in the process of dye decolourization and degradation. The photocatalytic activity is elevated with the increase until a certain temperature and then the activity becomes constant producing a plateau-like curve when plotted. The above phenomenon is attributed to the carbon substrate residue CTAB adhered on the surface of nanoparticles which affects the photocatalytic activity. The highest possible or efficient or ideal calcinationed temperature of 450 °C is preferred by the conducted experiments where the complete burning of CTAB happens [42].

5.4 Influence of Stirring

To study the effect of stirring on the rate of decolourization and dye degradation was studied under the above-mentioned ideal conditions (where the time 50 min, ideal pH 6, initial dye concentration of 10 mg/mL and the concentration of ZnO nanospheres 100 mg/L). The stirring results in the elevated rate of dye degradation and decolourization. Thus, the effect of rate of stirring is directly proportional to the increased dye photocatalytic activity. The stirring causes the release of trapped oxygen in between the nanosphere and removes hydroxyl radicals that are present, thus increasing in surface area for the adsorption of dye. To note, the stirring of the solution causes a decrease in the rate of equilibrium period hence increasing the rate of diffusion of dye molecules that adhere to the surface of photocatalysts [4].

5.5 Effect of Oxygen Concentration

The effect of concentration of oxygen in the photocatalyst has an impact on the rate of dye degradation and decolourization. A current of air containing oxygen and nitrogen was conducted through the solution to analyse the efficiency of ZnO photocatalyst. The procedure was conducted under the ideal condition (where the time 50 min, ideal pH 6, initial dye concentration of 10 mg/mL and the concentration of ZnO nanospheres 100 mg/L) and the gas containing nitrogen was passed through the solution for 2.5 min. The passage of nitrogen caused the removal of oxygen that decreased the activity of photocatalyst. Thus, it is determined that the presence of oxygen in the solution causes increased efficiency of photocatalyst nanospheres in dye degradation and decolourization. The free oxygen radicals carrying negative charge attributes to the oxidation of hydroxyl radicals resulting in the formation of hydrogen peroxide (H_2O_2) and molecular oxygen (O_2). Thus increases the efficiency of the photocatalytic ZnO nanospheres [28].

5.6 Recycling of the Photocatalytic ZnO Nanospheres

The recycling of photocatalytic ZnO nanospheres has its importance in the economic point of view. The reused photocatalytic ZnO nanospheres can be used by recovering them from the sources where the decolourization and degradation were already being conducted and can be utilized for further photocatalytic activities. To reuse the pre-used ZnO nanospheres the efficiency should be retrieved. The recovery of pure ZnO nanospheres by removal of the decolourized and degraded dye and impurities are done by subjecting them into multiple cycles of centrifugation under >4000 rpm. Thus, the impurities and degraded dyes are removed also by rinsing the centrifuged ZnO nanospheres using ethanol and bi-distilled water. The duly recovered and purified ZnO nanospheres can be reused in dye degradation. The efficiency or the capacity of ZnO nanoparticles that are recovered by recycling are comparatively lower than that of the freshly prepared or unused ZnO nanospheres. The recycled nanospheres also show a decrease in its activity of rate of decolourization and degradation processes in multiple usage [42] (Fig. 2).

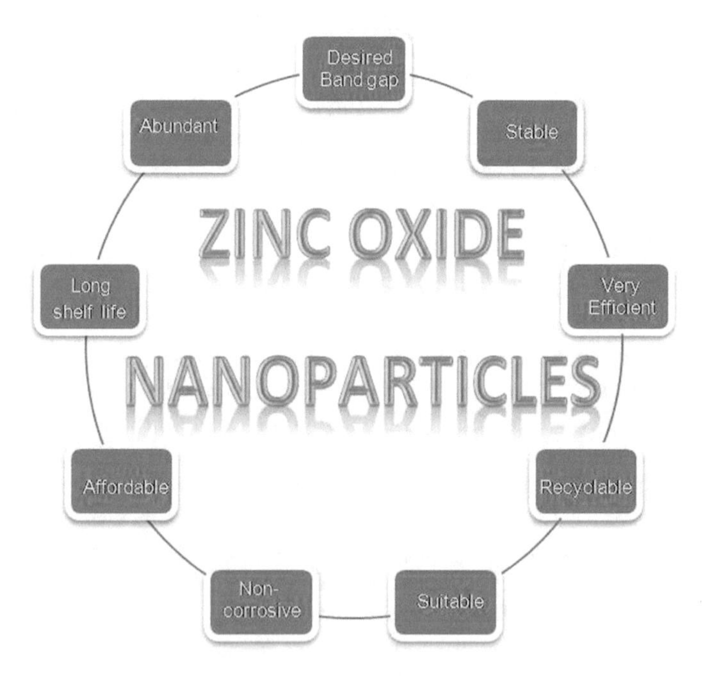

Fig. 2 Schematic representation of properties attributed to ZnO nanospheres

6 Discussions

Since ages, textile industries have been using a lot of variant dyes that are washed out as waste through outlets to the natural water bodies. The content of these outlets might be organic, inorganic, natural or synthetic but are hazardous to the biome present in the water bodies. Despite being hazardous they are also non-biodegradable. Different ways of approaches have been conducted till date to degrade the dyes, among which the use of nanospheres or nanoparticles are photocatalyts or sonocatalysts that are known to successfully decolourize or bleach and degrade the dyes. Few transition metal oxides are well known for their property of adsorption due to their capacity to produce lone-pair of electrons and holes by providing the dyes to adhere on its surface. Different transition metals like TiO_2, WO_3, Fe_2O_3, ZnO, MgO and other transition metal oxides are used for the preparation of photocatalytic nanospheres among which ZnO is known to be stable, efficient, affordable and environment-friendly catalyst. ZnO is a semiconductor metal oxide which has the capacity to hydrolyze aqueous solutions. When ZnO nanospheres are used to decolourize and degrade the dye pollutants present in the water under the influence of UV irradiation causes hydrolysis of water producing OH^- and H^+. These OH^- and H^+ might mask the ability of ZnO to act as photocatalyst by acting as electron scavengers.

To avoid this phenomenon of electron scavenging by OH^- and H^+ species, ZnO nanocatalysts are doped. Doping helps in the generation of charge carriers for the degradation of dye pollutants by recombinating photogenerated electron–hole pair. Thus, it helps to achieve high photocatalytic efficiency.

There are different studies that are carried out to dope ZnO with other narrow band gap semiconductors. This process is known as Co-doping as it attributes more qualities of photocatalysis when combined with ZnO nanospheres when compared to pure ZnO nanospheres. There are different ways of approach in dye degradation due to the multiple catalytic properties of ZnO or doped ZnO, they are photocatalysis, sonocatalysis under the influence of UV irradiation, IR radiation, natural light treatment, hydrothermal activity, combustion and calcination at high temperature.

The applications of nanoparticles are seen in diverse areas such as agriculture, animal husbandry, chemical sciences, medical sciences, biotechnology, electronic engineering veterinary science etc. They have been also used in medicinal treatments, fabrication of devices like solar cells, oxide fuel batteries for energy storage, as biosensors, cosmetic products, UV protection fabrics and so on. Antibacterial agents such as antiseptics, antidandruff shampoos, baby powder etc. contain ZnO particles. The antibacterial property and other biological functioning of ZnO nanoparticles depend on the concentration, morphology, size, healing, temperature and affinity between the bacterial cells and ZnO. On the other hand, Ag/NPs showed an effective antibacterial, antifungal and antiviral property.

Semiconductor-assisted photocatalytic degradation is found to be an effective method to degrade organic compounds present in the portable water sources. Apart from physiochemical methods like adsorption, nanofiltration and electrochemical methods, Advanced Oxidative Process (AOPs) is proved to be a good technique for the degradation of organic pollutants. Sonocatalysis is another emerging AOP to remove persistent organic compounds.

Sonocatalysts are active under ultrasonic irradiation and have a high rate of degradation of organic contaminants. TiO_2, ZnO, Cu_2S, CdS, ZnS, ZnO–NRs–AC and Cu: ZnS–NPs–AC, MIL–101(Cr)/RGO/$ZnFe_2O_4$ and $CoFe_2O_4$@ZnS are various catalysts that show good sonocatalytic activities. WO_3 is one of the promising photocatalysts for many opto-electronic applications etc. WO_3/ZnO nanorods synthesized via Hydrothermal-deposition technique are found to be highly effective in the degradation of 2,4-dichlorophenoxy acetic acid (2,4-D) under natural sunlight. At 600 °C calcination temperature WO_3/ZnO composites with 2 wt% concentration of WO_3 showed doubled photocatalytic activity as compared to pure ZnO. TiO_2 (3.37 eV) and ZnO (3.18 eV) are wide band gap semiconductor metal oxides, which have great importance due to their high catalytic efficiency. Noble metals like Ag/Pt can act as electron sinks thereby reducing the electron–hole recombination and charge separation. Noble metal/ZnO has the capacity to enhance the corrosion resistance, to improve the photocatalytic efficiency and photostability. Ag ions have attracted attention for their combined property of both photoactivity and antibacterial properties. Solvothermal method is another technique of preparing large-scale ZnO nanostructures. Cetyltrimethylammonium bromide (CTAB) has been used in this solvenothermal technique as organic dispersant. They interact with ZnO to form nanowires on the surface of ZnO by the synergic effect of the surfactant and the solvent (ethanol). Solvenothermal method helps in the deposition of noble-NPs on the ZnO–NWs and thereby obtains nanoparticles with a high surface to volume ratio, even distribution of noble metal etc. Flame aerosol synthesis is an alternate method used to prepare nanoscale materials commercially at a low cost. Flame Spray Pyrolysis (FSP) is used in particular for one-step drying synthesis of high surface area noble metal-laden catalysis. This method provides highly pure particles, control of particle size and good crystalline structure on one step.

Aquatic system is contaminated due to the growth of industrialization which releases chemicals like organic dyes, gases, oils, heavy metals which are hazardous to plants and animals. According to WHO, contaminated water caused many diseases like Giardiasis, Cholera, Viral gastroenteritis etc. which kills millions of people each year. The dye from textile industries, tanneries, kraft mills and other industrial wastes are non-biodegradable chemicals and contribute to the elevated levels of Biological Oxygen Demand (BOD), Chemical Oxygen Demand (COD), pH and colour. Hence there is a requirement for an efficient, harmless and environment-friendly technique to degrade the pollutants. Nanoparticles of size ranging from 1 to 100 nm have been studied in a wide range for their remarkable properties like catalytic degradation, antibacterial agents etc. Various studies have been conducted regarding the use of ZnO nanospheres as dye degrading components. Few among them are tabulated (Table 1).

Table 1 Various studies on dye degradation using different combinations of ZnO nanocomposites

Sl. no	Dye degraded	Nanocomposites used	Mode of degradation	Photo-spectrum	Type of degradation	References
1	Methylene blue	CuO/ZnO	Thermal decomposition	Visible light spectrum	Photocatalytic	[27]
2	Methylene blue	CuO	Carbothermal	Sunlight irradiation	Photocatalytic	[27]
3	Methylene blue	WO3	Precipitation grinding method	Sunlight	Photocatalytic	[1, 17, 56]
4	Methylene blue	N/ZnO	Hydrothermal decomposition	UV–Visible light range	Photocatalytic	[20, 24]
5	Methylene blue	N/ZnO	Combustion reaction	Visible light	Photocatalytic	[37, 46]
6	Methylene blue	Ni/ZnO	Co-gel/Co-precipitation techniques	UV–Visible spectrum	Photocatalytic	[11, 27]
7	Methylene blue and eosin Y	ZnO		UV irradiation (16 W lamp as source)	Photocatalytic	[7]
8	Methyl orange	CuO/ZnO	Thermal decomposition	Visible light spectrum	Photocatalytic	[27]
9	Methyl orange	CuO	Xe-Lamp	Photo-degradation	Photocatalytic	[27]
10	Azo Rhodamine b	CuO	Xe-Lamp	Photo-degradation	Photocatalytic	[27]
11	Brilliant blue	WO_3-ZnO	–	Ultrasonic irradiation	Sonocatalytic	[17]
12	Methylene blue	ZnO-green synthesized Ag	–	Sunlight	Photocatalytic	[19]
13	Methylene blue	ZnO, ZnO/Ag, ZnO/Pt	Mercury vapour lamp	UV light irradiation	Photocatalytic	[32]
14	Methylene blue	ZnO, Au/ZnO, Pt/ZnO	UV light 2 _ 15 W UV tube	UV irradiation	Photocatalytic	[36]
15	Methylene blue	Ag/ZnO	–	Visible irradiations	Photocatalytic	[34]
16	Acid red B and Rhodamine B	ZnO	–	Ultrasonic irradiation	Sonocatalytic	[52]
17	Coomassie brilliant blue R-250 dye	Green synthesized ZnO	–	Sunlight	Photocatalytic	[44]

7 Conclusion

Due to the release of toxic dyes from textile industry into the natural water bodies, there are many hazardous conditions being formed that spoils the biome and upsets the ecosystem [7, 47]. To solve this, various approaches are being done among which ZnO dye degradation is one of the reliable and efficient methods [11, 42]. To conclude with the above work, reviewing different experiments and comparison of dye degradation using ZnO as a basic semiconductor metal ion nanoparticle reflects that the use of ZnO is more appropriate because of its attributing properties being efficient and environment-friendly. As stability place an important role in any chemical reactions concerned with approaches of environmental application that helps in the betterment of conditions like pollution, wastewater treatment etc. ZnO (pure or doped form) [23, 35, 37] has excellent properties of remaining unaffected and also contributing to degradation. Natural solar radiations and ultraviolet radiations can be used to enhance the properties of ZnO in dye degradation. The usage of ultrasound to create higher efficiency of the nanoparticles results in sonocatalysis. These properties have positive contribution to the field of environmental application, ZnO nanoparticles are highly credited among other semiconductor nanoparticles.

Bibliography

1. Adhikari S, Sarkar D, Madras G (2015) Highly efficient WO_3–ZnO mixed oxides for photocatalysis. RSC Adv 5(16):11895–11904. https://doi.org/10.1039/c4ra13210f
2. Anandan S, Wu JJ (2014) Ultrasound assisted synthesis of TiO_2–WO_3 heterostructures for the catalytic degradation of Tergitol (NP-9) in water. Ultrason Sonochem 21(4):1284–1288. https://doi.org/10.1016/j.ultsonch.2014.01.014
3. Anju SG, Yesodharan S, Yesodharan EP (2012) Zinc oxide mediated sonophotocatalytic degradation of phenol in water. Chem Eng J 189–190:84–93. https://doi.org/10.1016/j.cej.2012.02.032
4. Bagheri AR, Ghaedi M, Asfaram A, Jannesar R, Goudarzi A (2017) Design and construction of nanoscale material for ultrasonic assisted adsorption of dyes: application of derivative spectrophotometry and experimental design methodology. Ultrason Sonochem 35:112–123. https://doi.org/10.1016/j.ultsonch.2016.09.008
5. Bustos-Torres KA, Vazquez-Rodriguez S, la Cruz AM de, Sepulveda-Guzman S, Benavides R, Lopez-Gonzalez R, Torres-Martínez LM (2017) Influence of the morphology of ZnO nanomaterials on photooxidation of polypropylene/ZnO composites. Mater Sci Semicond Process 68(May):217–225. https://doi.org/10.1016/j.mssp.2017.06.023
6. Carraway ER, Hoffman AJ, Hoffmann MR (1994) Photocatalytic oxidation of organic acids on quantum-sized semiconductor colloids. Environ Sci Technol 28(5):786–793. https://doi.org/10.1021/es00054a007
7. Chakrabarti S, Dutta BK (2004) Photocatalytic degradation of model textile dyes in wastewater using ZnO as semiconductor catalyst. J Hazard Mater 112(3):269–278. https://doi.org/10.1016/j.jhazmat.2004.05.013
8. Daneshvar N, Aber S, Seyed Dorraji MS, Khataee AR, Rasoulifard MH (2007) Photocatalytic degradation of the insecticide diazinon in the presence of prepared nanocrystalline ZnO powders under irradiation of UV-C light. Sep Purif Technol 58(1):91–98. https://doi.org/10.1016/j.seppur.2007.07.016

9. Dzsaber S, Negyedi M, Bernáth B, Gyüre B, Fehér T, Kramberger C, Pichler T, Simon F (2015) A Fourier transform Raman spectrometer with visible laser excitation. J Raman Spectrosc 46(3):327–332. https://doi.org/10.1002/jrs.4641
10. Ebrahimi R, Maleki A, Zandsalimi Y, Ghanbari R, Shahmoradi B, Rezaee R, Safari M, Joo SW, Daraei H, Harikaranahalli Puttaiah S, Giahi O (2019) Photocatalytic degradation of organic dyes using WO_3-doped ZnO nanoparticles fixed on a glass surface in aqueous solution. J Ind Eng Chem 73:297–305. https://doi.org/10.1016/j.jiec.2019.01.041
11. Hameed A, Montini T, Gombac V, Fornasiero P (2009) Photocatalytic decolourization of dyes on NiO–ZnO nano-composites. Photochem Photobiol Sci 8(5):677–682. https://doi.org/10.1039/b817396f
12. Havrdova M, Polakova K, Skopalik J, Vujtek M, Mokdad A, Homolkova M, Tucek J, Nebesarova J, Zboril R (2014) Field emission scanning electron microscopy (FE-SEM) as an approach for nanoparticle detection inside cells. Micron 67:149–154. https://doi.org/10.1016/j.micron.2014.08.001
13. Hollerith C, Wernicke D, Bühler M, Feilitzsch FV, Huber M, Höhne J, Hertrich T, Jochum J, Phelan K, Stark M, Simmnacher B, Weiland W, Westphal W (2004) Energy dispersive X-ray spectroscopy with microcalorimeters. Nucl Instrum Methods Phys Res Sect A 520(1–3):606–609. https://doi.org/10.1016/j.nima.2003.11.327
14. Hossaini H, Moussavi G, Farrokhi M (2017) Oxidation of diazinon in cns-ZnO/LED photocatalytic process: catalyst preparation, photocatalytic examination, and toxicity bioassay of oxidation by-products. Sep Purif Technol 174:320–330. https://doi.org/10.1016/j.seppur.2016.11.005
15. Huang Z, Maness PC, Blake DM, Wolfrum EJ, Smolinski SL, Jacoby WA (2000) Bactericidal mode of titanium dioxide photocatalysis. J Photochem Photobiol A 130(2–3):163–170. https://doi.org/10.1016/S1010-6030(99)00205-1
16. Hunge YM, Mahadik MA, Moholkar AV, Bhosale CH (2017) Photoelectrocatalytic degradation of oxalic acid using WO_3 and stratified WO_3/TiO_2 photocatalysts under sunlight illumination. Ultrason Sonochem 35:233–242. https://doi.org/10.1016/j.ultsonch.2016.09.024
17. Hunge YM, Yadav AA, Mathe VL (2018) Ultrasound assisted synthesis of WO_3–ZnO nanocomposites for brilliant blue dye degradation. Ultrason Sonochem 45(January):116–122. https://doi.org/10.1016/j.ultsonch.2018.02.052
18. Hunge YM, Yadav AA, Mohite BM, Mathe VL, Bhosale CH (2018) Photoelectrocatalytic degradation of sugarcane factory wastewater using WO_3/ZnO thin films. J Mater Sci Mater Electron 29(5):3808–3816. https://doi.org/10.1007/s10854-017-8316-1
19. Jadhav P, Shinde S, Suryawanshi SS, Teli SB, Patil PS, Ramteke AA, Hiremath NG, Prasad NR (2020) Green AgNPs decorated ZnO nanocomposites for dye degradation and antimicrobial applications. Eng Sci 12:79–94. https://doi.org/10.30919/es8d1138
20. Kabir R, Saifullah MAK, Ahmed AZ, Masum SM, Molla MAI (2020) Synthesis of n-doped zno nanocomposites for sunlight photocatalytic degradation of textile dye pollutants. J Compos Sci 4(2). https://doi.org/10.3390/jcs4020049
21. Kanade KG, Kale BB, Baeg JO, Lee SM, Lee CW, Moon SJ, Chang H (2007) Self-assembled aligned Cu doped ZnO nanoparticles for photocatalytic hydrogen production under visible light irradiation. Mater Chem Phys 102(1):98–104. https://doi.org/10.1016/j.matchemphys.2006.11.012
22. Kavitha MK, Pillai SC, Gopinath P, John H (2015) Hydrothermal synthesis of ZnO decorated reduced graphene oxide: understanding the mechanism of photocatalysis. J Environ Chem Eng 3(2):1194–1199. https://doi.org/10.1016/j.jece.2015.04.013
23. Khalid NR, Hammad A, Tahir MB, Rafique M, Iqbal T, Nabi G, Hussain MK (2019) Enhanced photocatalytic activity of Al and Fe co-doped ZnO nanorods for methylene blue degradation. Ceram Int 45(17):21430–21435. https://doi.org/10.1016/j.ceramint.2019.07.132
24. Khataee A, Karimi A, Arefi-Oskoui S, Darvishi Cheshmeh Soltani R, Hanifehpour Y, Soltani B, Joo SW (2015) Sonochemical synthesis of Pr-doped ZnO nanoparticles for sonocatalytic degradation of acid red 17. Ultrason Sonochem 22:371–381. https://doi.org/10.1016/j.ultsonch.2014.05.023

25. Kuriakose S, Avasthi DK, Mohapatra S (2015) Effects of swift heavy ion irradiation on structural, optical and photocatalytic properties of ZnO–CuO nanocomposites prepared by carbothermal evaporation method. Beilstein J Nanotechnol 6(1):928–937. https://doi.org/10.3762/bjnano.6.96
26. Lam SM, Sin JC, Abdullah AZ, Mohamed AR (2013) Investigation on visible-light photocatalytic degradation of 2,4-dichlorophenoxyacetic acid in the presence of MoO_3/ZnO nanorod composites. J Mol Catal A Chem 370:123–131. https://doi.org/10.1016/j.molcata.2013.01.005
27. Li B, Wang Y (2010) Facile synthesis and photocatalytic activity of ZnO–CuO nanocomposite. Superlattices Microstruct 47(5):615–623. https://doi.org/10.1016/j.spmi.2010.02.005
28. Litter MI (2005) Introduction to photochemical advanced oxidation processes for water treatment. In: Environmental photochemistry part II, vol 2, Issue September, pp 325–366. https://doi.org/10.1007/b138188
29. Lucas MS, Peres JA (2006) Decolorization of the azo dye reactive black 5 by Fenton and photo-Fenton oxidation. Dyes Pigm 71(3):236–244. https://doi.org/10.1016/j.dyepig.2005.07.007
30. Ma H, Cheng X, Ma C, Dong X, Zhang X, Xue M, Zhang X, Fu Y (2013) Synthesis, characterization, and photocatalytic activity of N-doped ZnO/ZnS composites. Int J Photoenergy 2013. https://doi.org/10.1155/2013/625024
31. Moroni M, Borrini D, Calamai L, Dei L (2005) Ceramic nanomaterials from aqueous and 1,2-ethanediol supersaturated solutions at high temperature. J Colloid Interface Sci 286(2):543–550. https://doi.org/10.1016/j.jcis.2005.01.097
32. Muñoz-Fernandez L, Sierra-Fernandez A, Milošević O, Rabanal ME (2016) Solvothermal synthesis of Ag/ZnO and Pt/ZnO nanocomposites and comparison of their photocatalytic behaviors on dyes degradation. Adv Powder Technol 27(3):983–993. https://doi.org/10.1016/j.apt.2016.03.021
33. Pal B, Sharon M (2002) Enhanced photocatalytic activity of highly porous ZnO thin films prepared by sol-gel process. Mater Chem Phys 76(1):82–87. https://doi.org/10.1016/S0254-0584(01)00514-4
34. Panchal P, Paul DR, Sharma A, Choudhary P, Meena P, Nehra SP (2020) Biogenic mediated Ag/ZnO nanocomposites for photocatalytic and antibacterial activities towards disinfection of water. J Colloid Interface Sci 563:370–380. https://doi.org/10.1016/j.jcis.2019.12.079
35. Pascariu P, Tudose IV, Suchea M, Koudoumas E, Fifere N, Airinei A (2018) Preparation and characterization of Ni, Co doped ZnO nanoparticles for photocatalytic applications. Appl Surf Sci 448:481–488. https://doi.org/10.1016/j.apsusc.2018.04.124
36. Pawinrat P, Mekasuwandumrong O, Panpranot J (2009) Synthesis of Au-ZnO and Pt-ZnO nanocomposites by one-step flame spray pyrolysis and its application for photocatalytic degradation of dyes. Catal Commun 10(10):1380–1385. https://doi.org/10.1016/j.catcom.2009.03.002
37. Prabakaran E, Pillay K (2019) Synthesis of N-doped ZnO nanoparticles with cabbage morphology as a catalyst for the efficient photocatalytic degradation of methylene blue under UV and visible light. RSC Adv 9(13):7509–7535. https://doi.org/10.1039/C8RA09962F
38. Prencipe I, Dellasega D, Zani A, Rizzo D, Passoni M (2015) Energy dispersive X-ray spectroscopy for nanostructured thin film density evaluation. Sci Technol Adv Mater 16(2). https://doi.org/10.1088/1468-6996/16/2/025007
39. Rahmani M, Kaykhaii M, Sasani M (2018) Application of Taguchi L16 design method for comparative study of ability of 3A zeolite in removal of Rhodamine B and Malachite green from environmental water samples. Spectrochim Acta Part A Mol Biomol Spectrosc 188:164–169. https://doi.org/10.1016/j.saa.2017.06.070
40. Reddy IN, Reddy CV, Shim J, Akkinepally B, Cho M, Yoo K, Kim D (2020) Excellent visible-light driven photocatalyst of (Al, Ni) co-doped ZnO structures for organic dye degradation. Catal Today 340:277–285. https://doi.org/10.1016/j.cattod.2018.07.030
41. Saini J, Garg VK, Gupta RK, Kataria N (2017) Removal of Orange G and Rhodamine B dyes from aqueous system using hydrothermally synthesized zinc oxide loaded activated carbon (ZnO-AC). J Environ Chem Eng 5(1):884–892. https://doi.org/10.1016/j.jece.2017.01.012

42. Saleh SM (2019) ZnO nanospheres based simple hydrothermal route for photocatalytic degradation of azo dye. Spectrochim Acta Part A Mol Biomol Spectrosc 211:141–147. https://doi.org/10.1016/j.saa.2018.11.065
43. Sing KSW (1989) The use of gas adsorption for the characterization of porous solids. Colloids Surf 38(1):113–124. https://doi.org/10.1016/0166-6622(89)80148-9
44. Singh K, Singh J, Rawat M (2019) Green synthesis of zinc oxide nanoparticles using Punica Granatum leaf extract and its application towards photocatalytic degradation of Coomassie brilliant blue R-250 dye. SN Appl Sci 1(6). https://doi.org/10.1007/s42452-019-0610-5
45. Šťastná M, Trávníček M, Šlais K (2005) New azo dyes as colored isoelectric point markers for isoelectric focusing in acidic pH region. Electrophoresis 26(1):53–59. https://doi.org/10.1002/elps.200406088
46. Sudrajat H, Babel S (2017) A novel visible light active N-doped ZnO for photocatalytic degradation of dyes. J Water Process Eng 16:309–318. https://doi.org/10.1016/j.jwpe.2016.11.006
47. Tanaka K, Padermpole K, Hisanaga T (2000) Photocatalytic degradation of commercial azo dyes. Water Res 34(1):327–333. https://doi.org/10.1016/S0043-1354(99)00093-7
48. Tang WZ, An H (1995) UV/TiO_2 photocatalytic oxidation of commercial dyes in aqueous solutions. Chemosphere 31(9):4157–4170. https://doi.org/10.1016/0045-6535(95)80015-D
49. Turan R, Perovic DD, Houghton DC (1996) Mapping electrically active dopant profiles by field-emission scanning electron microscopy. Appl Phys Lett 69(11):1593–1595. https://doi.org/10.1063/1.117041
50. Ullah R, Dutta J (2008) Photocatalytic degradation of organic dyes with manganese-doped ZnO nanoparticles. J Hazard Mater 156(1–3):194–200. https://doi.org/10.1016/j.jhazmat.2007.12.033
51. Velavan S, Amargeetha A (2018) X-ray diffraction (XRD) and energy dispersive spectroscopy (EDS) analysis of silver nanoparticles synthesized from erythrina indica flowers. Nanosci Technol Open Access 5(1):1–5. https://doi.org/10.15226/2374-8141/5/1/00152
52. Wang J, Jiang Z, Zhang Z, Xie Y, Wang X, Xing Z, Xu R, Zhang X (2008) Sonocatalytic degradation of acid red B and Rhodamine B catalyzed by nano-sized ZnO powder under ultrasonic irradiation. Ultrason Sonochem 15(5):768–774. https://doi.org/10.1016/j.ultsonch.2008.02.002
53. Xie J, Zhou Z, Lian Y, Hao Y, Liu X, Li M, Wei Y (2014) Simple preparation of WO_3–ZnO composites with UV-Vis photocatalytic activity and energy storage ability. Ceram Int 40(8 Part A):12519–12524. https://doi.org/10.1016/j.ceramint.2014.04.106
54. Yang TCK, Wang SF, Tsai SHY, Lin SY (2001) Intrinsic photocatalytic oxidation of the dye adsorbed on TiO_2 photocatalysts by diffuse reflectance infrared Fourier transform spectroscopy. Appl Catal B 30(3–4):293–301. https://doi.org/10.1016/S0926-3373(00)00241-1
55. Youssef Z, Colombeau L, Yesmurzayeva N, Baros F, Vanderesse R, Hamieh T, Toufaily J, Frochot C, Roques-Carmes T (2018) Dye-sensitized nanoparticles for heterogeneous photocatalysis: cases studies with TiO_2, ZnO, fullerene and graphene for water purification. Dyes Pigments (Elsevier Ltd.) 159:49–71. https://doi.org/10.1016/j.dyepig.2018.06.002
56. Yu C, Yang K, Shu Q, Yu JC, Cao F, Li X (2011) Preparation of WO_3/ZnO composite photocatalyst and its photocatalytic performance. Cuihua Xuebao/Chin J Catal 32(4):555–565. https://doi.org/10.1016/s1872-2067(10)60212-4
57. Zbair M, Anfar Z, Ait Ahsaine H, El Alem N, Ezahri M (2018) Acridine orange adsorption by zinc oxide/almond shell activated carbon composite: operational factors, mechanism and performance optimization using central composite design and surface modeling. J Environ Manag 206:383–397. https://doi.org/10.1016/j.jenvman.2017.10.058
58. Zhang L, Qi H, Yan Z, Gu Y, Sun W, Zewde AA (2017) Sonophotocatalytic inactivation of E. coli using ZnO nanofluids and its mechanism. Ultrason Sonochem 34:232–238. https://doi.org/10.1016/j.ultsonch.2016.05.045

Carbon Nitride Application on Advanced Oxidation Processes for Dye Removal

R. Suresh, Saravanan Rajendran, and Lorena Cornejo-Ponce

Abstract Water contamination by toxic organic dyes is a major concern in health and environmental safety. Numerous studies on the catalytic application of graphitic carbon nitride (g-C_3N_4) in advanced oxidation processes (AOPs) which are employed for degradation of organic dye pollutants have been reported. The g-C_3N_4 based catalysts have a good prospect in terms of sustainable environmental remediation techniques, due to their greater chemical stability, optical and metal free semiconducting property. In this book chapter, we compiled research articles related to g-C_3N_4 based catalysts, used in various AOPs exclusively for organic dye degradation reactions. This chapter contains four sections, in the first section, need of AOPs for organic dye contamination in water treatment was mentioned. Section 2 describes the fundamental properties such as structure, optical, and semiconducting behavior of pure g-C_3N_4 powder. Third section comprises research studies on g-C_3N_4 based catalysts in AOPs, particularly, catalytic activation of oxidants, photocatalysis, sonocatalysis, and Fenton-like process for effective dyes removal from water. Moreover, improvement strategies like doping process and composite formation adopted in g-C_3N_4 were highlighted. The catalytic mechanism of g-C_3N_4 based catalysts in each process was also discussed. Finally, summary and conclusion of the selected topics were given.

Keywords Waste water · Advanced oxidation process · Organic dyes · g-C_3N_4 · Composites · Doping process · Ozonation · Photocatalysis · Sonocatalysis · Fenton-like process

1 Introduction

Dyes are colored substances that will impart color to applied materials including fabrics, leather, paper, and food. They are usually organic compounds and readily

R. Suresh (✉) · S. Rajendran · L. Cornejo-Ponce
Laboratorio de Investigaciones Medioambientales de Zonas Áridas (LIMZA), Departamento de Ingeniería Mecánica, Facultad de Ingeniería, Universidad de Tarapacá, Avda. General Velásquez, 1775 Arica, Chile

S. S. Muthu and A. Khadir (eds.), *Advanced Oxidation Processes in Dye-Containing Wastewater*, Sustainable Textiles: Production, Processing, Manufacturing & Chemistry, https://doi.org/10.1007/978-981-19-0882-8_13

soluble in water. They can be broadly classified as natural and synthetic dyes (artificial dyes). Leaves, roots, flowers, vegetables, and fruits of many plants and insects have varieties of natural dyes [1]. Natural dyes are biocompatible and they have medicinal values too [2]. In 1856, aniline purple (Mauveine), a first artificial dye was synthesized by Henry Perkin. In the beginning of nineteenth century, vast number of artificial dyes or derivatives from natural dyes was synthesized. Compared to natural dyes, synthetic dyes can be prepared in large scale with low cost, due to the availability of heaper precursors. Nowadays, large quantities of synthetic dyes are usually utilized in manufacturing industries like fabrics, plastics, food, leather, and cosmetics [3]. Approximately, 10,000 dye varieties are being used by these industries with their annual production of over 0.7 million tons [4]. A study depicts that per year, ~20,000 tons of dyes are released with effluents into nearby environment owing to inadequate dyeing process [5]. Dyes in these effluents are stable and non-biodegradable in nature [6]. Besides, dyes show toxic effects such as reduction of photosynthesis in aquatic plants and phytoplankton and oxygen consumption thereby affect aquatic organisms [3]. Synthetic dyes also pollute surface and groundwater through surface run-off and percolation process [7]. Hence, dyes must be properly treated and then discharged in the environment.

Treatment methods like adsorption, coagulation-flocculation, ion exchange, membrane filtration, electrochemical methods, catalytic degradation, hydrodynamic cavitation, using subcritical and supercritical water and biological process (bacteria, algae, and enzymes) were developed for the treatment of various dyes contaminated effluents [8–11]. But, most of these methods suffer in the complete removal of dye effluents. Also, they need high capital investment, more time and energy, and may require additional treatments. Studies have also revealed that some dyes are refractory to current wastewater treatment plants [12]. Hence, complete mineralization of dyes in wastewater is a foremost challenge. In this context, AOPs were developed to achieve complete removal of aquatic organic dye contaminants.

In AOPs, dyes are decomposed by strong oxidizing agents like hydroxyl radicals ($^{\bullet}OH$) which are in-situ generated by ozonation, ultraviolet (UV), or catalytic activation of oxidants, sonocatalysis, electrochemical, Fenton reactions, photo-Fenton, electro-Fenton reactions, and photocatalysis (Fig. 1). These processes are commonly operated under specific temperature, UV light, pressure, pH, oxidant, and catalyst. Materials such as metal oxides and other compounds, pure metals, alloys, carbon-based materials, metal-organic frameworks, and hybrid materials were used as catalyst in light, temperature, ultrasound, and electricity assisted AOP for removal of dye pollutants [13].

In this chapter, utilization of g-C_3N_4 in catalytic activation of oxidizing agents, photocatalysis, sonocatalysis, and heterogeneous Fenton-like process for effective elimination of dyes were described in detail. Further, studies on improvement strategies in g-C_3N_4 were also explained. The pathway of production of free radicals by g-C_3N_4 based catalysts was also described.

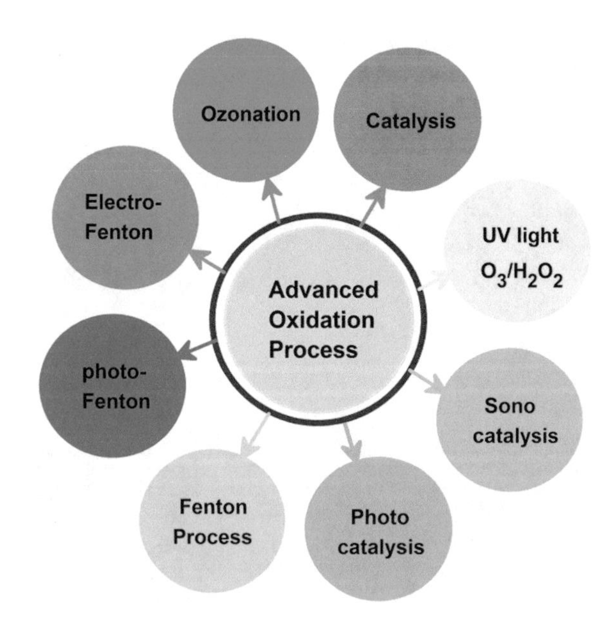

Fig. 1 Different types of advanced oxidation processes

2 Basics of g-C_3N_4

In 1834, Berzelius synthesized a polymeric melon (carbon nitride). In 1922, Franklin predicated the empirical composition of "melon" to be C_3N_4. In 1937, Pauling and Sturdivant predicted structure of C_3N_4 (tri-s-triazine). In 1940, Redemann and Lucas deduced that C_3N_4 has a graphite structure. So far, there are five allotropes of C_3N_4 viz. $\alpha - C_3N_4$, $\beta - C_3N_4$, cubic-C_3N_4, pseudo cubic-C_3N_4, and g-C_3N_4 [14]. Among them, g-C_3N_4 a stable allotrope of carbon nitride is used in various applications including energy production [15], environmental remediation [16], CO_2 sequestration [17], electrochemical devices [18], sensor [19], and hydrogen storage [20]. The vast applications of g-C_3N_4 are due to the following properties.

(a) Structure

The g-C_3N_4 has a similar structure to graphene, i.e., it has polymeric tri-s-triazines units connected through the tertiary amine group [21]. The two-dimensional layered polymeric structure contains s-triazine and tri-s-triazine units (Fig. 2) [22]. This flexible structure favors various metals and non-metal doping in g-C_3N_4.

(b) Optical property

Bulk g-C_3N_4 absorbs visible light (violet-blue; $\lambda = \sim 460$ nm) from solar light and the corresponding band gap is ~2.7 eV, this is due to the presence of higher energy nitrogen orbital. The absorption of violet-blue light (visible region) by C_3N_4 leads π–π^* transitions of the heterocyclic aromatic moieties [23]. It renders pale yellow to deep brown coloration to the g-C_3N_4 powder. Moreover, degree of polymer condensation also determines the color of g-C_3N_4 powder. During condensation, the

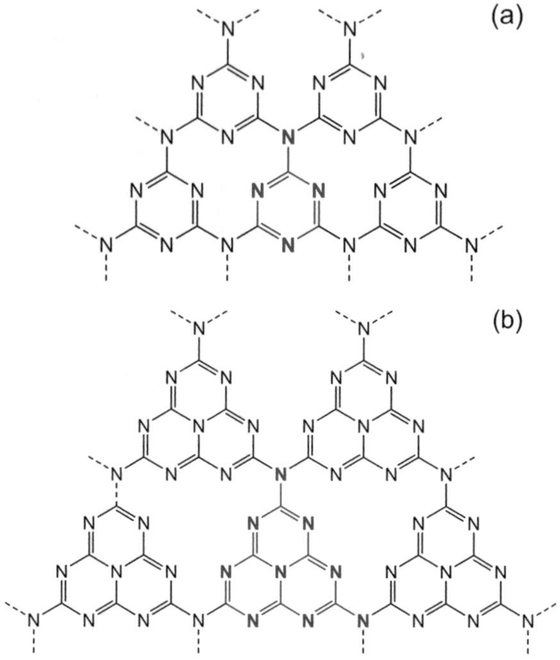

Fig. 2 The g-C_3N_4 structure contains **a** s-triazine and **b** tri-s-triazine units [22]

polytriazine and polyheptazine units become increasingly buckled. The non-bonded electrons present on nitrogen atoms undergo non-allowed n–π* transitions. It leads redshift in the absorption edge which gives brown color to the g-C_3N_4 powder [24].

(c) Stability

The g-C_3N_4 is a thermally stable compound, owing to the existing covalent linkage between C and N atoms. According to the thermogravimetric study, g-C_3N_4 was found to be stable at 600 °C in air [25]. Nonetheless, triazine unit will be decomposed at elevated temperature under pure oxygen atmosphere. Its high chemical stability is attributed to the Vander Waal's force of attraction [26].

(d) Semiconducting property

The g-C_3N_4 is a non-metallic semiconducting material. By using density functional theory calculations, Wang et al. [23] have determined band gap, i.e., energy gap between highest occupied molecular orbital (HOMO) and lowest unoccupied molecular orbital (LUMO) level of g-C_3N_4 (~2.7 eV). Like semiconductors, under light illumination, electrons are excited from valence to conduction band of g-C_3N_4. Here, lone pair electrons of P_Z orbitals of N atoms form valence band (lone pair valence band), while P_Z orbitals of C atoms form conduction band [27]. Later, El-kader et al. [28] have also established this observation by their experimental evidences.

(e) Easy synthesis

The g-C_3N_4 possesses only C and N atoms which are highly abundant in the Earth. It can be easily synthesized by simple thermal condensation approach using low-cost nitrogen-rich precursors like urea, dicyandiamide, and melamine [29]. During calcination step, precursor undergoes self-condensation, leading to the formation of g-C_3N_4.

3 Role of g-C_3N_4 in AOPs for Dye Removal

3.1 g-C_3N_4 Induced Activation of Oxidants

Ozone (O_3), an unstable molecule was employed as oxidant for decomposition of organic dyes due to its high oxidation potential ($E_0 = 2.07$ V versus NHE). In two ways, organics degradation reaction with ozone takes place (direct oxidation). At acidic condition, ozone directly attacks and breaks specific parts of dye molecule such as aromatic ring, unsaturated bonds, thereby degrading dye molecules [30]. In basic to neutral pH condition, ozone undergoes decomposition to form hydroxyl radical ($E_0 = 2.80$ V versus NHE) (Eq. 1) which led degradation reaction unspecifically with dye molecules (indirect oxidation). But, solution pH will be decreased owing to the production of acidic products. Consequently, the degradation pathway shifts toward direct oxidation mechanism. Indeed, ozonation leads to discoloration of dyes, but it does not afford complete mineralization.

$$O_3 + OH^- \rightarrow OH^\bullet \quad (1)$$

Hence, catalytic ozonation reaction was adopted for dye contaminated wastewater treatment. In presence of suitable solid catalyst such as metal oxides, carbon-based materials (heterogeneous ozonation), ozone decomposition is greatly increased, which produces more active radicals than ozone, thereby increasing the mineralization rate [31]. Additionally, catalyst is also promoting dye mineralization process due to its high surface area. Advantages of this method are effective utilization of ozone, less degradation process time, no sludge formation, and facile recovery of catalysts [32]. It should be mentioned that the application of pure g-C_3N_4 as catalyst in ozonation of dyes is rare [33]. However, metallic $Zn^{(0)}$ (5.4 wt%) immobilized g-C_3N_4 composite was used as catalyst for ozonation of degradation of atrazine in water [33]. Around 61.2% catalytic efficiency was achieved within 1.5 min. Higher catalyst dosage, solution pH (3.0 to 9.0) and initial concentration (1 to 10 mg/L) also affected atrazine degradation efficiency. Superior catalytic activity of Zn/g-C_3N_4 composite was attributed to high dispersibility, greater surface area, and conductivity. Moreover, $O_2^{\bullet-}$, $^\bullet OH$, and $^1O_2$ were involved in this catalytic degradation reaction.

Alternatively, sulfate radical ($SO_4^{\bullet-}$) based AOPs also gained much interest in dye removal. Oxidants such as peroxydisulfate ($S_2O_8^{2-}$, E° ($S_2O_8^{2-}/SO_4^{2-}$) =

2.01 V_{NHE}) and peroxymonosulfate (HSO_5^-, E∘ (HSO_5^-/SO_4^{2-}) = 1.75 V_{NHE}) are used for the production of sulfate radicals since they are cheaper, and having relatively long half-life period [16]. The g-C_3N_4 based catalysts are applied in sulfate radical-mediated dye degradation process. For example, g-C_3N_4 containing OMS-2 composite catalyst was also reported for activation of peroxymonosulfate for decomposition of acid orange 7 [34]. OMS-2 is a type of MnO_2 (2 × 2 edge-shared MnO_6 octahedral chains) [35]. The g-C_3N_4/OMS-2/peroxymonosulfate system showed 88% of degradation efficiency toward acid orange 7 within 30 min. Also, good degradation efficiencies were observed for degradation of reactive brilliant red X-3B, reactive brilliant blue KN-R, and methylene blue within 30 min in presence of g-C_3N_4/OMS-2/peroxymonosulfate system. As another example, Fe_3O_4@C/g-C_3N_4 catalyst was used for the activation of peroxymonosulfate in the decomposition of acid orange 7 [36]. This composite catalyst exhibits 97% of acid orange 7 degradation efficiency within 20 min (pH = 2–6). Both sulfate and hydroxyl radicals, generated through catalytic activation of peroxymonosulfate were involved in the decomposition of acid orange 7.

3.2 g-C_3N_4 Based Photocatalysts

Photocatalysis refers that the activation of photochemical reactions (oxidation or reduction) by semiconductors or plasmonic metals as photocatalysts. Photocatalysts are chemical substances which generate reactive oxygen species ($OH^•$ and $O_2^{•-}$) by absorbing light with appropriate wavelength in solar spectrum. Depending upon the band gap, photocatalysts are classified as ultraviolet active and visible or solar active materials. Among the available photocatalysts, g-C_3N_4 has gained huge interest as it is a metal free polymeric semiconductor with visible light harvesting capacity, structural stability, and ecofriendly nature [37]. Like semiconductor photocatalyst, visible light irradiation on g-C_3N_4 causes electronic excitation from valence to conduction band while creating holes at valence band. Water (H_2O) molecules interact with hole and form $OH^•$ radical, while dissolved oxygen molecule absorbs photoexcited electrons to form $O_2^{•-}$ radical. These radicals attack organic dye molecules subsequently until their decomposition [37]. However, bulk g-C_3N_4 does not show good photocatalytic behavior owing to its fast recombination of photogenerated charges, less visible light utilization, and worse dispersibility. To decrease recombination rate, thereby increasing photocatalytic efficiency, the following improvement strategies were developed.

(a) Synthesis of exfoliated g-C_3N_4

In comparison to bulk form of g-C_3N_4, exfoliated g-C_3N_4 shows improved photocatalytic properties [38]. The g-C_3N_4 with high porosity and active surface area was obtained by adopting thermal polymeric condensation method in water. Fattahimoghaddam et al. [39] prepared porous g-C_3N_4 (pore volume = 1.86 $cm^3\ g^{-1}$ and surface area = 408.8 $m^2\ g^{-1}$) which showed nearly 100% efficiency towards rhodamine B decomposition in presence of visible light (15 min). Furthermore, this catalyst also showed excellent stability (98%) even after five cycles.

(b) Using g-C_3N_4/oxidizing agent

In order to produce $OH^•$ radicals efficiently, oxidants such as hydrogen peroxide (H_2O_2) and persulfate ($S_2O_8^{2-}$) are along used with g-C_3N_4/visible light system. For example, enhanced bisphenol A degradation efficiency was observed in presence of persulfate/g-C_3N_4 nanosheet/visible light system [40]. The order of photocatalytic activity toward bisphenol A degradation under 90 min of visible light irradiation is as follows: (5 mM) persulfate/g-C_3N_4 nanosheet (100%) > g-C_3N_4 nanosheet (72.5%) > bulk g-C_3N_4 (22%). Photocatalysts dosage, pH, and persulfate concentration has also significant influence on bisphenol A degradation. Rather than sulfate and holes and superoxide radicals were participated in bisphenol A degradation. The generation of free radicals and bisphenol A degradation at g-C_3N_4 under visible light is given as Eqs. (2–6) [40]

$$\text{g-C}_3\text{N}_4 + \text{hv} \rightarrow \text{g-C}_3\text{N}_4 - \text{h}^+ + \text{g-C}_3\text{N}_4 - \text{e}^- \quad (2)$$

$$\text{g-C}_3\text{N}_4 - \text{e}^- + \text{O}_2 \rightarrow \text{O}_2^{\bullet-} \quad (3)$$

$$\text{g-C}_3\text{N}_4 - \text{e}^- + \text{S}_2\text{O}_8^{2-} \rightarrow \left[\text{S}_2\text{O}_8^{2-}\right]^- \quad (4)$$

$$\left[\text{S}_2\text{O}_8^{2-}\right]^- + \text{O}_2 \rightarrow \text{S}_2\text{O}_8^{2-} + \text{O}_2^{\bullet-} \quad (5)$$

$$\text{g-C}_3\text{N}_4 - \text{h}^+\text{or O}_2^{\bullet-} + \text{Bisphenol A} \rightarrow \text{Degradation products} \quad (6)$$

The g-C_3N_4 is also used in sulfate radical ($SO_4^{•-}$) based AOP in dye degradation. Wei et al. [41] have used g-C_3N_4/visible light system to generate sulfate radicals through activation of sulfite. The production pathway of sulfate radicals at g-C_3N_4/visible light system is represented in Fig. 3. The produced sulfate radicals are involved in the degradation of aqueous organic dye solution. Overall, the photocatalytic degradation efficiency of g-C_3N_4 depends on sulfite concentration, solution pH, and co-existence of other components (humic acid, chloride ion, etc.).

(c) Construction of homojunction composite

Interestingly, homojunction nanocomposite between g-C_3N_4 particles was prepared to decrease neutralization of photogenerated electrons and holes, so that achieved greater photocatalytic activity toward dye degradation. This g-C_3N_4/g-C_3N_4 homojunction composite was synthesized by hydrothermal polymerization method using two different precursors [42]. The band gap of g-C_3N_4, synthesized using urea and thiourea is found to be 2.72 and 2.59 eV respectively. This homojunction nanocomposite photocatalyst shows excellent degradation efficiency towards rhodamine B degradation reaction under LED (18 W) light illumination. The g-C_3N_4/g-C_3N_4

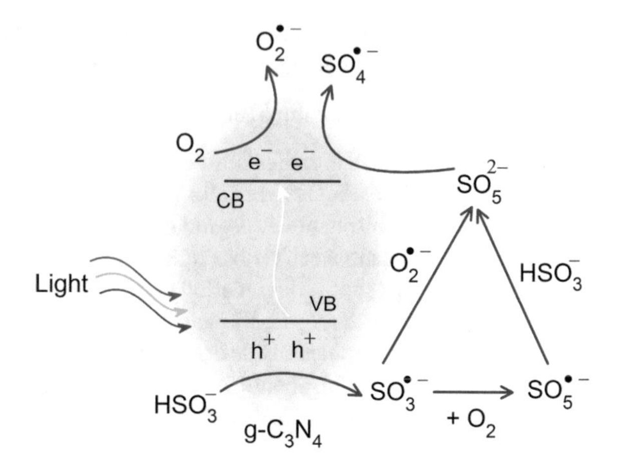

Fig. 3 Generation of $O_2^{\bullet-}$ and $SO_4^{\bullet-}$ free radicals by g-C_3N_4/HSO_3^-/visible light system. CB-conduction band; VB-valence band; h^+- hole and e^- - photoexcited electron [41]

shows two and 1.8 fold enhancement in degradation efficiency than the pristine g-C_3N_4 particles derived from urea and thiourea respectively. It was determined that the enhanced electron–hole pair separation at homojunction interface is responsible for the observed greater catalytic performances (Fig. 4).

(d) pH dependent activity

The effect of hydrion in the photocatalytic activity of g-C_3N_4 towards rhodamine B dye removal was investigated [43]. It was found that performance of g-C_3N_4 catalyst increased with increased hydrion concentration. Nearly, 95% efficiency was observed within 30 min of sunlight illumination.

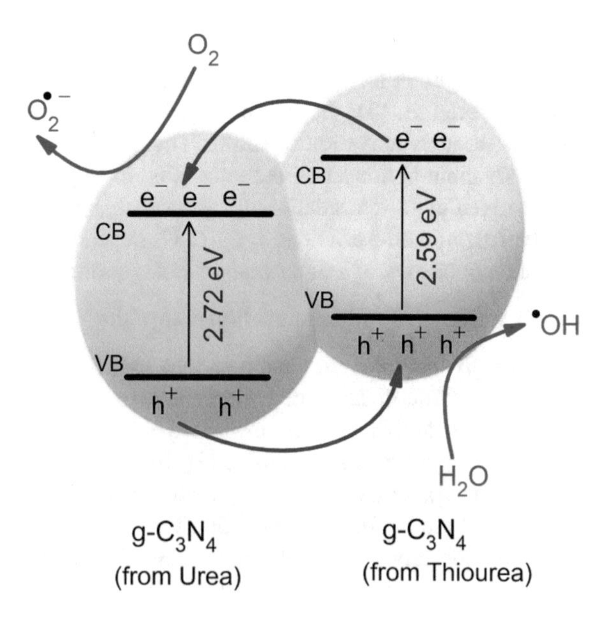

Fig. 4 Improved photocharge separation occurs at g-C_3N_4/g-C_3N_4 homojunction interface. CB-conduction band; VB-valence band; h^+- hole and e^- - photoexcited electron [42]

(e) Fabrication of g-C_3N_4 based composites

To reduce recombination of photogenerated electrons and holes, construction of composite photocatalyst is a most widely adopted strategy. At the same time, quality of interfaces between two or more photocatalysts determines their photocatalytic activity [44]. Hence, selection of appropriate photocatalyst for combination to g-C_3N_4 is an important step. With this concern, numerous g-C_3N_4 containing other suitable counter photocatalysts were reported. Few examples of g-C_3N_4 based composite photocatalyst and their photocatalytic activities toward dyes degradation have been listed in Table 1.

The inference from Table 1 is discussed below:

(a) Metal compounds such as metal oxide, iodide, sulfide, molybdate, selenide, and ferrite were coupled with g-C_3N_4 particles to achieve better photocatalytic performances. Depending on the counter parts, g-C_3N_4 based binary composites show photocatalytic activity under UV or visible light source.
(b) The mechanism of production of active free radicals by g-C_3N_4 based binary composite photocatalysts is also different. For example, g-C_3N_4/AgI binary composite adopts conventional type II mechanism, i.e., in presence of visible light source, AgI and g-C_3N_4 are excited instantaneously [45]. The photogenerated electrons in the conduction band of g-C_3N_4 are shifted to the conduction band of AgI while holes at valence band of AgI are shifted to the valence band of g-C_3N_4 nanosheets, thereby achieving effective separation of photocharges. The radical trapping experimental results are well supported by the above mentioned conclusions. Miao et al. [46] have discussed Z-scheme mechanism for production of free radicals by FeSiB metallic glass/g-C_3N_4 composite photocatalyst in the acid orange 7 degradation reaction. The excited electrons in conduction band of g-C_3N_4 are jumped to valence band of FeSiB glass and neutralized with holes. It leads to accumulation of holes at valence band of g-C_3N_4 while electrons are accumulated at conduction band of FeSiB glass. Therefore, g-C_3N_4 and FeSiB glass act as oxidation and reduction center respectively.
(c) The photocatalytic performance of g-C_3N_4 based binary composites is also dependent on the composition between g-C_3N_4 and counter photocatalyst.

Besides, g-C_3N_4 based ternary composite photocatalysts for removal of organic dyes have also been reported in literature (Table 2)

(f) Doping into g-C_3N_4

Doping, i.e., addition of impurity elements in a semiconductor is an important strategy to improve its functions like enhanced visible light harvesting property, surface area, mobility and lifetime of charges and decrease recombination rate of photocharges [47]. As g-C_3N_4 is a semiconductor, doping process enhances the catalytic activity of g-C_3N_4 particles [48]. Further, doping process also changes the texture of g-C_3N_4. Hence, different alkali and transition metal ions were used as dopants in g-C_3N_4 for enhancement of photocatalytic organic dye degradation reactions [49]. Similar

Table 1 Organic dye degradation performance of C_3N_4 based binary composite photocatalysts

Photocatalyst	Photocatalyst dosage	Light source	Organic dye (volume)	Concentration of dye (mg L^{-1})	Degradation efficiency (%)	Light irradiation time (min)	References
g-C_3N_4/AgI	20 mg	500 W Xe lamp	rhodamine B (30 mL)	20	~100.0	60	[45]
$CdMoO_4$/g-C_3N_4	50 mg	35WXe arc lamp	methylene blue (50 mL)	10	93.7	150	[50]
CdS/g-C_3N_4	0.5 g L^{-1}	W lamp (100 W)	methyl orange (10 mL)	2	85.0	60	[51]
$CoFe_2O_4$/ mesoporous graphitic mpg-C_3N_4	0.08 g L^{-1}	UV-C lamp (16 W)	Malachite green	10	93.4	120	[52]
FeSiB metallic glass/g-C_3N_4	1 g L^{-1}	UV–visible light	acid orange 7 (250 mL)	20	80.0	60	[46]
NiSe/g-C_3N_4	–	Visible light	methyl orange (20 mL)	20	92.2	5	[53]
g-C_3N_4/SnO_2	25 mg	UV 5 Watts LED lamp (visible light)	rhodamine B (50 ml)	10	100.0 98.5	240 240	[54]

Table 2 Photocatalytic dye degradation performance of C_3N_4 based ternary composites

Photocatalyst	Photocatalyst dosage	Light source	Organic dye (volume)	Dye Concentration	Degradation efficiency (%)	Light irradiation time (min)	References
Ag/g-C_3N_4/$LaFeO_3$	20 mg	Xe arc lamp (300 W; λ > 420 nm)	Methylene blue (60 mL)	10 mg L^{-1}	98.9	90	[55]
Ag/g-C_3N_4/reduced graphene oxide	30 mg	Xe lamp (300 W; λ > 420 nm)	Rhodamine 6G (35 mL)	10^{-3} M	93.0	240	[56]
AgCl/Au/g-C_3N_4	50 mg	Xe lamp (200 W λ < 420 nm)	Rhodamine B (50 mL)	10 mg L^{-1}	93.1	25	[57]
g-C_3N_4@Ag-ZnO	0.01 g	Sunlight	Methylene blue (200 mL)		98.0	80	[58]
g-C_3N_4/carbon dot/FeOCl	0.1 g	Visible light	Rhodamine B (250 mL)	1×10^{-5} M	~100.0	60	[59]
g-C_3N_4/polyaniline-polypyrrole	15 mg	Xe lamp (8 W)	Methylene blue (100 mL)	5×10^{-5} M	~100.0	70	[60]

to metal doping, non-metal doping process was also attempted in g-C_3N_4. In fact, doping of non-metal is highly preferred than metal ions since it produces completely metal free catalyst as well as no thermal variation of chemical states. Non-metals such as boron (B), phosphorous (P), oxygen (O), sulfur (S) were used as dopants in g-C_3N_4 [47, 61]. For instance, Yan et al. [62] have prepared B-doped g-C_3N_4 photocatalyst for decomposition of rhodamine B and methyl orange dyes. Recently, Li et al. [63] have also prepared (0.87%) P-doped g-C_3N_4 photocatalyst with greater specific surface area (202.9 cm^2 g^{-1}). This photocatalyst shows 99.5% efficiency toward decomposition of rhodamine B dye.

3.3 g-C_3N_4 Based Sonocatalysis

Sonolysis is another important AOP which is used for organic dyes degradation. In this treatment, ultrasound waves are used to grow bubbles in water rapidly. Subsequently, those bubbles collapsed and increase heat and pressure tremendously. Consequently hydroxyl radicals are generated by the following way (Eqs. 7–10) [64]: (a) sonication causes production of hydrogen ($H^•$) radicals, (b) $H^•$ radical instantly reacts with O_2 to form hydroperoxyl radicals ($HOO^•$) and (c) $^•OH$ radical also combined to form H_2O_2 [64].

$$H_2O + \text{Sonication})))))) \rightarrow {}^{\bullet}H + {}^{\bullet}OH \tag{7}$$

$$^{\bullet}H + O_2 \rightarrow HOO^{\bullet} \tag{8}$$

$$2\,HOO^{\bullet} \rightarrow H_2O_2 + O_2 \tag{9}$$

$$2^{\bullet}OH \rightarrow H_2O_2 \tag{10}$$

The in-situ generated H_2O_2 will decompose dye pollutants in water. However, sonolysis has the following disadvantages: (a) huge amounts of energy in the form of ultrasound need to supply for dye decomposition and (b) complete mineralization of dyes rarely occurred. But, these shortcomings can be eliminated by sonocatalytic process. Ultrasonic waves are used to excite electrons from valence to conduction band of catalyst (sonocatalyst) and thus generate electron–hole pairs. The elevation of local temperature by ultrasound stimulates the thermal excitation of electrons in catalyst [65]. Ultrasound induced electrons and holes react with oxygen and H_2O to form superoxide and hydroxyl radicals respectively. The degradation efficiency of sonocatalyst is dependent on appropriate catalyst. In this context, g-C_3N_4 has been considered in sonocatalysis.

(a) Pure g-C_3N_4

Pure g-C_3N_4 was applied as sonocatalyst in the degradation of aqueous solution of methylene blue [66]. In this process, 100 mg pure g-C_3N_4 was suspended in 100 mL of 10 mg/L methylene blue aqueous solution (pH = 7). Ultrasound with frequency of = 40 $_{kHz}$ (power = 80 W) was used to irradiate the suspension. Noticeably, photocatalytic degradation efficiency is dependent on synthesis temperature of g-C_3N_4. Photocatalytic efficiency of g-C_3N_4 increases from 450 to 600 °C. Higher calcination temperature leads to structural defects in g-C_3N_4 [67]. Generally, defects cause adsorption of pollutant on catalyst's surface efficiently that accelerate the degradation reaction. Pure g-C_3N_4 showed 96% degradation efficiency toward methylene blue degradation reaction under optimized experimental conditions.

(b) g-C_3N_4 based composites

Pure g-C_3N_4 has some restrictions including less photocharge separation efficiency and low active surface area, thus it exhibits poor efficiency in dye removal process. This drawback could be eliminated by combining g-C_3N_4 with suitable catalysts. For example, $SrTiO_3$/mpg-C_3N_4 sonocatalyst was fabricated for degradation of basic violet 10 dye [68]. This composite shows 80% sonocatalytic efficiency under the following experimental condition: sonocatalyst concentration = 0.3 g L^{-1}; initial concentration = 10 mg L^{-1}; ultrasonic power = 240 W; irradiation time = 120 min and pH = 5. The degradation of dye at the surface of $SrTiO_3$/mpg-C_3N_4 sonocatalyst was described by Eqs. 11–17 [68].

$$\text{Basic violet 10 dye}+))))(\text{ultrasound}) \rightarrow \text{Basic violet 10 dye}^* \quad (11)$$

$$\text{g - }C_3N_4/SrTiO_3 + \text{Basic violet 10 dye}^* +)))) \rightarrow \text{Basic violet 10 dye}^+ + SrTiO_3(e^-) \quad (12)$$

$$\text{g - }C_3N_4/SrTiO_3 +))))(\text{ultrasound}) \rightarrow \text{g - }C_3N_4/SrTiO_3(h^+ + e^-) \quad (13)$$

$$\text{g - }C_3N_4(e^-) \rightarrow SrTiO_3(e^-) \quad (14)$$

$$SrTiO_3\,(h^+) \rightarrow \text{g - }C_3N_4(h^+) \quad (15)$$

$$SrTiO_3\,(e^-) + O_2 \rightarrow O_2^{\bullet -} \quad (16)$$

$$SrTiO_3\,(h^+) + OH^- \rightarrow {}^{\bullet}OH \quad (17)$$

(c) Sonophotocatalysis

The integration of sonocatalysis with light (sonophotocatalysis) could further improve the degradation efficiency of dye pollutants, i.e., in addition to ultrasound waves, light is also used as irradiation source to catalyst that generates electron–hole pairs. Sonophotocatalysis produces vast amount of active free radicals that are quite enough to decompose dye molecules [69]. The processes involved in sonophotocatalysis are activation of catalyst, improvement of the pollutant transport and breakage of aggregation. The g-C_3N_4 based ternary composite was reported as sonophotocatalyst for methylene blue removal. The g-C_3N_4 containing Cu_2O and 1-butyl-3-methylimidazolium acetate (ionic liquid) catalyst was utilized in methylene blue along with 10 mM H_2O_2 as oxidant [69]. Heterojunction between g-C_3N_4/Cu_2O interface promotes the charge separation while ionic liquid traps photogenerated electrons. Hence, reduced recombination rate was observed for this composite catalyst under the optimized experimental conditions: light source = UV (6 W); frequency = 20 kHz and pH = 7. Within thirty minutes of dual irradiation (ultrasound and UV light), almost complete removal dye was achieved.

3.4 Heterogeneous Fenton-Like Reaction

Fenton reaction, a homogenous process is a commonly used removal method for organic dye pollutants in water. This is due to its degradation capability over contaminants by producing $OH^{\bullet}$ radicals from H_2O_2. The $OH^{\bullet}$ radicals were formed by two steps. In step - 1, Fe^{2+} ions was oxidized to Fe^{3+} ions, while in step-2, Fe^{3+} ions again reduced to Fe^{2+}. Importantly, rate constant of Fe^{2+} oxidation is much faster than reduction of Fe^{3+} ions. However, practical utilization of Fenton reaction in wastewater treatment is suffered by the following reasons:

(a) It will work only in narrow pH range (pH = 2–3).
(b) Easy loss of iron ions
(c) It causes secondary pollution and
(d) It generates large amount of sludge.

To overcome the above said drawbacks, heterogeneous Fenton-like reaction was developed. The production pathway of free radicals by Fenton-like process is similar to the conventional homogenous Fenton process. Usually, iron-based catalysts were employed as heterogeneous Fenton catalysts, owing to their biocompatibility, mixed oxidation states, cheap, narrow band gap, surface area, and abundant in earth crust. Examples of such catalysts are Fe^0, Fe_2O_3, Fe_3O_4, FeOOH and FeOCl [70]. Like, iron, elemental copper also shows similar redox capability (Cu^+/Cu^{2+}). In fact, Cu^{2+} ion will readily reduce to Cu^+ ion during the Fenton-like process. Hence, copper-based Fenton-like catalysts also have huge attention in water pollution minimization [71]. Merits of Fenton-like process are easy operation, requirement of mild experimental conditions and cheap catalysts. Nevertheless, performance of iron-based catalysts is low due to the less surface area, slow conversion of Fe^{3+}/Fe^{2+} redox couple, interference of co-existing species, and leaching of iron at low pH (highly acidic). Further, Cu_2O is not stable during Fenton process.

Hence, several researches have been conducted to improve the activity of Fenton catalysts. In this context, g-C_3N_4 was used into the Fenton-like system. The occurrence of N sites makes it an appropriate support material for the immobilization of iron-based Fenton catalyst. The following strategies were adopted to enhance performance of g-C_3N_4 based Fenton –like catalysts

(a) Construction of three-dimensional g-C_3N_4

Facile recovery from the reaction aliquot is an important requirement of any catalyst. But, due to finer size and thinness, two-dimensional structured g-C_3N_4 powder is difficult to recover from the suspension after completing Fenton process. In case of three-dimensional porous g-C_3N_4 particles, adsorption and diffusion of organic pollutants is greatly enhanced and recovery of the catalyst is also highly feasible. Theoretically, it was determined that Fe_2O_3 immobilized on three-dimensional porous g-C_3N_4 catalyst will be an efficient way to achieve better degradation efficiency in organic dye degradation reactions [72].

(b) Doping in g-C_3N_4

Studies on doped g-C_3N_4 as catalyst in Fenton-like reaction have been explored. For example, g-C_3N_4 doped with Na and Fe was used as catalyst in the methylene blue degradation reaction [73]. The ratio between Fe and Na determines the catalytic performance. The obtained degradation efficiency of the optimized catalyst (Fe/Na = 1.5/4.5) is 98.4%. The mechanism of methylene blue degradation in presence of Fe/Na-g-C_3N_4/H_2O_2 was described as follows: (a) adsorption of methylene blue on catalyst's surface through electrostatic attraction, (b) H_2O_2 activation followed $^{\bullet}OH$ radical generation through Fe^{3+}/Fe^{2+} redox couple, (c) production of $O_2^{\bullet-}$ radical and (d) both $^{\bullet}OH$ and $O_2^{\bullet-}$ radicals get involved in the decomposition of dye molecules.

(c) Using light (photo-Fenton process)

Fenton-like reaction is accelerated by using suitable light energy (photo-Fenton reaction). This process has the following advantages: (a) it doesn't generate sludge, (b) it doesn't cause secondary pollution and (c) catalyst can be easily recovered and reused. Hence, g-C_3N_4 based systems have been used in this process. In typical g-C_3N_4 based binary photo-Fenton heterojunction catalysis, the generation of $^{\bullet}OH$ radical takes place through reduction of Fe^{3+} ions by photogenerated electrons (Fig. 5). In addition, H_2O_2 also undergoes decomposition in presence of light energy to form $^{\bullet}OH$ radicals. Thus, light irradiation could also improve the degradation efficiency of the catalyst used in Fenton-like reaction. For example, g-C_3N_4/$ZnFe_2O_4$ composite was synthesized through sol–gel followed calcination route and this composite was applied in degradation of 20 mg L^{-1} of mixture of methylene blue and rhodamine B dyes under natural solar light illumination [74]. A complete degradation was observed within shorter reaction time (35 min). In another study, g-C_3N_4/diatomite/Fe_3O_4 ternary composite showed more than 90% efficiency toward rhodamine B dye decomposition under visible light illumination with H_2O_2 [75]. Excellent activity of this ternary catalyst is due to decreased recombination efficiency of photocharges and synergistic

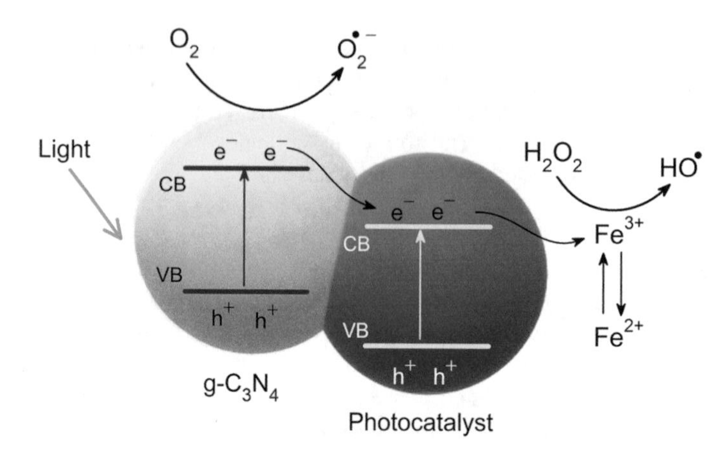

Fig. 5 Upon irradiation of light, superoxide ($O_2^{\bullet-}$) and hydroxyl ($^{\bullet}OH$) free radicals are generated by g-C_3N_4/photocatalyst/H_2O_2/Fe^{2+} system through photo-Fenton process. CB-conduction band; VB-valence band; h^+- hole and e^- - photoexcited electron

effect of Fenton reaction and photocatalysis. Further, this ternary composite catalyst can be readily separated by an external magnet. Radical trapping experiments assure that both $^{\bullet}OH$ and $O_2^{\bullet-}$ radicals participated in this degradation reaction. In certain cases, some dye molecules can be excited under visible light to produce photosensitizer molecules, which can further stimulate the reduction of Fe^{3+} ions [76].

4 Conclusions

In AOPs for organic dyes removal, g-C_3N_4 has huge consideration since it has favorable structural and physicochemical properties. But, the catalytic activity of pure g-C_3N_4 is experimentally found to be low in various organic dye degradation reactions. Therefore, strategies like construction of composites and doping process have been adapted to enhance the performance of pure g-C_3N_4. Overall, g-C_3N_4 based catalysts were used in activation of oxidant, photocatalysis, sonocatalysis, and Fenton-like reactions for abatement of aqueous solution of organic dyes. The following points are summarized from the literature survey.

(i) In activation of oxidants, g-C_3N_4 based composite catalyst shows capability to decompose oxidants like ozone to form active free radicals. They also induced sulfate radicals from sulfite and peroxymonosulfate reagents. Surface property of g-C_3N_4 is the key factor in these processes.
(ii) In photocatalysis, mesoporous structure, pH, fabrication of homojunction, heterojunction, and doping process promote photocatalytic efficiency of pure g-C_3N_4 toward targeted aqueous dye solution, since these strategies improve

specific surface area, band gap reduction, and reduced recombination rate of photocharges.

(iii) In sonocatalysis, g-C_3N_4 based composite catalysts showed improved photocatalytic efficiency toward various organic dyes, as they show improved charge separation efficiency under ultrasonic irradiation.

(iv) In Fenton-like process, g-C_3N_4 composites containing iron-based compounds exhibit greater performance than pure iron-based catalysts. In addition, g-C_3N_4 also improves the activity of copper-based Fenton catalysts.

(v) Sonocatalysis and Fenton-like catalysis are also significantly influenced by using UV and visible light irradiation. The g-C_3N_4 based catalysts play vital role in these coupled techniques (sonophotocatalysis and photo-Fenton-like process). Hence, to conclude, g-C_3N_4 based catalysts in treatment of organic dyes will become a great option in the future.

However, most of the aforementioned studies are performed in laboratory using pure dye solutions. Hence, in order to implement in wastewater treatment plants, more studies using real dye effluents need to be done.

Acknowledgments The authors wish to thank Solar Energy Research Center, SERC-Chile (FONDAP/ANID/15110019).

References

1. Aggarwal S (2021) Indian dye yielding plants: Efforts and opportunities. Nat Resour Forum 45:63–86. https://doi.org/10.1111/1477-8947.12214
2. Yusuf M, Shabbir M, Mohammad F (2017) Natural colorants: historical, processing and sustainable prospects. Nat Prod Bioprospect 7:123–145. https://doi.org/10.1007/s13659-017-0119-9
3. Sharma J, Sharma S, Soni V (2021) Classification and impact of synthetic textile dyes on aquatic flora: a review. Reg Stud Mar Sci 45:101802. https://doi.org/10.1016/j.rsma.2021.101802
4. Wang CC, Li JR, Lv XL, Zhang YQ, Guo G (2014a) Photocatalytic organic pollutants degradation in metal-organic frameworks. Energy Environ Sci 7:2831–2867. https://doi.org/10.1039/C4EE01299B
5. Chequer FMD, Oliveira GARD, Ferraz ERA, Cardoso JC, Zanoni MVB, Oliveira DPD (2013) Textile dyes: Dyeing process and environmental impact. Book: Eco-Friend Text Dye Finish edited by Melih Gunay,13. https://doi.org/10.5772/53659
6. Zille A, Gornacka B, Rehorek A, Cavaco A (2005) Degradation of azo dyes by *Trametes villosa laccase* over long periods of oxidative conditions. Appl Environ Microbiol 71:6711–6718. https://doi.org/10.1128/AEM.71.11.6711-6718.2005
7. O'Neill C, Hawkes FR, Hawkes DL, Lourenço ND, Pinheiro HM, Delée W (1999) Colour in textile effluents-sources, measurement, discharge consents and simulation: a review. J Chem Technol Biotechnol 74:1009–1018. https://doi.org/10.1002/(SICI)1097-4660(199911)74:11%3c1009::AID-JCTB153%3e3.0.CO;2-N
8. Das S, Bhat AP, Gogate PR (2021) Degradation of dyes using hydrodynamic cavitation: Process overview and cost estimation. J Water Process Eng 42:102126. https://doi.org/10.1016/j.jwpe.2021.102126

9. Javaid R, Qazi UY, Ikhlaq A, Zahid M, Alazmi A (2021) Subcritical and supercritical water oxidation for dye decomposition. J Environ Manage 290:112605. https://doi.org/10.1016/j.jenvman.2021.112605
10. Samsami S, Mohamadizaniani M, Sarrafzadeh MH, Rene ER, Firoozbahr M (2020) Recent advances in the treatment of dye-containing wastewater from textile industries: overview and perspectives. Process Saf Environ Prot 143:138–163. https://doi.org/10.1016/j.psep.2020.05.034
11. Wong JKH, Tan HK, Lau SY, Yap PS, Danquah MK (2019) Potential and challenges of enzyme incorporated nanotechnology in dye wastewater treatment: a review. J Environ Chem Eng 7:103261. https://doi.org/10.1016/j.jece.2019.103261
12. Vikrant K, Giri BS, Raza N, Roy K, Kim KH, Rai BN, Singh RS (2018) Recent advancements in bioremediation of dye: current status and challenges. Bioresour Technol 253:355–367. https://doi.org/10.1016/j.chemosphere.2019.05.205
13. Kurian M (2021) Advanced oxidation processes and nanomaterials—a review. Cleaner Eng Technol 2:100090. https://doi.org/10.1016/j.clet.2021.100090
14. Inagaki M, Tsumura T, Kinumoto T, Toyoda M (2019) Graphitic carbon nitrides (g-C_3N_4) with comparative discussion to carbon materials. Carbon 141:580–607. https://doi.org/10.1016/j.carbon.2018.09.082
15. Jiang R, Lu G, Liu J, Wu D, Yan Z, Wang Y (2021) Incorporation of π-conjugated molecules as electron donors in g-C_3N_4 enhances photocatalytic H_2-production. Renew Energy 164:531–540. https://doi.org/10.1016/j.renene.2020.09.040
16. Hasija V, Nguyen VH, Kumar A, Raizada P, Krishnan V, Khan AAP, Singh P, Lichtfouse E, Wang C, Huong PT (2021) Advanced activation of persulfate by polymeric g-C_3N_4 based photocatalysts for environmental remediation: A review. J Hazard Mater 413:125324. https://doi.org/10.1016/j.jhazmat.2021.125324
17. Jiang XX, Hu XD, Tarek M, Saravanan P, Alqadhi R, Chin SY, Khan MMR (2020) Tailoring the properties of g-C_3N_4 with CuO for enhanced photoelectrocatalytic CO_2 reduction to methanol. J CO_2 Util 40:101222. https://doi.org/10.1016/j.jcou.2020.101222
18. Ma J, Tao XY, Zhou SX, Song XZ, Guo L, Wang Y, Zhu YB, Guo LT, Liu ZS, Fan HL, Wei XY (2019) Facile fabrication of Ag/PANI/g-C_3N_4 composite with enhanced electrochemical performance as supercapacitor electrode. J Electroanal Chem 835:346–353. https://doi.org/10.1016/j.jelechem.2018.12.025
19. Yang ZH, Wu XY, Liu XC, Xu MM (2021) One-step bridging of g-C_3N_4 and graphene oxide by successive electrolysis for constructing electrochemical sensor of Pb^{2+}. Chin J Anal Chem 49:e21179–e21186. https://doi.org/10.1016/S1872-2040(21)60115-9
20. Panigrahi P, Kumar A, Karton A, Ahuja R, Hussain T (2020) Remarkable improvement in hydrogen storage capacities of two-dimensional carbon nitride (g-C_3N_4) nanosheets under selected transition metal doping. Int J Hydrog Energy 45:3035–3045. https://doi.org/10.1016/j.ijhydene.2019.11.184
21. Yao Z, Liu J, Liang J, Jaroniec M, Qiao SZ (2012) Graphitic carbon nitride materials: controllable synthesis and applications in fuel cells and photocatalysis. Energy Environ Sci 5:6717–6731. https://doi.org/10.1039/C2EE03479D
22. Wang Y, Wang X, Antonietti M (2014b) Graphitic carbon nitride as a heterogeneous catalyst: from photochemistry to multipurpose catalysis to sustainable chemistry. Angew Chem Int Ed Engl 51:68–89. https://doi.org/10.1002/anie.201101182
23. Wang X, Maeda K, Thomas A, Takanabe K, Xin G, Carlsson JM, Domen K, Antonietti M (2009) A metal-free polymeric photocatalyst for hydrogen production from water under visible light. Nat Mater 8:76–80. https://doi.org/10.1038/nmat2317
24. Jorge AB, Martin DJ, Dhanoa MTS, Rahman AS, Makwana N, Tang J, Sella A, Cora F, Firth S, Darr JA, McMillan PF (2013) H_2 and O_2 evolution from water half-splitting reactions by graphitic carbon nitride materials. J Phys Chem C 117:7178–7185. https://doi.org/10.1021/jp4009338
25. Wang X, Siegfried B, Markus A (2012) Polymeric graphitic carbon nitride for heterogeneous photocatalysis. ACS Catal 2:1596–1606. https://doi.org/10.1021/cs300240x

26. Dong GP, Zhang YH, Pan QW, Qiu JR (2014) A fantastic graphitic carbon nitride (g-C_3N_4) material: Electronic structure, photocatalytic and photoelectronic properties. J Photochem Photobio C 20:33–50. https://doi.org/10.1016/j.jphotochemrev.2014.04.002
27. Fanchini G, Tagliaferro A, Conway NMJ, Godet C (2002) Role of lone-pair interactions and local disorder in determining the interdependency of optical constants of a−CN: H thin films. Phys Rev B 66:195415. https://doi.org/10.1103/PhysRevB.66.195415
28. El-kader FHA, Moharram MA, Khafagia MG, Mamdouh F (2012) Influence of the nitrogen content on the optical properties of CNx films. Spectrochim Acta A Mol Biomol Spectrosc 97:1115–1119. https://doi.org/10.1016/j.saa.2012.07.126
29. Ahmaruzzaman M, Mishra SR (2021) Photocatalytic performance of g-C_3N_4 based nanocomposites for effective degradation/removal of dyes from water and wastewater. Mater Res Bull 143:111417. https://doi.org/10.1016/j.materresbull.2021.111417
30. Agustina TE, Ang HM, Vareek VK (2005) A review of synergistic effect of photocatalysis and ozonation on wastewater treatment. J Photochem Photobiol C: Photochem Rev 6:264–273. https://doi.org/10.1016/j.jphotochemrev.2005.12.003
31. Wang J, Chen H (2020) Catalytic ozonation for water and wastewater treatment: recent advances and perspective. Sci Total Environ 704:135249. https://doi.org/10.1016/j.scitotenv.2019.135249
32. Hordern KB, Ziołek M, Nawrocki J (2003) Catalytic ozonation and methods of enhancing molecular ozone reactions in water treatment. Appl Catal B Environ 46:639–669. https://doi.org/10.1016/S0926-3373(03)00326-6
33. Yuan X, Qin W, Lei X, Sun L, Li Q, Li D, Xu H, Xia D (2018) Efficient enhancement of ozonation performance via ZVZ immobilized g-C_3N_4 towards superior oxidation of micropollutants. Chemosphere 205:369–379. https://doi.org/10.1016/j.chemosphere.2018.04.121
34. Li J, Fang J, Gao L, Zhang J, Ruan X, Xu A, Li X (2017) Graphitic carbon nitride induced activity enhancement of OMS-2 catalyst for pollutants degradation with peroxymonosulfate. Appl Surf Sci 402:352–359. https://doi.org/10.1016/j.apsusc.2017.01.129
35. Nyutu EK, Chen CH, Sithambaram S, Crisostomo VMB, Suib SL (2008) Systematic control of particle size in rapid open-vessel microwave synthesis of K-OMS-2 nanofibers. J Phys Chem C 112:6786–6793. https://doi.org/10.1021/jp800672m
36. Guo F, Lu J, Liu Q, Zhang P, Zhang A, Cai Y, Wang Q (2018) Degradation of acid orange 7 by peroxymonosulfate activated with the recyclable nanocomposites of g-C_3N_4 modified magnetic carbon. Chemosphere 205:297–307. https://doi.org/10.1016/j.chemosphere.2018.04.139
37. Xing Y, Wang X, Hao S, Zhang X, Wang X, Ma W, Zhao G, Xu X (2021) Recent advances in the improvement of g-C_3N_4 based photocatalytic materials. Chin Chem Lett 32:13–20. https://doi.org/10.1016/j.cclet.2020.11.011
38. Lin QY, Li L, Liang SJ (2015) Efficient synthesis of monolayer carbon nitride 2D nanosheet with tunable concentration and enhanced visible-light photocatalytic activities. Appl Catal B Environ 163:135–142. https://doi.org/10.1016/j.apcatb.2014.07.053
39. Fattahimoghaddam H, Shamsabadi TM, Lee BK (2021) Efficient photodegradation of rhodamine B and tetracycline over robust and green g-C_3N_4 nanostructures: Supramolecular design. J Hazard Mater 403:123703. https://doi.org/10.1016/j.jhazmat.2020.123703
40. Liu B, Qiao M, Wang Y, Wang L, Gong Y, Guo T, Zhao X (2017) Persulfate enhanced photocatalytic degradation of bisphenol A by g-C_3N_4 nanosheets under visible light irradiation. Chemosphere 189:115–122. https://doi.org/10.1016/j.chemosphere.2017.08.169
41. Wei Y, Zou Q, Ye P, Wang M, Li X, Xu A (2018) Photocatalytic degradation of organic pollutants in wastewater with g-C_3N_4/sulfite system under visible light irradiation. Chemosphere 208:358–365. https://doi.org/10.1016/j.chemosphere.2018.06.006
42. Phang SJ, Goh JM, Tan LL, Lee WPC, Ong WJ, Chai SP (2021) Metal-free n/n–junctioned graphitic carbon nitride (g-C_3N_4): a study to elucidate its charge transfer mechanism and application for environmental remediation. Environ Sci Pollut Res 28:4388–4403. https://doi.org/10.1007/s11356-020-10814-z
43. Shi W, Fang WX, Wang JC, Qiao X, Wang B, Guo X (2021) pH-controlled mechanism of photocatalytic RhB degradation over g-C_3N_4 under sunlight irradiation. Photochem Photobiol Sci 20:303–313. https://doi.org/10.1007/s43630-021-00019-9

44. Li CC, Gao MY, Sun XJ, Tang HL, Dong H, Zhang FM (2020a) Rational combination of covalent-organic framework and nano TiO_2 by covalent bonds to realize dramatically enhanced photocatalytic activity. Appl Catal B 266:118586. https://doi.org/10.1016/j.apcatb.2020.118586
45. Huang H, Li YX, Wang HL, Jiang WF (2021) In situ fabrication of ultrathin-g-C_3N_4/AgI heterojunctions with improved catalytic performance for photodegrading rhodamine B solution. Appl Surf Sci 538:148132. https://doi.org/10.1016/j.apsusc.2020.148132
46. Miao F, Wang Q, Zhang LC, Shen B (2021) Magnetically separable Z-scheme FeSiB metallic glass/g-C_3N_4 heterojunction photocatalyst with high degradation efficiency at universal pH conditions. Appl Surf Sci 540:148401. https://doi.org/10.1016/j.apsusc.2020.148401
47. Patnaik S, Sahoo DP, Parida K (2021) Recent advances in anion doped g-C_3N_4 photocatalysts: A review. Carbon 172:682–711. https://doi.org/10.1016/j.carbon.2020.10.073
48. Ma Z, Cui Z, Lv Y, Sa R, Wu K, Li Q (2020) Three-in-One: Opened Charge-transfer channel, positively shifted oxidation potential, and enhanced visible light response of g-C_3N_4 photocatalyst through K and S Co-doping. Int J Hydrog Energy 45:4534–4544. https://doi.org/10.1016/j.ijhydene.2019.12.074
49. Chen Z, Zhang S, Liu Y, Alharbi NS, Rabah SO, Wang S, Wang X (2020) Synthesis and fabrication of g-C_3N_4-based materials and their application in elimination of pollutants. Sci Total Environ 731:139054. https://doi.org/10.1016/j.scitotenv.2020.139054
50. Gandamalla A, Manchala S, Anand P, Fu YP, Shanker V (2021) Development of versatile $CdMoO_4$/g-C_3N_4 nanocomposite for enhanced photoelectrochemical oxygen evolution reaction and photocatalytic dye degradation applications. Mater Today Chem 19:100392. https://doi.org/10.1016/j.mtchem.2020.100392
51. Pourshirband N, Ejhieh AN, Mirsattari SN (2021) The CdS/g-C_3N_4 nano-photocatalyst: Brief characterization and kinetic study of photodegradation and mineralization of methyl orange. Spectrochim Acta A Mol Biomol Spectrosc 248:119110. https://doi.org/10.1016/j.saa.2020.119110
52. Hassani A, Eghbali P, Ekicibil A, Metin Ö (2018) Monodisperse cobalt ferrite nanoparticles assembled on mesoporous graphitic carbon nitride ($CoFe_2O_4$/mpg-C_3N_4): A magnetically recoverable nanocomposite for the photocatalytic degradation of organic dyes. J Magn Magn Mater 456:400–412. https://doi.org/10.1016/j.jmmm.2018.02.067
53. Chen Z, Gao Y, Chen F, Shi H (2021) Metallic NiSe cocatalyst decorated g-C_3N_4 with enhanced photocatalytic activity. Chem Eng J 413:127474. https://doi.org/10.1016/j.cej.2020.127474
54. Wang X, He Y, Xu L, Xia Y, Gang R (2021a) SnO_2 particles as efficient photocatalysts for organic dye degradation grown in-situ on g-C_3N_4 nanosheets by microwave-assisted hydrothermal method. Mater Sci Semicond Process 121:105298. https://doi.org/10.1016/j.mssp.2020.105298
55. Zhang W, Ma Y, Zhua X, Liu S, An T, Bao J, Hu X, Tian H (2021a) Fabrication of Ag decorated g-C_3N_4/$LaFeO_3$ Z-scheme heterojunction as highly efficient visible-light photocatalyst for degradation of methylene blue and tetracycline hydrochloride. J Alloys Compd 864:158914. https://doi.org/10.1016/j.jallcom.2021.158914
56. Song Y, Peng Y, Long NV, Huang Z, Yang Y (2021) Multifunctional self-assembly 3D Ag/g-C_3N_4/RGO aerogel as highly efficient adsorbent and photocatalyst for R6G removal from wastewater. Appl Surf Sci 542:148584. https://doi.org/10.1016/j.apsusc.2020.148584
57. Zhang W, Xu D, Wang F, Chen M (2021b) AgCl/Au/g-C_3N_4 ternary composites: efficient photocatalysts for degradation of anionic dyes. J Alloys Compd 868:159266. https://doi.org/10.1016/j.jallcom.2021.159266
58. Sher M, Khan SA, Shahid S, Javed M, Qamar MA, Chinnathambi A, Almoallim HS (2021) Synthesis of novel ternary hybrid g-C_3N_4@Ag-ZnO nanocomposite with Z-scheme enhanced solar light-driven methylene blue degradation and antibacterial activities. J Environ Chem Eng 9:105366. https://doi.org/10.1016/j.jece.2021.105366
59. Khaneghah SA, Yangjeh AH, Seifzadeh D, Chand H, Krishnan V (2021) Visible-light-activated g-C_3N_4 nanosheet/carbon dot/FeOCl nanocomposites: photodegradation of dye pollutants and tetracycline hydrochloride. Colloids Surf A Physicochem Eng Asp 617:126424. https://doi.org/10.1016/j.colsurfa.2021.126424

60. Munusamy S, Sivaranjan K, Sabhapathy P, Narayanan V, Mohammad F, Sagadevan S (2021) Enhanced electrochemical and photocatalytic activity of g-C_3N_4-PANI-PPy nanohybrids. Synth Met 272:116669. https://doi.org/10.1016/j.synthmet.2020.116669
61. He F, Wang Z, Li Y, Peng S, Liu B (2020a) The nonmetal modulation of composition and morphology of g-C_3N_4-based photocatalysts. Appl Catal B: Environ 269:118828. https://doi.org/10.1016/j.apcatb.2020.118828
62. Yan SC, Li ZS, Zou ZG (2010) Photodegradation of Rhodamine B and methyl orange over boron-doped g-C_3N_4 under visible light irradiation. Langmuir 26:3894–3901. https://doi.org/10.1021/la904023j
63. Li Z, Chen Q, Lin Q, Chen Y, Liao X, Yu H, Yu C (2020b) Three-dimensional P-doped porous g-C_3N_4 nanosheets as an efficient metal-free photocatalyst for visible-light photocatalytic degradation of Rhodamine B model pollutant. J Taiwan Inst Chem Eng 114:249–262. https://doi.org/10.1016/j.jtice.2020.09.019
64. Anandan S, Ponnusamy VK, Ashokkumar M (2020) A review on hybrid techniques for the degradation of organic pollutants in aqueous environment. Ultrason Sonochem 67:105130. https://doi.org/10.1016/j.ultsonch.2020.105130
65. Qiu P, Park B, Choi J, Thokchom B, Pandit AB, Khim J (2018) A review on heterogeneous sonocatalyst for treatment of organic pollutants in aqueous phase based on catalytic mechanism. Ultrason Sonochem 45:29–49. https://doi.org/10.1016/j.ultsonch.2018.03.003
66. Song L, Zhang S, Wu X, Wei Q (2012) A metal-free and graphitic carbon nitride sonocatalyst with high sonocatalytic activity for degradation methylene blue. Chem Eng J 184:256–260. https://doi.org/10.1016/j.cej.2012.01.053
67. Zhang JS, Chen XF, Takanabe K, Maeda K, Domen K, Epping JD, Fu XZ, Antonietti M, Wang XC (2010) Synthesis of a carbon nitride structure for visible-light catalysis by copolymerization. Angew Chem Int Ed 49:441–444. https://doi.org/10.1002/anie.200903886
68. Eghbali P, Hassani A, Sündü B, Metin Ö (2019) Strontium titanate nanocubes assembled on mesoporous graphitic carbon nitride ($SrTiO_3$/mpg-C_3N_4): Preparation, characterization and catalytic performance. J Mol Liq 290:111208. https://doi.org/10.1016/j.molliq.2019.111208
69. Eshaq G, ElMetwally AE (2019) Bmim[OAc]-Cu_2O/g-C_3N_4 as a multi-function catalyst for sonophotocatalytic degradation of methylene blue. Ultrason Sonochem 53:99–109. https://doi.org/10.1016/j.ultsonch.2018.12.037
70. Wang J, Tang J (2021b) Fe-based Fenton-like catalysts for water treatment: catalytic mechanisms and applications. J Mol Liq 332:115755. https://doi.org/10.1016/j.molliq.2021.115755
71. Wang C, Cao Y, Wang H (2019) Copper-based catalyst from waste printed circuit boards for effective Fenton-like discoloration of Rhodamine B at neutral pH. Chemosphere 230:278–285. https://doi.org/10.1016/j.chemosphere.2019.05.068
72. Liu D, Li C, Ni T, Gao R, Ge J, Zhang F, Wu W, Li J, Zhao Q (2021) 3D interconnected porous g-C_3N_4 hybridized with Fe_2O_3 quantum dots for enhanced photo-Fenton performance. Appl Surf Sci 555:149677. https://doi.org/10.1016/j.apsusc.2021.149677
73. Wang X, He M, Nan Z (2021c) Effects of adsorption capacity and activity site on Fenton-like catalytic performance for Na and Fe co-doped g-C_3N_4. Sep Purif Technol 256:117765. https://doi.org/10.1016/j.seppur.2020.117765
74. Palanivel B, Jayaraman V, Ayyappan C, Alagiri M (2019) Magnetic binary metal oxide intercalated g-C_3N_4: Energy band tuned p-n heterojunction towards Z-scheme photo-Fenton phenol reduction and mixed dye degradation. J Water Process Eng 32:100968. https://doi.org/10.1016/j.jwpe.2019.100968
75. Xiong C, Ren Q, Liu X, Jin Z, Ding Y, Zhu H, Li J, Chen R (2021) Fenton activity on RhB degradation of magnetic g-C_3N_4/diatomite/Fe_3O_4 composites. Appl Surf Sci 543:148844. https://doi.org/10.1016/j.apsusc.2020.148844
76. He L, Yang SS, Bai SW, Pang JW, Liu GS, Cao GL, Zhao L, Feng XC, Ren NQ (2020b) Fabrication and environmental assessment of photo-assisted Fenton-like Fe/FBC catalyst utilizing mealworm frass waste. J Clean Prod 256:120259. https://doi.org/10.1016/j.jclepro.2020.120259

Combination of Photocatalysis and Membrane Separation for Treatment of Dye Wastewater

Veronice Slusarski-Santana, Leila D. Fiorentin-Ferrari, Samara D. P. Massochin, Keiti L. Maestre, Carina C. Triques, and Monica L. Fiorese

Abstract The dye wastewaters damage the environment, aquatic systems, and human beings, due to their high organic load and dissolved salts, and to the toxic, carcinogenic, and mutagenic character of dyes. Because of these characteristics, this kind of effluent must be adequately treated before being released into water resources or reused in the industry. Several methods can be used to treat dye wastewater, such as adsorption, biological treatment, membrane separation, and the advanced oxidation processes, and among this latter, Fenton, Photo-Fenton, ozonation, peroxidation, electrochemical oxidation, sonolysis, and heterogeneous photocatalysis. These methods can be used individually or combined. One combination that is highlighted is heterogeneous photocatalysis and membrane separation. Thus, this chapter presents (1) a general overview about basic concepts of heterogeneous photocatalysis, and membrane separation; (2) the different configuration combinations of these techniques; (3) a case study using the sequential combination to treat raw dye wastewater; and (4) applications of different combinations to treat dye wastewater. The combination of photocatalysis and membrane separation can be performed through sequential combination or hybrid/one-pot combination, being that, in this last configuration, the photocatalyst can be suspended together with the submerged membrane in the reactional mixture or immobilized in the membrane, constituting the photocatalytic membrane. Most studies use the hybrid/one-pot combination with photocatalytic membrane because the presence of the photocatalyst in the membrane promotes self-cleaning and the fouling reduction in the membrane due to the increase of membrane hydrophilicity.

V. Slusarski-Santana (✉) · L. D. Fiorentin-Ferrari · S. D. P. Massochin
Department of Chemical Engineering, State University of West Parana, Toledo 85903–000, Brazil
e-mail: veronice.santana@unioeste.br

L. D. Fiorentin-Ferrari
e-mail: leila.ferrari@unioeste.br

K. L. Maestre · C. C. Triques · M. L. Fiorese
Program of Post-Graduation in Chemical Engineering, State University of West Parana, Toledo 85903-000, Brazil

S. S. Muthu and A. Khadir (eds.), *Advanced Oxidation Processes in Dye-Containing Wastewater*, Sustainable Textiles: Production, Processing, Manufacturing & Chemistry, https://doi.org/10.1007/978-981-19-0882-8_14

Keywords Photocatalyst · Polymeric membrane · Phase inversion · Ultrafiltration · Fouling · Sequential combination · Hybrid/one-pot combination · Photocatalytic membrane · Slurry photocatalytic reactor · Wastewater degradation

1 Introduction

The necessity to increase production in all industrial sectors, mainly in the last decades, made the discharge of polluting liquid residuals into water bodies considerably increase [13, 88]. Water pollution directly affects, and often irreversibly, the quality of life of human beings depending on the exposition time and pollutants concentration [16, 34]. In this sense, among all kinds of pollutants, the residuals generated by the textile industries, dyeing, paper and cellulose, and leather dyeing considerably contribute to degrading the environment, for being sectors that produce large volumes of effluents with intense color, high organic load and toxicity, high concentration of dissolved salts, besides their carcinogenic and mutagenic nature due to the presence of dyes, regardless of which one [7, 15, 29, 49, 55, 60, 68, 88, 119].

The treatments currently used to remove dyes from wastewater and effluents include, among others, physicochemical methods (coagulation/flocculation, filtration, electrocoagulation, and adsorption), oxidative methods (advanced oxidation processes and chemical oxidation), biological methods (aerobic and anaerobic activated sludge), and membrane separation techniques (microfiltration, ultrafiltration, nanofiltration, and reverse osmosis) [13, 16, 29, 60, 75, 133].

The choice of the adequate method to treat effluents and dye wastewaters is based on specific characteristics of each technique, and also in practical aspects such as easiness of operation, low cost, quick separation, easy scale-up, and efficiency to remove and degrade compounds. In addition, all mentioned characteristics must be associated with the use of ecologically correct and viable technologies. In this sense, the advanced oxidation processes and membranes separation have been highlighted in the scientific community regarding degradation and rejection of pollutants, especially dyes molecules from several classes presented in industrial wastes [17, 39, 132]. These techniques can be used individually and present satisfactory results, however, depending on the characteristic of the raw dye wastewater and mainly on the desired treated effluent, combining photocatalysis and membrane separation is necessary. This union combines the high efficiency of photocatalytic degradation with the selective capacity of membranes to retain dye molecules [33, 72, 94]. The combination of photocatalysis and membranes separation can be performed through a sequential combination process, where both processes are coupled in series, or through hybrid/one-pot combination, where both processes are inside the same equipment: the photocatalytic membrane reactor. In the hybrid/one-pot combination, the photocatalyst may be suspended in solution together with the submerged membrane, or immobilized under/inside the membrane, constituting the photocatalytic membrane. The system with photocatalytic membrane has been highlighting because it does not require the photocatalyst recovery and for the synergistic effect between the photocatalyst and the membrane that promotes self-cleaning and reduced membrane fouling [46, 90, 92, 112].

2 Dye Wastewater Treatment Processes

The advance of environmental laws and the consensus on environmental preservation encouraged the researchers to develop technologies that aim to reduce pollution in an industrial process using efficient wastewater treatments and based on the principles of green chemistry [51]. The dye wastewaters, especially the ones from textile industries, need to be adequately treated since their presence in the environment, besides being a visual problem due to intense color, causes serious damages to the aquatic and terrestrial ecosystems because it reduces light penetration and microbial growth, and to human health due to the toxic, mutagenic, and carcinogenic character of dyes [6, 105].

In general, dyes are classified according to the structure of the chromophore group responsible for fixing the color and the auxochrome group that complements the chromophore, turning the dye molecule soluble in water [115]. The main dyes used are the acid, reactive, dispersed, sulfur, direct, and alkaline ones, which can be used in wool, silk, cotton fibers, cellulose, wood, inks, and leather dyeing [15, 115]. Due to the lack of fixation of the chromophore group in synthetic fibers, dyes are often released in liquid wastewaters from textile industries [15, 141].

In this sense, physicochemical processes such as adsorption, ion exchange, coagulation/flocculation, and sedimentation, biological processes using fungus, bacteria, enzymes, seaweed, and activated sludge, advanced oxidation processes, such as heterogeneous photocatalysis, photolysis, Fenton, photo-Fenton, ozonation, sonolysis, electrochemical oxidation, and peroxidation, and the process of membrane separation that involves microfiltration, ultrafiltration, nanofiltration, and reverse osmosis can be used to treat dye wastewaters [46, 52, 75, 105, 106, 121, 142]. Among all these processes, heterogeneous photocatalysis and membranes separation deserve to be highlighted because they can degrade and retain a great variety of organic and inorganics pollutants, especially dyes.

2.1 Heterogeneous Photocatalysis

Most advanced oxidation processes (AOPs) occur at room temperature and use energy to produce highly reactive intermediates with high oxidation/reduction potential. These reactive species generated attack and destroy a variety of organic compounds in a relatively quick and non-selective way, promoting their total mineralization into CO_2, H_2O, and inorganic ions [40].

Several processes can generate the reactive species, mainly the hydroxyl radicals (•OH) and superoxides ($\bullet O_2^-$), and are considered advanced oxidation processes.

However, the way that each one of these processes generates these reactive species ranges according to the oxidant or promoting agent, for example photolysis uses the ultraviolet (UV), visible (Vis), or solar radiation; Ozonation uses ozone (O_3); Peroxidation employs hydrogen peroxide (H_2O_2); Fenton reaction uses ferrous ions and hydrogen peroxide; Sonication employs ultrasound (US); Heterogeneous photocatalysis uses a semiconductor and light radiation (UV, Vis or solar); and Electro-oxidation (EO) employs electron beam from the passage of electrical current through the electrodes. These processes can be used individually, but when combined (UV/O_3; UV/H_2O_2; UV/Fe^{+2}/H_2O_2; UV/US; UV/US/H_2O_2; semiconductor/O_3/UV; semiconductor/H_2O_2/UV; UV/EO; EO/Fe^{+2}/H_2O_2), the treatments have their efficiency potentialized, being able to treat different kinds of effluents [6, 40, 44, 91]. The advanced oxidation processes can be an option of pre-treatment to biological methods because they are able to reduce the effluent toxicity in a way that does not inactivate the microorganisms, and also to membranes separation because it can reduce the number of species that may cause membrane fouling [30]. Another option to use advanced oxidation processes is as a post-treatment, aiming to eliminate recalcitrant compounds that persist in the effluent turning it able to be properly disposed of in water resources or to be reused in the industrial process.

Among these oxidation processes, heterogeneous photocatalysis deserves to be highlighted for being a process that combines photochemistry and catalysis, from the activation of a semiconductor by appropriate luminous radiation (UV, Vis, or solar). The activation of the semiconductor occurs by the absorption of photons with equal or higher energy than the energy of its band gap (distance between the valence band and the conduction band), which promotes excitation of the electrons from the valence band to the conduction band with consequent formation of the electron/hole pair. From the formation of the electron/hole pair, several chain reactions are developed, promoting the in situ generation of reactive oxygen species (•OH, $•O_2^-$, HO^-, HOO^-, HOO•, H_2O_2, and $^1O_2$) which are able to convert toxic and non-biodegradable organic compounds into relatively innocuous products, the green products [4, 30, 58, 99, 107, 111].

The main advantages of heterogeneous photocatalysis are as follows: reasonably fast reaction at room temperature; capacity to oxidize several organic and inorganic compounds; capacity to completely degrade pollutants and intermediates; system of easy operation and maintenance; possibility to act in several media (gas phase, pure organic liquid phase, or aqueous solutions); relatively low cost of implementation and operation, especially when solar radiation is used; does not require the addition of chemical product; does not generate sludge after the treatment; the possibility to recycle the photocatalyst, which reduces operating costs [96, 112, 136, 145]. The disadvantages of industrial application of heterogeneous photocatalysis are the rapid recombination rate of photogenerated electrons and holes, the possible formation of subproducts by partial oxidation that can be less reactive and more toxic than the initial contaminant, incapacity of absorbing visible and solar radiation of some catalysts, the necessity of an additional step to separate little particles of catalysts dispersed in the effluent after the treatment, the difficulty to treat high concentrations or extremely low concentrations (trace levels) of organic pollutants, lower viability to treat great volumes of pollutants at low concentrations and low efficiency to treat effluents that contain chemical species, such as chloride and carbonate ions, which act as scavengers of reactive radicals or centers of recombination [30, 46, 72, 96, 111, 124, 136].

Several materials, such as metal oxides (TiO_2, ZnO, Nb_2O_5, Fe_2O_3, CuO), metal sulfides (CdS, ZnS), compound oxides (N-ZnO/CdS/Graphene oxide, $Fe_3O_4/SiO_2/CeO_2$, Co_3O_4/Fe_2O_3), carbon nanotubes (CNT/TiO_2, g-C_3N_4/CNTs/Ag_3PO_4), dendrimers (TiO_2 magnetic-cored dendrimers), and polymer composites (TiO_2/PES, ZnO/PMMA), can be employed as a catalyst in photocatalytic processes [11, 35, 43, 45, 63, 67, 71, 108, 138, 146]. The catalysts can be classified, among others, according to their structure and compositions as [19, 72, 111, 126]:

- Bulk catalysts: in which the active phase corresponds to the whole material. Usually, these catalysts present small particle size and are dispersed in the effluent to be treated;
- Doped catalysts: in which a metal or a metallic compound is added to the structure or surface of the bulk catalyst to improve its photocatalytic activity, stability, and selectivity to a desired reaction. Doping can dislocate or enlarge the range of light absorption by the catalyst, allowing the use of solar radiation in its activation;
- Supported or immobilized catalysts: in which the active phase of the catalyst is immobilized or impregnated into an adequate support matrix, aiming to facilitate the recovery and reuse of the catalysts. Photocatalytic activity is improved due to the synergistic effect generated between the active phase and the support;
- Composites and/or hybrid catalysts: in which two or more materials, semiconductors or not, are combined in the composition of the catalyst, aiming to delay the recombination process of the electron/hole pairs and increase photoactivity.

The development of these structures and composition of catalysts as well as new catalysts was, and has still been, encouraged by the search for more efficient, sustainable, and eco-friendly materials and processes. In the 70 s and 80 s, most studies of photocatalysis were performed with pure oxides, especially TiO_2 and ZnO, which could be found dispersed in solution inside a slurry-type photoreactor. This kind of photoreactor presents advantages such as simple configuration, high adsorption and degradation rate, and low limitation of mass transfer; however, slurry photoreactor presents the inconvenience that the recovery of the catalyst particles is necessary [20, 33, 50, 57, 104].

Since TiO_2 and ZnO are usually activated with ultraviolet radiation, there was an urgent need for studies about the synthesis of doped catalysts, aiming to use visible/solar radiation and consequently reduce costs [3, 26, 85]. Therefore, trying to make the photocatalytic process more ecologically and economically viable from the recovery and reutilization of the catalysts, the supported catalysts were developed. The problem with this kind of catalyst is that in some kinds of support, especially the porous ones, limitations of the mass transfer can occur, which reduces the process efficiency [18, 23, 89]. Thus, up to now, new catalysts are being developed, both in micrometer and nanometer scale, especially from the combination of different materials to create the composites/nanocomposites, hybrid catalysts, or single-atom catalysts in order to achieve catalysts that are highly photoactive also in the visible region to use to remediate effluents containing persistent and recalcitrant compounds. Among the materials used to prepare the composites and hybrid catalysts, there are metal oxides, metal hydroxides, metals, semimetals, nonmetals, bimetallic nanoparticles, graphene oxide, reduced graphene oxide, graphite such as g-C_3N_4, carbon nanotubes, nanoclusters, and nanodots [31, 45]. As examples of single-atom catalysts, there are monodisperse silver atoms and carbon quantum dots, co-loaded with ultrathin g-C_3N_4, single-atom dispersed Ag mesoporous g-C_3N_4 hybrid [62].

With the development of catalysts with smaller and smaller particle sizes and the need to reuse these catalysts, the coupling of photocatalysis with the system of membranes separation aiming the confinement and recovery of photocatalysts inside the treatment plant, is a promising alternative [33, 72, 94].

2.2 *Membrane Separation*

In industrial processes that involve separation technologies, the use of synthetic membranes has been highlighted due to the versatility of separating solutes with varied types and sizes. This characteristic occurs because of the morphology of the membranes that are divided into two big groups: symmetrical and asymmetrical [32, 97]. The symmetrical structure can have uniform internal morphology with similar pore sizes, as it is the case of microporous membranes; but also it does not necessarily present pores, such as in the case of dense membranes. In the asymmetrical membranes, the main characteristics are the non-uniformity of the pores and the formation of a thin surface layer supported on a porous sublayer. When this thin layer, often called skin, is constituted by small pores, it is called asymmetric type Loeb-Sourirajan membrane [14, 97].

According to the morphology that results in different ranges of pore size, the membranes used for liquid permeation possess different classifications, as shown in Fig. 1. Microfiltration and ultrafiltration membranes with pore size higher than 50 nm and between 50 and 3 nm, respectively, can retain molecular species in the range of 300 to 1106 Da, including microparticles such as protozoa and colloidal solids to the up to macromolecules such as viruses, proteins, and enzymes (Fig. 1) [25]. Nanofiltration, classified between 0.5 and 10 nm, retain substances with molar mass in the range of 1000 to 3000 Da, such as ions and organic molecules [98]. In the case of reverse osmosis, in which the membranes possess pore sizes lower than 200 Da and are considered dense membranes, inorganic ions and organic matter with a molar mass lower than 1000 Da are retained [25, 118].

The morphologies of the membranes depend on the kind of material and the fabrication technique used. Membranes can be fabricated from hydrophobic and hydrophilic polymeric materials (cellulose acetate, polyvinyl chloride, polyvinylidene fluoride, polytetrafluoroethylene, polypropylene, polycarbonate, polysulfone, polyethersulfone, polyamide, and polyester) or ceramic materials (ceramics, carbon,

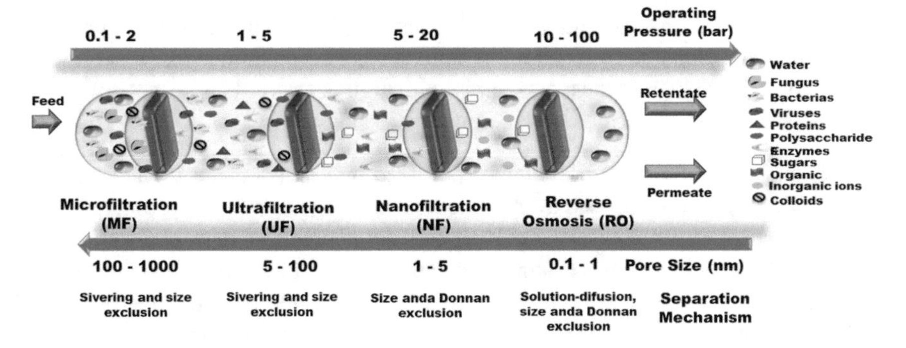

Fig. 1 The schematic classification of pressure-driven membrane

metal oxides, metals, alumina, borosilicate glass, and zirconia) [1, 73]. Ceramic membranes manufactured with inorganic materials present greater mechanical, thermal, and chemical resistance, and longer service life than polymeric membranes. However, the cost involved in the utilization of these inorganic materials is higher than that of polymeric membranes [5]. Lee and Kim [77] claim that the ceramic membranes compete with the polymeric ones regarding potable water treatment. What can be claimed based on studies from the literature is that a direct relationship between the kind of particulate material present in the fluid that will be filtrated and the material that constitutes the membrane exists.

Generally, synthetic porous membranes of microfiltration produced from polymeric materials can be manufactured from sintering, stretching, track-etching, template leaching, and phase inversion techniques, while the production of ultrafiltration membranes is by phase inversion. A succinct description of each one of these techniques that are used to manufacture the polymeric membranes classified in the range of micro and ultrafiltration is presented below [97]:

- Sintering: consists of compressing the material that will be used in the membrane with a certain size and sintering at elevated temperature. Both organic, such as polymers, and inorganic materials, such as metals, graphite, and glass, can be used. This technique produces membranes with pore sizes in the range of 0.1 to 10 μm, and porosity between 10 and 20%;
- Stretching: consists of stretching a polymeric film or sheet until breakage occurs that constitutes the porous structure with pore sizes in the range of 0.1 to 3 μm;
- Track-etching: consists of, initially, submitting a film or sheet to high energy particle radiation to create tracks and then, inserting the film by immersion into acid or alkaline bath to form the pores. The pore sizes can be in the range of 0.02 to 10 μm;
- Template leaching: consists of lixiviating one of the components presented in the structure of the membrane with the aid of an acid or base, obtaining pore sizes around 0.005 μm;

- Phase inversion: this technique is used when the phase of a polymeric solution is desired to be changed from liquid to solid state. This technique allows to obtain open and dense membranes, preferably with high porosity. This process of phase inversion can be performed by precipitation techniques by:

- Solvent evaporation: the polymer is dissolved into an adequate solvent and the polymeric solution is spread over an inert surface. The solvent is evaporated at an inert atmosphere or a non-solvent bath with something else than a solvent;
- Vapor phase: in this case, a casting film, composed of the polymer and the solvent is put into contact with a vapor phase saturated with a non-solvent. The formation of the porous membrane can occur through the diffusion of the non-solvent inside the casting film;
- Evaporation controlled: the polymer is dissolved into a mixture of solvent and non-solvent, which is then evaporated;
- Thermally-induced: the polymeric solution is cooled to occur phase separation, forming membranes with a thin layer, called skin. Usually, it results in microfiltration membranes;
- Immersion: the polymeric solution is spread into an inert supporter and put in a coagulation bath containing a non-solvent.

Since the different techniques originate membranes with distinct characteristics, the technology employing polymeric porous membranes has its application consolidated in all processes that involve separation, purification, and concentration of substances of the most varied chemical species. Reverse osmosis, which presents morphology with small pores or is often dense, is mainly applied in the desalination of brackish or seawater [38], and also in the production of ultra-pure water used to produce drugs and cosmetics. Porous membranes of microfiltration and ultrafiltration possess a wide range of use, such as concentration of products aiming to add value to the food, cheese processing, separation of proteins and lactose, clarification, and juice concentration, separation of oil-water emulsions, biotechnology area, drugs filtration, water recovery and recycle in plants, and treatment of effluents [14, 22, 47, 69, 143, 144]. Porous membranes of microfiltration and ultrafiltration are satisfactorily used to treat water and industrial effluents containing dyes, due to the high retention that these membranes present when particulate materials in the effluent possess sizes relatively bigger than the membrane porous, as it is the case of dye molecules [14, 16, 34].

Industrially, the transport of solutes through membranes can be performed via dead-end or cross-flow filtration modules. Considering dead-end filtration, the fluid is perpendicularly permeated through the membrane with continuous deposition of solutes into the surface, often hampering the performance of continuous processes. In the cross-flow filtration, the flow occurs adjacently to the membrane surface, favoring less deposition of solutes, being the most industrially used process due to continuous filtration [41, 97]. To enable the transport of different chemical species and different molar sizes, a driving force is necessary; in this case, the transmembrane pressure.

In the dead-end filtration, the transmembrane pressure is determined by the difference between the feed pressure and the permeate pressure, while in the cross-flow filtration, the pressure is defined by the mean between the feed pressure and the permeate pressure subtracted by the permeate pressure [129]. The lower the size of the membrane pore, the higher the necessary driving force to perform the separation. Thus, to have an adequate transport of solute, it is necessary to apply a transmembrane pressure in the range of 1 to 3 bar for microfiltration, 3 to 10 bar for ultrafiltration, 10 to 40 bar for nanofiltration, and osmotic pressures for reverse osmosis (Fig. 1), independently of the filtration being dead-end or cross-flow [73].

With the deposition of solute into the membrane surface during the separation of liquid solutions, the pores of the membranes become obstructed, which can occur by complete, internal, and random blocking of the pore and/or formation filter cake (Fig. 2). In the complete and internal blocking, the pores are completely obstructed by continuous deposition of the particles from the effluent. In random blocking, the deposition of the particles occurs in different regions, causing random accumulation of the retentates particles. Regarding the formation of filter cake, deposition continuously occurs causing an additional resistance to mass transfer of the solvent through the membrane. This phenomenon is also called polarization of concentration [14, 97]. From the development of these phenomena that cause continuous flow decrease after the stabilization of the polarization of concentration, fouling occurs, which can lead to such low flows that the membrane cannot be used anymore.

Fouling can occur reversibly with a reduction of the permeability at a short period due to the deposition of solutes in the surface. In the irreversible fouling, by its turn, the loss of permeability occurs more slowly, because the solutes are sipped by the membrane structure. Thus, irreversible fouling is considered a phenomenon that irreversibly reduces the service time of membranes [5, 62, 78, 97, 113]. After the reduction of permeability via reversible fouling to values that are not industrially enough, hydraulic backwash and/or chemical cleaning are necessary.

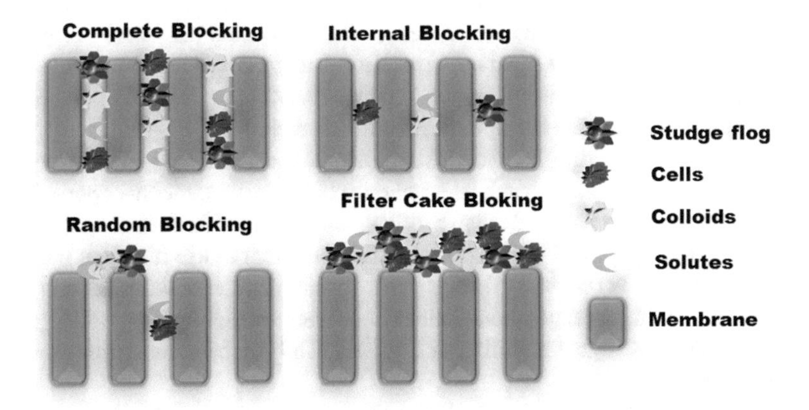

Fig. 2 Types of pore blocks in pressure-driven membrane

Physical or hydraulic cleaning consists of the permeation of continuous water flow at high pressures, forcing the removal of solutes adhered to the surface and/or inner part of the membrane. The backwashing method, which consists of performing the filtration with the membrane inverted, is the most common method to remove fouling, especially in membranes with bigger pores as in the case of microfiltration and ultrafiltration [24, 131]. Due to its higher efficiency, the chemical cleaning causes the recuperation of the membrane by filtrating acid or alkaline solutions, chemically destabilizing internal incrustation and the layer adhered in the membrane surface, often called filtration pie. When the flow reduction is caused mainly by reversible fouling, physical cleaning can recover the initial flow of the membrane. However, for the irreversible fouling, depending on the nature of incrustation, the combination of chemical and hydraulic cleaning by backwashing is necessary [124, 131]. It should also be noted that, in addition to cause the filtration process to stop, the hydraulic cleaning process uses considerable volumes of water, and the chemical cleaning generates solutions with varied pH [97].

Membranes should present high mechanical, thermal, and chemical resistance in order to have a good performance regarding permeability, cleaning, and flow recovery. These characteristics can be obtained during the manufacture according to the technique employed, but structural and superficial modifications in the membranes can still be performed to improve its performance regarding flow and anti-fouling properties. Among the modification techniques that are based on the traditional manufacturing techniques, it can be highlighted [12, 16, 34, 42, 61, 72, 79, 88, 110, 124]:

- Blending: consists of the mixture of different mass proportions of polymers and/or a mixture of particulate materials of the most varied species;
- Surface coating: this method is employed mainly in the structural modification of commercial membranes. Solutions, suspensions, colloids, or other polymeric solutions can be inserted into the surface of the membranes;
- Sol-Gel: consists of hydrolyzing the precursor, usually an alkoxide, forming the hydrolyzed products (active monomers), which are then polymerized, forming an oxide network. This method also allows a good dispersion of the particles/nanoparticles under the support/membrane. Membranes with pore sizes in the range of nanometers are obtained using this technique;
- Surface coating or dipping: in this technique, the scattering of nanocrystals, nanofibers, nanoparticles, such as graphene oxide, carbon, metal-organic frameworks, and metal/metal oxide is performed on the surface of membrane produced from any traditional technique. It should be noted that to use it for the treatment of residual waters containing dyes, negatively charged membranes present better separation. In this case, graphene oxide nanosheets are highlighted, which present negative charge, high flexibility, and excellent physicochemical and transport properties;
- Anodizing: the modification occurs from the addition of polyelectrolytes that bond themselves to the species that are desired to be removed forming a macromolecular

complex that can be more easily retained by the membrane. In this process, it is usual to retain anions or cations;

- Electrospinning: in this technique, a polymeric solution can be pulverized using an electric field potential to obtain particles or fibers. In this system, nanoparticles can be inserted into the morphological structure of the membranes.

The technology of membrane separation is adapting itself to new possibilities of changes of morphological configuration with the addition of compounds that aid in the obtention of greater industrial performances. However, depending on the kind and concentration of the particulate material present in the effluent, it is not always possible to meet the effluent discharge standards established by legislation. In these cases, the combination of different treatment processes for dye wastewater can optimize the results regarding high pollutant degradation with a minimum time of treatment process.

3 Combination of Heterogeneous Photocatalysis and Membrane Separation

Processes that combine different treatment technologies constitute the concept of process intensification and green technology, which aim to minimize operating and capital costs and the mitigation of environmental impacts due to the reduction of waste generation. These reductions are possible from the use of technologies that are cleaner, more efficient, that occupy lower industrial structure (reduction of unit operation and processes scale), and that allow the flexibilization of the process, as well as the synergistic effect created by the coupling of these technologies [9, 64, 87, 100]. Some combination of processes to treat dye wastewaters can be mentioned: photolysis and ozonation [123]; photolysis and peroxidation [52, 123]; photocatalysis and membrane separation [21, 28, 102, 114]; photocatalysis, and sonolysis [87, 120]; photocatalysis, microfiltration, and sonolysis [120]; photocatalysis, adsorption, and membrane separation [143, 144]; biological treatment and membrane separation [95]; biological treatment, membrane separation, photolysis, and ozonation [95]; photocatalysis and biological treatment, among others [27, 105].

The combination of the processes of photocatalysis and membrane separation, in particular, can be applied in the remediation of several kinds of pollutants, such as drugs [46, 86], dissolved organic matter (humic and fulvic acid) [59, 83, 109, 122], microorganisms [53, 66], oil wastewater [100, 101], primary and secondary effluents [66, 93, 128], and dyes [21, 51, 54, 102, 114, 143, 144].

When coupling the processes of photocatalysis and membrane separation in the remediation of dye wastewaters, as in the case of effluents generated by the textile industry, the high efficiency of photocatalytic degradation is combined with the selective capacity of the membranes to retain ions and/or molecules. In this combined process, the photocatalyst performs the role of degrading the organic molecules of dye, while the membrane rejects organic and inorganic contaminant species,

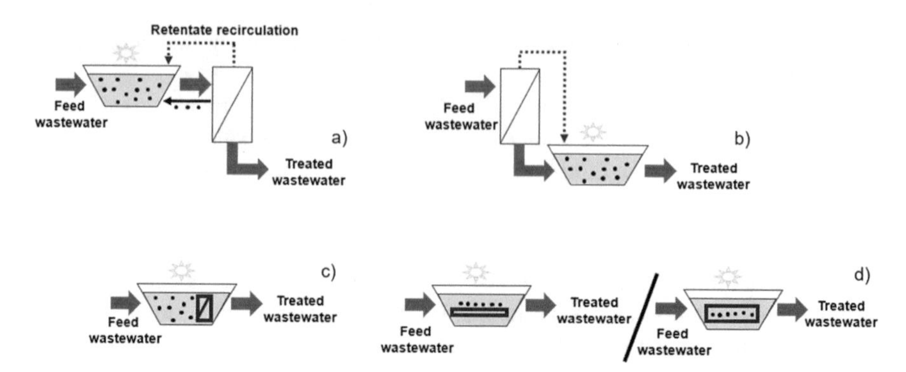

Fig. 3 Scheme of the possible combinations of heterogeneous photocatalysis and membrane separation

thus increasing the global efficiency of the process. In addition, the membrane also performs the important role of retaining the photocatalyst for its later recovery and reuse. The combination of those two processes is also favored by the similarity of some operating conditions, such as operation in liquid phase and continuous system, as well as relatively low temperature and pressure, and low concentration of the chemicals [33, 72, 94].

The combination of the processes of photocatalysis and membrane separation can be performed in different configurations (Fig. 3), which can be divided into two main groups [9, 33, 37, 46, 64, 72, 79, 90, 92, 94, 112].

(1) Sequential combination: in this kind of combination, the modules of each process are coupled in series, heterogeneous photocatalysis followed by membrane separation, according to Fig. 3a, or membrane separation followed by heterogeneous photocatalysis, according to Fig. 3b, being this kind of combination called split-type system. The order in which the processes are arranged depends on the objective to be achieved, so, photocatalysis can be used as a pre-treatment stage and also as a post-treatment stage. As a pre-treatment to the feed flow of the membrane, photocatalysis aims to degrade pollutants that cause incrustations and blockage of the pores of the membrane, thus reducing the fouling. Photocatalysis as a stage of post-treatment aims to degrade the organic microcompounds that were not rejected by the membrane and highly-concentrated organic compounds in the retentate flow, allowing the adequate disposal of the treated effluent into the environment. In the same way, membrane separation can act as a pre- or as a post-treatment. The membrane separation can be used as a pre-treatment to retain part of the pollutants to increase photocatalysis efficiency, or also as a post-treatment, in which the membrane acts in the separation and recovery of the photocatalyst, allowing its return to the photoreactor and reuse, as well as in the retention of remaining pollutants. In some cases, the effluent is recirculated between the two modules until it reaches the desired quality to be adequately released;

(2) Hybrid/one-pot combination: in this combination, both systems (heterogeneous photocatalysis and membrane separation) are inside a sole module and the oxidation and the separation/rejection of pollutants simultaneously occur. This combination can be subdivided into other two configurations, a system with suspended photocatalyst and a system with the photocatalytic membrane. In the system with suspended photocatalyst (Fig. 3c), the membrane is submerged into a reactional mixture that contains the suspended photocatalyst, so that the photocatalysis and the membrane separation act independently, that is, the photocatalyst causes the generation of the reactive oxygen species that degrade the pollutants, while the membrane acts rejecting pollutants after degradation, selective separating the desired product, avoiding undesired reactions, and also as a barrier to keep the photocatalyst confined inside the photoreactor. In the configuration with the photocatalytic membrane (Fig. 3d), the photocatalysis and membrane separation act in an intrinsically integrated way, that is, the membrane presents photocatalytic activity due to the presence of photoactive material (photocatalyst) in its constitution, performed by adding photocatalyst in the polymeric solution, or by depositing the photocatalyst in the membrane surface. Thus, the membrane acts as a photocatalyst when activated by luminous radiation and generating the reactive oxygen species and also as a selective barrier to non-degraded pollutants and intermediate compounds from partial degradation. This kind of combination is in accordance with the principles of process intensification that aim at the design of innovative equipment and the development of more compact and efficient processes.

Despite being different, the configurations sequential combination and hybrid/one-pot combination present the same basic operating principles of oxidation and separation. Because of this, the photoreactors used in those two configurations can receive the same denomination. In the literature, the denominations found for these photoreactors include incorporated photocatalytic membrane reactor (IPMR), slurry photocatalytic membrane reactor (SPMR), slurry membrane photoreactor (SMPR), submerged membrane photocatalysis reactor (SMPR), photocatalytic membrane reactor (PMR), membrane photocatalytic reactor (MPR), and integrated/hybrid membrane reactor (HMR). However, among all denominations, photocatalytic membrane reactor is the one that prevails [9, 33, 46, 64, 72, 74, 79, 90, 92, 112, 135].

It should be noted that, regardless of the choice of the kind of configuration (sequential combination or hybrid/one-pot combination), several parameters affect the efficiency of the photocatalytic membrane reactors, among them, it can be mentioned [9, 33, 79, 90, 92, 145].

- Photocatalyst: kind, state (dispersed or immobilized), particle size, surface area, concentration/charge, and mainly the energy of the band gap;
- Effluent: pH, composition/presence of organic ions, and concentration and size of the pollutant molecule;
- Aeration: concentration of dissolved oxygen;

- Radiation: light source typology, wavelength, time, intensity, and position (above/inside the feed tank or membrane unit or additional tank);
- Membrane: material, porosity, roughness, mechanical resistance, stability, configuration (hollow fiber and flat sheet membrane), operating mode (dead-end and cross-flow), permeate flow rate, transmembrane pressure, and anti-fouling property.

Thus, the choice of the optimum operating parameters and the configuration of adequate combination will be determinants for the efficiency of removal and degradation of colored compounds found in industrial effluents.

3.1 Sequential Combination

The sequential combination of the processes photocatalysis and membrane separation aims to increase the efficiency of the dye wastewaters treatment because it combines characteristics and advantages from both processes. In this kind of configuration, the effluent can be first subjected to the photocatalytic treatment, and then to the membrane separation process, or vice versa. The choice of the adequate sequence of treatment depends on characteristics of the raw wastewater, but mainly on the characteristics that the treated effluent should present to be adequately disposed into the environment without causing environmental impacts or still that could be reused by the industry itself.

Despite the possibility of using photocatalysis as pre-treatment or as post-treatment, most studies used the sequential combination of photocatalysis and membrane separation with recycling of the permeate effluent between the two modules until reaching the quality to be adequately discarded; or the process of membrane separation after the photocatalysis in a continuous system with the main aim of retaining and recovering the small particles of the photocatalyst.

Some studies can be highlighted regarding the ones that promoted the recirculation of the dye solution and simulated effluent between the modules of photocatalysis and membrane separation until its complete discoloration, as mentioned in the sequence. Grzechulska-Damszel et al. [51] used the combination of photocatalysis (ultraviolet radiation and TiO_2 P25 dispersed in solution) and ultrafiltration separation with a flat-sheet polysulfone membrane to remove azo dyes (Acid Red 18, Direct Green 99, and Acid Yellow 36) from water. The sequential system was operated with permeate recirculation to the photoreactor feed until the complete discoloration of the dye solution, which also promoted the removal of 48-73% of total organic carbon. The ultrafiltration process was able to separate the TiO_2 particles from the treated solution and still retain some products of partial degradation. Jiang et al. [65] used a TiO_2 slurry photocatalytic reactor with submerged ultraviolet light lamp followed by an ultrafiltration unity containing a polyvinylidene difluoride or polyacrylonitrile membrane operated in the recirculation mode during the treatment to degrade simulated wastewater containing azo Acid Red B dye. The authors verified that the

TiO_2 particles were recovered by ultrafiltration, the permeate flow reduction in the combined system was negligible compared to simple ultrafiltration and the combination of photocatalysis and ultrafiltration promoted the greatest rate of discoloration of the simulated effluent. Buscio et al. [21] evaluated the degradation of the solution of Disperse Red 73 dye into a photocatalytic membrane reactor by recirculating the de solution between the filtration module containing a polysulfone membrane and the photocatalytic module containing TiO_2 dispersed in solution and 99% of degradation was reached after 180 min of treatment. The authors verified that the presence of the catalyst contributed to membrane fouling only at the beginning of the experiment, possibly by the formation of a porous cake layer on the membrane surface due to the deposition of catalyst particles.

The degradation of dye solution and textile effluent using sequential combination of photocatalysis and membrane separation in a continuous system was also reported in the literature. Wang et al. [137] investigated the degradation of the Reactive Brilliant Red X-3B dye in a photocatalytic membrane reactor using TiO_2 nanotubes as the photocatalyst, ultraviolet radiation, and polyethersulfone (PES) ultrafiltration membrane to recover the photocatalyst. The authors obtained 94.6% of discoloration rate after 75 min of treatment and 100% of TiO_2 nanotubes recovery rate. Desa et al. (2019) investigated the degradation of industrial textile wastewater in a photocatalytic membrane reactor with sequential combination of photocatalysis and ultrafiltration. Initially, the effluent was degraded by the action of the dispersed ZnO capped with polyethylene glycol photocatalyst and ultraviolet radiation, being that the luminous source was submerged in the photocatalytic module. Then, the effluent treated by photocatalysis was subjected to the filtration module containing a polypiperazine-amide tight ultrafiltration membrane. Excellent results, 100% of discoloration and 97% of chemical oxygen demand reduction, were obtained at pH 11 when coupling photocatalysis to membrane separation. The authors identified that the catalyst particles were recovered and that membrane fouling was the cake-filtration-model type.

A sequential combination using membrane separation as a pre-treatment to photocatalysis was evaluated by [114]. These authors prepared a photocatalytic membrane from the incorporation of TiO_2 to the polymer polyvinylidene difluoride and used this membrane to degrade Remazol Turquoise Blue dye in both the filtration system and in the photocatalytic system as the photocatalyst (photocatalytic membrane). Initially, the dye solution was filtered in the cross-flow ultrafiltration module that contained the photocatalytic membrane, and then, in a photocatalytic module, the permeate was subjected to ultraviolet radiation in the presence of the same photocatalytic membrane. The filtration process was responsible for 72% of discoloration and 40% of total organic carbon reduction, and after the combination with photocatalysis, better results were achieved (94 and 90%, respectively), especially regarding mineralization.

The combination of photocatalysis and membrane separation performed continuously presents certain inconveniences, such as the possibility of deposition of the photocatalyst in the pipes between both processes in the sequential combination, flow variation of photocatalyst, and, consequently, photocatalyst concentration inside the reactor, and still, in some cases, the low efficiency of the process due to the short residence time of the effluent in the system [72, 112, 145]. Thus, a study regarding all factors that can influence the efficiency of sequential combination processes is necessary.

Besides, a complete study in the literature that compares the efficiency of the two configurations of the sequential combinations was not found; that is, a study that uses photocatalysis as pre-treatment and also as post-treatment for the same dye wastewater. Thus, a case study to evaluate the treatment of one real dye wastewater employing these two configurations will be presented.

3.1.1 Case Study

This case study aimed to compare both configurations of sequential combination (photocatalysis followed by membrane separation and vice versa) to treat one real dye wastewater, which was collected after the dyeing stage in the process of fish skin tanning. The Acid Green 16 ($C_{31}H_{33}N_2NaO_6S_2$) dye was used in this process, which is also used in textile industries. The real effluent presented green coloration due to the presence of the dye ($\approx$24 mg L^{-1}) with maximum absorption wavelength at 646 nm, pH of 5, electric conductivity of 438 μS cm^{-1}, and organic matter of 700 mg O_2 L^{-1}.

In photocatalysis, a supported catalyst (10%TiO_2_P25/MOR) was used in the experiments. This catalyst was prepared by the wet impregnation method, in which 10% (%wt) of TiO_2 (P25 Degussa/Evonik) was supported on mordenite zeolite (pre-treated with NaCl solution and with a diameter between 0.589 and 0.833 mm) and the material was dried at 100°C/24 h and calcinated at 500ºC/5 h [125]. The catalyst 10%TiO_2_P25/MOR was characterized by N_2 physisorption (BET method), X-ray diffraction, photoacoustic spectroscopy, point of zero charge (pH_{pzc}), and scanning electron microscopy. The photocatalytic assays lasted 3 h and were performed under UV and solar radiation. In the assays with UV radiation, a mercury vapor lamp with 250 W was used located 15 cm above the reactional mixture, while the assays with solar radiation were performed between 11 a.m. and 2 p.m. during summer. In the photoreactor, 0.3 g of the 10%TiO_2_P25/MOR catalyst were suspended in 300 mL of effluent and subjected to UV or solar radiation for 3 h. Samples of the effluent were collected at regular time intervals and centrifuged. The degradation of the effluent was evaluated between 200 and 800 nm in UV-Vis spectrophotometer (UV-1800 SHIMADZU), while color removal was evaluated at 646 nm. pH and conductivity of the initial and final samples were measured.

In the membrane separation process, 12% of polyethersulfone (PES) (%wt) were dissolved in N'N-dimethylformamide (DMF) (12%PES/DMF) under stirring and controlled temperature of 50 °C to fabricate the membrane. The polymeric solution was then spread under a plane surface and the membrane was obtained applying the phase inversion technique [42]. The membrane was superficially characterized by scanning electron microscopy and contact angle. Regarding membrane permeability, permeation assays with distilled water to determine the initial flow and with the real effluent were performed in a dead-end filtration cell with a membrane of 2.5 cm^2. After the effluent permeation, the membrane was cleaned by the backwashing method and the fouling was determined regarding the initial flow of the membrane. All permeation and membrane cleaning assays were performed in the dead-end filtration cell with a pressure of 1 bar.

The characterization analysis of the 10%TiO_2_P25/MOR catalyst evidenced that it is a porous material with a mean diameter of 17.5 Å and crystalline with peaks referring to the anatase and rutile phases of TiO_2, besides the peaks that are characteristics from mordenite zeolite. This catalyst can be activated by Vis/solar radiation, because its activation occurs at 488 nm according to the analysis of photoacoustic spectroscopy, and presents pH_{PZC} at 6.8. Thus, the catalyst, when in contact with an effluent with a pH lower than the pH_{PZC}, will have its surface positively charged [116], which will favor the degradation of anionic solutions, as in the case of the effluent under study. The presence of TiO_2 agglomerates in the surface of the mordenite zeolite is seen from the micrography image of the 10%TiO_2_P25/MOR catalyst (Fig. 4a), evidencing that the TiO_2 immobilization preferably occurred in the external surface of the support.

Regarding the characterizations of the membrane fabricated with 12%PES/DMF, it presented thermal stability up to 300 °C, high hydrophilicity with a contact angle lower than 90°, and a mean pore diameter of 950 nm [42]. From the micrography of the membrane surface (Fig. 4b), a morphology that is typical of asymmetrical membranes is observed with the presence of macrovoids that facilitate the permeation through the membrane. Usually, membranes that present macrovoids possess a high flow, regardless of the kind of permeate contaminant.

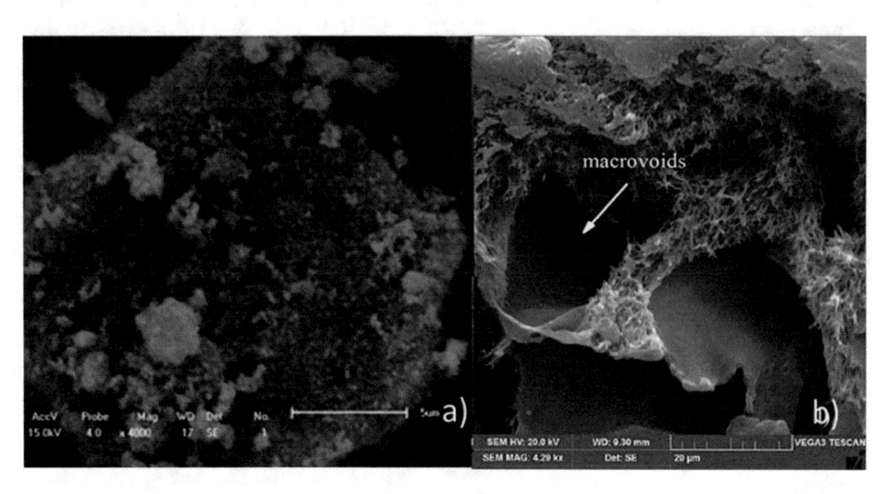

Fig. 4 Micrographs of **a** 10%TiO_2_P25/MOR catalyst with ampliation of 4.0 kx; **b** 12%PES/DMF membrane with ampliation of 4.3 kx

The effluent was subjected, initially, to the individual processes of photocatalysis and membrane separation and the results are presented in Fig. 5. Regarding photocatalysis, the discoloration kinetics of the effluent using the 10%TiO_2_P25/MOR catalyst at concentration 1 g L^{-1} under UV and solar radiation is presented in Fig. 5a. The effluent discoloration occurred according to a pseudo-first-order Langmuir-Hinshelwood kinetic. Considering the kinds of radiation, the discoloration is observed to have occurred mostly in the first 60 min of solar irradiation, evidencing the capacity of the supported catalyst to be activated also by Vis/solar radiation, corroborating with the photoacoustic spectroscopy analysis result. The high intensity of UV light promoted a more effective discoloration of the effluent during all the time of the assay. After the photocatalytic treatment, the pH did not considerably change pH (5.0-5.3) and a small reduction of electric conductivity was observed (345–350 μS cm^{-1}), regardless of the kind of radiation. The effluent discoloration was 56 and 40%, and the chemical oxygen demand reduction was 38% and 23% after 180 min of UV and solar irradiation, respectively. This result was attributed to the coloration and to the presence of particles in the effluent that hampered the light penetration and the catalyst activation.

Regarding the effluent treatment by the membrane separation process (Fig. 5b), a high steady permeate flow is observed for the water filtration (600 kg h^{-1} m^{-2}) and also for the effluent filtration (100 kg h^{-1} m^{-2}). This result corroborates with the classification of the 12%PES/DMF membrane as being porous and with the formation of macrovoids (Fig. 4b). The permeate flow with distilled water presented a higher fall compared to the permeate flow with effluent in the first 40 min of operating (Fig. 5b). This fact is associated with the period of membrane compaction that occurs due to the movement of the polymeric chains caused by the applied pressure of 1 bar. However, comparing the steady state permeate flows, a reduction of 58% of the steady flow with effluent compared to the flow with water is verified. This flow reduction occurs due to the deposition of organic matter and remaining pollutants from the effluent in the surface of the membrane, which is more accentuated in the dead-end filtration. Despite the increase in the concentration of solutes in the membrane surface causing

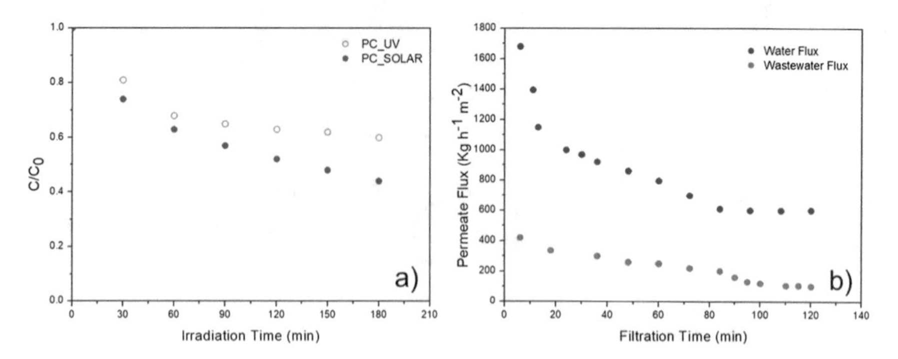

Fig. 5 **a** Profile of effluent discoloration using 1 g L^{-1} of the 10%TiO_2_P25/MOR catalyst under UV and solar radiation; **b** filtration of the effluent with the 12%PES/DMF membrane at 1 bar

a diffusive movement of the solute in the sense of returning to the solution [97, 14], this movement was not enough to avoid permeate flow reduction with effluent. After the permeation of the effluent in the membrane, neither the pH nor the electric conductivity of the effluent changed, and the 12%PES/DMF membrane enabled 32% of color and 75% of chemical oxygen demand reduction.

After the effluent permeation, the membrane was subjected to a cleaning procedure, using a backwashing with distilled water and acid solution, and the flow returned to the initial value. This result indicated that no irreversible fouling occurred in the 12%PES/DMF membrane, only reversible fouling of 85% with the predominance of pore-clogging by random blockage (intermediate blockage) according to the concepts developed by the Hermia models [56].

Generally analyzing the individual application of the methods photocatalysis and membrane separation to treat the dye wastewater of fish skin tanning containing Acid Green 16 dye, satisfactory results regarding color and chemical oxygen demand reduction were not reached. Thus, the effluent was subjected to the sequential combination of the processes photocatalysis and membrane separation, aiming to reach an effluent able to be disposed of according to standards of effluent release. Two sequences were tested, photocatalysis followed by membrane separation process (PC + MSP) as sequence 1, and membrane separation process followed by photocatalysis (MSP + PC) as sequence 2, under both UV and solar radiation (Fig. 6).

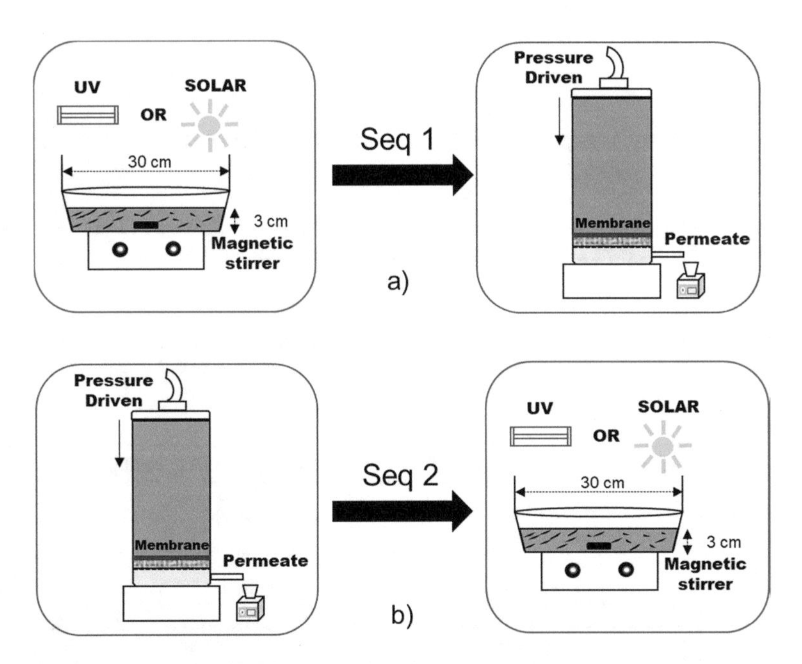

Fig. 6 Scheme of sequential combination used to treat dye wastewater **a** Sequence 1 – photocatalysis followed by membrane separation process (PC + MSP) and **b** Sequence 2 – membrane separation process followed by photocatalysis (MSP + PC)

The profiles of discoloration and permeate flow, the discoloration and the chemical oxygen demand reduction of the effluent for Sequences 1 and 2 under UV and solar radiation are presented in Fig. 7 and Tables 1 and 2, respectively. The sequential combination of membrane separation process followed by photocatalysis (Sequence 2) under solar radiation (Fig. 7a) presented the same behavior and efficiency in the effluent discoloration as individual photocatalysis (40 and 42%, Table 1), evidencing that the determinant factor, in this case, was the solar radiation. However, when using UV radiation, the previous filtration of the effluent made the process more efficient regarding discoloration (63 and 56%, Table 1), but mainly regarding the organic load (52 and 38%, Table 2). This is due to the retention of particulate materials in the effluent that hampered the penetration of UV radiation, allowing a more efficient activation of the catalyst and favoring the degradation of more diluted solution, since the number of organic species decreased to a constant quantity of catalyst and luminous intensity [84].

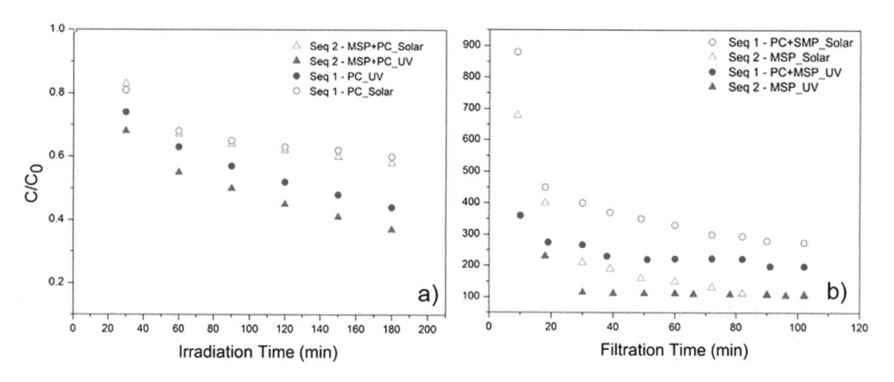

Fig. 7 **a** Profile of effluent discoloration; **b** permeate flow of the effluent with the 12%PES/DMF membrane at 1 bar for the sequences 1 and 2 under UV and solar radiation

Table 1 Discoloration of the effluent after the treatment by photocatalysis (PC) and membrane separation process (MSP) and the membrane fouling for sequences 1 and 2 under UV and solar radiation

Sequence 1				
Radiation	Dis_PC (%)	Dis_MSP (%)	Fouling (%)	Dis_Total process (%)
UV	56	3	48	57
Solar	40	18	50	51
Sequence 2				
Radiation	Dis_MSP(%)	Fouling (%)	Dis_PC (%)	Dis_Total process (%)
UV	35	86	63	76
Solar	34	85	42	62

[a]Dis_PC: discoloration by photocatalysis; Dis_MSP: discoloration by membrane separation process; Dis_Total Process: total discoloration of the combined processes

Table 2 Chemical oxygen demand (COD) reduction after the treatment by photocatalysis (PC) and the membrane separation process (MSP) for sequences 1 and 2 under UV and solar radiation

Sequence 1			
Radiation	COD_PC (%)	COD_MSP (%)	COD_Total process (%)
UV	38	2	39
Solar	23	9	30
Sequence 2			
Radiation	COD_MSP (%)	COD_PC (%)	COD_Total process
UV	77	52	89
Solar	75	36	84

[a]COD: chemical oxygen demand; COD_PC: COD reduction by photocatalysis; COD_MSP: COD reduction by membrane separation process; COD_Total Process: total COD reduction of the combined processes

According to Fig. 7b, the combination sequence of the processes influenced the steady flow value, which was 100 kg h^{-1} m^{-2} for Sequence 2 independent of radiation and 210 and 280 kg h^{-1} m^{-2} for Sequence 1 under UV and solar radiation, respectively. The lower flow obtained in Sequence 2 is associated with the deposition of organic matter and remaining pollutants present in the effluent under the membrane that occurs when the raw wastewater was permeated by the membrane. When the raw wastewater was previously subjected to photocatalysis, the organic molecules were broken, which facilitated the permeation in the membrane and promoted a lower reduction of the flow. Since the molecules of the effluent became smaller, their rejection by the membrane for Sequence 1 was lower than Sequence 2, for both color and organic matter (Tables 1 and 2). The highest efficiency of the 12%PES/DMF membrane was in the rejection of organic matter (Table 2), possibly because the size of these molecules is higher than the ones of the dye. When the steady flows of sequences 1 and 2 are compared with the initial flow obtained with water (600 kg h^{-1} m^{-2}), a decrease of 83% of the flow is verified for Sequence 2 and of 53-65% for Sequence 1. This process clearly evidences the polarization by concentration with the deposition of particles and macromolecules in the membrane surface, with posterior development of fouling [97].

Just like in the individual process, the membrane was cleaned and the permeate flow was re-established, evidencing that in the combined process, the reversible fouling also occurred. Another important point is that the fouling mechanism continued to be random blockage [62, 70]. Despite the membrane reversible fouling for Sequence 2 having been much higher than for sequence 1 (Table 1), this membrane can be used to treat dye wastewater.

Evaluating the global efficiency of the combined processes, the sequence of membrane separation process followed by photocatalysis (Sequence 2) promoted a higher color (76 and 62%) and chemical oxygen demand reduction (89 and 84%) under UV and solar radiation. Thus, the sequential combination of membrane separation process and heterogeneous photocatalysis was the most efficient to treat the effluent from the process of fish skin tanning employing the Acid Green 16 dye. Thus, this kind of combination is an alternative treatment of dye wastewater.

3.2 Hybrid/One-Pot Combination

Another way of coupling photocatalysis to membrane separation is by hybrid/one-pot combination. In this kind of combination, the photocatalyst can be suspense in solution or immobilized. The process is called hybrid/one-pot combination with suspended photocatalyst when the photocatalyst is suspended in the reactional medium and remains confined inside a photoreactor by a barrier promoted by the membrane that is submerged in the reactional medium. When the photocatalyst is immobilized in the membrane surface or is added to the polymeric solution, the membrane starts to present photocatalytic activity, and then, there is the called hybrid/one-pot combination with photocatalytic membrane [46, 90, 114, 145].

3.2.1 Hybrid/One-Pot Combination with Suspended Photocatalyst

In the hybrid/one-pot combination with suspended photocatalyst, the membrane is submerged, while the photocatalyst is suspended in the reactional medium, all inside a single piece of equipment: the photocatalytic membrane reactor. The radiation source can also be submerged or suspended in the center of the photocatalytic module. This kind of combination is favored for simultaneously promoting oxidation and separation of pollutants in a single piece of equipment, maintaining the photocatalyst confined inside the photoreactor, and with this, controlling its concentration/optimum dosage, improving mass transfer between the pollutant and photocatalyst, avoiding additional steps to separate and recover the photocatalyst, reducing the installation area, and enabling low load loss in the pipes, controlling the time length of permanence of the compounds in the photoreactor and with this enabling several industrial applications [46, 72, 90, 92, 114, 145].

In contrast, in this kind of configuration, depending on the ratio between the particle size of the photocatalyst and the size of the membrane pore, the photocatalyst particles can cause mechanical damages to the membrane, as well as be deposited under the surface of the membrane, encrust, and block the membrane pores, causing fouling, especially in membranes of microfiltration and ultrafiltration. When photocatalyst nanoparticles are used, the agglomeration with the formation of cake layer on the membrane surface can be intensified [9, 46, 72, 92, 94, 112].

This blockage of the pores causes a reduction of the permeate flow, being necessary to stop the process to perform the cleaning. The longer the time and frequency of chemical and/or physical cleaning, the higher will be the operating cost of the process. It should also be noted that the operating cost will be even more increased if the membranes need to be changed, due to irreversible blockage of the pores [24, 97, 145]. However, such inconvenience can be minimized by the building of specific systems of stirring and aeration inside the photocatalytic membrane reactor, however, there will be an additional energy consumption [145].

As mentioned, several parameters influence the efficiency of photocatalytic membrane reactors, and in the case of the hybrid/one-pot combination with suspended photocatalyst, two parameters deserve special attention: the concentration/dosage of the photocatalyst and the type/intensity of luminous radiation. It is extremely important to determine the optimum concentration of photocatalyst to be used as a function of the photoreactor configuration, because a high concentration of photocatalyst particles in suspension can favor the phenomenon of particle agglomeration, reducing the effective surface area, and still partially blocking the incidence of radiation inside the photoreactor, hampering the activation of the photocatalyst and consequently reducing the efficiency of the photocatalytic process [72, 135, 145]. Due to the need to apply adequate luminous radiation to activate the photocatalyst, the time of exposition of the membrane to UV radiation and the intensity of this light should be performed carefully. Depending on the polymer or bend that was used to manufacture the membrane, UV radiation and/or photogenerated reactive oxygen species can damage the morphological structure of the membrane, due to its low chemical stability [112]. In an attempt to mitigate this problem, a protective apparatus can be placed between the submerged luminous source and the membrane, but this can decrease the area of light coverage, reducing the photonic efficiency of the system [145]. Thus, when using the hybrid/one-pot combination, it is essential to manufacture membranes that present high chemical, physical, and thermal stability [112].

Most literature reports use the hybrid/one-pot combination with suspended photocatalyst to degrade dye solution, being that few studies evaluated the degradation of real dye wastewaters. Regarding the removal of dye, [33] evaluated the degradation of Reactive Black 5 dye in a membrane photocatalytic reactor using TiO_2 suspended in solution, two UV-C light lamps submerged in the reactional mixture, together with a microfiltration module put in the center of the membrane photocatalytic reactor containing a polytetrafluoroethylene membrane. The best results were 98–100% of discoloration, 75–85% of total organic carbon reduction and 85-95% of chemical oxygen demand reduction using 0.5 mg L^{-1} of TiO_2 and 125 mg L^{-1} of dye during 700 h of treatment. [74] used a ZnO slurry membrane photoreactor in continuous system to degrade the Reactive Black 5 dye being that the flat polyvinylidene fluoride membrane module was submerged in solution as well as both two ultraviolet light lamps. The authors verified high efficiency in the discoloration (higher than 95%) and in the reduction of chemical oxygen demand (50-80%) during 120 h of treatment. The lowest chemical oxygen removal (50%) occurred around 80 h, being attributed to the fixation of the ZnO particles to the reactor wall, however, the efficiency was recovered when the reactor wall was cleaned and the ZnO particles re-suspended. [135] evaluated the degradation of Rhodamine B dye in a continuous pilot-scale submerged membrane photocatalysis reactor. In this reactor, the TiO_2 nanoparticles were suspended in solution, while an apparatus protected the polyvinylidene fluoride microfiltration membrane from the submerged luminous source (four low-pressure UV-C lamps of 6 W). The authors optimized the operating parameters from a central composite design and determined that the optimum conditions to degrade 95% of Rhodamine B dye were: 0.1 g/L TiO_2, 3 lamps, pH 8, and 100 L/h m^2 Rhodamine B solution flux.

Still regarding the use of hybrid/one-pot combination with suspended photocatalyst to degrade dye solution, other techniques can be coupled to the photocatalysis and membrane separation. [120] evaluated the degradation of the Reactive Orange 29 dye using the hybrid combination of photocatalysis and microfiltration and also photocatalysis, microfiltration, and also those two processes (photocatalysis and microfiltration), together with sonolysis. TiO_2 dispersed in solution was used as the photocatalyst, and the polymeric membrane was constituted of polyvinylidene fluoride. The authors observed that there was a synergistic effect in the combination of photocatalysis and microfiltration that favored the removal of dye (60%) compared to the use of these processes solely, and the application of sonolysis to this combination potentiated the degradation of the dye (100%).

Regarding the use of hybrid/one-pot combination with suspended photocatalyst in the treatment of raw textile wastewaters, [36] evaluated the degradation of raw textile wastewater using a hybrid system composed of ZnO slurry photocatalytic reactor with six UVA or UVC light lamps and ultrafiltration module submerged in the photoreactor containing a flat polyethersulfone membrane. The authors reported that the concentration of the effluent was the preponderant factor in the process efficiency, the membrane fouling was reduced by the photocatalytic action of the small quantity of ZnO that was deposited on the surface of the membrane under UVA radiation and that the ZnO particles were kept confined inside the photoreactor by the membrane. [117] developed a pilot-scale photocatalytic membrane bioreactor (20 L) to remediate an effluent of textile dyeing containing a mixture of dyes, which originated a black color. The photocatalytic membrane bioreactor is constituted by a photocatalytic module in which WO_3 and 1%GO/WO_3 alginate beads were used as photocatalyst employing visible radiation, and one microfiltration module with polyethersulfone membrane. The authors verified that the hybrid combination of photocatalysis and microfiltration favored fouling reduction, being that the reversible fouling of the membrane prevailed, and the best results (76% chemical oxygen demand removal and 70% color removal) were obtained with 1%GO/WO_3 alginate beads photocatalyst. This study evidenced the possibility to apply a photocatalytic membrane bioreactor in the treatment of real textile effluent on a pilot-scale, a necessary stage to industrial implementation. Besides dye removal, the hybrid/one-pot combination with suspended photocatalyst was also used by [76] in the removal of 1,4-dioxane from textile wastewater effluent after biological treatment, and good results were obtained under solar and UV radiation, however, periodic backwashes were necessary to remove TiO_2 particles that became encrusted in the membrane.

3.2.2 Hybrid/One-Pot Combination with Photocatalytic Membrane

In the hybrid/one-pot combination with photocatalytic membrane, the photocatalyst can be incorporated into the polymeric matrix or immobilized in the membrane surface, being that the membrane can be fully constituted of the photoactive material or a substrate revested with the photoactive material. Thus, the membranes can act only as a support to the photocatalyst or even as the photocatalyst and selective barrier to pollutant compounds, being denominated, in this case, photocatalytic membranes. Photocatalytic membranes can be classified into inorganic membranes (ceramic, nonporous carbon, and metallic oxides) and inorganic polymeric hybrid membranes (combination of inorganic and polymeric materials), and each one presents advantages and disadvantages. Polymeric membranes are easy to manufacture and relatively cheap, but they are less resistant to UV radiation, chemical agents, and photogenerated reactive oxygen species, while inorganic membranes, the ceramic in particular, present high chemical stability and mechanical resistance, however, they possess a high manufacture cost [72, 94, 112].

In this kind of combination, there is no need to recover the photocatalyst and the hydroxyl radicals generated by the activation of the membrane oxidize the pollutants present in the external and internal surface of the membrane, promoting the self-cleaning and the reduction of membrane fouling, because there are fewer incrustations in the membrane surface due to alteration of its wettability (increased hydrophilicity). As for disadvantages of photocatalytic membranes, it can be highlighted: the low resistance of polymeric membranes to UV radiation, difficulty to control the photocatalyst dosage that is really incorporated into the membrane, impossibility to adjust the photocatalyst concentration as a function of the effluent characteristics, reduction of photocatalytic activity of the membrane with usage, and lower photocatalytic activity of the membrane compared to the system of photocatalysts in suspension. This lower activity of photocatalytic membranes is due to limitations of mass transfer between the pollutant and the photocatalyst, short service time of hydroxyl radicals, and accessibility of the active phase to luminous radiation [46, 72, 79, 92, 94, 112, 114, 145].

Among the photocatalysts that can be immobilized or incorporated in the membrane, pure or doped TiO_2 and ZnO are the most used ones. The majority of the studies about hybrid/one-pot combination with photocatalytic membrane performed the studies in the batch system and in a partial way, being that some studies evaluated only the activity of the photocatalytic membrane in the degradation of dyes solutions, mainly under visible radiation, while other studies evaluated only the anti-fouling and self-cleaning ability properties of the photocatalytic membrane. Some studies about the activity of the photocatalytic membranes in the removal of dyes can be observed in Table 3.

The anti-fouling potential and self-cleaning ability of the photocatalytic membranes were evaluated by several authors using, mainly, solutions of humic acid and bovine serum albumin. In these studies, the incorporation of materials, such as TiO_2, ZnO, Ni-TiO_2, SiO_2, Fe_3O_4, Ni^+, Ag^+, and Ni-ZnO, favored the permeate flow increase and the reduction of cake layer formation under the membrane surface, due to alterations of hydrophilicity, morphology, pore size, and photocatalytic action of materials [2, 8, 48, 80–82, 127, 130, 140, 143, 144]. Li et al. [80] verified that the permeability and the anti-fouling property of the membrane were improved from

Table 3 Summary of some studies about the activity of photocatalytic membranes to remove dyes

Materials	Processing conditions	Removal efficiency	Comments	References
Copper-iron bimetal modified polyacrylonitrile nanofibrous membrane (BM-PANNM membrane)	Preparation of PAN nanofibrous membrane using electrospinning method and immersion on Cu-Fe metal salt solutions; 0.2 g of membrane; 25–100 ppm of Reactive Blue 19, Reactive Red 195 and Acid Orange 7 dyes; UV light (365 nm); 60 min of irradiation	Photocatalytic capacity for the three dyes above 99% at 60 min	Good reusability of BM-PANNM membrane after 5 cycles of reuse for Reactive Blue 19 dye	[139]
Fe^{3+} doped ZnO on polyester fabric membrane (Fe^{3+}@ZnO/PMR membrane)	Functionalization using seeding and hydrothermal methods; 5 cm^2 of Fe^{3+}@ZnO/PMR membrane with 60 μg of catalyst; Artificial daylight (72 W); Reactive Black 5 dye degradation using response surface methodology and central composite design	98.3% degradation at optimized conditions: Fe^{3+}@ZnO/PMR membrane, 30 ppm of Reactive Black 5 dye, pH 3, 30 mM of H_2O_2, and 3 h of irradiation	Excellent reusability of Fe^{3+}@ZnO/PMR membrane with an efficiency reduction of 10% between 8 and 15 cycles of reuse	[10]
N-vinylcaprolactam-TiO_2-acrylic acid blended polysulfone membrane (VCL-TiO_2-AA/PSF membrane)	Preparation using phase inversion method; 0.1 g of membrane; 10 ppm of Methylene Blue dye; UV light (100 W) and dark conditions; 1 h of irradiation Humic acid filtration was used to evaluate the anti-fouling properties	Maximum dye removal was 40% (in the dark) and 94% (under UV light) at alkaline pH due to the difference of surface charge between membrane and dye	Rejection of humic acid was 97.7% with a flux recovery ratio of 98.9%, reversible fouling below 10%, and total fouling below 25%	[124]

(continued)

Table 3 (continued)

Materials	Processing conditions	Removal efficiency	Comments	References
Polyvinylidene fluoride membrane incorporated with TiO_2 (TiO_2/PVDF membrane)	TiO_2 nanoparticles were synthesized using solution combustion synthesis (TiO_2_SCS) and hydrothermal synthesis (TiO_2_S) and incorporated in membrane fabricated by phase inversion technique; 15 cm × 10 cm of membrane; 50 ppm of Congo Red and Reactive Yellow 145 dyes; UV light (8 W) and sunlight; 300 min of irradiation	Discoloration and mineralization of 84 and 71% (Congo Red) and 100 and 87% (Reactive Yellow 145) under UV light, and 67 and 54% (Congo Red) and 77 and 55% (Reactive Yellow 145) under sunlight using TiO_2_S/PVDF membrane	TiO_2_S/PVDF membrane had reusability after 4 cycles with excellent long-term stability (6 months)	[39]
Ln(Sm, La, Er) doped ZnO blended polyvinylpyrrolidone and polyvinylidene fluoride membrane (PVDF/PVP/ZnO:Ln(Sm,La,Er) membrane)	Preparation using electrospinning method; 0.1 g of membrane; 10 ppm of Methylene Blue dye and 5 ppm of Rhodamine B dye; Visible light (400 W); 360 min of irradiation	Highest efficiency PVDF/PVP/ZnO:La (37%) membrane, with color removal of 96.33% (Methylene Blue) and 93.36% (Rhodamine B)	PVDF/PVP/ZnO:La (37%) membrane had reusability after 10 cycles as well as excellent desorption under sunlight for 24 h	[103]
Polytetrafluoroethylene membrane with TiO_2/ZnO nanolayer (TiO_2/ZnO/PVDF membrane)	Modification using atomic layer deposition; 19.6 cm^2 of TiO_2/ZnO/PVDF membrane; 10^{-5} M of Methylene Blue dye; Visible light (200 W); 30 min of irradiation Filtration of humic acid solution with and without irradiation was used to evaluated anti-fouling	Color removal above 90% and 84% total organic carbon removal with TiO_2:ZnO ratio of 1:3 in the membrane	TiO_2/ZnO/PVDF membrane with TiO_2:ZnO ratio of 1:3 had good reusability after 5 cycles of reuse and presented excellent anti-fouling property associated with high hydrophilicity under visible light irradiation and dark conditions	[80]

(continued)

Table 3 (continued)

Materials	Processing conditions	Removal efficiency	Comments	References
Polyvinylidene fluoride membrane coated with polyvinyl alcohol and TiO_2 (PVA/TiO_2/PVDF membrane)	The membrane was fabricated using phase inversion and dip-coating methods; 5 cm × 5 cm of membrane; 50 ppm of Reactive Blue, Rhodamine B, and Methyl Orange dyes; UV light (15 W); 150 min irradiation Dye solution and bovine serum albumin filtration were used to evaluate rejection and anti-fouling, respectively	Maximum photodegradation of 44.4% (Reactive Blue), 45.8% (Rhodamine B), and 47.8% (Methyl Orange) was obtained using 3%PVA/1%TiO_2/16%PVDF membrane	Greater membrane rejection of Reactive Blue (85%) dye compared to Rhodamine B (55%) and Methyl Orange (35%) dyes. The smaller the molecular weight of the dye, the higher the photocatalytic degradation but the lower the rejection by the membrane. Flow recovery after bovine serum albumin filtration was 76.2% for 3%PVA/1%TiO_2/16%PVDF membrane with lower irreversible fouling factor	[114]

the incorporation of three-dimensional TiO_2/ZnO composites layers on polytetrafluoroethylene membrane surface and pore walls. Sun et al. [127] observed that the anchor of magnetic nanoparticles of Ni-TiO_2 onto the polymeric polyethersulfone membrane surface improved the anti-fouling performance of the membrane, being that the higher the content of nanoparticles, the higher the permeate flow. These photocatalytic membranes were efficient to degrade the Methylene Blue dye and the high rates of flow recovery (between 75.4 and 99.56%) evidenced the ability of self-cleaning of the membrane due to the presence of nanoparticles in the membrane surface. Liu et al. [82] attributed the anti-fouling performance to the interaction energy between the species adhered in the membrane surface and the photocatalytic activity of Ni-ZnO particles deposited under the surface of the polytetrafluoroethylene membrane.

Few are the studies that really used the hybrid/one-pot combination with photocatalytic membrane to treat dye solutions and wastewaters. Rajeswari et al. 2016 used the sol-gel method to synthesize nano ZnO and solution dispersion blending and phase inversion methods to prepare the nano ZnO blended cellulose acetate-polyurethane membrane (CA-PU/ZnO). This photocatalytic membrane was used to degrade solutions of the Reactive Red 11 and Reactive Orange 84 dyes under sunlight in a continuous system, ranging the initial concentration of the dye solutions (50–250 mg L^{-1}), pH (3–11), and irradiation time (10–60 min). When analyzing the membrane permeate, the optimum conditions to degradation (93-98%) and mineralization (15%) of the dyes were: 50 mg L^{-1} of initial concentration, pH 7, and 40 min of sunlight irradiation. The pure water flux was 93% and 66% for the CA-PU/ZnO and CA-PU membranes, respectively and this higher flow was attributed to the hydrophilic character of the CA-PU/ZnO membrane. Vatanpour et al. [134] prepared TiO_2 blended polyvinylidene fluoride membranes using the phase inversion method and evaluated the degradation of the Rhodamine B (RhB) dye in a continuous pilot-scale submerged photocatalytic membrane reactor using response surface methodology and Box-Behnken method. Optimum conditions to 70% dye removal were: 18% polyvinylidene fluoride, 10% TiO_2 in membrane, 40 °C, pH 5, 1.6 L min^{-1} aeration, 0.1 ^{g} L^{-1} TiO_2 suspended in solution and 320 min of UV irradiation. These optimum conditions were applied in the textile wastewater degradation and 60% degradation was obtained.

4 Conclusion

This chapter approached the combination of photocatalysis and membrane separation as a treatment for dye wastewaters. Basic concepts were presented and then, the available options of combination, sequential, and hybrid/one-pot combination, were exposed and explained. Each kind of combination can still be divided into subcategories; for example, the sequential combination can be performed with photocatalysis as a pre- or as a post-treatment, and in the hybrid/one-pot combination, the possibilities are the photocatalyst can be suspended in the reactional mixture with the membrane, or incorporated inside/under the membrane, constituting the photocatalytic membrane. Considering the lack of complete studies comparing some possible combinations, a case study combining photocatalysis and membrane separation as a sequential process to treat dye wastewater was performed and showed that, among the evaluated combinations, the membrane separation process followed by photocatalysis enabled the highest color (76 and 62%) and chemical oxygen demand reduction (89 and 84%) of the dye wastewater under UV and solar radiation. From this chapter, it was found that the photocatalytic membrane reactor with the photocatalyst incorporated in the membrane is the most advantageous due to the photocatalytic membrane characteristics of self-cleaning and anti-fouling and to the intrinsic performance of both two techniques in this combination. Since the majority of studies were performed with dye solution and in a partial way, complete studies regarding photocatalytic activity and anti-fouling ability of the membranes are necessary, mainly aiming its application to treat industrial effluents, because several factors influence the efficiency of the photocatalytic membrane reactors. Thus, this chapter elucidated the different options to treat dye wastewater using the combination of photocatalysis and membrane separation, contributing to the application of sustainable and eco-friendly technologies according to the concepts of process intensification.

Acknowledgements The authors acknowledge the Coordination of Improvement of Higher Level Personnel—Brazil (Capes)—Finance Code 001, Araucária Foundation of Support to the Scientific and Technological Development of the State of Paraná.

References

1. Abdallah H (2017) A review on catalytic membranes production and applications. Bull Chem React Eng Catal 12:136–156. https://doi.org/10.9767/bcrec.12.2.462.136-156
2. Ahmad AL, Abdulkarim AA, Shafie ZMHM, Ooi BS (2017) Fouling evaluation of PES/ZnO mixed matrix hollow fiber membrane. Desalination 403:53–63. https://doi.org/10.1016/j.desal.2016.10.008
3. Ahmed F, Kumar S, Arshi N, Anwar MS, Koo BH (2012) Morphological evolution between nanorods to nanosheets and room temperature ferromagnetism of Fe-doped ZnO nanostructures. CrystEngComm 14:4016–4026. https://doi.org/10.1039/c2ce25227a
4. Ahmed S, Rasul MG, Martens WN, Brown R, Hashib MA (2010) Heterogeneous photocatalytic degradation of phenols in wastewater: A review on current status and developments. Desalination 261:3–18. https://doi.org/10.1016/j.desal.2010.04.062
5. Alresheedi MT, Barbeau B, Basu OD (2019) Comparisons of NOM fouling and cleaning of ceramic and polymeric membranes during water treatment. Sep Purif Technol 209:452–460. https://doi.org/10.1016/j.seppur.2018.07.070
6. Amaterz E, Tara A, Bouddouch A, Taoufyq A, Bakiz B, Benlhachemi A, Jbara O (2020) Photo-electrochemical degradation of wastewaters containing organics catalysed by phosphate-based materials: a review. Rev Environ Sci Biotechnol 19:843–872. https://doi.org/10.1007/s11157-020-09547-9
7. Amutha K (2017) Sustainable chemical management and zero discharges, sustainable fibres and textiles. Elsevier Ltd. https://doi.org/10.1016/B978-0-08-102041-8.00012-3
8. Ananth A, Arthanareeswaran G, Ismail AF, Mok YS, Matsuura T (2014) Effect of bio-mediated route synthesized silver nanoparticles for modification of polyethersulfone

membranes. Colloids Surf A Physicochem Eng Asp 451:151–160. https://doi.org/10.1016/j.colsurfa.2014.03.024
9. Argurio P, Fontananova E, Molinari R, Drioli E (2018) Photocatalytic membranes in photocatalytic membrane reactors. Processes 6. https://doi.org/10.3390/pr6090162
10. Ashar A, Bhatti IA, Ashraf M, Tahir AA, Aziz H, Yousuf M, Ahmad M, Mohsin M, Bhutta ZA (2020) Fe3+@ZnO/polyester based solar photocatalytic membrane reactor for abatement of RB5 dye. J Clean Prod 246. https://doi.org/10.1016/j.jclepro.2019.119010
11. Asif SAB, Khan SB, Asiri AM (2014) Efficient solar photocatalyst based on cobalt oxide/iron oxide composite nanofibers for the detoxification of organic pollutants. Nanoscale Res Lett 9:1–9. https://doi.org/10.1186/1556-276X-9-510
12. Bai L, Liu Y, Ding A, Ren N, Li G, Liang H (2019) Surface coating of UF membranes to improve antifouling properties: A comparison study between cellulose nanocrystals (CNCs) and cellulose nanofibrils (CNFs). Chemosphere 217:76–84. https://doi.org/10.1016/j.chemosphere.2018.10.219
13. Baig U, Uddin MK, Gondal MA (2020) Removal of hazardous azo dye from water using synthetic nano adsorbent: facile synthesis, characterization, adsorption, regeneration and design of experiments, Colloid Surf A Physicochem Eng Asp Elsevier B.V. https://doi.org/10.1016/j.colsurfa.2019.124031
14. Baker RW (2004) Membrane technology and applications. John Wiley & Sons, New York
15. Benkhaya S, M'Rabet S, El Harfi A (2020a) A review on classifications, recent synthesis and applications of textile dyes. Inorg Chem Commun 115:107891. https://doi.org/10.1016/j.inoche.2020.107891
16. Benkhaya S, M'Rabet S, Hsissou R, El HA (2020b) Synthesis of new low-cost organic ultrafiltration membrane made from Polysulfone/Polyetherimide blends and its application for soluble azoic dyes removal. J Mater Res Technol 9:4763–4772. https://doi.org/10.1016/j.jmrt.2020.02.102
17. Bhat AP, Jadhav AJ, Holkar CR, Pinjari DV (2021) Doped-TiO_2 and doped-mixed metal oxide-based nanocomposite for photocatalysis. In: Handbook of nanomaterials for wastewater treatment. Elsevier, pp. 155–180. https://doi.org/10.1016/B978-0-12-821496-1.00018-0
18. de Brites-Nóbrega FF, Polo ANB, Benedetti AM, Mô MDL, Slusarski-Santana V, Fernandes-Machado NRC (2013) Evaluation of photocatalytic activities of supported catalysts on NaX zeolite or activated charcoal. J Hazard Mater 263:61–66. https://doi.org/10.1016/j.jhazmat.2013.07.061
19. Brites-Nóbrega FF, Lacerda IA, Santos SV, Amorim CC, Santana VS, Fernandes-Machado NRC, Ardisson JD, Henriques AB, Leão MMD (2015) Synthesis and characterization of new NaX zeolite-supported Nb, Zn, and Fe photocatalysts activated by visible radiation for application in wastewater treatment. Catal Today 240:168–175. https://doi.org/10.1016/j.cattod.2014.06.036
20. Brites FF, Santana VS, Fernandes-Machado NRC (2011) Effect of support on the photocatalytic degradation of textile effluents using Nb_2O_5 and ZnO: Photocatalytic degradation of textile dye. Top Catal 54:264–269. https://doi.org/10.1007/s11244-011-9657-2
21. Buscio V, Brosillon S, Mendret J, Crespi M, Gutiérrez-Bouzán C (2015) Photocatalytic membrane reactor for the removal of C.I. disperse red 73. Materials (Basel). 8:3633–3647. https://doi.org/10.3390/ma8063633
22. Carter BG, Cheng N, Kapoor R, Meletharayil GH, Drake MA (2021) Invited review: Microfiltration-derived casein and whey proteins from milk. J Dairy Sci 104:2465–2479. https://doi.org/10.3168/jds.2020-18811
23. Castañeda-Juárez M, Martínez-Miranda V, Almazán-Sánchez PT, Linares-Hernández I, Santoyo-Tepole F, Vázquez-Mejía G (2019) Synthesis of TiO_2 catalysts doped with Cu, Fe, and Fe/Cu supported on clinoptilolite zeolite by an electrochemical-thermal method for the degradation of diclofenac by heterogeneous photocatalysis. J Photochem Photobiol A Chem 380. https://doi.org/10.1016/j.jphotochem.2019.04.045
24. Chang H, Liang H, Qu F, Liu B, Yu H, Du X, Li G, Snyder SA (2017) Hydraulic backwashing for low-pressure membranes in drinking water treatment : a review. J. Memb. Sci. 540:362–380. https://doi.org/10.1016/j.memsci.2017.06.077

25. Charcosset C (2012) Membrane processes in biotechnology and pharmaceutics, 1st edn. Elsevier
26. Chauhan R, Kumar A (2012) Photocatalytic studies of silver doped ZnO nanoparticles synthesized by chemical precipitation method. J Sol-Gel Sci Technol 546–553. https://doi.org/10.1007/s10971-012-2818-3
27. Chebli D, Fourcade F, Brosillon S, Nacef S, Amrane A (2011) Integration of photocatalysis and biological treatment for azo dye removal-application to AR183. Environ Technol 32:507–514. https://doi.org/10.1080/09593330.2010.504236
28. Chen H, Zhang YJ, He PY, Li CJ, Li H (2020) Coupling of self-supporting geopolymer membrane with intercepted Cr(III) for dye wastewater treatment by hybrid photocatalysis and membrane separation. Appl Surf Sci 515. https://doi.org/10.1016/j.apsusc.2020.146024
29. Chowdhury MF, Khandaker S, Sarker F, Islam A, Rahman MT, Awual MR (2020) Current treatment technologies and mechanisms for removal of indigo carmine dyes from wastewater: a review. J Mol Liq 318:114061. https://doi.org/10.1016/j.molliq.2020.114061
30. Coha M, Farinelli G, Tiraferri A, Minella M, Vione D (2021) Advanced oxidation processes in the removal of organic substances from produced water: potential, configurations, and research needs. Chem Eng J 414:128668. https://doi.org/10.1016/j.cej.2021.128668
31. Costa SIG, Cauneto VD, Fiorentin-Ferrari LD, Almeida PB, Oliveira RC, Longo E, Módenes AN, Slusarski-Santana V (2020) Synthesis and characterization of $Nd(OH)_3$-ZnO composites for application in photocatalysis and disinfection. Chem Eng J 392:123737. https://doi.org/10.1016/j.cej.2019.123737
32. Cui ZF, Muralidhara HS (2010) Membrane technological a practical guide to membrane technology and applications in food and bioprocessing. Elsevier
33. Damodar RA, You SJ, Ou SH (2010) Coupling of membrane separation with photocatalytic slurry reactor for advanced dye wastewater treatment. Sep Purif Technol 76:64–71. https://doi.org/10.1016/j.seppur.2010.09.021
34. Dasgupta J, Singh A, Kumar S, Sikder J, Chakraborty S, Curcio S, Arafat HA (2016) Poly (sodium-4-styrenesulfonate) assisted ultrafiltration for methylene blue dye removal from simulated wastewater: optimization using response surface methodology. J Environ Chem Eng 4:2008–2022. https://doi.org/10.1016/j.jece.2016.03.033
35. Mauro AD, Cantarella M, Nicotra G, Pellegrino G, Gulino A, Brundo MV, Privitera V, Impellizzeri G (2017) Novel synthesis of ZnO/PMMA nanocomposites for photocatalytic applications. Sci Rep 7:1–12. https://doi.org/10.1038/srep40895
36. Doruk N, Yatmaz HC, Dizge N (2016) Degradation efficiency of textile and wood processing industry wastewater by photocatalytic process using in situ ultrafiltration membrane. Clean: Soil, Air, Water 44:224–231. https://doi.org/10.1002/clen.201400203
37. El Baraka N, Laknifli A, Saffaj N, Addich M, Ait Taleb A, Mamouni R, Fatni A, Ait Baih M, 2020. Study of coupling photocatalysis and membrane separation using tubular ceramic membrane made from natural Moroccan clay and phosphate. E3S Web Conf 150:1–8. https://doi.org/10.1051/e3sconf/202015001007
38. Eljaddi T, Mejia Mendez DL, Favre E, Roizard D (2021) Development of new pervaporation composite membranes for desalination: membrane characterizations and experimental permeation data. Data Br 35:106943. https://doi.org/10.1016/j.dib.2021.106943
39. Erusappan E, Thiripuranthagan S, Radhakrishnan R, Durai M, Kumaravel S, Vembuli T, Kaleekkal NJ (2021) Fabrication of mesoporous TiO_2/PVDF photocatalytic membranes for efficient photocatalytic degradation of synthetic dyes. J Environ Chem Eng 9:105776. https://doi.org/10.1016/j.jece.2021.105776
40. Esplugas S, Giménez J, Contreras S, Pascual E, Rodríguez M (2002) Comparison of different advanced oxidation processes for phenol degradation. Water Res 36:1034–1042. https://doi.org/10.1016/S0043-1354(01)00301-3
41. Field RW, Bekassy-Molnar E, Lipnizki F, Vatai G (2017) Engineering aspects of membrane separation and application in food processing, 1st edn. CRC Press
42. Fiorentin-Ferrari LD, Celant KM, Gonçalves BC, Teixeira SM, Slusarski-Santana V, Módenes AN (2021) Fabrication and characterization of polysulfone and polyethersulfone membranes

applied in the treatment of fish skin tanning effluent. J Clean Prod 294. https://doi.org/10.1016/j.jclepro.2021.126127
43. Fischer K, Schulz P, Atanasov I, Latif AA, Thomas I, Kühnert M, Prager A, Griebel J, Schulze A (2018) Synthesis of high crystalline TiO2 nanoparticles on a polymer membrane to degrade pollutants from water. Catalysts 8. https://doi.org/10.3390/catal8090376
44. Galindo C, Jacques P, Kalt A (2000) Photodegradation of the aminoazobenzene acid orange 52 by three advanced oxidation processes: UV/H_2O_2, UV/TiO_2 and VIS/TiO_2. J Photochem Photobiol A Chem 130:35–47. https://doi.org/10.1016/s1010-6030(99)00199-9
45. Ganie AS, Bano S, Khan N, Sultana S, Rehman Z, Rahman MM, Sabir S, Coulon F, Khan MZ (2021) Nanoremediation technologies for sustainable remediation of contaminated environments: recent advances and challenges. Chemosphere 275. https://doi.org/10.1016/j.chemosphere.2021.130065
46. Ganiyu SO, Van Hullebusch ED, Cretin M, Esposito G, Oturan MA (2015) Coupling of membrane filtration and advanced oxidation processes for removal of pharmaceutical residues: a critical review. Sep Purif Technol 156:891–914. https://doi.org/10.1016/j.seppur.2015.09.059
47. García R, Naves A, Anta J, Ron M, Molinero J (2021) Drinking water provision and quality at the Sahrawi refugee camps in Tindouf (Algeria) from 2006 to 2016. Sci Total Environ 780:146504. https://doi.org/10.1016/j.scitotenv.2021.146504
48. Ghaemi N, Madaeni SS, Daraei P, Rajabi H, Zinadini S, Alizadeh A, Heydari R, Beygzadeh M, Ghouzivand S (2015) Polyethersulfone membrane enhanced with iron oxide nanoparticles for copper removal from water: application of new functionalized Fe_3O_4 nanoparticles. Chem Eng J, Elsevier B.V. 263:101-112. https://doi.org/10.1016/j.cej.2014.10.103
49. Gola D, Kriti A, Bhatt N, Bajpai M, Singh A, Arya A, Chauhan N, Srivastava SJ, Tyagi PK, Agrawal Y (2021) Silver nanoparticles for enhanced dye degradation. Curr Res Green Sustain Chem 4:100132. https://doi.org/10.1016/j.crgsc.2021.100132
50. Gouvêa CAK, Wypych F, Moraes SG, Durán N, Nagata N, Peralta-Zamora P (2000) Semiconductor-assisted photocatalytic degradation of reactive dyes in aqueous solution. Chemosphere 40:433–440. https://doi.org/10.1016/S0045-6535(99)00313-6
51. Grzechulska-Damszel J, Mozia S, Morawski AW (2010) Integration of photocatalysis with membrane processes for purification of water contaminated with organic dyes. Catal Today 156:295–300. https://doi.org/10.1016/j.cattod.2010.06.033
52. Guimarães JR, Maniero MG, de Araújo RN (2012) A comparative study on the degradation of RB-19 dye in an aqueous medium by advanced oxidation processes. J Environ Manage 110:33–39. https://doi.org/10.1016/j.jenvman.2012.05.020
53. Guo B, Pasco EV, Xagoraraki I, Tarabara VV (2015) Virus removal and inactivation in a hybrid microfiltration-UV process with a photocatalytic membrane. Sep Purif Technol 149:245–254. https://doi.org/10.1016/j.seppur.2015.05.039
54. Hairom NHH, Mohammad AW, Kadhum AAH (2014) Effect of various zinc oxide nanoparticles in membrane photocatalytic reactor for Congo red dye treatment. Sep Purif Technol 137:74–81. https://doi.org/10.1016/j.seppur.2014.09.027
55. Haque MS, Nahar N, Sayem SM (2021) Industrial water management and sustainability: development of SIWP tool for textile industries of Bangladesh. Water Resour Ind 25:100145. https://doi.org/10.1016/j.wri.2021.100145
56. Hermia J (1982) Constant pressure blocking filtration law: Application to power law non-Newtonian fluids. Trans Inst Chem Eng 60:183–187
57. Herrmann JM (1999) Heterogeneous photocatalysis: fundamentals and applications to the removal of various types of aqueous pollutants. Catal Today 53:115–129. https://doi.org/10.1115/IMECE200743738
58. Herrmann JM (2010) Fundamentals and misconceptions in photocatalysis. J Photochem Photobiol A Chem 216:85–93. https://doi.org/10.1016/j.jphotochem.2010.05.015
59. Ho DP, Vigneswaran S, Ngo HH (2009) Photocatalysis-membrane hybrid system for organic removal from biologically treated sewage effluent. Sep Purif Technol 68:145–152. https://doi.org/10.1016/j.seppur.2009.04.019

60. Holkar CR, Jadhav AJ, Pinjari DV, Mahamuni NM, Pandit AB (2016) A critical review on textile wastewater treatments: possible approaches. J Environ Manage 182:351–366. https://doi.org/10.1016/j.jenvman.2016.07.090
61. Homem NC, Beluci NCL, Amorim S, Reis R, Vieira AMS, Vieira MF, Bergamasco R, Amorim MTP (2019) Surface modification of a polyethersulfone microfiltration membrane with graphene oxide for reactive dyes removal. Appl Surf Sci 486:499–507. https://doi.org/10.1016/j.apsusc.2019.04.276
62. Huang B, Wu Z, Zhou H, Li J, Zhou C, Xiong Z, Pan Z, Yao G, Lai B (2021) Recent advances in single-atom catalysts for advanced oxidation processes in water purification. J Hazard Mater 412:125253. https://doi.org/10.1016/j.jhazmat.2021.125253
63. Huo P, Zhou M, Tang Y, Liu X, Ma C, Yu L, Yan Y (2016) Incorporation of N-ZnO/CdS/Graphene oxide composite photocatalyst for enhanced photocatalytic activity under visible light. J Alloys Compd 670:198–209. https://doi.org/10.1016/j.jallcom.2016.01.247
64. Iglesias O, Rivero MJ, Urtiaga AM, Ortiz I (2016) Membrane-based photocatalytic systems for process intensification. Chem Eng J 305:136–148. https://doi.org/10.1016/j.cej.2016.01.047
65. Jiang H, Zhang G, Huang T, Chen J, Wang Q, Meng Q (2010) Photocatalytic membrane reactor for degradation of acid red B wastewater. Chem Eng J 156:571–577. https://doi.org/10.1016/j.cej.2009.04.011
66. Jiang L, Zhang X, Choo KH (2018) Submerged microfiltration-catalysis hybrid reactor treatment: photocatalytic inactivation of bacteria in secondary wastewater effluent. Sep Purif Technol 198:87–92. https://doi.org/10.1016/j.seppur.2017.01.018
67. Jiang T, Zhang L, Ji M, Wang Q, Zhao Q, Fu X, Yin H (2013) Carbon nanotubes/TiO_2 nanotubes composite photocatalysts for efficient degradation of methyl orange dye. Particuology 11:737–742. https://doi.org/10.1016/j.partic.2012.07.008
68. Katheresan V, Kansedo J, Lau SY (2018) Efficiency of various recent wastewater dye removal methods: a review. J Environ Chem Eng 6:4676–4697. https://doi.org/10.1016/j.jece.2018.06.060
69. Kavetsou E, Koutsoukos S, Daferera D, Poslisiou MG, Karagiannis D, Perdikis DC, Detsi A (2019) Encapsulation of mentha pulegium essential oil in yeast cell microcarriers: an approach to environmentally friendly pesticides. J Agric Food Chem 67:4746–4753. https://doi.org/10.1021/acs.jafc.8b05149
70. Khan IA, Lee YS, Kim JO (2020) A comparison of variations in blocking mechanisms of membrane-fouling models for estimating flux during water treatment. Chemosphere 259:127328. https://doi.org/10.1016/j.chemosphere.2020.127328
71. Kim LJ, Jang JW, Park JW (2014) Nano TiO_2-functionalized magnetic-cored dendrimer as a photocatalyst. Appl Catal B Environ 147:973–979. https://doi.org/10.1016/j.apcatb.2013.10.024
72. Koe WS, Lee JW, Chong WC, Pang YL, Sim LC (2020) An overview of photocatalytic degradation: photocatalysts, mechanisms, and development of photocatalytic membrane. Environ Sci Pollut Res 27:2522–2565. https://doi.org/10.1007/s11356-019-07193-5
73. Kuvarega AT, Mamba BB (2016) Photocatalytic membranes for efficient water treatment. In: Semiconductor photocatalysis-materials, mechanisms and applications. pp 523–539
74. Laohaprapanon S, Matahum J, Tayo L, You SJ (2015) Photodegradation of Reactive Black 5 in a ZnO/UV slurry membrane reactor. J Taiwan Inst Chem Eng 49:136–141. https://doi.org/10.1016/j.jtice.2014.11.017
75. Laohaprapanon S, Matahum J, Tayo L, You SJ (2015) Photodegradation of Reactive Black 5 in a ZnO/UV slurry membrane reactor. J Taiwan Inst Chem Eng 49:136–141. https://doi.org/10.1016/j.jtice.2014.11.017
76. Lee KC, Beak HJ, Choo KH (2015) Membrane photoreactor treatment of 1,4-dioxane-containing textile wastewater effluent: performance, modeling, and fouling control. Water Res 86:58–65. https://doi.org/10.1016/j.watres.2015.05.017
77. Lee SJ, Kim JH (2014) Differential natural organic matter fouling ofceramic versus polymeric ultrafiltration membranes. Water Res 48:43–51. https://doi.org/10.1016/j.watres.2013.08.038

78. Lei Z, Dzakpasu M, Li Q, Chen R (2020) Anaerobic membrane bioreactors for domestic wastewater treatment. In: Current developments in biotechnology and bioengineering advanced membrane separation processes for sustainable water and wastewater management anaerobic membrane bioreactor processes and technologies, pp 143–165
79. Li N, Lu X, He M, Duan X, Yan B, Chen G, Wang S (2021) Catalytic membrane-based oxidation-filtration systems for organic wastewater purification: a review. J Hazard Mater 414:125478. https://doi.org/10.1016/j.jhazmat.2021.125478
80. Li N, Tian Y, Sun ZZJ, Zhao J, Zhang J, Zuo W (2017) Precisely-controlled modification of PVDF membranes with 3D TiO_2/ZnO nanolayer: enhanced anti-fouling performance by changing hydrophilicity and photocatalysis under visible light irradiation. J Memb Sci 528:359–368. https://doi.org/10.1016/j.memsci.2017.01.048
81. Lin J, Ye W, Zhong K, Shen J, Jullok N, Sotto A, Van der Bruggen B (2016) Enhancement of polyethersulfone (PES) membrane doped by monodisperse Stöber silica for water treatment. Chem Eng Process-Process Intensif 107:194–205. https://doi.org/10.1016/j.cep.2015.03.011
82. Liu Y, Shen L, Lin H, Yu W, Xu Y, Li R, Sun T, He Y (2020) A novel strategy based on magnetic field assisted preparation of magnetic and photocatalytic membranes with improved performance. J Memb Sci 612:118378. https://doi.org/10.1016/j.memsci.2020.118378
83. Ly QV, Kim HC, Hur J (2018) Tracking fluorescent dissolved organic matter in hybrid ultrafiltration systems with TiO_2/UV oxidation via EEM-PARAFAC. J Memb Sci 549:275–282. https://doi.org/10.1016/j.memsci.2017.12.020
84. Mangrulkar PA, Kamble SP, Joshi MM, Eshram JS, Labhsetwar NK, Rayalu SS (2012) Photocatalytic degradation of phenolics by N-doped mesoporous titania under solar radiation. Int J Photoenergy 2012. https://doi.org/10.1155/2012/780562
85. Marques RG, Ferrari-Lim AM, Slusarski-Santana V, Fernandes-Machado NRC (2017) Ag_2O and Fe_2O_3 modified oxides on the photocatalytic treatment of pulp and paper wastewater. J Environ Manage 195:242–248. https://doi.org/10.1016/j.jenvman.2016.08.034
86. Martínez F, López-Muñoz MJ, Aguado J, Melero JA, Arsuaga J, Sotto A, Molina R, Segura Y, Pariente MI, Revilla A, Cerro L, Carenas G (2013) Coupling membrane separation and photocatalytic oxidation processes for the degradation of pharmaceutical pollutants. Water Res 47:5647–5658. https://doi.org/10.1016/j.watres.2013.06.045
87. May-Lozano M, Lopez-Medina R, Mendoza Escamilla V, Rivadeneyra-Romero G, Alonzo-Garcia A, Morales-Mora M, González-Díaz MO, Martinez-Degadillo SA (2020) Intensification of the Orange II and Black 5 degradation by sonophotocatalysis using Ag-graphene oxide/TiO2 systems. Chem. Eng. Process. - Process Intensif. 158. https://doi.org/10.1016/j.cep.2020.108175
88. Modi A, Bellare J (2020) Amoxicillin removal using polyethersulfone hollow fiber membranes blended with ZIF-L nanoflakes and cGO nanosheets: Improved flux and fouling-resistance. J Environ Chem Eng 8. https://doi.org/10.1016/j.jece.2020.103973
89. Mogal SI, Gandhi VG, Mishra M, Tripathi S, Shripathi T, Joshi PA, Shah DO (2014) Single-step synthesis of silver-doped titanium dioxide: Influence of silver on structural, textural, and photocatalytic properties. Ind Eng Chem Res 53:5749–5758. https://doi.org/10.1021/ie404230q
90. Molinari R, Lavorato C, Argurio P, Szymański K, Darowna D, Mozia S (2019) Overview of photocatalytic membrane reactors in organic synthesis, energy storage and environmental applications. Catalysts 9. https://doi.org/10.3390/catal9030239
91. Moraes SG, Freire RS, Durán N (2000) Degradation and toxicity reduction of textile effluent by combined photocatalytic and ozonation processes. Chemosphere 40:369–373. https://doi.org/10.1016/S0045-6535(99)00239-8
92. Mozia S (2010) Photocatalytic membrane reactors (PMRs) in water and wastewater treatment: a review. Sep Purif Technol 73:71–91. https://doi.org/10.1016/j.seppur.2010.03.021
93. Mozia S, Darowna D, Szymański K, Grondzewska S, Borchert K, Wróbel R, Morawski AW (2014) Performance of two photocatalytic membrane reactors for treatment of primary and secondary effluents. Catal Today 236:135–145. https://doi.org/10.1016/j.cattod.2013.12.049

94. Mozia S, Darowna D, Wróbel R, Morawski AW (2015) A study on the stability of polyethersulfone ultrafiltration membranes in a photocatalytic membrane reactor. J Memb Sci 495:176–186. https://doi.org/10.1016/j.memsci.2015.08.024
95. Mozia S, Janus M, Bering S, Tarnowski K, Mazur J, Szymański K, Morawski AW (2020) Hybrid system coupling moving bed bioreactor with UV/O3 oxidation and membrane separation units for treatment of industrial laundry wastewater. Materials (Basel). 13. https://doi.org/10.3390/ma13112648
96. Muggli DS, Falconer JL (1998) Catalyst design to change selectivity of photocatalytic oxidation. J Catal 175:213–219. https://doi.org/10.1006/jcat.1998.2008
97. Mulder M, Mulder J (1996) Basic principles of membrane technology, 2edn. Kluwer Academic Publishers
98. Nasir A, Masood F, Yasin T, Hameed A (2019) Progress in polymeric nanocomposite membranes for wastewater treatment: preparation, properties and applications. J Ind Eng Chem 79:29–40. https://doi.org/10.1016/j.jiec.2019.06.052
99. Nosaka Y, Nosaka AY (2013) Identification and roles of the active species generated on various photocatalysts. In: Pichat P (ed) Photocatalysis and water purification: from fundamentals to recent applications. Weinheim, pp 1–24
100. de Oliveira CPM, Viana MM, Silva GR, Frade Lima LS, Paula EC, Amaral MCS (2020) Potential use of green TiO2 and recycled membrane in a photocatalytic membrane reactor for oil refinery wastewater polishing. J Clean Prod 257:120526. https://doi.org/10.1016/j.jclepro.2020.120526
101. Ong CS, Lau WJ, Goh PS, Ng BC, Ismail AF (2014) Investigation of submerged membrane photocatalytic reactor (sMPR) operating parameters during oily wastewater treatment process. Desalination 353:48–56. https://doi.org/10.1016/j.desal.2014.09.008
102. Ou W, Zhang G, Yuan X, Su P (2015) Experimental study on coupling photocatalytic oxidation process and membrane separation for the reuse of dye wastewater. J Water Process Eng 6:120–128. https://doi.org/10.1016/j.jwpe.2015.04.001
103. Pascariu P, Cojocaru C, Samoila P, Olaru N, Bele A, Airinei A (2021) Novel electrospun membranes based on PVDF fibers embedding lanthanide doped ZnO for adsorption and photocatalytic degradation of dye organic pollutants. Mater Res Bull 141:111376. https://doi.org/10.1016/j.materresbull.2021.111376
104. Peralta-Zamora P, Moraes SG, Pelegrini R, Freire M, Reyes J, Mansilla H, Durán N (1998) Evaluation of ZnO, TiO_2 and supported ZnO on the photoassisted mediation of black liquor, cellulose and textile mill effluents. Chemosphere 36:2119–2133. https://doi.org/10.1016/S0045-6535(97)10074-1
105. Pérez-Osorio G, Hernández-Gómez FDR, Arriola-Morales J, Castillo-Morales M, Mendoza-Hernández JC (2020) Blue dye degradation in an aqueous medium by a combined photocatalytic and bacterial biodegradation process. Turkish J Chem 44:180–193. https://doi.org/10.3906/kim-1902-33
106. Piaskowski K, Świderska-Dąbrowska R, Zarzycki PK (2018) Dye removal from water and wastewater using various physical, chemical, and biological processes. J AOAC Int 101:1371–1384. https://doi.org/10.5740/jaoacint.18-0051
107. Pickard JM, Zeng MY, Caruso R, Núñez G (2017) Gut microbiota: Role in pathogen colonization, immune responses, and inflammatory disease. Immunol Rev 279:70–89. https://doi.org/10.1111/imr.12567
108. Pirkanniemi K, Sillanpää M (2002) Heterogeneous water phase catalysis as an environmental application: a review. Chemosphere 48:1047–1060. https://doi.org/10.1016/S0045-6535(02)00168-6
109. Rajca M (2020) NOM (HA and FA) reduction in water using nano titanium dioxide photocatalysts (P25 and P90) and membranes. Catalysts 10. https://doi.org/10.3390/catal10020249
110. Ray S, Raychaudhuri U, Chakraborty R (2016) An overview of encapsulation of active compounds used in food products by drying technology. Food Biosci 13:76–83. https://doi.org/10.1016/j.fbio.2015.12.009

111. Rengifo-Herrera JA, Kiwi J, Pulgarin C (2009) N, S co-doped and N-doped Degussa P-25 powders with visible light response prepared by mechanical mixing of thiourea and urea: reactivity towards *E. coli* inactivation and phenol oxidation. J Photochem Photobiol A Chem 205:109–115. https://doi.org/10.1016/j.jphotochem.2009.04.015
112. Romay M, Diban N, Rivero MJ, Urtiaga A, Ortiz I (2020) Critical issues and guidelines to improve the performance of photocatalytic polymeric membranes. Catalysts. https://doi.org/10.3390/catal10050570
113. Sadr SMK, Saroj DP (2015) Membrane technologies for municipal wastewater treatment. In: Advances in membrane technologies for water treatment materials, processes and applications woodhead publishing series in energy, pp 443–463
114. Sakarkar S, Muthukumaran S, Jegatheesan V (2020) Polyvinylidene fluoride and titanium dioxide ultrafiltration photocatalytic membrane: fabrication, morphology, and its application in textile wastewater treatment. J Environ Eng 146:04020053. https://doi.org/10.1061/(asce)ee.1943-7870.0001716
115. Samsami S, Mohamadi M, Sarrafzadeh MH, Rene ER, Firoozbahr M (2020) Recent advances in the treatment of dye-containing wastewater from textile industries: overview and perspectives. Process Saf Environ Prot 143:138–163. https://doi.org/10.1016/j.psep.2020.05.034
116. Saraf S, Vaidya VK (2016) Elucidation of sorption mechanism of *R. arrhizus* for reactive blue 222 using equilibrium and kinetic studies. J Microb Biochem Technol 8:236–246. https://doi.org/10.4172/1948-5948.1000292
117. Sathya U, Keerthi P, Nithya M, Balasubramanian N (2021) Development of photochemical integrated submerged membrane bioreactor for textile dyeing wastewater treatment. Environ Geochem Health 43:885–896. https://doi.org/10.1007/s10653-020-00570-x
118. Selatile MK, Ray SS, Ojijo V, Sadiku R (2018) Recent developments in polymeric electrospun nanofibrous membranes for seawater desalination. RSC Adv 8:37915–37938. https://doi.org/10.1039/C8RA07489E
119. Sharma J, Sharma S, Soni V (2021) Classification and impact of synthetic textile dyes on aquatic flora: a review. Reg Stud Mar Sci 45. https://doi.org/10.1016/j.rsma.2021.101802
120. Sheydaei M, Zangouei M, Vatanpour V (2019) Coupling visible light sono-photocatalysis and sono-enhanced ultrafiltration processes for continuous flow degradation of dyestuff using N-doped titania nanoparticles. Chem Eng Process-Process Intensif 143:107631. https://doi.org/10.1016/j.cep.2019.107631
121. Shindhal T, Rakholiya P, Varjani S, Pandey A, Ngo HH, Guo W, Ng HY, Taherzadeh MJ (2021) A critical review on advances in the practices and perspectives for the treatment of dye industry wastewater. Bioengineered 12:70–87. https://doi.org/10.1080/21655979.2020.1863034
122. Shon HK, Phuntsho S, Vigneswaran S (2008) Effect of photocatalysis on the membrane hybrid system for wastewater treatment. Desalination 225:235–248. https://doi.org/10.1016/j.desal.2007.05.032
123. Shu HY, Chang MC (2005) Decolorization effects of six azo dyes by O_3, UV/O_3 and UV/H_2O_2 processes. Dye Pigment 65:25–31. https://doi.org/10.1016/j.dyepig.2004.06.014
124. Singh R, Sinha MK, Purkait MK (2020) Stimuli responsive mixed matrix polysulfone ultrafiltration membrane for humic acid and photocatalytic dye removal applications. Sep Purif Technol 250:117247. https://doi.org/10.1016/j.seppur.2020.117247
125. Slusarski-Santana V, Ribeiro MVS, Fiorenti-Ferrari LD, Modenes AN, Caldato AAF, Sales Junior E, Olsen-Scaliante MHN, Fernandes-Machado NRC (2017) ZnO supported on zeolites: an efficient catalyst in the photodegradation of fish skins tanning process wastewater. In: Howell F (ed) Eutrophication: causes, mechanisms and ecological effects-series: marine science and technology. Nova Science Publishers, New York, pp 105–124
126. Su W, Zhang Y, Li Z, Wu L, Wang X, Li J, Fu X (2008) Multivalency iodine doped TiO_2: preparation, characterization, theoretical studies, and visible-light photocatalysis. Langmuir 24:3422–3428. https://doi.org/10.1021/la701645y

127. Sun T, Liu Y, Shen L, Xu Y, Li R, Huang L, Lin H (2020) Magnetic field assisted arrangement of photocatalytic TiO_2 particles on membrane surface to enhance membrane antifouling performance for water treatment. J Colloid Interface Sci 570:273–285. https://doi.org/10.1016/j.jcis.2020.03.008
128. Szymański K, Morawski AW, Mozia S (2018) Effectiveness of treatment of secondary effluent from a municipal wastewater treatment plant in a photocatalytic membrane reactor and hybrid UV/H_2O_2–ultrafiltration system. Chem Eng Process-Process Intensif 125:318–324. https://doi.org/10.1016/j.cep.2017.11.015
129. Tamime AY (2013) Membrane processing dairy and beverage applications. Blackwell Publishing Ltd.
130. Teow YH, Ooi BS, Ahmad AL, Lim JK (2021) Investigation of anti-fouling and UV-Cleaning properties of PVDF/TiO_2 mixed-matrix membrane for humic acid removal. Membranes (Basel) 11:1–22. https://doi.org/10.3390/membranes11010016
131. Thombre NV, Gadhekar AP, Patwardhan AV, Gogate PR (2020) Ultrasound induced cleaning of polymeric nanofiltration membranes. Ultrason Sonochem 62:104891. https://doi.org/10.1016/j.ultsonch.2019.104891
132. Thuyavan YL, Arthanareeswaran G, Ismail AF, Goh PS, Shankar MV, Lakshmana Reddy N (2020) Treatment of synthetic textile dye effluent using hybrid adsorptive ultrafiltration mixed matrix membranes. Chem Eng Res Des 159:92–104. https://doi.org/10.1016/j.cherd.2020.04.005
133. Torres NH, Souza BS, Ferreira LFR, Lima ÁS, Santos GN, Cavalcanti EB (2019) Real textile effluents treatment using coagulation/flocculation followed by electrochemical oxidation process and ecotoxicological assessment. Chemosphere 236. https://doi.org/10.1016/j.chemosphere.2019.07.040
134. Vatanpour V, Darrudi N, Sheydaei M (2020) A comprehensive investigation of effective parameters in continuous submerged photocatalytic membrane reactors by RSM. Chem Eng Process-Process Intensif 157:108144. https://doi.org/10.1016/j.cep.2020.108144
135. Vatanpour V, Karami A, Sheydaei M (2017) Central composite design optimization of Rhodamine B degradation using TiO_2 nanoparticles/UV/PVDF process in continuous submerged membrane photoreactor. Chem Eng Process-Process Intensif 116:68–75. https://doi.org/10.1016/j.cep.2017.02.015
136. Vučić MDR, Mitrović JZ, Kostić MM, Velinov ND, Najdanović SM, Bojić DV, Bojić AL (2020) Heterogeneous photocatalytic degradation of anthraquinone dye reactive blue 19: optimization, comparison between processes and identification of intermediate products. Water SA 46:291–299. https://doi.org/10.17159/wsa/2020.v46.i2.8245
137. Wang L, Xiong W, Yao L, Wang Z (2016) Novel Photocatalytic Membrane Reactor with TiO_2 Nanotubes for Azo Dye Wastewater Treatment. MATEC Web Conf 67:06020. https://doi.org/10.1051/matecconf/20166706020
138. Wang W, Wang L, Li W, Feng C, Qiu R, Xu L, Cheng X, Sha G (2019) Fabrication of a novel g-C_3N_4/Carbon nanotubes/Ag_3PO_4 Z-scheme photocatalyst with enhanced photocatalytic performance. Mater Lett 234:183–186. https://doi.org/10.1016/j.matlet.2018.09.098
139. Yi S, Sun S, Zhang Y, Zou Y, Dai F, Si Y (2020) Scalable fabrication of bimetal modified polyacrylonitrile (PAN) nanofibrous membranes for photocatalytic degradation of dyes. J Colloid Interface Sci 559:134–142. https://doi.org/10.1016/j.jcis.2019.10.018
140. Yu W, Liu Y, Xu Y, Li R, Chen J, Liao BQ, Shen L, Lin H (2019) A conductive PVDF-Ni membrane with superior rejection, permeance and antifouling ability via electric assisted in-situ aeration for dye separation. J Memb Sci 581:401–412. https://doi.org/10.1016/j.memsci.2019.03.083
141. Yuan H, Chen L, Cao Z, Hong FF (2020) Enhanced decolourization efficiency of textile dye Reactive Blue 19 in a horizontal rotating reactor using strips of BNC-immobilized laccase: Optimization of conditions and comparison of decolourization efficiency. Biochem Eng J 156:107501. https://doi.org/10.1016/j.bej.2020.107501
142. Zeng Q, Liu Y, Shen L, Lin H, Yu W, Xu Y, Li R, Huang L (2021) Facile preparation of recyclable magnetic Ni@filter paper composite materials for efficient photocatalytic degradation

of methyl orange. J Colloid Interface Sci 582:291–300. https://doi.org/10.1016/j.jcis.2020.08.023

143. Zhang J, Tong HT, Pei W, Liu W, Shi F, Li Y, Huo Y (2021) Integrated photocatalysis-adsorption-membrane separation in rotating reactor for synergistic removal of RhB. Chemosphere 270:129424. https://doi.org/10.1016/j.chemosphere.2020.129424
144. Zhang J, Zhu L, Zhao S, Wang D, Guo Z (2021) A robust and repairable copper-based superhydrophobic microfiltration membrane for high-efficiency water-in-oil emulsion separation. Sep Purif Technol 256:117751. https://doi.org/10.1016/j.seppur.2020.117751
145. Zheng X, Shen ZP, Shi L, Cheng R, Yuan DH (2017) Photocatalytic membrane reactors (PMRs) in water treatment: configurations and influencing factors. Catalysts 7. https://doi.org/10.3390/catal7080224
146. Ziaadini F, Mostafavia A, Shamspura T, Fathiradb F, 2019. Photocatalytic degradation of methylene blue from aqueous solution using Fe_3O_4@SiO_2@CeO_2 core-shell magnetic nanostructure as an effective catalyst. Adv Environ Technol 5:127–132. https://doi.org/10.22104/aet.2020.4137.1204

Enhancement of Anaerobic Digestion and Photodegradation Treatment of Textile Wastewater Through Adsorption

Seth Apollo and John Kabuba

Abstract Textile industry wastewater contains toxic dyes which are difficult to biodegrade using conventional biological methods. Zeolites can be used to improve biodegradation of textile dye due to their good adsorption properties. Alternatively, photodegradation using TiO_2 photocatalyst can be used for effective degradation of textile dyes. Adsorption on zeolite can be used to improve the biodegradation or TiO_2 photodegradation process. This study used zeolite to improve anaerobic digestion and photodegradation of methylene blue (MB) dye. It was established that zeolite could concentrate MB dye on its surface through adsorption thus improving biomass and photocatalyst activity. During the anaerobic digestion process zeolite adsorbed 23% of the dye on its surface leading to a threefold increase in biogas production due to the close contact between the dyes and the immobilized micro-organisms. Increase in zeolite amount from 0 g/l to 50 g/l led to an increase in COD reduction from 50 to 65%. Adsorption on zeolite also enhanced the efficiency of photodegradation since 66% of the dye could be concentrated on the composite catalyst surface through adsorption leading to 85% photodegradation efficiency. The optimal TiO_2 to zeolite ratio was found to be at 15% TiO_2. This study showed the significance of zeolite in improving anaerobic and photodegradation processes used in the treatment of textile wastewater.

Keywords Anaerobic digestion · Photodegradation · Zeolite · Dyes

S. Apollo (✉) · J. Kabuba
Department of Chemical Engineering, Vaal University of Technology, Private Bag X021, Vanderbijlpark 1900, South Africa
e-mail: sethapollo@gmail.com

S. Apollo
Department of Physical Sciences, University of Embu, Embu 60100, Kenya

S. S. Muthu and A. Khadir (eds.), *Advanced Oxidation Processes in Dye-Containing Wastewater*, Sustainable Textiles: Production, Processing, Manufacturing & Chemistry, https://doi.org/10.1007/978-981-19-0882-8_15

1 Introduction

Wastewater from textile industries is considered among the most industrial polluters in the world [8]. Textile wastewater is a major pollutant because of its toxicity caused by residual recalcitrant dyes, high salt and heavy metal content [3]. The polluting nature of the textile effluent is further aggravated by the fact that about 15% of the dye used in textile processing remains unreacted and constitutes the wastewater [20, 32]. The dyes not only increase chemical oxygen demand to the receiving waters, but their presence in water even at low concentration has negative aesthetic impact. The dyes, due to their colour, hampers light penetration in the receiving water bodies leading to reduction in dissolved oxygen concentration.

Anaerobic digestion has been studied for the treatment of textile dyes, the major pollutants in textile wastewater. However, low degradation efficiency has often been recorded because the dyes present in textile wastewater are toxic to microorganisms and therefore cannot be efficiently biodegraded [2, 24]. However, to achieve better degradation a higher retention time and use of robust digesters is recommended [27]. Alternatively, the textile dyes can be degraded using advanced oxidation processes such as photodegradation [28]. Adsorption using suitable adsorbents has been used to improve the performance of biodegradation and photodegradation processes used in wastewater treatment.

Zeolite, being a good adsorbent [25], can be used to improve the efficiency of both the anaerobic and photodegradation processes. In one way, it can be used in bioreactor to limit microbial washout thereby ensuring high microbial concentration during the anaerobic process [22]. Also, due to its good adsorptive property, it helps in concentrating the pollutants on its surface therefore bringing the pollutants in close proximity to the microorganism colonies [31]. At the same time, zeolite can be used in the photoreactor as a catalyst support, and this ensures integration of adsorption and photocatalysis as well as facilitating post-treatment catalyst recovery [30]. This work focused on the use of zeolite as an adsorbent to improve the performance of anaerobic digestion and photodegradation processes used in the degradation of methylene blue dye.

2 Methodology

2.1 *Equipment and Materials*

The experiments conducted included anaerobic digestion, adsorption, and photodegradation. South African natural zeolite with characteristics reported in our previous work [5] was used as the adsorbent. The zeolite was used to improve the anaerobic digestion process and the photodegradation process. Preliminary anaerobic digestion and photodegradation were carried out in 100 ml conical flasks. A 500 ml anaerobic reactor packed with zeolite was used to predict the behaviour of anaerobic digestion using a fixed bed reactor packed with zeolite. Titanium dioxide (Technical grade, 99% purity) was used as a catalyst, while methylene blue dye was used as a model pollutant present in textile wastewater.

2.2 *Experimental Procedure*

Preliminary anaerobic digestion using zeolite was carried out in 100 ml flasks covered with balloons for gas collection and incubated in a thermostatic shaker at 37 °C. Inoculum used was obtained from breweries industry and municipal wastewater treatment plant. Zeolite mass in the flasks were varied from 0 to 200 g/l, while methylene blue dye concentration was varied from 1000 to 4000 ppm. This was followed by investigating the effect of zeolite on packed bed reactor using a 500 ml reactor maintained at 37 °C using a water bath, and biogas was collected using inverted graduated cylinder through water displacement method [1].

Photodegradation was carried using a composite TiO_2/zeolite catalyst in a 100 ml conical flasks illuminated with 30 W UV-C lamp. The concentration of MB dye degraded was 20 ppm, and mixing was achieved in a shaker. The composite catalyst was prepared by supporting the TiO_2 on zeolite. The titanium dioxide was supported on zeolite using solid state dispersion method where a required proportions of zeolite and TiO_2 were mixed thoroughly with ethanol, heated to evaporate the solvent then calcined for 6 h at 450 °C [4, 9]. Factors such as catalyst to zeolite ratio, composite amount and effect of pH on the catalyst performance were investigated. For either the anaerobic digestion or the photodegradation process, the potential of zeolite to adsorb the pollutants on its surface was investigated through adsorption studies.

2.3 *Chemical Analyses and Catalyst Characterization*

Standard analytical method was applied for COD analysis using dichromate as the oxidant in a closed reflux method, the colour of the oxidized sample was analysed using Nanocolour colorimeter. Colour analyses for the MB dye were carried out using UV-vis spectrophotometer (HACH, model DR 2800) at maximum absorption wavelength of 664 nm [10]. The volume of biogas produced was determined from the gas collected in the inverted measuring cylinder [1]. Biogas methane composition was determined from a gas chromatograph (GC, SRI 8610C). The GC chromatograph had a thermal conductivity detector (TCD). The morphology of TiO2/zeolite was studied from a scanning electron microscope (SEM, FEI NOVANANO 230), and results can be found in our previous work [5].

3 Results and Discussion

3.1 Effect of Concentration on Adsorption and Biodegradation

Investigating the effect of dye concentration on the anaerobic digestion and adsorption of dye was carried out using dye concentration range of 1000 to 4000 ppm. Dye and inoculum were added in the biodegradation experiments, while for the adsorption experiments only dye was used. Equal amount of zeolite was used in both the biodegradation and adsorption experiments. The experiments were carried out in duplicates, and the mean COD values are presented in Fig. 1.

The COD removal efficiencies was always higher for biodegradation than for adsorption in all concentrations studied. At low concentrations of 1000 ppm adsorption was capable of removing 25% of the dye while biodegradation could remove about 38% of the dye. The biodegradation potential of dye was found to be low because dyes are toxic to bacteria. The hazardous potential of textile dyes on bacterial strands has been reported, and this could have hindered the biodegradation process [29].

In all cases, it can be seen that COD removal by adsorption and anaerobic digestion decreased with an increase in substrate concentration. The decrease in adsorption with concentration may be due to the fact that since the mass of zeolite used was constant; therefore, there were a fixed number of adsorptive sites. This number of the active sites could only hold a specific amount of substrate. It therefore means that for higher concentration the percentage adsorbed was negligible compared to the total amount of substrate available. It therefore follows that at high concentration of 4,000 ppm, no appreciable removal by adsorption could be obtained.

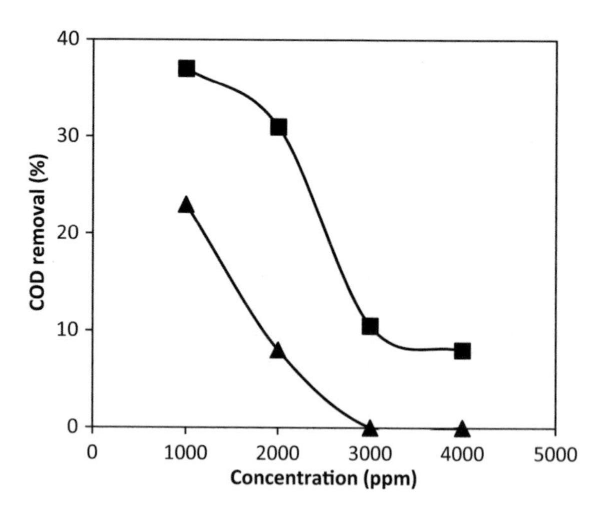

Fig. 1 Effect of substrate concentration on anaerobic degradation with zeolite (■) and adsorption (▲)

3.2 *Effect of Zeolite Dosage on Anaerobic Degradation*

Effect of zeolite dosage was studied varying zeolite concentration in different flasks then purging with nitrogen after which the flasks were incubated for 42 days. Varying the amount of zeolite from 0 to 200 g/l showed a remarkable increase in colour and COD reduction during the anaerobic process. The colour reduction could have been due to combined biodegradation and adsorption on zeolite. The general increase in biodegradation when zeolite is used compared to cases without zeolite shows that zeolite offers large surface area for microbial attachment [13]. In addition, zeolite is reportedly a good adsorbent and therefore can concentrate the pollutants on its surface for easy degradation by the microorganisms attached on its surface. Excessive use of zeolite does not result in increase in degradation due to increase viscosity of the solution that hinders mass transfer [21]. A study using Saponite and Esmectite as microbial support material in the treatment of molasses wastewater reported a higher degradation than in cases where no support material was used [15]. It has been explained that the high performance of zeolite as support material is due to the high capacity for the immobilization of microorganisms [12].

A similar study investigated the effect of addition of zeolite on batch thermophilic anaerobic digestion of piggery waste, doses of 0 to 12 g/l zeolite were evaluated, and high degradation was observed at dosage of 8 g/l [17]. Another study investigated zeolite dosage in the range of 0.2–10 g/l, and a maximum degradation was obtained with dosage between 2 and 4 g/l [21]. Dosage above this values leads to a decrease in degradation due to mass transfer hindrance (Fig. 2).

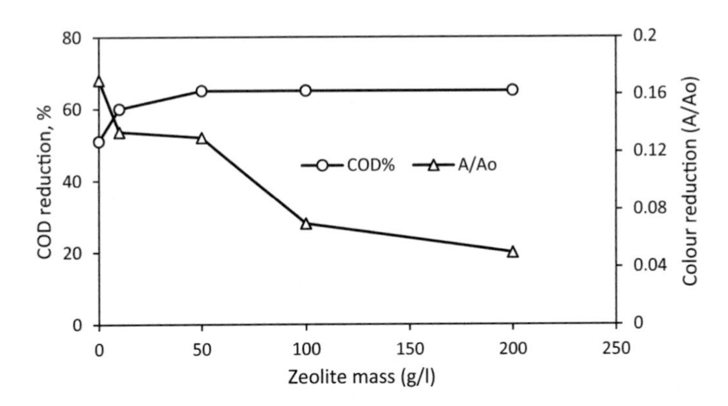

Fig. 2 Effect of zeolite dosage in COD and colour reduction during anaerobic treatment of MB dye, COD (○) and colour (Δ)

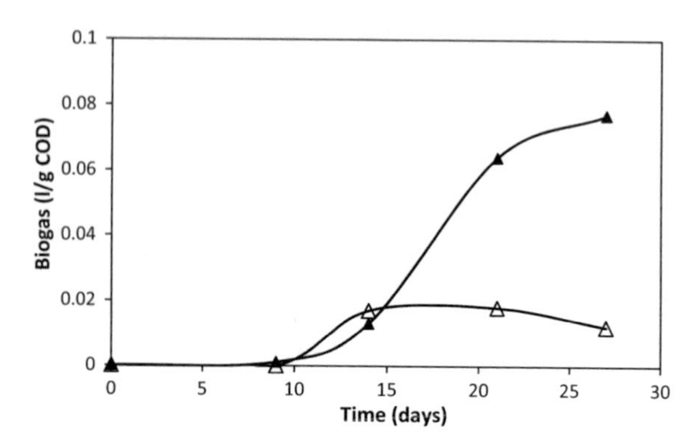

Fig. 3 Effect of zeolite on methane yield in reactor with zeolite (▲) and reactor without zeolite (Δ)

3.3 *Effect of Zeolite on Biogas Production*

Biogas yield of a 500 ml reactor packed with zeolite and applied to treat 5 g/l of MB dye was compared with that of similar reactor without zeolite. A threefold increase in biogas yield was recorded in the reactor with zeolite compared to that without zeolite (Fig. 3). This is because zeolite has high capacity for immobilization of microorganisms; this makes the microorganisms to have a faster growth rate [22]. Similar observations were made when zeolite was used in the anaerobic digester treating swine wastewater [17].

4 Photodegradation of MB Dye

4.1 *Adsorption and Photodegradation*

From the anaerobic digestion studies of the MB dye, it is evidenced that in cases where zeolite is not used the efficiency of degradation of the dye is significantly low because of the chemical toxicity of the dye. Photodegradation offers an alternative path for treatment of biorecalcitrant organic compounds [26]. The efficiency of photodegradation can be improved by using appropriate adsorbents like zeolites [7, 18]. This is because the adsorbents are capable of concentrating the pollutants on their surface where catalyst is attached, and this increases the rate of photodegradation [19]. Figure 4 shows the comparison between the removal of methylene blue by the adsorption on TiO_2/zeolite surface and that of the photodegradation. The study of methylene blue adsorption was performed at 298 K in the dark. It is observed that photodegradation achieved higher colour removal than adsorption. This is due to the fact that the adsorption of the dye takes place up to a maximum level where equilibrium is achieved and thereafter no more adsorption can take place while for photodegradation there is no equilibrium to be attained as the MB molecules are photodecomposed as soon as they are adsorbed on the TiO_2/zeolite surface.

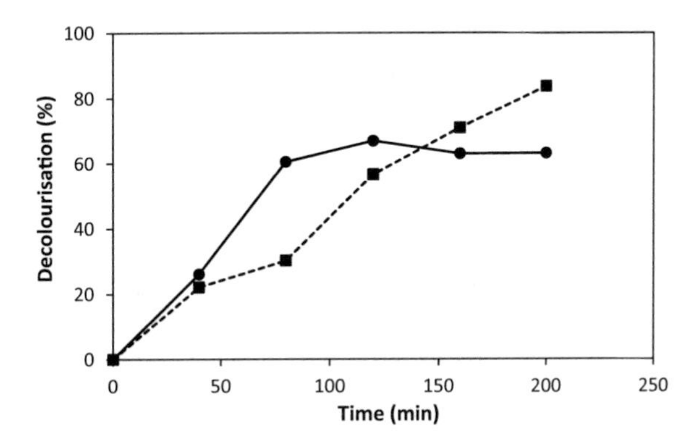

Fig. 4 Colour removal by; photodegradation (■) and adsorption (●)

However, adsorption is observed to take place faster than photodegradation within the first 100 min. This may be an indication that in the adsorptive photodegradation process for the treatment of wastewater containing MB, adsorption of the molecules on the surface of the adsorbent takes place first followed by photodegradation.

It can also be pointed out in Fig. 4 that adsorption removed some 65% of the dye while photodegradation removed 85%. Although adsorption can be considered as effective in dye removal, it is better when adsorption is integrated with photodegradation because adsorption only transfer the pollutants from the solution to the adsorbent surface while photodegradation achieves mineralization of the pollutants [16]. Adsorption of MB dye of 68% on zeolite and a subsequent photodegradation removal of 95% on TiO2/zeolite composite photocatalyst has been reported [19].

4.2 *Effect of TiO_2 to Zeolite Ratio*

To obtain optimal synergy between adsorption and photodegradation it is important to determine the appropriate ratio of the catalyst to the adsorbent to avoid using too much catalyst that can block the adsorption sites of the adsorbent [5, 14]. The study found that a composite catalyst of 10 wt% TiO_2 achieved low dye degradation than 15 wt% and 20 wt% TiO_2 as shown in Fig. 5. This is because the composite with 10 wt% TiO_2 had less catalyst active sites compared to the cases with higher TiO_2 composition. The removal efficiency for 15 wt% and 20 wt% TiO_2 was nearly the same indicating that loading of the catalyst on the adsorbent to a value above the optimal amount do not result in an appreciable increase in degradation [9]. The failure of 20 wt% to produce higher degradation than 15 wt% as may have been expected might have been due to the fact that at this catalyst loading, the TiO_2 became excess and started falling off from the zeolite surface thereby making the solution turbid and consequently blocking the penetration of UV rays into the solution. Similar observation has been reported in the photodegradation of textile dye using TiO_2/zeolite composite catalyst [19].

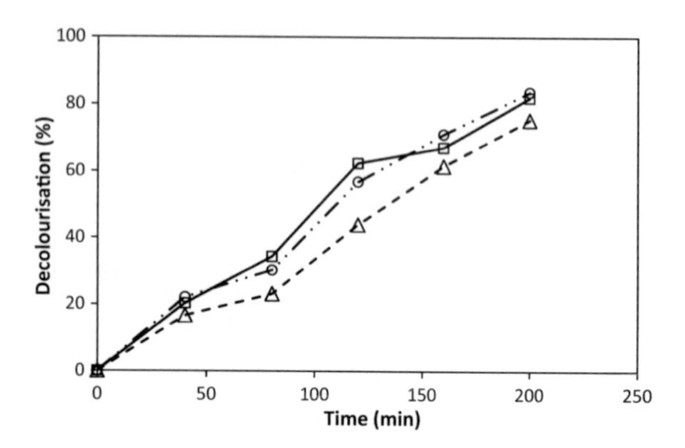

Fig. 5 The effect of TiO_2 loading on Zeolite, 10% (Δ), 15% (□) and 20% (○)

4.3 *Effect of Catalyst Dosage*

The effect of the amount of the composite catalyst (TiO_2/zeolite) was studied using various amounts of catalyst (25 g/l, 50 g/l and 75 g/l) at constant MB concentration of 20 ppm while catalyst loading was maintained at 15 wt% TiO_2. In Fig. 6 a loading of 50 g/l performed better than both 25 and 75 g/l after 160 min of degradation. The increase in degradation when catalyst loading was increased from 25 to 50 g/l could be due to an increase in the number of adsorption site and catalyst active sites [18]. Higher loading of above 50 g/l did not result in any appreciable increase in degradation because of a likely decrease in diffusion of the dye onto the catalyst surface because the high catalyst loading hindered the mass transfer process [23, 23]. In addition, the high catalyst loading hindered the penetration of UV radiation.

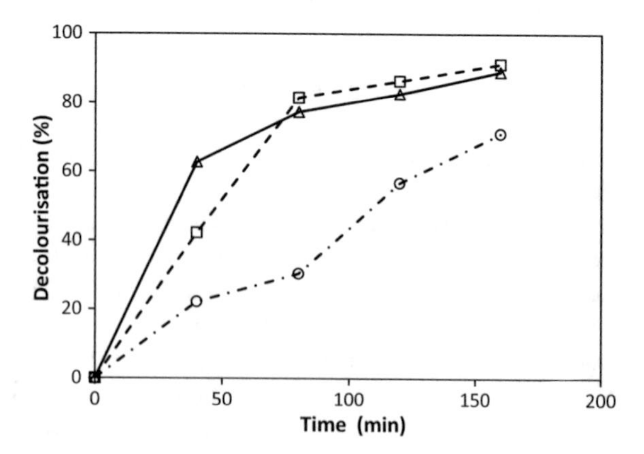

Fig. 6 The effect of catalyst (TiO_2-zeolite) dosage, 25 g/l (○), 50 g/l (□), and 75 g/l (Δ)

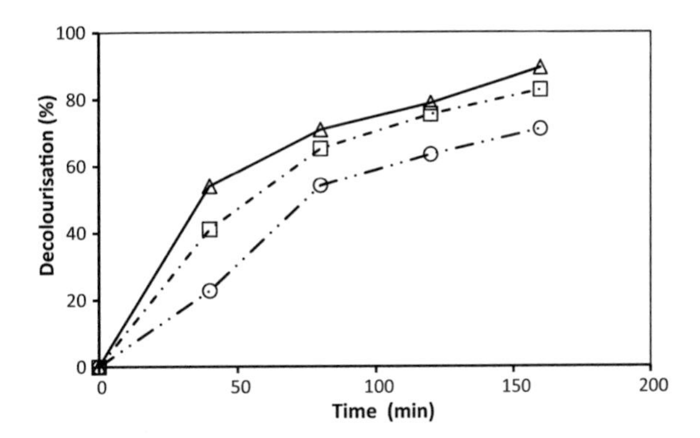

Fig. 7 The effect of pH on photodegradation of MB using zeolite as catalyst support, pH 2 (○), pH 4 (□) and pH 9 (Δ)

4.4 Effect of Solution pH Value

The rate of degradation on the composite catalyst increased with an increase in solution pH as shown Fig. 7. This variation is due to the surface properties of the TiO_2/zeolite photocatalyst The zero point charge of TiO_2 is reported as 4.7 [6]; therefore, the catalyst is positively charged below the zero point charge and negatively charged above the zero point charge value. Since methylene blue (MB) had a cationic configuration, its adsorption is favoured in alkaline solution and this increased photodegradation on the catalyst surface [11]. On the other hand the adsorption of MB on the catalyst surface in acidic medium is poor because the surface of the catalyst was positively charged and therefore could not easily absorb the cationic dye. Similar observations have been reported [7]. It is generally accepted that the pH-dependent photodecomposition was mainly attributed to the variations of surface charge properties of a photocatalyst. Consequently, this changed the adsorption behaviour of a dye on catalyst surface.

5 Conclusions

The dyes that are found in textile effluent have low biodegradability and are not effectively degraded using the conventional biological treatment methods. Methods that can be used to improve biodegradation of the textile dyes include application of zeolite in biological treatment. This is because zeolite is both a good adsorbent and an effective biomass support material. Alternatively, photodegradation using TiO_2 can be used for effective degradation of textile dyes. Adsorption on zeolite can be used to improve the biodegradation or TiO_2 photodegradation process. This study used zeolite to improve anaerobic digestion and and photodegradation of methylene blue (MB) dye. During anaerobic digestion zeolite adsorbed 23% of dye on its surface leading to a threefold increase in biogas production due to the close contact between the dyes and the immobilized micro-organisms. Increase in zeolite amount from

0 g/l to 50 g/l leads to an increase in COD reduction from 50 to 65%. Adsorption on zeolite also enhanced the efficiency of photodegradation since 66% of the dye could be concentrated on the composite catalyst surface through adsorption leading to 85% photodegradation efficiency. The optimal TiO_2 to zeolite ratio was found to be at 15% TiO_2. This study showed the significance of zeolite in improving anaerobic and photodegradation processes used in the treatment of textile wastewater.

References

1. Adamu A (2015) Effect of substrate on biogas yield. Glob J Eng Res 13:35. https://doi.org/10.4314/gjer.v13i1.4
2. Al-Momani F, Touraud E, Degorce-Dumas J, Roussy J, Thomas O (2002) Biodegradability enhancement of textile dyes and textile wastewater by VUV photolysis. J Photochem Photobiol A Chem 153:191–197. https://doi.org/10.1016/S1010-6030(02)00298-8
3. Alinsafi A, Evenou F, Abdulkarim EM, Pons MN, Zahraa O, Benhammou A, Yaacoubi A, Nejmeddine A (2007) Treatment of textile industry wastewater by supported photocatalysis. Dye Pigment 74:439–445. https://doi.org/10.1016/j.dyepig.2006.02.024
4. Apollo S, Onyango MS, Ochieng A (2013) An integrated anaerobic digestion and UV photocatalytic treatment of distillery wastewater. J Hazard Mater 261. https://doi.org/10.1016/j.jhazmat.2013.06.058
5. Apollo S, Onyango S, Ochieng A (2014) UV/H_2O_2/TiO_2/Zeolite hybrid system for treatment of molasses wastewater. Iran J Chem Chem Eng 33:107–117
6. Azeez F, Al-Hetlani E, Arafa M, Abdelmonem Y, Nazeer AA, Amin MO, Madkour M (2018) The effect of surface charge on photocatalytic degradation of methylene blue dye using chargeable titania nanoparticles. Sci Rep 8:1–9. https://doi.org/10.1038/s41598-018-25673-5
7. Chenab KK, Sohrabi B, Jafari A, Ramakrishna S (2020) Water treatment: functional nanomaterials and applications from adsorption to photodegradation. Mater Today Chem 16:100262. https://doi.org/10.1016/j.mtchem.2020.100262
8. Delée W, O'Neill C, Hawkes FR, Pinheiro HM (1998) Anaerobic treatment of textile effluents: a review. J Chem Technol Biotechnol 73:323–335. https://doi.org/10.1002/(SICI)1097-4660(199812)73:4%3c323::AID-JCTB976%3e3.0.CO;2-S
9. Durgakumari V, Subrahmanyam M, Subba RK, Ratnamala A, Noorjahan M, Tanaka K (2002) An easy and efficient use of TiO_2 supported HZSM-5 and TiO2+HZSM-5 zeolite combinate in the photodegradation of aqueous phenol and p-chlorophenol. Appl Catal A Gen 234:155–165. https://doi.org/10.1016/S0926-860X(02)00224-7
10. Eskizeybek V, Sari F, Gülce H, Gülce A, Avci A (2012) Preparation of the new polyaniline/ZnO nanocomposite and its photocatalytic activity for degradation of methylene blue and malachite green dyes under UV and natural sun lights irradiations. Appl Catal B Environ 119–120:197–206. https://doi.org/10.1016/j.apcatb.2012.02.034
11. Faghihian H, Bahranifard A (2011) Application of TiO_2–zeolite as photocatalyst for photodegradation of some organic pollutants. Iran J Catal 1:45–50
12. Fernández N, Montalvo S, Borja R, Guerrero L, Sánchez E, Cortés I, Colmenarejo MF, Travieso L, Raposo F (2008) Performance evaluation of an anaerobic fluidized bed reactor with natural zeolite as support material when treating high-strength distillery wastewater. Renew Energy 33:2458–2466. https://doi.org/10.1016/j.renene.2008.02.002
13. Fernández N, Montalvo S, Fernández-Polanco F, Guerrero L, Cortés I, Borja R, Sánchez E, Travieso L (2007) Real evidence about zeolite as microorganisms immobilizer in anaerobic fluidized bed reactors. Process Biochem 42:721–728. https://doi.org/10.1016/j.procbio.2006.12.004

14. Huang M, Xu C, Wu Z, Huang Y, Lin J, Wu J (2008) Photocatalytic discolorization of methyl orange solution by Pt modified TiO2 loaded on natural zeolite. Dye Pigment 77:327–334. https://doi.org/10.1016/j.dyepig.2007.01.026
15. Jiménez AM, Borja R, Martín A, Raposo F (2006) Kinetic analysis of the anaerobic digestion of untreated vinasses and vinasses previously treated with Penicillium decumbens. J Environ Manage 80:303–310. https://doi.org/10.1016/j.jenvman.2005.09.011
16. Jing J, Liu M, Colvin VL, Li W, Yu WW (2011) Photocatalytic degradation of nitrogen-containing organic compounds over TiO_2. J Mol Catal A Chem 351:17–28. https://doi.org/10.1016/j.molcata.2011.10.002
17. Kotsopoulos TA, Karamanlis X, Dotas D, Martzopoulos GG (2008) The impact of different natural zeolite concentrations on the methane production in thermophilic anaerobic digestion of pig waste. Biosyst Eng 99:105–111. https://doi.org/10.1016/j.biosystemseng.2007.09.018
18. Leal Marchena C, Lerici L, Renzini S, Pierella L, Pizzio L (2016) Synthesis and characterization of a novel tungstosilicic acid immobilized on zeolites catalyst for the photodegradation of methyl orange. Appl Catal B Environ 188(23):30. https://doi.org/10.1016/j.apcatb.2016.01.064
19. Liu ZF, Liu ZC, Wang Y, Li YB, Qu L, E L, Ya J, Huang PY (2012) Photocatalysis of TiO_2 nanoparticles supported on natural zeolite Mater Technol 27:267–271. https://doi.org/10.1179/1753555712Y.0000000011
20. Maas R, Chaudhari S (2005) Adsorption and biological decolourization of azo dye Reactive Red 2 in semicontinuous anaerobic reactors. Process Biochem 40:699–705. https://doi.org/10.1016/j.procbio.2004.01.038
21. Milán Z, Sánchez E, Weiland P, Borja R, Martín A, Ilangovan K (2001) Influence of different natural zeolite concentrations on the anaerobic digestion of piggery waste. Bioresour Technol 80:37–43. https://doi.org/10.1016/S0960-8524(01)00064-5
22. Montalvo S, Guerrero L, Borja R, Sánchez E, Milán Z, Cortés I, Angeles de la la Rubia M (2012) Application of natural zeolites in anaerobic digestion processes: a review. Appl Clay Sci 58:125–133. https://doi.org/10.1016/j.clay.2012.01.013
23. Nawi MA, Sabar S (2012) Photocatalytic decolourisation of Reactive Red 4 dye by an immobilised TiO_2/chitosan layer by layer system. J Colloid Interface Sci 372:80–87. https://doi.org/10.1016/j.jcis.2012.01.024
24. Oller I, Malato S, Sánchez-Pérez JA (2011) Combination of Advanced Oxidation Processes and biological treatments for wastewater decontamination–a review. Sci Total Environ 409:4141–4166. https://doi.org/10.1016/j.scitotenv.2010.08.061
25. Onyango M, Kittinya J, Hadebe N, Ojijo V, Ochieng A (2011) Sorption of melanoidin onto surfactant modified zeolite. Chem Ind Chem Eng Q 17:385–395. https://doi.org/10.2298/CICEQ110125025O
26. Otieno BO, Apollo SO, Naidoo BE, Ochieng A (2017) Photodecolorisation of melanoidins in vinasse with illuminated TiO_2-ZnO/activated carbon composite. J Environ Sci Heal Part A Toxic/Hazard Subst Environ Eng 52. https://doi.org/10.1080/10934529.2017.1294963
27. Pirsaheb M, Mohamadi S, Rahmatabadi S, Hossini H, Motteran F (2018) Simultaneous wastewater treatment and biogas production using integrated anaerobic baffled reactor granular activated carbon from baker's yeast wastewater. Environ Technol (United Kingdom) 39:2724–2735. https://doi.org/10.1080/09593330.2017.1365939
28. Rauf MA, Meetani MA, Khaleel A, Ahmed A (2010) Photocatalytic degradation of Methylene Blue using a mixed catalyst and product analysis by LC/MS. Chem Eng J 157:373–378. https://doi.org/10.1016/j.cej.2009.11.017
29. Rizzo L (2011) Bioassays as a tool for evaluating advanced oxidation processes in water and wastewater treatment. Water Res 45:4311–4340. https://doi.org/10.1016/j.watres.2011.05.035
30. Shan AY, Ghazi TIM, Rashid SA (2010) Immobilisation of titanium dioxide onto supporting materials in heterogeneous photocatalysis: A review. Appl Catal A Gen 389:1–8. https://doi.org/10.1016/j.apcata.2010.08.053

31. Weiß S, Zankel A, Lebuhn M, Petrak S, Somitsch W, Guebitz GM (2011) Investigation of mircroorganisms colonising activated zeolites during anaerobic biogas production from grass silage. Bioresour Technol 102:4353–4359. https://doi.org/10.1016/j.biortech.2010.12.076
32. Zhu H, Jiang R, Fu Y, Guan Y, Yao J, Xiao L, Zeng G (2012) Effective photocatalytic decolorization of methyl orange utilizing TiO_2/ZnO/chitosan nanocomposite films under simulated solar irradiation. Desalination 286:41–48. https://doi.org/10.1016/j.desal.2011.10.036

Printed in the United States
by Baker & Taylor Publisher Services